DYNAMICS

OF

OIL AND GAS ACCUMULATIONS

This is the entirely revised and updated text of the French edition :

PERRODON, A. : Géodynamique pétrolière. Genèse et répartition des gisements d'hydrocarbures. — Elf Aquitaine et Masson, 1980.

Translated by Nissim Marshall.

ISBN 2-901026-12-5
ISSN 0181-0901

Elf Aquitaine, Documentation Centre Micoulau, 64018 Pau CEDEX, France

DYNAMICS
OF
OIL AND GAS ACCUMULATIONS

BY

ALAIN PERRODON

Former Head of the Geology Division of Elf Aquitaine

BULLETIN DES CENTRES DE RECHERCHES EXPLORATION-PRODUCTION ELF-AQUITAINE,
MEM. 5

PAU 1983

CONTENTS

PART II — HYDROCARBON FIELDS

PART III — PETROLEUM PROVINCES

PREFACE

Alain Perrodon's Géologie du Pétrole *(Petroleum Geology) was published in 1966.*

This new book that he offers us fourteen years later could have been a mere updating of the earlier work. In fact, an interval of nearly fifteen years easily justifies such an endeavor, both from the standpoint of methods and techniques as well as results and regional descriptions, especially in a field so rich in innovations and discoveries (1) *as that of petroleum.*

Alain Perrodon has not failed to rise to the occasion, and his new book not only brings his Géologie du Pétrole *perfectly up-to-date, but goes much further. Considering the particular nature of petroleum, which makes of it a sediment in the most widespread acceptance of the term, but a sediment that is complex, unstable and eminently wandering, he rewrites its entire history and presents it as the reflection of its depositional environment and of the geodynamics of its environment. These considerations lead him to reposition hydrocarbon fields within the 'profound logic of the geologic mechanism and events' that presided over their genesis.*

The title of his new book is " Géodynamique Pétrolière ".

The author's orientation is perfectly reflected by the subject, the genesis and distribution of hydrocarbon accumulations.

But let us make no mistake: this light illuminates the new philosophy of all petroleum exploration, and Alain Perrodon's work is perfectly timed to participate in the renovation of certain outdated concepts of prospecting. Following the empirical stage of wells drilled haphazardly or near oil shows, the success of exploration has been based on the virtually systematic identification of all varieties of traps, considered essentially within the framework of the inventory of their geometric reality. We know that this approach has been advantageous, and that it has led to the discovery of most of the hydrocarbons known today, but we also know that with the passage of time, the rate of success has been steadily declining, and that access to the reserves that have evaded the methodical 'combing' practised hitherto is liable to be compromised seriously if prospecting is pursued exclusively in this manner. This is why it is vital to reconsider some of our concepts, and to exploit the progress achieved in the understanding of the geology of petroleum, this 'dynamic sediment', to shift our traditional exploration methods towards an approach that lays greater stress, within the framework of the knowledge of sedimentary basins, on the dynamics of hydrocarbon formation, of their migrations, and of their accumulation in the form of pools. The importance of Perrodon's study in such a context cannot be over-emphasized.

The time has come when geologists are faced with a serious challenge, the need to discover the final hydrocarbon reserves which will serve in the coming century, to provide a relay with the new energies without which the future of our civilization is in jeopardy, and Perrodon's work is a timely arrival to fuel the imaginative effort necessary for the discovery of the vast quantities of additional hydrocarbons that the world demands.

No other person is better suited to undertake the task and accomplish it successfully. The author, confronted for more than twenty years with all the exploration problems of the State-owned BRP-ERAP Group, which became Elf-Aquitaine in 1976, and, as Head of the Geology Division, specially responsible for all activities associated with scientific and technical research in matters of exploration, the enrichment of the knowledge of our teams and the renovation of our prospecting concepts, has compiled and updated a vast body of information and experience, with a ceaseless concern to guarantee the broadest circulation. The new book that he gives us offers an outstanding demonstration of his mastery in this field.

Raymond G. LÉVY

Former Exploration Manager
BRP-ERAP Group

Paris, January 1980.

(1) Over half of all the oil ever discovered in the world has been discovered since 1964.

FOREWORD TO THE ENGLISH EDITION

The many encouragements received, especially from a number of American and English petroleum geologists, led us to publish this English translation of Géodynamique Pétrolière, *published in 1980.*

Considering the time elapsed and the different remarks and criticisms received, this edition has been amended slightly and, whenever possible, brought up to date in comparison with the 1980 French edition. In particular, the beginning of Chapter 3 of Part 3, Global Geotectonic Sketch, Mexico, and the end of the Conclusion have been re-written. A glance at the petroleum geology of Western Europe has also been added.

My special thanks for their encouragement go to the late N.B. Vassoevitch, to Gerald M. Friedman and H. Douglas Klemme, and in particular to A.A. Meyerhoff. I am very grateful to A.W. Bally for his comments and his Preface to this edition.

I also wish to thank Nissim Marshall for his translation, and the geologists J. Borocco and A. Mitchell, for their thorough review of the final text.

Paris, Summer 1982

A. PERRODON

PREFACE TO THE ENGLISH EDITION

To explore is to synthesize. In a succession of exploration cycles, that is only occasionally rewarded by discoveries, earth scientists find themselves involved in a perennial learning process, that ultimately will deepen our understanding of the earth and its resources. We are fortunate to have Alain Perrodon as a most eloquent chronicler of this process and of the current state-of-the-art. His book updates the technical thinking that is essential for a good exploration campaign today. W.E. Pratt's motto, "Where oil is first found, in the final analysis, is in the minds of men", is the theme of this book.

Hydrocarbon exploration is a demanding task. Given propitious political and economic conditions, the explorer of today has to absorb the information obtained with an ever increasing array of new, often esoteric, geophysical and geochemical tools. They are described in numerous manuals, books, and courses that focus on the "nuts and bolts" of modern reflection seismic, remote sensing, logging and geochemical technology. However, books that provide a modern overview of the whole subject of exploration geology are scarce.

Today we have many different specialists who provide the background for decisions made by technically trained managers and entrepreneurs. Teams of experts have taken over, where until the late forties individuals reigned, such as the giants of exploration E. L. de Golyer, W. E. Pratt and others. Thus exploration today risks becoming a patchwork of different specialties that often obscures the nature of the total exploration play and its critical limiting factors. Perrodon's book offers us a logic for the big picture. He views hydrocarbon accumulations in the context of a dynamic basin evolution, which ultimately determines all exploration strategies. These, of course, need to be reconciled with economic, political and environmental boundary conditions.

Perrodon illustrates the linkages between the occurrence of hydrocarbons, the chemistry of their source beds, reservoir genesis and distribution, the importance of seals, migration and trapping mechanisms. The perception of new exploration opportunities is often inspired by similarities of the new play with better understood analogs, and therefore the author provides us with concise sketches of a number of petroleum provinces.

Most importantly, however, our French author shows a deep appreciation for the reasons behind the great effectiveness of exploration efforts in North America. Reasons that include the desire to follow a flexible exploration path, that always adapts to new insights and the willingness to drill on the basis of high quality geology and geophysics. Thus, the intent of gaining new knowledge is combined with the tenacity and imagination needed to develop and test new concepts.

In times when explorers are burdened by excessive economic fine-tuning, by the creation of spurious and bureaucratic obstacles of various governments, and by naive claims that the same goals may be achieved by statistical drilling, it is good to be reminded that exploration is primarily a geologic concern that feeds on sound geologic thinking, the same kind of thinking that also can help in the sound management of the environment.

This book was originally written in French. Its splendid translation into English does not intend to hide the fact that the spirit of the book is French, as reflected by the many quotes from writers, philosophers and scholars. It is this difference in perspective which makes Perrodon's contribution particularly interesting to the English speaking reader. For some, exploration for hydrocarbons may appear to be a mundane enterprise. Alain Perrodon, however, treats exploration as the exciting intellectual adventure it really is.

A. W. BALLY.

INTRODUCTION

« Il conviendra d'écarter de ceux qui sont orientés vers ce que Levorsen appelle la « géologie créatrice », les tracas et les avalanches de travaux de routine dans lesquels beaucoup de nos chefs géologues sont trop souvent empêtrés. N'oublions pas que l'avenir d'une société de prospection est faite de la valeur de ses géologues. »

('It is essential to spare those who lean towards what Levorsen calls' "creative geology", the trouble and the avalanche of routine work in which many of our leading geologists are too often bogged down. Let us remember that the future of an exploration company is founded on the competence of its geologists.')

D. Schneegans, 1947.

The approach to the truth follows complex paths, in which logic and intellectual satisfaction sometimes experience rude shocks. Knowledge has advanced in rapid strides and also interminable detours which sometimes look like reversals. But seen in the right perspective, these contradictions, which are more apparent than real, are so often merely components of vaster and more complex overall views of the image that we try to develop, and the model, which we believed to be fixed and unchanging, is ceaselessly enriched.

The teacher cannot be unaware of this long and arduous approach. He must go through all the twists and turns at the risk of changing his statements, and even of contradicting himself. All teaching must be familiar with the most recent acquisitions, and must also rely on the most traditional, but not necessarily the most intangible, facts and concepts.

In this context, the first and easiest attitude is to accumulate notes and data sheets, and to gain a sufficient overview to make a solid exposition. This may take time. Yet the unparalleled progression of the oral exposé soon needs to be shored up by the written account, which through its intrinsic nature and the reflection that it imposes — form often clarifies content — opens new horizons, and furthers the advance of knowledge.

A quarter century since the first course, over twenty years after the first publication, I felt the need to bring it up-to-date, and to rewrite the work in the form of this "Dynamics of Oil and Gas Accumulations". Rather than a simple rewriting, necessarily entirely reshaped in a number of chapters, I felt that it would be simpler and more interesting to write a new account of the same old story of petroleum, using other illustrations.

This is not the treatise that "Géologie du Pétrole" attempted to be. And, in particular, this text does not contain a number of examples and basic data, often elementary, which appeared in the earlier work.

The scale assigned to the different chapters, sub-chapters and presentations is not always balanced to match their position in nature (or literature), and the accent is frequently placed on concepts and examples which are relatively recent, and hence necessarily less familiar to the geological community.

Nor does the book offer developments on methods and techniques, some of them the subject of recent publications, including those issued by the Elf Aquitaine Group. Petroleum exploration is a field rich enough for its illustration to be profoundly altered, and I have attempted to use, as examples, the description of a number of recent places and results of exploration.

Hence rather than an update, this is essentially a second book that supplements the first, without claiming to replace it completely.

I felt the need to stress the fundamentally mobile and dynamic character of the different mechanisms that lead to the formation and preservation of hydrocarbon accumulations and provinces. Like any geological process, this adventure of oil and gas is a story rooted in the earliest manifestations of subsidence, unfolding throughout the history of the sedimentary basin.

The study of the evolution of geological occurrences, especially the shifts and deformations of sediments, whose movements are comparable successively to spreadings, flows, and intumescences, like the aquifers that play a tremendously important yet so often underestimated role, is certainly a matter of geodynamics.

The entire Earth's crust is continuously subjected to forces and stresses trying to find a perpetual yet unattainable equilibrium. Matter shifts and is deformed continuously, and these movements are observed on all levels, from the lithospheric plates to the migrations of solutions, molecules and ions which, by aggradation and degradation, succeed in transforming the most unalterable rocks.

While the living species are no more than a sort of clumsy codification, at the present moment, of a long evolving pattern, the mineral and geological species are also no more than 'snapshots' of an unrelenting series of transformations and mutations.

The representation we have to make of these occurrences forces us to 'freeze' them and to classify them in simple and rigid structures, although their reality actually consists of imperceptible gradations and incessant variations. Thus we can understand the plaint of Baudelaire :

"*Je hais le mouvement qui déplace les lignes.*"

("I hate the movement that disturbs the lines.")

Yet we must not shrink from describing, analyzing and representing if we wish to understand and to make understood, and to transcribe the kinematics of our representations, lines are our only resource.

In an exposition that leans to the pedagogical, this approach implies backward glances and repetitions. A number of geological and geodynamic events necessarily occur a little like cycles, or more precisely incipient cycles, like Ravel's Boléro, as suggested by P. Routhier. Geological history never repeats itself exactly, any more than the history of mankind.

This means that on various occasions, repetitions are inserted in complementary presentations. Obviously a book of this type cannot be read as a novel, even if petroleum geodynamics has all the trappings of a story, and it seemed wiser to write each part, and even each chapter, as a unit as coherent as possible.

Like all the elements, oil and gas are found everywhere in trace amounts, but very rarely in large accumulations. This false paradox is the law of all mineral concentrations : the greater their number, the poorer and smaller their condition.

The exceptional character of concentrations, and of giant fields in particular, appears as the culmination of a long and complex history, and especially the history of episodes favoring the preservation and transformation of organic matter, the expulsion and concentration of the hydrocarbons formed, the protection and replenishment of the accumulations. At the end of this story, we have a source rock, a reservoir, a trap, closely associated in space and time. In other words, every hydrocarbon accumulation finally lies at the crossroads of strictly defined paleogeographies and chronologies, the terminus of a succession of narrowly dependent landscapes and events, the result of a geodynamics of sediments and basins.

In this view, Part 1, which is short, reviews the general geological framework of sedimentary basins, the setting of the whole story of the formation of petroleum and its accumulations.

Part 2 describes the general mechanisms giving rise to the genesis of hydrocarbons and to their concentration in accumulations, or to their dispersion. This series of processes goes on to the description of a number of types of traps and pools. It illustrates the particularly dynamic character of petroleum-related sedimentation.

Part 3 is devoted to petroleum provinces, families of hydrocarbon fuels. It is at once — and successively — synthetic and descriptive. The analysis of certain examples of petroleum provinces and systems illustrates the different phases of these general processes by means of concrete cases. It does not attempt to reduce reality to a few models or identikits, but offers a broad illustration of the best-known provinces. Reality is too rich to be consigned to a few stereotypes, and the detailed knowledge of certain petroleum systems, through an attempt to discover their logic, is the best means to guide our action.

This classic presentation intimately combines the analysis of the mechanisms of formation of hydrocarbons and pools, which are still frequently based on assumptions, and descriptions of fields and provinces representing the synthesis of a host of convergent geological occurrences.

These two viewpoints also guide the conduct of exploration activities, with the observations and measurements of increasingly powerful techniques. But as accurate as they may be, the data furnished by these tools must be interpreted and integrated within a global geological synthesis on the scale of the sedimentary basin.

One difficulty facing this procedure is that the knowledge of a province only unfolds gradually, on the completion of costly geophysical and drilling operations. A basin is only really well known when most of its petroleum wealth has been revealed. The most fervent hopes of success are also those incurring the highest risks. This means that throughout exploration, data must be gathered and processed continuously, and must be used to compile new hypotheses and more reliable models. In short, petroleum geologists have to analyze and compile syntheses, compare and select, remain prudent yet know when to take risks, build a scientific edifice and yet take gambles. But above all, in the words of the late Daniel Scheegans, who died prematurely just thirty years ago, "*il faut que le géologue digne de ce nom reste un homme de science et qu'il ne soit pas un simple 'automate' à faire des contours*" ("the geologist worthy of the name must remain a man of science and not a simple 'robot' plotting contours").

Acknowledgements

A work of this scope obviously reflects the knowledge and experience of a team, and I take great pleasure in thanking the experts and specialists who were kind enough to criticize areas falling specifically within their competence.

Among those whose help I have enlisted, I wish to thank in particular for their participation J.P. Cassan, A. Chiarelli, J. Claret, J. Connan, H. Coustau, R. Dame, F. Duffaud, J. Le Fournier, P. Masse, P. Robert and J. du Rouchet. I must add to this list of researchers H. Oertli, a specialist in micropaleontology and head of publications, who kindly reviewed the text with his experience and methodical attention.

Discussions held on many occasions always proved to be fertile experiences. These discussions emphasized the gaps that may subsist on many points. They must constantly inspire us to greater modesty and illustrate the need for synthetic works. While the search and acquisition of new knowledge are absolutely indispensable to make headway, these new conquests of the mind would be in vain, if we do not try together to transmit the message to those who shoulder the responsibility for selecting new license areas, and for drilling new wells and interpreting them.

From the material standpoint, I wish to thank the typists and draftsmen who gave these texts and images their form, in many cases final. Without their efforts, this book would not be what it is today.

My special thanks go to R.G. Levy, former Exploration Manager of the Elf-Aquitaine Group, who kindly reviewed all these chapters and wrote the Preface. Having been one of his close collaborators for nearly twenty years, a measure of his own experience and knowledge has also gone into these pages.

My thanks also go to the Exploration Department, especially its Manager, H. Radier, to the management of Scientific and Technical Research, and to the executive management of SNEA(P), who authorized this publication.

MEASUREMENT UNITS

The measurement units and symbols used in this book are those of the international system (SI). The presentation is generally given by exponents (for example t km^{-2} for t/km^2).

To avoid breaking with deeply-rooted custom, certain data are also given in more familiar units. A conversion table is given below.

	SI UNITS	MULTIPLES AND SUB-MULTIPLES DECIMALS			CONVERSION FACTORS TO SI	
LENGTH	meter (m)	angström	(Å)	10^{-10}	inch (in) foot (ft) mile	2.54×10^{-2} 3.048×10^{-1} 1.609
AREA	square meter (m^2)				acre square mile	0.405×10^4 2.59×10^{-6}
VOLUME	cubic meter (m^3)	liter	(l)	10^{-3}	barrel (bbl) cubic foot (ft^3) (cf) gallon (US) (gal) cubic mile	0.159 0.028 3.785×10^{-3} 4.15 km^3
MASS	kilogram (kg)	ton	(t)	10^3	pound (lb)	0.4536 kg
PRESSURE	pascal (Pa)	kilogram per square centimeter	(kg cm^{-2})		atmosphere (atm)	1.013×10^5
		bar	(bar)	10^5	pounds per square inch (lb/in^2)	6.894×10^3 $\cong$ 0.07 bar continued
WORK ENERGY HEAT	joule (J)	erg	(erg)	10^{-7}	watt hour (Wh) calorie (cal) thermie (th) British thermal unit (BTU)	3.600 4.185 4.185×10^6 1.054×10^3 J 0.252 kcal

Decimal multiples and sub-multiples adopted in this text also comply with SI, with the usual symbols and including :

— k kilo or thousand (10^3)
— M mega or million (10^6)
— G giga or billion (10^9)
— T tera or trillion (10^{12})
— d deci (10^{-1})
— c centi (10^{-2})
— m milli (10^{-3})
— μ micro (10^{-6})
— n nano (10^{-9}).

The term giant applies to any field with initial reserves equal to or greater than 70 Mt ($500 \cdot 10^6$ or Mbbl) or 70 Gm^3 ($2500 \cdot 10^9$ or Gcf), and the term 'major' applies to fields of more than 14 Mt ($100 \cdot 10^6$ or Mbbl) or gas equivalent.

Other equivalences

— 1 barrel per day (bopd)	≅ 50 t per year (for API density 0.86-32°)
— 1 cubic foot per day scuftd or cuftd	≅ 10 m^3 per year
— 1 ton oil	≅ 1000 m^3 natural gas
	≅ 10,000 thermies
— 1 m^3 liquefied natural gas LNG	≅ 600 m^3 natural gas
— 1 barrel per acre	≅ 40 m^3 per square kilometer
— GOR 100 cubic feet per barrel	≅ 17.7 m^3 m^{-3}

1 °F per foot	1.8 °C per meter
2 °F per foot	3.65 °C per meter
3 °F per foot	5.5 °C per meter

Oil density

°API	Specific gravity	Barrels per ton	°API	Specific gravity	Barrels per ton
10	1.00	6.4	40	0.82	7.6
15	0.96	6.6	45	0.80	8.0
20	0.93	6.8	50	0.78	
25	0.90	7.0	55	0.76	
30	0.88	7.2	60	0.74	
35	0.85	7.4	65	0.72	

PART I

THE FRAMEWORK : SEDIMENTARY BASINS AND MECHANISMS

« ... Les périodes de remplissage des grands bassins deviennent dans ce cas les espaces-temps orogéniques par excellence, tandis que la classification des phases les considère comme des « périodes de repos » entre les phases orogéniques.

Ce fait marque un déplacement de l'intérêt, typique pour la psychologie du « spectateur » : c'est-à-dire, ce n'est que l'arrivée des coureurs qui intéresse la majorité des assistants et non pas la course bien qu'elle dure beaucoup plus longtemps. »

('... The filling periods of the great basins become orogenic space/time continua par excellence in this case, while the classification of phases considers them as "periods of rest" between orogenic phases.

This fact marks a shift in interest typical of the psychology of the "spectator". In other words, it is only the finish line that interests the crowd, rather than the race, although the latter lasts much longer. ')

E. Wegmann.

INTRODUCTION

Oil and gas fields represent the culmination of many long sedimentary, structural and hydrogeological processes, ordered in accordance with the inherent characteristics of sedimentary basins, and above all the subsidence movements at the source of every basin. No basin, no oil. Hence to alleviate the ensuing discussion, it is wise to recall briefly a few sedimentary processes that play a more direct role in the distribution of hydrocarbon accumulations.

Sedimentary basins and petroleum provinces

Petroleum geology is closely related to major geological and geodynamic events, which may be classified roughly in three orders of magnitude, in decreasing order :

— global tectonics,
— basin geodynamics,
— sedimentation mechanisms.

On the scale of lithospheric plates, first order units, the kinetics can be characterized generally in terms of divergence and convergence plates and transform faults. We know that the passive margins associated with divergence zones are distinguished by vast sedimentary prisms, deformed very little or not at all. Convergence or subduction areas are illustrated, in accordance with the continental or oceanic character of the plates involved, by the genesis of Cordilleras, folded ranges or island arcs, namely complex, often volcanic sedimentary packages. Finally, the transform fault areas associated with major shear fault zones are often marked by strike slip basins.

Sedimentary basins appear to be deeply marked by plate geodynamics, but nevertheless strictly reflect this general framework. They occur as second order units, showing both a certain relationship with the overall setting of global tectonics, but also a real independence, reflected by a more complex mosaic and by the convergence of forms.

For instance, a number of basins of the rift type are observed, associated with major plate spreading movements, like the passive margin basins and Tertiary deltaic basins, as well as compression basins of the miogeosyncline type, closely associated with subduction zones. On the other hand, rift valleys are familiar just as well in intracratonic areas and certain parts of convergence zones. Many rifts and grabens display tangential features that materialize compression phases.

Sedimentary occurrences are situated on a third scale, more limited in space as well as time. Certain basin types undeniably infer a sedimentological and paleogeographic tendency, to the extent that their characteristics are often mixed; for example, platform series and basin, geosyncline series. But frequent similarities are observed in most cases, like the paralic series in subsident as well as platform areas, together with reef building in stable and graben type basins.

The petroleum provinces, substantially of the same order of magnitude as the basins, are the result of the geodynamics and sedimentary filling of the latter. Their definition and classification are distinct from those of the basins, to account for the sedimentary factor.

In this context, we shall deal successively in Part 1 with some general aspects and others of specific interest from the petroleum standpoint of the sedimentary basins, and the sedimentary environments and mechanisms. The study of petroleum provinces occupies the bulk of Part 3.

1. – SEDIMENTARY BASINS

« Les volumes, les surfaces, les lignes, en un mot, les structures qui composent un édifice tectonique ne sont pas tout : il y a le mouvement qui anima, qui anime encore ces choses, car l'histoire continue. »

('Volumes, surfaces and lines, in short, the structures that make up a tectonic edifice, are not all that count: there is the movement that created and continues to create these elements, because the story goes on.')

E. Argand
The Tectonics of Asia, 1922.

A sedimentary basin may be defined, as by Bally (1975) as "realms of subsidence with thicknesses of sediments — commonly exceeding one kilometer — that are today still preserved in a more or less coherent form". This definition places basins between, and also outside :

— stable shield areas with a predominantly positive tendency,

— highly tectonized units which are deeply attacked by erosion.

Sedimentary basins often appear to us in a post-orogenic situation that guarantees them a certain permanence in our eyes and makes their features more visible. But they are equally developed in a pre-orogenic situation, giving birth, after long convulsions, to folded and faulted structures which obliterate and more or less completely mask their inherent features, rendering their analysis more difficult, if not impossible.

A basin is born from the meeting of a sedimentary deposit and a more or less pronounced concavity in the basement. The basal deposits, which represent the start of a new sedimentary adventure, generally mark a certain break with the past. Hence the term discordant basin, used occasionally, but actually widely applicable. The isopachs of the sedimentary volume display a closed unit, one which could correspond to a sedimentary prism, if the depositional environment crosses from coastal zones to deep areas.

The Earth's surface exhibits a wide variety of sedimentary basins. Their study reveals them to be in different stages of evolution, at distinct ages, on the one hand, and in varied geodynamic contexts, on the other. In other words, their distribution as well as their characteristics obey a twofold logic :

— geodynamic situational logic,

— historical logic.

Within the framework of the theory of global tectonics, which today represents the best working hypothesis for understanding the logic of the distribution and evolution of sedimentary basins, the latter are naturally organized in accordance with their situation in relation to the lithospheric plates.

Most of the basins, which are mobile zones by definition, are encountered at the plate boundaries. However, some of them, particularly the most extensive basins, are situated on the plates themselves. These are the cratonic and intracratonic basins :

— platform, internal and intracontinental basins,

— rift type basins.

The most mobile basins are located on the margins of the major plates :

— pull-apart basins of the passive margins, particularly on the periphery of the Atlantic Ocean,

— strike slip basins of the transform or shear zones, for example along the Californian coast,

— tectonized basins of the active or convergent margins of the megasutures (Bally & Snelson, 1980), and especially of the subduction zones (fore arc, back arc, etc.).

From this standpoint, we shall now examine the chief types of basin in succession, and then discuss their history.

1.1. MAIN BASIN TYPES

On consideration of their dominant mechanisms alone, the following four major classes of basin may be distinguished in the simplest manner :

— rift type basins, in cratonic areas,
— stable area basins, in continental areas,
— passive margin basins, in pericratonic situations,
— mobile basins, in orogenic belts.

It may be considered roughly that :

— the first are encountered in a cratonic or intracratonic situation and in pre-tectonic periods,
— the second are in a cratonic or intracratonic position and, from the chronological standpoint, in an a-tectonic situation,
— the third are in a pericratonic situation,
— the fourth are in an orogenic and late-tectonic area.

It is important to insist on the very general character of this classification of entities as complex and varied as sedimentary basins. In particular, it is important to analyze carefully the geological characteristics of a basin, to clarify its history and its degree of evolution, before attempting to classify it.

In fact, it is not unusual for tensile and compressive forces to act at the same time in different areas. It frequently occurs that distinct basins relay each other somewhat in time, superimposing, nesting, matching and adapting to each other in accordance with variations in the geodynamic context.

Basin classifications are now frequently encountered, with few variations (Perrodon, 1971; Bally, 1975; Poulet *et al.,* 1975; Klemme, 1975; Dickinson & Yarborough, 1976, are among the most recent).

The classification that we propose here is above all a framework for the identification of petroleum provinces.

1.1.1. Basins of the rift type in cratonic areas

These basins correspond to true continental fractures, often following an arching movement. They also form the seat of a sudden and active subsidence in an area that previously displayed great stability (Figure 1).

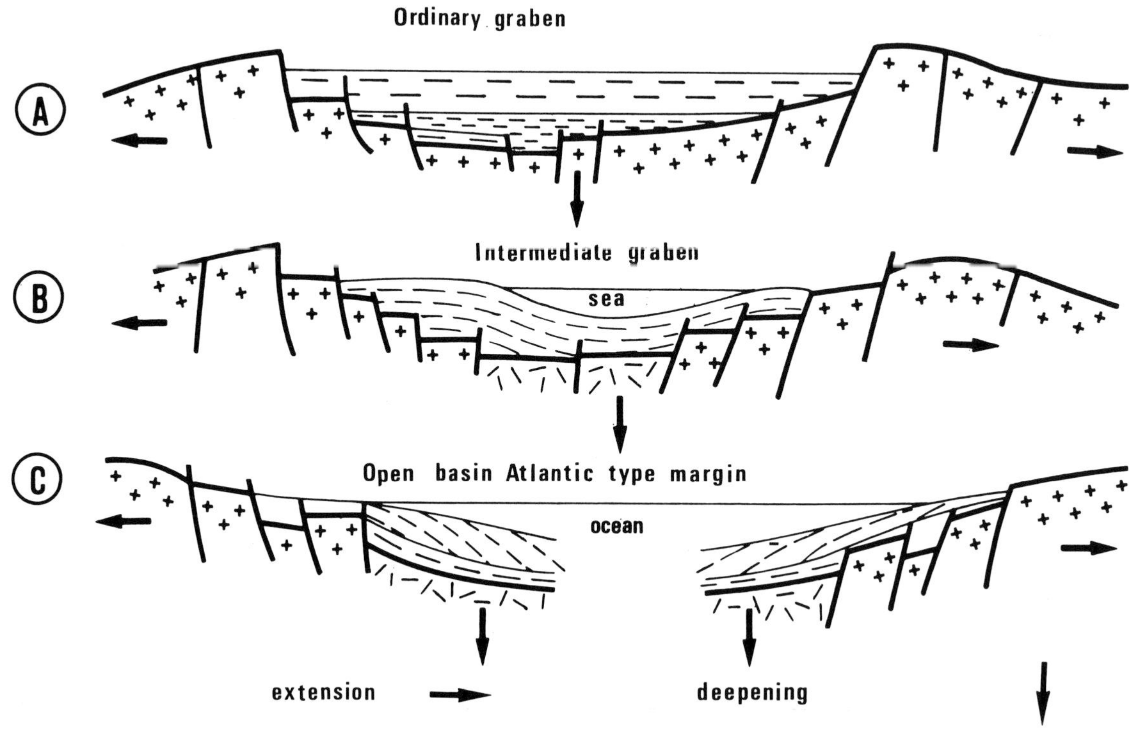

Fig. 1. — Schematic cross-section of rift basins and oceanic drift.

They are distinguished more specifically by :

— distension forces and negative vertical movements, causing considerable subsidence, often acting along listric faults,
— rapid sedimentation, capable of giving rise to undercompaction and shale diapirism, especially in cases of large inflows (deltaic basins),
— generally high geothermal gradients, often associated with volcanic occurrences.

These basins correspond to rift valleys and classic grabens. They generally exhibit an elongated or triangular form and pronounced subsidence. They appear as the result of distension movements, and often also shear movements.

The rate of sedimentation is generally lower than the rate of subsidence, giving rise to a starved basin stage. At the end of this distensive period, the trough is progressively filled by transgressive series which mask the deep flexures and lead to the platform basin pattern.

These basins are numerous, especially in the Permian and the Tertiary, and often make interesting petroleum provinces. Among these petroleum basins are the following :

— in the late Paleozoic, the Dnepr-Donetz-Pripyat troughs in the USSR (Figure 2),
— in the Mesozoic, the North Sea Viking Graben (Figure 183), the Gifhorn Trough in Germany, the internal basin of Gabon (Figure 153), the Reconcavo Basin in Brazil (Figure 151), and the Sirte Basin in Libya (Figure 207).
— in the Cenozoic, the Rhine Graben, and the Gulf of Suez Basin (Figure 150).

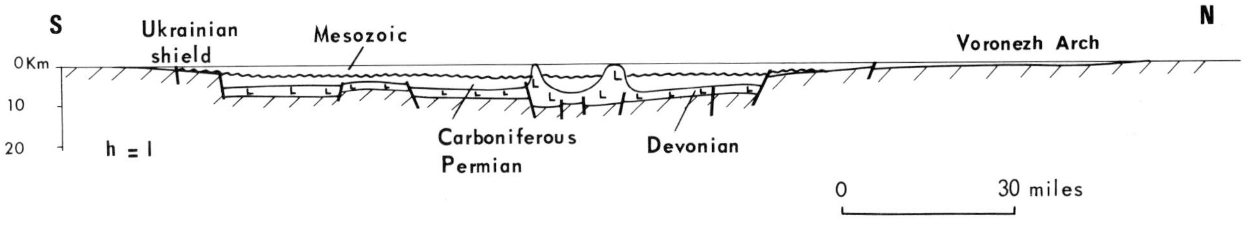

Fig. 2. — Schematic cross-section of Dnepr/Donets Trough, USSR (Andreyev & Chirvinskaya, 1961).

Depending on their evolution and the geodynamic context, rifts may be a prologue to the other great basin families :

— overall stabilization leads to the formation of platform basins,
— the acceleration of distension forces yields grabens of the intermediate type, characterized by the appearance of the oceanic crust on part of the basin,
— a paroxysm of spreading movement gives rise to the formation of passive margin basins,
— divergence followed by a convergent movement leads to the formation of orogenic belt basins.

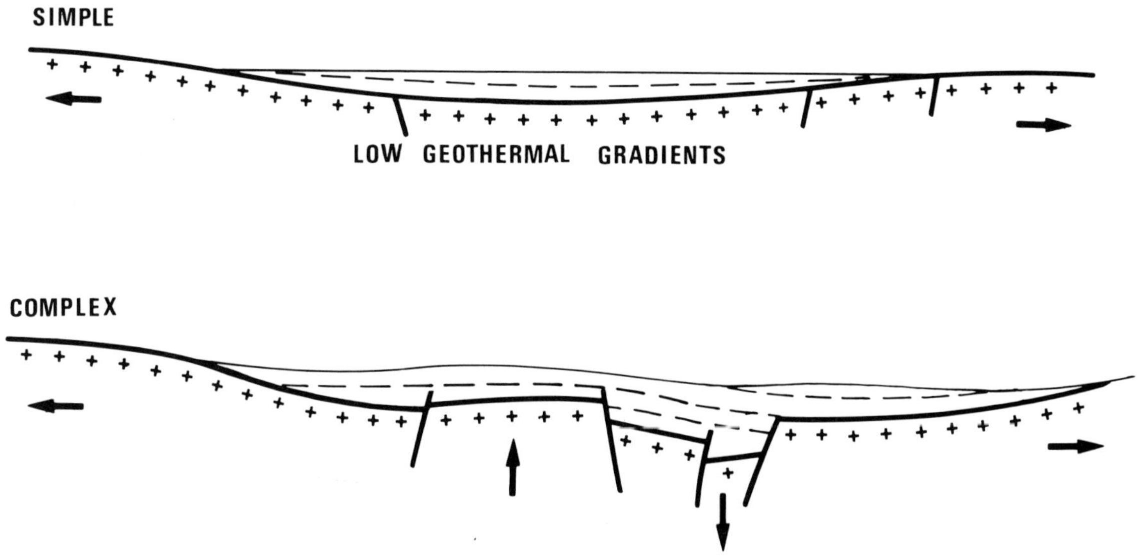

Fig. 3. — Schematic cross-section of Cratonic basins.

1.1.2. **Stable area basins in continental regions** (Figures 3 and 4)

The characteristics of these basins, of which the Paris Basin is a good example, reflect those of the overall geodynamic framework in which they are situated :

— low subsidence,
— indistinct structural deformations, broad arches or fractures in low amplitude horsts and grabens, caused by predominant distension movements,
— relatively thin, homogeneous sedimentation of the neritic type,
— generally low geothermal gradients,
— predominantly hydrostatic regime.

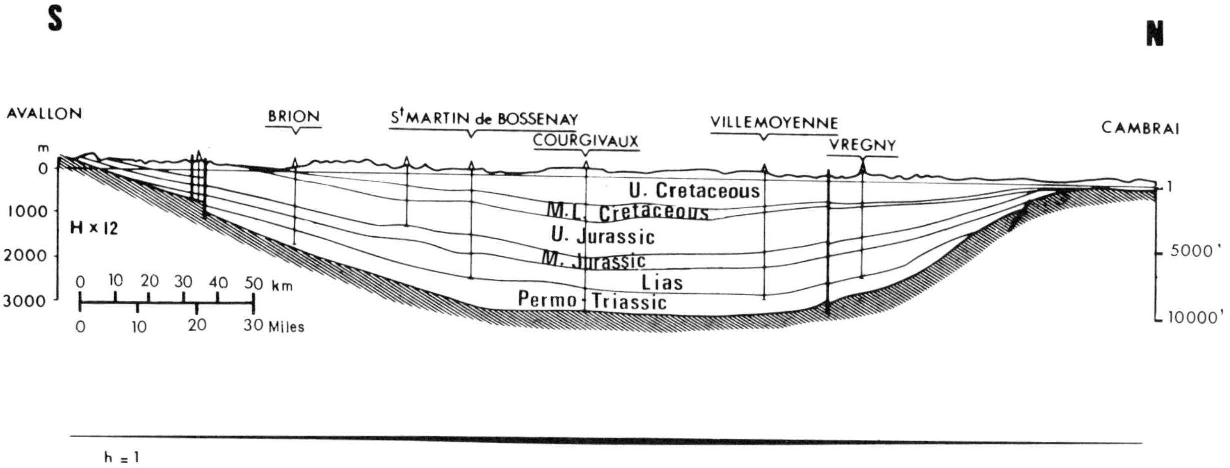

Fig. 4. — Section across the Paris Basin from Avallon to Cambrai (Heritier & Villemin, 1971).

The simplest basins, in a cratonic situation, are generally filled by a single megasequence. In some cases (Chad, Michigan), they may result partly from the formation of peripheral uplifts (Burke, 1976). They frequently appear to start with the formation of grabens.

The more extensive basins, usually in an intracratonic situation, are also more complex and may include sub-basins, "nesting" basins, internal troughs, in which the slow movements of subsidence areas give rise to several adjacent and successive sedimentary megasequences.

The Saharan basins of Algeria and western Libya (Figure 203 et seq.), the Great Artesian Basin in Australia, the West Siberian Basin (Figure 178), the Mid-Continent Basin in the United States (Figure 190), and the Upper Paleozoic Basin of Northern Europe, offer excellent illustrations of this family.

However, certain platform basins may exhibit significant variants, especially large sediment thicknesses.

The Sverdrup Basin in the Canadian Arctic Isles offers a special example of a cratonic basin characterized by high subsidence and the absence of orogenic activity (Figure 5).

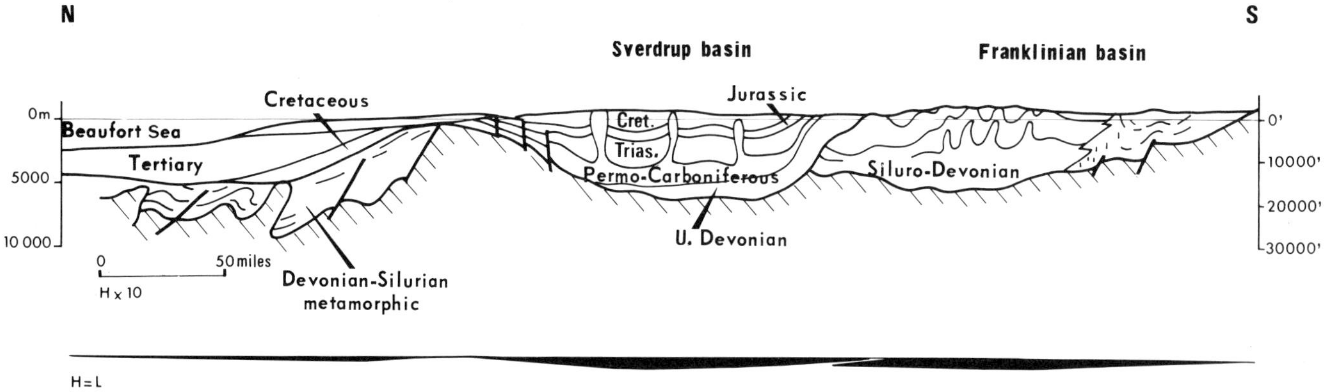

Fig. 5. — Section across Sverdrup and Franklinian Basins (Canadian Arctic) (Elf-Aquitaine document).

It is vast in area (500,000 km^2) but, above all, its depth is exceptional for this type of basin, around 13 km, for a duration of 285 million years from the early Carboniferous to the late Cretaceous. In accordance with a fairly general rule, its history began with a system of grabens. Subsidence was rapid and the environment relatively deep in the Carboniferous and the Permian. During the Triassic the axial zone received a considerable thickness of fine detrital materials; the sedimentation rate was slower in the Jurassic and the Neocomian. From the Aptian, the deposits largely transgressed the edges of the basin, forming vast continental shelves. Until the end of the Cretaceous, sandstone series alternated with transgressive black shales (from Balkwill & Kenneth, 1979).

1.1.3. **Passive margin basins**

Passive margin or pull-apart basins occur in a pericratonic situation. They are well-known on the edges of the Atlantic and Indian Oceans, hence their alternative name of Atlantic margin basins. They display a highly dissymetric profile, and two clearly distinct sedimentary cycles, corresponding to the rifting and post-rifting periods (Figure 6).

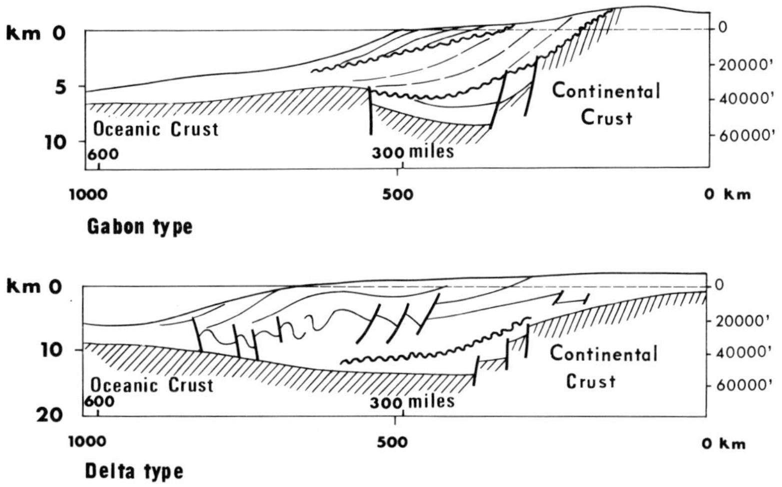

Fig. 6. — Schematic cross-sections of passive margin basins, simple type (above), delta type (below).

The rifting stage, in intracratonic position, is characteristic of the rift type, with predominantly vertical horst/graben movements and highly diversified sedimentation, with shaly and detrital series corresponding mainly to lacustrine and fluviatile environments.

The period corresponding approximately to the beginning of drifting is generally characterized by the formation of evaporites, often in combination with black shales and marls rich in organic matter.

The post-rifting stage, in a platform boundary or pericratonic position, witnesses the progressive development of marine troughs and often increasingly deep facies, from the sedimentary, and the structural standpoints, the appearance of deposition dips inclined towards the slope, often passing into progradating formations. These deposits become thinner offshore, where they often rest on the oceanic crust. In case of massive inflows, especially deltaic, complex sedimentary prisms may be formed, ranging up to 10 km in thickness. The effects of gravity and undercompaction are reflected by shale ridges and a network of listric faults.

From the petroleum standpoint, excellent illustrations of passive margin basins are the basins of Equatorial Africa (Gabon, Congo and Angola) (Figure 153), of Sergipe Alagoas and Campos in Brazil, Newfoundland and Labrador in Canada, Dampier (Figure 155) and Gippsland (Figure 156) in Australia. Added to these are the deltaic provinces of the Niger delta (Figure 209) and the Gulf Coast (Figure 139). They are all discussed in the chapter on petroleum provinces (Part 3).

1.1.4. **Orogenic belt basins**

The basins of this family, situated in a particularly unstable area and often qualified as mobile, are generally distinguished by a wide variety of types and by greater structural and very often stratigraphic complexity, since tectonically unstable zones are also very mobile during periods of deposition.

A distinction may be drawn roughly between two chief groups :

— basins with predominantly vertical or shear movements, corresponding more or less to the term of intermontane basin,

— basins strongly marked by convergent movements and consisting largely of basins of the miogeosyncline or foredeep type, since geosynclines usually show a tendency to occur more in the form of folded and eroded ranges than specific basins.

1.1.4.1. ***Intermontane type basins***

These are basically fault basins, situated, unlike rifts, in orogenic belts or cratonized areas, although many transition forms link these two cases.

Depending on the respective share of vertical and horizontal movements, two chief groups may be distinguished :

— classic fault basins or grabens, intersected by large fracture, like the Vienna Basin (Figure 159), and the Cook Inlet Basin in Alaska (Figure 160),

— 'strike slip' basins : bounded by the great shear features which often induce *en échelon* folds in relation to the trough axis, clearly illustrated by the Californian basins (Figures 161, 162 and 193).

Others are found in subduction and island arc areas. Finally, even in cratonic areas, some grabens may subsequently be uplifted, folded and eroded, as shown by the Dutch Jurassic and Cretaceous troughs.

These different basins often exhibit good petroleum characteristics, and some of them represent major petroleum provinces (Chapters 2 and 3 of Part 3).

1.1.4.2. ***Miogeosyncline or foredeep type basins***

These basins are basically characterized by convergent and compressive movements generating folded structures and reverse faults (Figure 7). They generally exhibit a very sharply asymmetric transverse profile, with a mobile folded margin often affected by complex tectonics, corresponding to the most subsident zone, and a stable, more or less extended margin, crossing to the deeper central area through a zone of flexures and facies variations. These basins generally have a more diversified geological history than that of the previous family, and the latest events of their history are characterized by dominant compressive forces which impart a determining feature. Nevertheless, the essential phases of their formation primarily obey tectonic mechanisms that give rise to subsidence, of which the features are often subsequently more or less masked by later compression and folding phases.

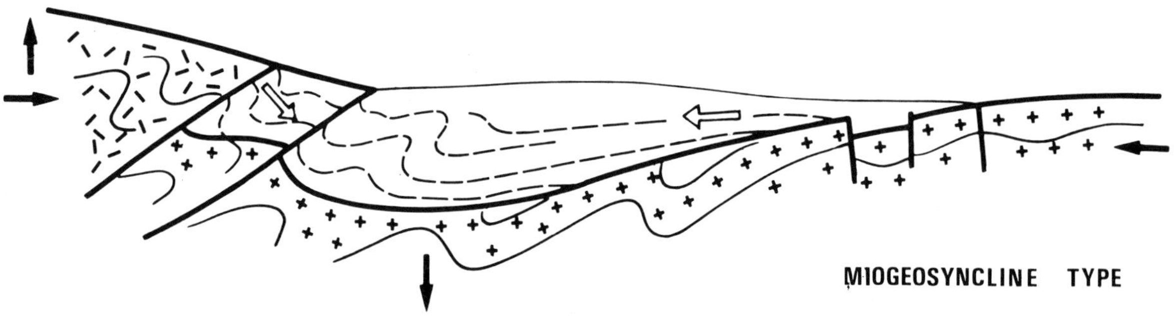

Fig. 7. — Schematic cross-section of tectonic compression basins.

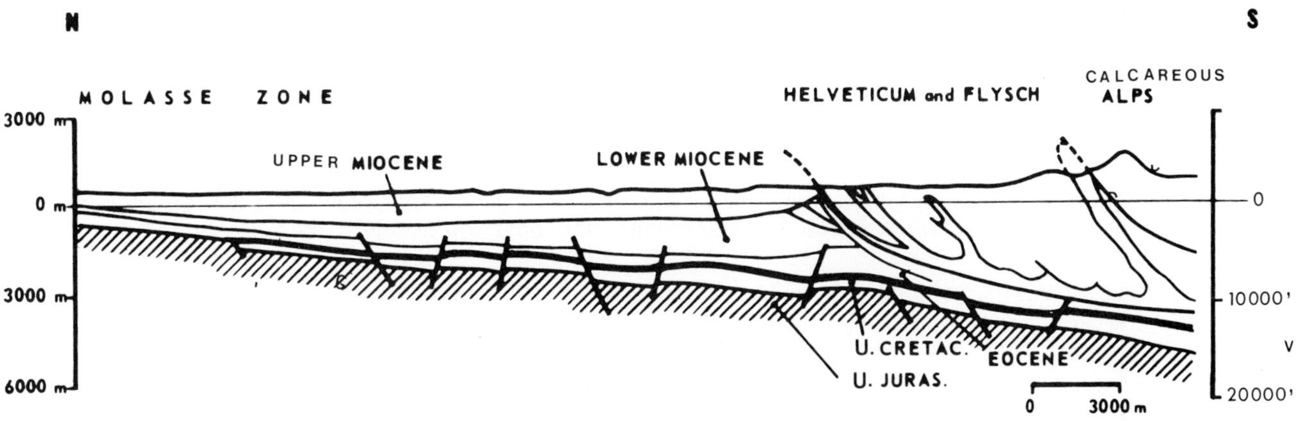

Fig. 8. — Cross-section of Molassic/Austrian Basin (Guide to Excursion A.C. 26, 23rd Int. Geol. Cong., 1968).

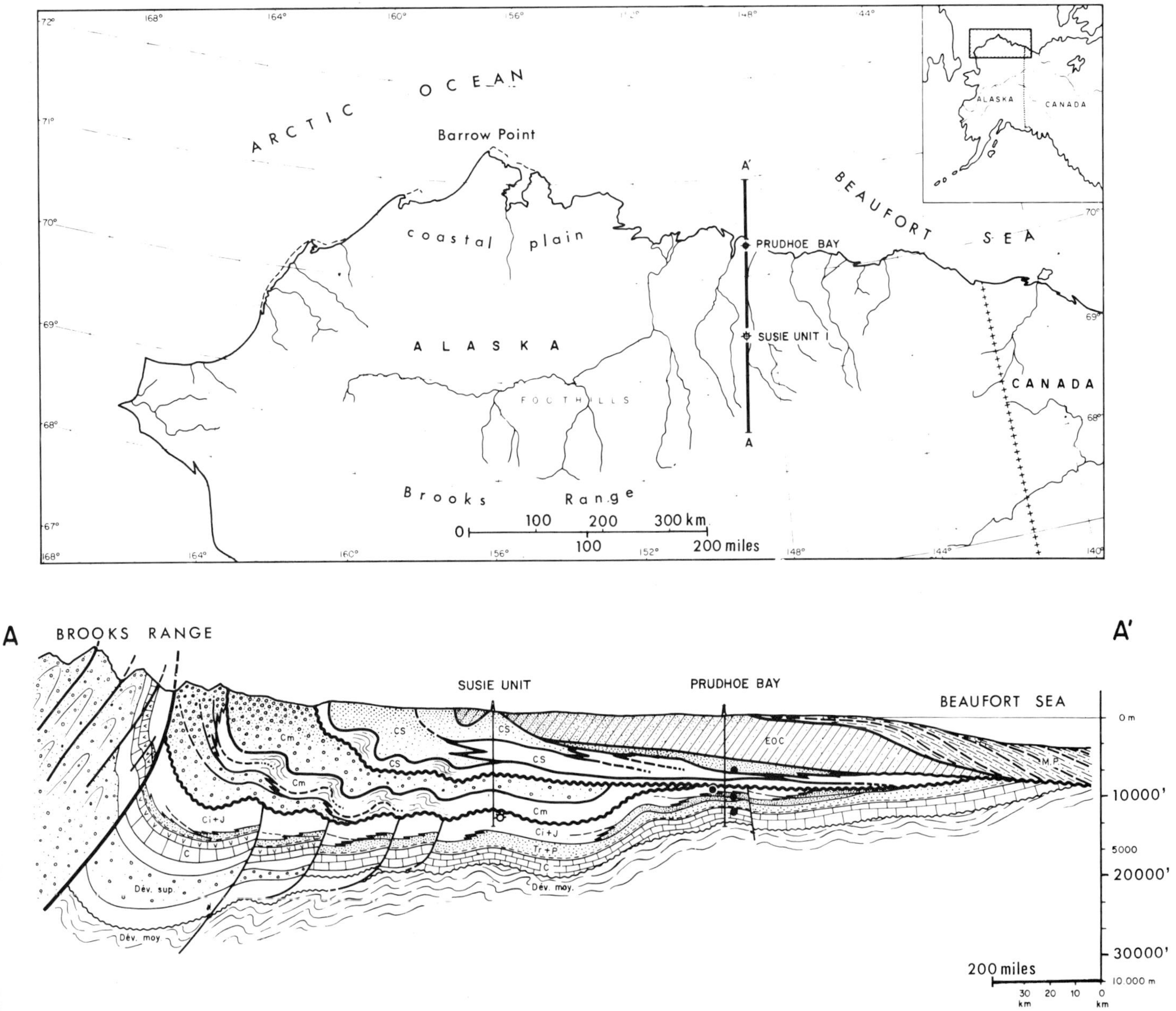

Fig. 9. — Map and section of the 'Arctic Slope' North Alaska Basin (Bouillot, 1972).

The basins of this group appear more specifically characterized by :

— folds and tectonic deformations of the tangential type, possibly by a piling up of nappes in the most subsident zone, as shown by the extra-Alpine Basin (Figure 8),

— a thick sedimentary series, at least in the most subsident and most unstable belt, exhibiting a sequence with a negative tendency,

— possibly by volcanic events.

Depending on their geodynamic situation, in the case of convergent zones between continental and oceanic or pericratonic plates, one speaks of foredeep (the best represented), back-deep and intra-deep basins.

This group of basins, generally displaying a large and diversified sedimentary volume, as well as many early anticline structures, forms a remarkable group of rich petroleum provinces, of which the leading ones are analyzed below (Chapters 2 and 3 of Part 3).

Some of these basins may display a shift of the most subsident zones in the form of relays in space and time, identifying sorts of hinges or permanent highs, as illustrated by the Arctic Basin in Alaska (Figure 9), and the Mzab Arch in the Algerian Sahara (Figure 144).

Troughs of the island arc type are individualized at the convergence of two oceanic plates, of which the outermost one corresponds to a neo-oceanization zone. At the boundary of these two plates towards the internal zone, the island arc may generate compression basins with volcano-detrital sedimentation and high energy carbonate intervals, in a relatively low thermal gradient context. Tension basins with greater sedimentation and high thermal gradients are developed on the outside (Figure 10). Basins of this type are generally of lesser petroleum interest than the former.

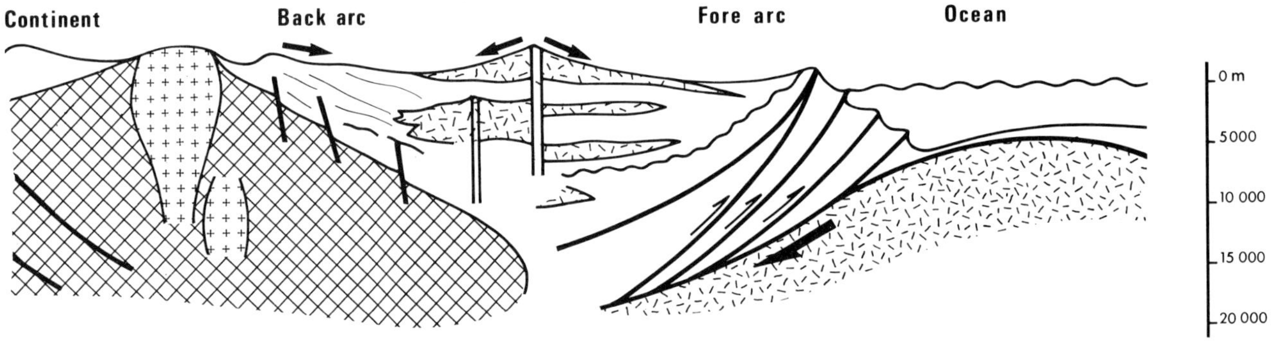

Fig. 10. — Section of basins associated with orogeny zones, island arc type

The geographic distribution of these great basin families is not random. It satisfies a global dynamic logic, which concurs with the relationships between extension areas and compression areas. "*Intracontinental deformation, especially extension deformation of the rift type, is organized in accordance with its location, and the duration and nature of the convergent zones between plates*" (P. Masse, 1979, unpublished). Rifts, like 'en échelon' tension gashes on a smaller scale, indicate zones associated with converging plates. The opening of the Red Sea and the action of the Great Levant Fault appear as the logical counterpart of the Zagros Fold, and that of the Rhine Graben the complement of the Alpine Arc.

1.2. GEODYNAMICS OF SEDIMENTARY BASINS

The present image of the earth offers us, in a sort of snapshot on the geological scale, a complete sampling of images depicting the advance through time of the history of the basins. This means that one can find all the stages of a common evolution in the different types of basin known today on the Earth's surface and subjected to various tensile, shear and compressive forces.

When the lithospheric crust reacts elastically to an overload, the sedimentary basins form the seat of a series of transformations that occur in a logical order, and which may be schematized very roughly by three major successive stages (Figure 11) :

— juvenile stage,

— mature stage,

— final stage.

(Perrodon, 1969, 1977)

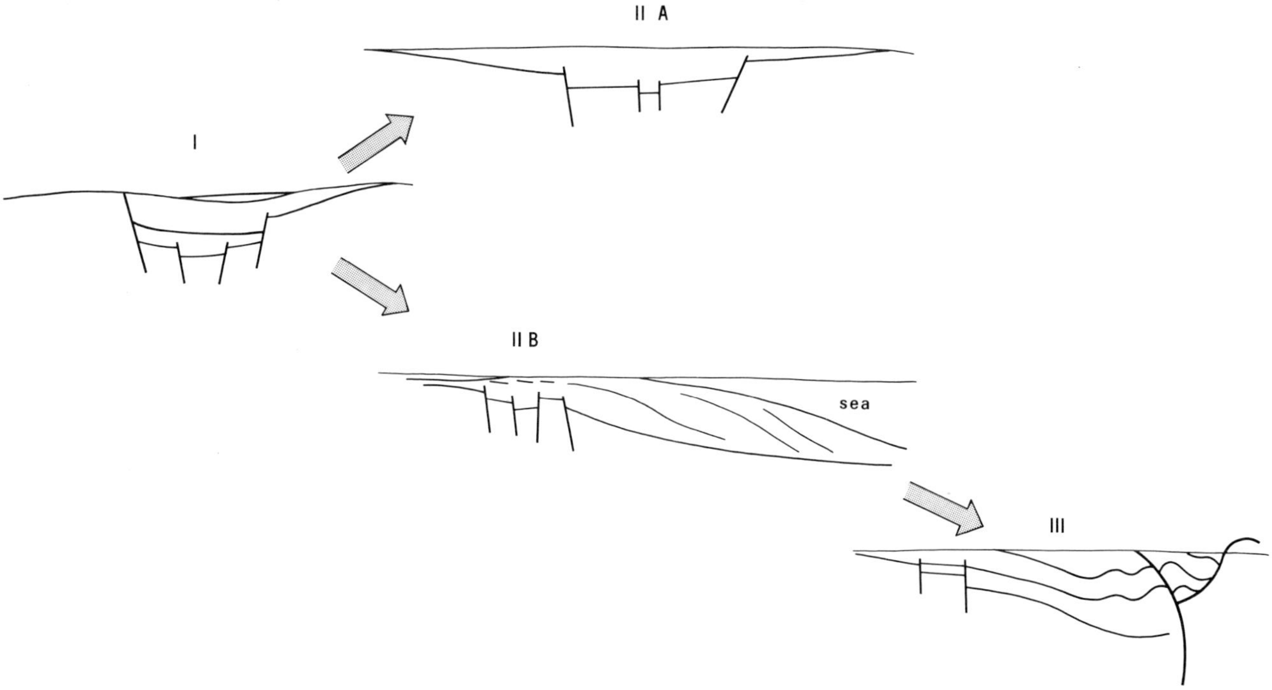

Fig. 11. — Diagrams of sedimentary basin kinematics.

1.2.1. **Juvenile stage**

The history of a sedimentary basin begins with a stage of active subsidence of the rift type. This youthful stage, the most important in the basin's history, appears to be more specifically characterized by :

- — high thermal gradients,
- — high subsidence, acting along faults or active marginal flexures, in a usually distensive tectonic framework,
- — variable sedimentation, primarily continental and mainly lacustrine and palustrine, with shales often rich in organic matter and detrital fans, possibly with evaporites in a desert climate,
- — centrifugal hydrodynamic gradients related to the progressive compaction of the deposits when these reach a sufficient volume, namely at the end of this period,
- — fracture tectonics in monocline panels usually sloping towards the basin margins and bounded by normal faults.

These geological characteristics evidently undergo complete evolution during this active subsidence phase. This appears to begin with a fracturation phase, brought about by the action of basement compartments, and materialized by normal faults. This is generally followed by a flexure stage, which results in the response of the basement to the sedimentary load.

This filling, which is generally discontinuous in space and time, causes incessant structural readjustments (Walcott, 1972). Some authors (Sweeney, 1976) estimate that 70 % of final subsidence could result from the response of the lithosphere to the sedimentary load, while compaction mechanisms cause a delay for the surface layers. The response time of the lithosphere is estimated at around one million years (Watts & Ryan, 1976).

The speed of subsidence may be a source of relatively deep facies initially, including turbidites corresponding to the starved basin phase (Aubouin, 1959). The environment becomes progressively shallower by an intensification of sedimentation and a gradual filling of the basin, and very rapid sedimentation rates also cause the formation of growth faults and rollovers. The deposits of these periods are distinguished by rapid variations in thickness, gradually diminishing upward in the series, and by undercompaction mechanisms, often resulting in shaly intumescences, as well as mud volcanoes.

During this initial period, the hydrogeological processes play a major role, closely combined with diagenetic mechanisms (see Section 1.2.2.). The interstitial water is progressively expelled from the deepest zones subjected to the highest pressures, and migrates towards zones with a weaker potential, laterally

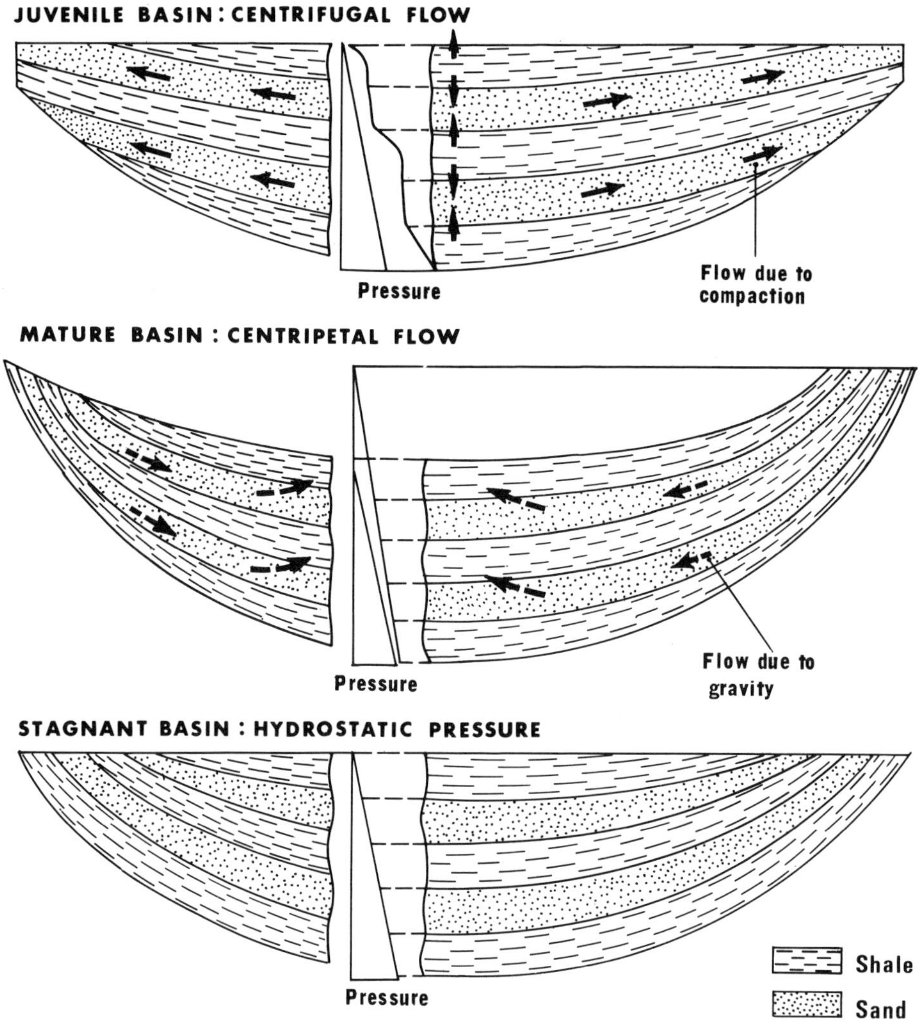

Fig. 12. — The three hydrodynamic types of sedimentary basin (Coustau *et al.*, 1975).

towards the edges and vertically through the beds into the surface layers. These movements, amounting to a few centimeters per year, thus obey a roughly centrifugal pattern (Figure 12).

To this force resulting from compaction is added the movement caused by the expansion of the deep aquifers under the effect of temperature. This flow occurs from the warmer zones to the colder zones, namely generally from the depths to the surface and the edges, and is added to the normal movement due to compaction.

With a geothermal gradient of 25 °C km^{-1}, the specific volume of the water increases from 1 at atmospheric pressure to 1.10 $cm^3 g^{-1}$ at a pressure of 800 bar, corresponding to a burial depth of 7500 m, or an increase of 10 % for a burial of 7500 m.

The expansion of the rock grains is not negligible, but generally less than that of the water. The ratio of the volume of water to the volume of rock grains at this depth is more than about 1/15 (corresponding to more than 6 % pore volume). The effect of expansion of the water exceeds that of the expansion of the rock, so that the water is set in motion. In the Gulf Coast, this 6 % porosity occurs above 7200 m.

From the chemical standpoint, the water expelled from the rocks undergoing compaction, and especially the shaly series, is generally subjected to electrofiltration mechanisms, and the solutions extracted, which generally flow into the most permeable drains, tend to display lower salinity than the solutions that remain imprisoned.

However, the compacted shales cannot withstand excessively high pressure differences and tend to fracture, so that the aquifers on each side of the membrane are placed in communication and differences in salinity are eliminated.

1.2.2. **Mature stage**

In intracratonic areas, following a certain evolution corresponding, from the diastrophism standpoint, to a progressive diminution in the flexures and in subsidence, and to a gradual transgression on the margins, and from the sedimentary standpoint, to a positive megasequence, the basin reaches a stage of consolidation and maturity comparable to the characteristics of platform basins.

The general framework remains distinguished by a general extension movement marked by the outward shifting and gradual attenuation of the marginal flexures, causing a lowering of the margins and a raising of the central area. Some action of horsts and grabens may be manifested, and sometimes volcanic occurrences also. Sedimentation becomes more homogeneous, the environment shallower, and chemical deposits may appear, the transgression on the basin margins preventing the flow of sediments into the basin. The geothermal gradients are lower, and the aquifers are in a hydrostatic régime.

This phase, which corresponds to a platform basin, is chiefly characterized by :

— great structural stability, always in a distension or shear pattern,

— relatively low geothermal gradients,

— shallow homogeneous facies, with a transgressive tendency in a positive sedimentary sequence.

A good illustration of this general evolution is the history of the Miocene Basin of Chélif, Algeria (Perrodon, 1969), the Tortonian Basin of Ploesti (Small, 1959), the Great Mesozoic West Siberian Basin (Rudkevich, 1976), and the Gippsland Basin (Weeks & Hopkins, 1967). It is also found in the Jurassic cycle of the French basins, illustrated by the French Jurassic Atlas.

This stage may mark the (temporary) end of the evolution of the basin, which is somewhat fossilized, or the onset of a new pull-apart stage, which may or may not be followed by convergence movements.

In pericratonic areas, the rift stage may evolve into passive margin or pull-apart basins. These basins assume a profoundly dissymmetric profile, extending from the emerged margin to the deep facies of the continental rise, consisting of a succession of deposits corresponding to the progressive spreading of an oceanic area : Continental rift formations at the base, passing gradually to open sea facies, often through confined environments contemporary with the drifting.

1.2.3. **Final stage**

Diastrophic activity may resume in a third stage, which may be qualified as the final stage and the death of the basin. The marginal flexures resume their action around progressively smaller areas, while compression mechanisms emerge, often associated with convergence movements. The sedimentary sequence, frequently comparable to molasses, corresponds to a negative megasequence. The basin fills up, while the marginal areas are folded and eroded.

Under the action of the reliefs, the aquifers are stressed by gravity and flow towards low potential zones, which are often the low points of the emergence of nappes. From these internal, low potential zones, the water can escape vertically through the beds. Here one can speak of predominantly centripetal movements. In these aquifers, moreover, the 'aggressive' meteoric waters occupy an important place. They can penetrate deeply into the basin along certain especially permeable levels, particularly along overthrust and unconformable areas, giving rise to a whole sequence of weathering and degradation, especially of organic matter and hydrocarbons.

The burial of shaly series, corresponding to rapid sedimentation rates, under sand masses, a frequent case of regressive cycles, generates great instability in the series, which is usually reflected by growth faults and by diapiric processes going as far as overthrusts (Chapman, 1974).

In a short time the basin may yield a folded range, which itself combats erosion for a while and then finally disappears, with its elements flowing into a new sedimentary basin.

This evolutionary scheme emerges from the analysis of a number of basin histories. It is illustrated in particular by the publications of Pruvost (1939) on the Saint-Etienne Coal Basin, Natland & Kuenen (1951) on the Pliocene Basin of Ventura, Renz *et al.* (1958) on the Orinoco Basin, and Sweeney (1976) on the Sverdrup Basin.

These different stages correspond to the progressive evolution from the 'basin of deposition' into the 'structural basin'. This history is more or less complete, and eventually may be subdivided by variable time intervals. It is depicted by the 'S' curves in the diagrams showing cumulative subsidence as a function of time (especially Dickinson & Yarborough, 1976) (Figure 13).

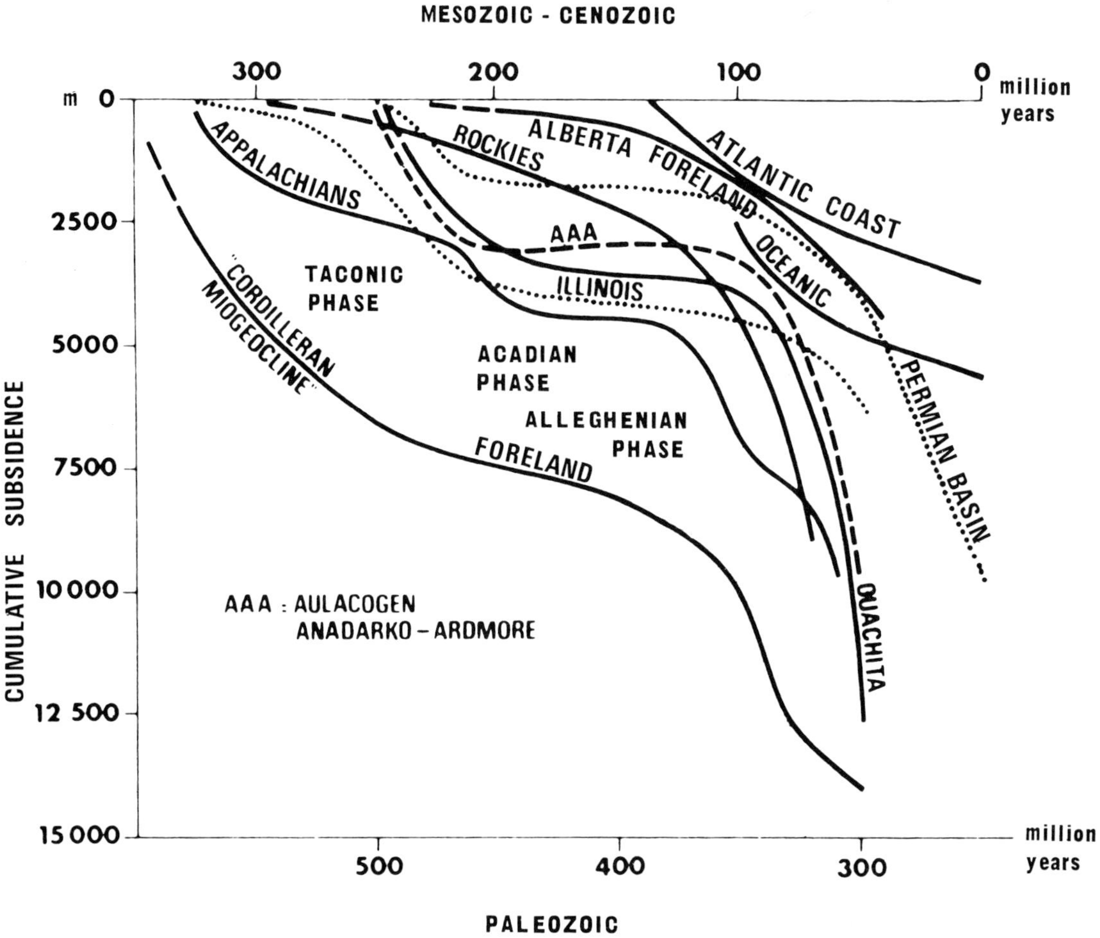

Fig. 13. — Diagram of cumulative subsidence versus time (Dickinson & Yarborough, 1976).

Of these three stages, the first, which corresponds to the active subsidence phase, is undeniably the most important, both with respect to the inherent geology of the basins and also that of petroleum provinces. This stage also appears to be the mandatory prelude to the different types of basin. Depending on the intensity, continuity and direction of the forces acting on the trough, the latter may evolve towards a platform or passive margin basin, possibly an ocean, or it may remain a simple graben.

From the petroleum standpoint, it is during this initial phase that most of the sediments accumulate, especially the basic sedimentary formations, the most deeply buried, which may perform the role of source rocks.

The mature stage often corresponds to favorable periods for the deposition of source rocks and seals, as well as the formation of reservoirs and stratigraphic traps.

The filling stage, which belongs to juvenile or final periods, generally favors the maturation of hydrocarbon source rocks. At this stage, what we tend to call 'mort terrains' or 'overburden' from the petroleum standpoint may also play an active role in the formation of oil accumulations.

If the lithospheric crust follows this pattern, not in an elastic manner but by a visco-elastic reaction, the sediments tend to be deposited in accordance with a generally regressive pattern, and the most recent deposits are confined to the central area.

Hence the sedimentary basins provide the geological framework for petroleum provinces, and the structural, stratigraphic and hydrogeological characteristics of the former closely condition the main economic features of the latter.

2. – SEDIMENTARY ENVIRONMENTS AND MECHANISMS AND THEIR PETROLEUM CONSEQUENCES

« Cette vue consiste à accepter le discontinu partout où il se présente et à le réintégrer dans le continu des échelles supérieures. »

('This view consists of accepting discontinuity wherever it occurs and reintegrating it within the continuity of a greater scale.')

E. Argand,
The Tectonics of Asia, 1922.

The sedimentary mechanisms which specifically condition petroleum geology and geodynamics are chiefly organized around subsidence and sedimentation mechanisms, which occur at different scales in time. The sediments constitute the 'flesh' represented here or there by the source rocks, and the hydrocarbons that they generate, the reservoirs where these may collect, and the seals that protect them. On another scale, more continuous in space and time, subsidence allows the accumulation of sediments and their transformations during burial.

Initially the dynamics of depositional environments allow, let us say, the fabrication of an assortment of elementary and varied bricks, and in a second stage, the basin dynamics use these materials and arrange them in a unique architectural pattern.

In accordance with the natural mechanisms, we shall deal briefly with the following :
— the characteristics of certain depositional environments,
— the main mineral transformations associated with burial,
preceded by a short review of certain general relationships between sedimentation and paleogeography.

The transformations of organic matter which are dealt with in detail in "Hydrocarbon Genesis" (Chapter 1 of Part 2), are only presented here for reference purposes.

On the basin scale, the great movements of the sea often dictate the mode of sedimentation and indirectly the petroleum characteristics. As a rule, one may note that the great transgressive periods correspond in the basins to a certain stability, and to relatively low sedimentation rates, as the bulk of the detrital sediments is halted in the alluvial plain zones. On the other hand, regressive intervals are characterized by rapid subsidence and sedimentation rates.

Sand bodies in the form of bars may however accumulate on the platforms during transgressive periods, often making excellent stratigraphic traps. The advance of the sea, materialized by marly, shaly and sometimes evaporitic levels, offers excellent seals that effectively protect the underlying detrital series. These formations often appear as the key component of a petroleum province.

2.1. CHARACTERISTICS AND RATES OF CERTAIN SEDIMENTARY ENVIRONMENTS

Depositional environments, which determine the lithological facies and main stratigraphic features, closely condition the petroleum characteristics of the sediments. These are basically the dynamic units resulting from the interaction of a substantially constant factor on the scale of the deposition process, and especially of the paleotopography, with short periodicity, hydrodynamic, biodynamic and climatic variables. The energy of the depositional environment thus appears to be fundamental criterion from the petroleum standpoint.

Low energy environments principally give rise to thick beds of carbonates, marls, and shales in an open marine environment, and sequences of similar lithology, but often rich in organic matter and evaporites in a confined environment, in other words, in most cases, rocks with slight or very slight porosity and permeability.

These environments, often corresponding to basins which are closed or silled, witness the accumulation of shales, marls and carbonates rich in organic matter, mostly of algal origin. Hence they offer a rich oil-generating potential and many effective seals.

Among environments with very low or zero energy, special mention should be made of the confined environments that may combine evaporites and sapropelic deposits, namely, in petroleum terms, seals and source rocks of high quality hydrocarbons, although the evaporitic environments only represent one possibility of the deposition of source rocks.

The black shales rich in algal or planktonic organic matter, which usually occur in the form of laminites, are actually outstanding for their petroleum-generating potential. They generally result from intense surface activity, usually planktonic, characteristically very limited in the number of species, and the absence of a benthic population.

These environments are frequently formed within basins or depressions of varying depths, containing perfectly stagnant and stratified water in a carbonate platform or archipelago landscape, particularly associated with reefs (Busson, 1978).

The higher energy environments, subjected in particular to currents, are the source of extensive sandstone deposition either from detrital or carbonate inflows, or often built up by animal or vegetable organisms, in the absence of inflows. In both cases, the sediments generally exhibit good porosity and permeability characteristics.

From the climatic standpoint, medium precipitation in a moderately humid climate (60 to 120 cm per year) offers the most favorable conditions, notably a sufficient quantity of shale, displaying favorable characteristics concerning their affinity for organic matter (Tanner, 1978).

*
* *

Since oil does not generally accumulate in the places of its birth, the transition zones through these different environments are of particular interest. Of special importance are the zones of facies change between confined deposits and reservoir rocks, and the facies associated by small angle unconformities or by faults. In other words, the positions of the different deposits within the basin, both by their morphology and by their history, are factors of fundamental importance.

2.2. MINERAL TRANSFORMATIONS RELATED TO SUBSIDENCE

Subsidence allows the accumulation of large volumes of sediments which are progressively buried and compacted under the weight of new deposits. As this burial proceeds, the deposits are subjected to increasing temperatures and pressures which engender a number of physical and chemical mechanisms that alter them progressively and thoroughly. These transformations are also closely related to the characteristics of the depositional environment, as we shall see further (Chapter 1 of Part 2). Their analysis leads to a separate examination of the actions of pressure, temperature, and also time, although these different factors are closely interrelated in reality.

2.1.1. External physical parameters

The action of pressure

In the course of burial, the sediments are subjected to growing pressures which compact them and reduce their volume, this process occurring more rapidly with finer-grained materials. As a rule, the rock grains form the mini-arches that support the structure. The water which saturates the intergranular spaces is then subjected to hydrostatic pressure, which can be explained by the weight of the water head from the loading point to the measurement point (1 bar per 10 m in the case of fresh water, 1.09 bar with very saline water).

Various types of detrital sedimentation, generally formed of inherited materials, occur at all depths, from the alluvial plains to the deep detrital fans and turbidites, through delta and platform deposits. In these conditions, the sediments generally exhibit variable facies and thicknesses. They display many unconformities and a relatively limited extension, giving them a diachronous arrangement that often makes stratigraphic correlations difficult. The rapid accumulation does not allow the sorting of deposits and rarely offers good porosity and permeability characteristics. Deposits of turbidites in the large deep detrital fans may nevertheless display good continuities.

However, in accordance with a complex and changing geography, better individualized environments may occur, locally displaying facies rich in organic matter, usually originating in the higher plants of the continent, as well as environments favorable to the development of porosity, in particular sandy bodies of the channel sands type, beach deposits and offshore bars.

With the exception of turbidite zones, sedimentation of the 'oceanic' type, characterized by :

— slight thicknesses and very slow accumulation rates,

— pelitic materials, mainly oxidized,

— unconformities and hiatuses,

are generally of little or no petroleum interest.

The burial rate of the sediments also plays an important role from the petroleum standpoint. In the absence of a perfectly reducing depositional environment, a high burial rate guarantees the protection of the organic deposits. Combined with high subsidence, it also offers the advantage, from the petroleum standpoint, of carrying these beds relatively quickly beyond the action of bacteria, which largely contribute to the destruction of the organic molecules.

Low burial rates combined with high energy levels are often reflected by resedimentation and the absence of organic matter. Rapid sedimentation rates and very low energies do not allow the deposition of reservoirs, but favor the accumulation and preservation of organic matter. In the case of sediments liable to high compaction, such as shales, the pressure causes a considerable and rapid decrease in the volume of pores. The water that occupied the bulk of the pores is expelled, and very large quantities may thus be released.

In the case of shales, an approximately linear relationship is observed between burial depth and the logarithm of their porosity. This means that the largest quantities of water are expelled in the first hundred meters of burial. In thick uniformly shaly series, without sand intercalations, the water is chiefly expelled vertically upwards. In series with sand intercalations, the flow is directed sideways towards the margins of the basin. In the case of the Tertiary of the Gulf Coast, where a shale/detrital unit is superimposed on an unchanging shaly series, vertical flows occur at deeper levels, and horizontal transfers above (Magara, 1976).

Compaction is thus reflected primarily by the mechanical expulsion of interstitial water from the sediment. As the water is expelled, compaction brings saline filtration mechanisms into play, the shaly levels performing the role of semi-permeable barriers as long as they are not fractured. This results in the addition of chemical action to the mechanical action, and the water expelled has a lower salinity than the initial sediment water.

The compaction of clastic sediments is dominated by mechanical and physical rearrangements of the grains. The compaction of carbonates and evaporites primarily obeys chemical mechanisms. In the first hundreds of meters of burial, carbonate compaction is substantially similar to that of detrital rocks. Chemical processes predominate below this level, especially recrystallizations.

In brief, by compacting the sediment, increased pressure :

— reduces its porosity, permeability and water content,

— increases its density, resistivity and speed of sound transmission,

— alters its chemical and mineralogical properties.

But the latter point also involves the temperature factor, which is intimately associated with the increase in pressure during burial. Finally, compaction is an irreversible process, at least beyond a burial depth of about 1200 m for shales, corresponding to a phyllite pore diameter less than 50 A. The compacted shale is modified very little of at all in case of erosion of the sedimentary load, and this serves to estimate the maximum degree of burial of a sedimentary series.

The increase in pressure with depth often reveals subnormal anomalies, and more often abnormal anomalies. The overpressures are generally caused by diagenetic or metamorphic transformations, as well as undercompaction mechanisms, which are analyzed further in this chapter.

The action of temperature

Since Arrhenius, the general effect of temperature is known to accelerate chemical reaction kinetics very substantially. Heat transfer by conduction in the earth's crust, dQ, is roughly governed by Fourier's Law :

$$dQ = -KS \frac{d\theta}{dz} dt$$

where :

K is the thermal conductivity of the medium,

S is the area crossed by the quantity of heat dQ during time d*t*,

$\frac{d\theta}{dz}$ is the temperature gradient or geothermal gradient.

The heat flow θ is given by the equation :

$$\theta = \frac{1}{S}\frac{dQ}{dt} = -K\frac{d\theta}{dz}$$

The geothermal factors, particularly the geothermal gradients, result from the heat flows and the thermal conductivity of the rocks.

Geothermal flows partly originate in rock radioactivity, and to an equal extent in a deep sub-crustal flow which may be strongly amplified locally by the rise of basic magmas. The heat evolved by certain chemical reactions, and the conversion into heat of the mechanical energy of orogenic deformations, are also active to a lesser degree. They are expressed in joules per square meter per second ($J\,m^{-2}\,s^{-1}$) and more routinely in calories per square meter per second ($cal\,m^{-2}\,s^{-1}$). These flows are relatively constant, on average, on the scale of the globe, at around 1.2 to 1.4 $\mu cal\,cm^{-2}\,s^{-1}$. Locally, heat transfers may be increased very substantially by vertical flows of water.

They generally exhibit very high values :

— in rift and fracture zones, and zones generally subjected to extension tectonics,

— in periods of distension and volcanic occurrences, namely, often during basin downwarping periods.

The thermal conductivity of the rocks depends on their lithological and petrophysical characteristics, and especially on their density. It is expressed by the measurement of the flow as the number of calories crossing an area of one square centimeter in one second, with a temperature gradient of 1 °C per meter, or in calories per square meter per second per degree Celsius per meter of thickness : $cal\,m^{-2}\,s^{-1}$ ($°C\,m^{-1}$), or, in SI, in joules per square meter per second per degree Celsius per meter of thickness : $J\,m^{-2}\,s^{-1}$ ($°C\,m^{-1}$).

It is very high in rock salt, medium in sandstones and carbonates, low in shaly series. The following values are given for compacted series located at a depth of about 2000 m :

— anhydrite and salt	13 to $14 \cdot 10^{-3}$	$cal\,m^{-2}\,s^{-1}$ ($°C\,m^{-1}$)
— compact sandstone	8	—
— dolomitic limestone	7	—
— limestone	5 to 6	—
— porous sandstone, granite, basalt	5 to 8	—
— shale	4 to 6	—
— undercompacted shale	3.5	—
— coal	1 to 2	—

Salt diapirs thus play the role of vents exhibiting, at equivalent depth, temperatures which are substantially higher than the adjacent series. Gradients of 35 to 50 °C km^{-1} were observed in the North Sea in the vicinity of salt domes and thick Tertiary series (Cooper, 1978). Generally and schematically, it may be considered that seals act as insulators, and reservoirs act as good conductors.

Geothermal gradients result mainly from these two factors. They exhibit a mean world value of 25 °C per kilometer (14 °F per 1000 ft), with variations from less than half to more than double. They often vary substantially in space and time. In the same basin, gradients may spread from 30 to 70 °C km^{-1}, as in the Rhine Graben and the Indonesian Basins, from 20 to 45 in the Gippsland Basin (southeast Australia), and from 17 to 27 in the Niger Delta. In the North Sea, for example, they vary by more than 5 °C km^{-1} with generally high values, 30 to 40 °C km^{-1} for the Tertiary, low for the Cretaceous, 10 to 30 °C km^{-1}, and are again high in the Jurassic, 40 to 100 °C km^{-1} (Carstens *et al.*, 1979). They are higher along the extensive fault and fracture zones in particular. In South Texas, for example, where the average gradient ranges from 20 to

30 °C km^{-1}, values of 55 to 100 °C km^{-1} have been recorded along certain faults (Cooper, 1978). Very low gradients are measured in the carbonate provinces, and gradients of around 5 °C km^{-1} have been measured in the Bahamas.

Especially low temperatures prevail in the Arctic regions, such as Siberia, where the average ground temperature today is —10 °C. It is also necessary to consider the proximity of the last glacial periods, which deeply extended the Arctic climate southward. In the great gas fields of northwestern Siberia, temperatures of 2 to 8 °C have been recorded at 500 m, and 22 to 30 °C at 1000 m.

It is difficult to reconstruct variations occurring during geological or paleogeothermal history. However, it is possible to observe and to measure certain properties of mineral or organic particles which have been affected by an elevation in temperature, and which have preserved these transformations as a recorder or a maximum thermometer.

Nor is it easy to establish relationships between the measured parameters and the temperatures which cause them to exist. The time factor is an important one and prevents the calibration of field measurements. Nomographs have nevertheless been compiled. But the data gathered, especially on organic particles, provide direct information on the maturation processes of organic matter and its conversion into hydrocarbons, which is generally the information sought, making it unnecessary to take account of paleotemperatures (Robert, 1980).

The following are among the most widely used markers;

— vitrinite reflectance (R_v),

— Thermal Alteration Index (TAI) estimated as the light transmitted by sporopollenin,

— T_m or Maximum Peak Temperature of a pyrolysis in standard conditions, a marker that is entering widespread use.

The following parameters are also exploited :

— magnetic resonance or Electron Paramagnetic Resonance (EPR) which offers the advantage of requiring a very small sample, furnishes parameters of which the most reliable is Ng, the number of free radicals, depending on temperature or rank, with a peak at 500 °C and R_v of 2.5 %,

— the Level of Organic Metamorphism (LOM) (Hood *et al.*, 1975) which offers the advantage of providing a linear relationship between temperature variations, in case of a constant geothermal gradient in the reference basin : this is a scale of corresponding values.

Mineral markers, particularly in the smectite → mixed layer shale → illite series, are not only sensitive to the temperature parameter, but also to the pressure and chemism or chemical reaction mechanisms of the sediment. Also used is the illite crystallinity index or the variation in the proportion of swelling mixed-layer shales. The transformation of smectite into illite today appears to be the mineral criterion most sensitive to an elevation in temperature (Kübler *et al.*, 1979).

It is quite feasible to consider present temperatures as peak values in basins subjected to continuous and recent subsidence, particularly in Cenozoic basins. In the Neogene basins of Los Angeles

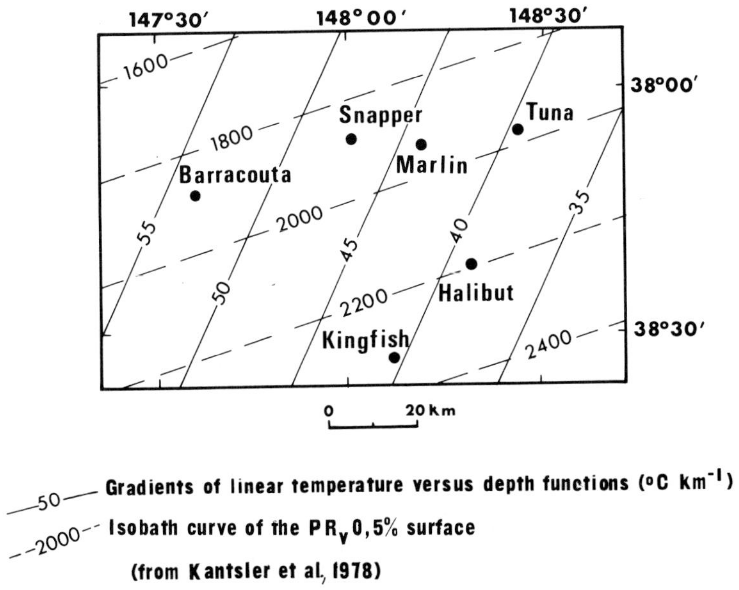

Fig. 14. — Map of Gippsland Basin.

and Ventura, for example, geothermal gradients of 24 to 35 °C per kilometer correspond to R_v gradients of 0.033 to 0.090 ‰. This comparison must nevertheless be made with caution, and significant discrepancies may appear between the paleogradients given by R_v and the gradients that may be observed in some areas, as in the Australian Gippsland Basin (Figure 14).

Moreover, past and present geothermal gradients vary significantly according to the framework of the basin. Higher temperatures occurring over longer periods in geological eras give rise to more profound diagenetic transformations. The action of heat thus tends to obliterate the imprint and the logic of the conditions of deposition and supplants them by its own.

This action of heat is especially important, and initially favorable, as we shall see below in speaking of hydrocarbon genesis, although it is negative for reservoir quality, except for the formation of secondary porosity. It also plays a major role, one which is probably less well known, in the expansion of aquifers and in their movement in case of a sufficiently rapid temperature rise, and finally in the petroleum wealth of a province.

Action of the time factor

Although chemical reactions occur more rapidly at higher temperatures, time may partly offset the heat flow. This has been observed in particular for the maturation of organic matter and the vitrinite reflectance (Figure 15). The latter requires higher temperatures in the younger series. This has also been observed in the results of laboratory investigations, in which organic matter is 'degraded' into oil and gas at temperatures from 400 to 500 °C in a few minutes.

These two parameters, time and temperature, are relatively indissociable, and any diagenetic transformation is an integral function of both of these variables, something that 'resembles' their product. The action of temperature cannot be defined independently of time, and vice versa.

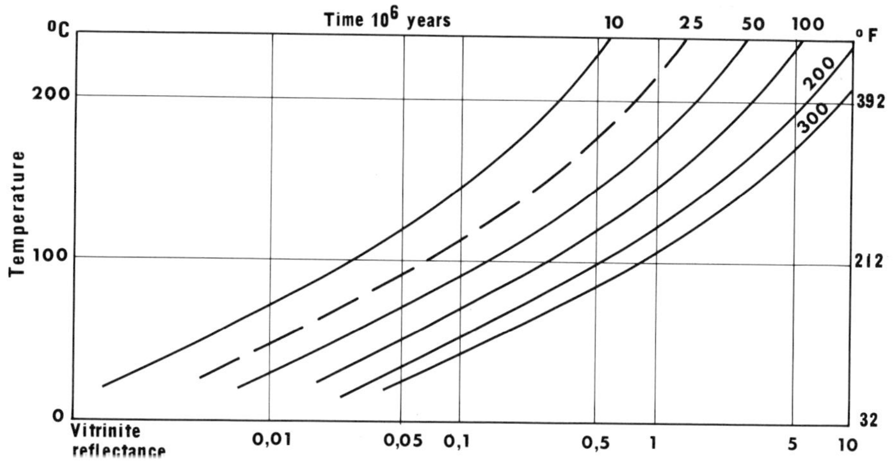

Fig. 15. — Diagram showing the variation in vitrinite reflectance versus temperature and time (Cooper, 1978).

2.2.2. **Transformation mechanisms (diagenesis, catagenesis and metagenesis stages)**

The action of temperature, pressure and solutions on sediments undergoing burial is reflected by a series of geochemical and mineralogical transformation mechanisms, which are particularly marked in argillaceous sediments. The term diagenesis is applied to transformations occurring within a vitrinite reflectance range of 0 to 0.5 ‰, corresponding approximately to maximum burial depths around 1200 to 1800 m. The term catagenesis applies to R_v between 0.5 and 2 ‰, or depths up to 4000 to 6000 m for average geothermal gradients. This level is the boundary of the realms of anchimetamorphism and metamorphism, which toll the knell of the sedimentary events.

However, it is important to note that these transformations are progressive, that these limits vary in accordance with the materials and the time of 'cooking', and are consequently very imprecise. It is also worth noting that diagenesis and catagenesis, both inorganic and organic, as well as hydrochemism, are only three aspects of the same ensemble (Scholle & Schluger, 1979).

2.2.2.1. *Diagenesis stage*

"Diagenesis is the sum of all physical, biochemical and physicochemical processes by which changes in a sediment (freshly deposited) are brought about after its deposition but before its final lithification (rock formation) » (Encyclopedia Britannica). Biological action, chiefly the action of bacteria, normally occurs in the first tens and exceptionally hundreds of meters of burial. It is first reflected by a strong reduction of the environment, which differs very soon from the aquatic environment in which the sediments are deposited. It also attacks complex organic molecules to yield CO_2, H_2O and CH_4.

Up to a depth of about 500 m, which corresponds to a sudden loss of porosity and the expulsion of large volumes of water, the characteristics of the rock, particularly its porosity and permeability, appear to be determined by the grain size, composition of clay mineral, and the concentration of electrolytes in the interstitial water, which plays an important role in mineral rearrangement.

In this first interval, the sediment is generally very porous and the waters with the salinity of the depositional environment flow freely within the pores. By bacterial action, this water is relatively acidic, and the Al and Fe occurring in solution in the upper portions are present in the form of amorphous hydroxides in the lower portion. This transformation causes increases in the speed of sound transmission and the resistivity of the sediment, which are added to the effects of compaction (van Elsberg, 1978).

Below a depth of around 500 m, temperature begins to play a predominant role simultaneous with the decrease in porosity and permeability of shaly sediments, and the various components are redistributed. The water is enriched with calcium and depleted of sulfates and carbonates, while the feldspars are dissolved and the pH becomes more alkaline.

These changes in the environment favor the appearance of sulfides, the incipient mobilization of silica, and the formation of CO_2, CH_4, H_2S and a complex polymerized organic substance, kerogen. The largest volumes of water are expelled in this phase, and the salinity of the water in place increases progressively with depth.

2.2.2.2. *Catagenesis stage*

This stage corresponds to the end of the principal compaction phase and marks the predominant action of the heat factor. As a rule, the sedimentary materials tend to reach equilibrium with the temperature conditions that they encounter, and these transformations are facilitated by the greater flow of water which guarantees the reversible character of the reactions. Most of the clay minerals are encountered in this zone.

Throughout these transformations, the minerals, and especially the clays, are kept constantly in chemical equilibrium with the interstitial water. This is the area in which smectite is transformed into mixed-layer minerals and illite in an alkaline environment, and where kaolinite can be transformed into dickite in an acidic environment.

In accordance with the increased proportion of potassium in the interstitial water, for example at temperatures around 100 °C and burial depths ranging from 2500 to 4500 m, smectite is converted from disordered to ordered mixed-layer clay minerals. If the potassium content of the water reaches high levels, illite represents the stable mineral. This transformation occurs with the elimination of extremely pure water at the start, which at this stage of catagenesis could play an important role in the primary migration of hydrocarbons. This expulsion of water appears to be widespread in catagenetic occurrences, and the quantities thus released may be as high as 15 to 17 % of the volume of the sediment (Price, 1976). It could contribute actively to hydraulic fracturation mechanisms.

Large amounts of silica are simultaneously dissolved, subsequently playing a major role in the cementation of sandstones (Dunoyer, 1970).

In the advanced catagenesis stages corresponding to a burial depth generally exceeding 4 to 6 km, pressures of 1000 to 1500 bar and temperatures of 150 to 200 °C, the environment becomes restricted, and in the absence of fractures the clays act as osmotic membranes which contribute to the concentration of brines. The argillaceous minerals are chiefly represented by illite and chlorite, of which the degree of order, and especially the crystallinity, increases with temperature and pressure. The organic substances, which are particularly fragile and unstable, are very sensitive to increases in temperature. This applies especially to carbonaceous particles, of which one of the properties, the vitrinite reflectance, is used as a marker of catagenetic evolution. It increases on the average by 0.15 % per 1000 m.

This phase witnesses the 'maturation' of the organic matter and its conversion into both coal and oil, and then into wet gas and finally dry gas (methane), with the proportions of these different constituents varying according to the composition of the initial organic matter. The reaction corresponding to this transformation gives rise to two types of product, one fluid, and the second condensed, more aromatic than the original organic matter.

2.2.2.3. *'Anchizone' or metagenesis stage*

The term anchizone, defined by Kübler (1967), is characterized by an illite crystallinity index of 4 mm, namely by a mineral criterion. The term metagenesis, as used by Tissot and Welte (1978), applies to series exhibiting an R_v equal to or greater than 2 %. These two words are hence not strictly comparable. These zones, the vestibule of the metamorphic area, are chiefly distinguished by the predominance of illite and chlorite in the clays, these minerals displaying high crystallinity, and by highly evolved organic matter, as the only organic substances capable of being present are anthracite and methane. The base of the anchizone corresponds approximately to the appearance of schistosity and a reflectance greater than 4 (Dunoyer, 1970).

In brief, these three zones can be broken down roughly into the following intervals of the main markers of the conversion of organic matter, R_v and TAI, and their equivalence in LOM.

Stage	*R_v %*	*TAI*	*LOM*
Diagenesis	0 to 0.5	0 to 2.5	0 to 7
Catagenesis	0.5 to 2	2.5 to 4.5	7 to 13.5
Metagenesis	2 to 4	$>$ 4.5	13.5 to 19

(Tissot & Welte, 1978)

2.2.3. **Burial anomalies : undercompacted series**

In the case of thick and rapid clay sedimentation, in which the particles have a large surface area per unit mass, and in the absence of detrital intercalations that play the role of drains, a delay occurs in compaction under the action of temperature, with the formation of undercompacted beds, under abnormally high pressures. Other geological events may also cause high pressures, at least locally :

— dehydration of smectite, which mainly appears to add its effects to those of undercompaction,
— osmotic processes triggered by differences in salinity,
— tectonic action (geodynamic pressure),
— carbonization of organic matter and the genesis of gas (Bradley, 1975; Magara, 1978).

From the mineralogical standpoint, these clay deposits are often characterized by the abundance of smectite. As in the present Niger Delta, this relationship may be explained by a convergence of particle size sortings of smectite muds with spores and pollens, and even algae (du Rouchet, unpublished report, 1974). Undercompacted shales thus consist of particularly fine clay particles, exhibiting large surface areas.

In these conditions, the interstitial water, as well as diagenetic water possibly derived from smectite, cannot escape regularly as burial continues. It remains imprisoned, preventing the normal reduction in porosity and transmitting a large share of the sedimentary load, thus giving rise to a sometimes considerable increase in pressure. In this case one speaks of lithostatic or geostatic pressures. The pressure gradients are extremely high, with increases of up to 2 bar per 10 m. This situation limits or prevents certain chemical conversions. The undercompacted shales play the role of a seal and thermal insulation, corresponding to high gradients and preserving high temperatures below. They exhibit abnormally high porosity and internal pressure, giving them considerable plasticity, and are distinguished by decreases in the speed of sound transmission, density and resistivity.

These formations appear to form very early during sedimentation. They may be preserved for tens of millions of years. Under the action of new sedimentary loads, and detrital deposits in particular — the case of progradating series — they are deformed into a series of sedimentary faults and folds, resulting in shaly intumescences which are perfectly comparable to the tectonics of salt, and which may even split at the surface to yield mud volcanoes. This process is called shale diapirism. These facies are encountered in particular in Tertiary deltaic formations, but also in Cretaceous and Jurassic formations, together with deposits associated with salt series. In Louisiana, for instance, abnormal pressure zones are encountered :

— locally, at the base of growth faults,
— regionally, in belts parallel to the paleoshorelines, in formations that contain the least sand.

They have also been identified in Nigeria — Akata Formation — in marine shale series situated below the sand/shale facies, and in the Mackenzie Delta Basin, in the Uppermost Cretaceous shales (Bituminous zone).

Undercompacted shales are often rich in organic matter and can thus become hydrocarbon source rocks. They can also act as a heat shield and hence delay hydrocarbon genesis, in comparison with normally compacted formations. The hydrocarbons formed can also migrate more easily in the transition zone, but this remains to be clearly demonstrated (J. Claret, oral communication).

Summary and conclusion

Sedimentary basins are crucibles in which hydrocarbons are produced from organic matter, within the framework of multifaceted sedimentation and diagenetic mechanisms. Hence the knowledge of sedimentary mechanisms forms the very basis of petroleum geology. The sedimentary logic falls within the broader framework of basin geodynamics, while obeying its own laws, just as the logic of the basins reflects, with a degree of independence, the grandiose framework of global tectonics. The same forces produce the same effects on different scales, from the basin to the simple tension split, while the sedimentary processes follow their own logic, like the stones that combine to produce the architecture of a building.

Like all other geological units, the main basin types offer a first approximation of the various steps in their history : rift valley or juvenile stage, mature stage and platform or stable area basins, senile stage and troughs in orogenic belts. In this way, sometimes with repetitions and convergences, each stage displays its successions of landscapes, its sequences of facies, some of them culminating in porous and permeable rocks, others in formations remarkably rich in organic matter, while yet others give rise to impermeable levels with high internal pressure. The development of these deposits then transforms these elements, subjects them to forces that fracture and fold them, facilitating or thwarting the movement of the liquids over long distances.

With its accumulating and transforming deposits, the sedimentary basin thus provides the setting in which the history of oil and gas accumulations is acted out.

REFERENCES

AUBOUIN, J. (1959). – A propos d'un centenaire : les aventures de la notion de géosynclinal. – *Rev. Géogr. phys. Géol. dyn.*, **2**, 3, 135-188, 28 fig.

AUBOUIN, J. (1972). – Chaînes liminaires et chaînes géosynclinales. – *Int. geol. Congr., 24th Sess., Canada (Montreal)*, **3**, 438-461, 6 fig.

BALKWILL, H. R. & KENNETH, J. H. (1979). – Sverdrup Basin response to cratonic events. – *Bull. amer. Assoc. Petroleum Geol.*, **63**, 3, 413 (Abstract).

BALLY, A. W. (1975). – A geodynamic scenario for hydrocarbon occurences. – *Proc. 9th World Petroleum Congr. (Tokyo)*, **2**, 33-44, 8 fig.

BALLY, A. W. & SNELSON, S. (1980). – Realms of subsidence. – *Mem. canad. Soc. Petroleum Geol.*, **6**, 9-94.

BEAUMONT, C. (1978). – The evolution of sedimentary basins on a viscoelastic lithosphere : theory and examples. – *Geophys. J. R. astron. Soc.*, **55**, 471-497.

BERNARD, A. (1976). – Metallogénie et tectonique des plaques. – *Géologues*, **37**, 45-52.

BRADLEY, J. S. (1975). – Abnormal formation pressure. – *Bull. amer. Assoc. Petroleum Geol.*, **59**, 6, 957-973, 23 fig., 4 tables.

BUSSON, G. (1978). – L'unité des faciès confinés en milieu de plateforme carbonatée. – Livre jubil. J. FLANDRIN, 87-112, 4 fig., Lyon.

CARSTENS, H. & FINSTAD, K. G. (1979). – Geothermal investigation of northern North Sea. – *Bull. amer. Assoc. Petroleum Geol.*, **63**, 3. 430 (abstract).

CHAPMAN, R. E. (1974). – Diapirism and overthrust faulting. – *Bull. geol. Soc. Amer.*, **85**, 1597-1602, 5 fig.

CHIARELLI, A. (1973). – Etude des nappes aquifères profondes. Thèse Bordeaux I. – Publ. Elf, 187 p.

COLEMAN, M. L., CURTIS, C. D. & IRWIN, H. (1979). – Burial rate : a key to source and reservoir potential. – *World Oil*, March, 83-92, 4 fig.

COOPER, B. S. (1978). – Estimation of the maximum temperatures attained in sedimentary rocks. – *In :* HOBSON, G. D. (ed.) : *Developments in petroleum geology*, **1**, 127-146, 6 fig. Applied Science Publ., London.

COUSTAU, H., RUMEAU, J.L., SOURISSE, C., CHIARELLI, A. & TISON, J. (1975). – Classification hydrodynamique des bassins sédimentaires. – *Proc. 9th World Petroleum Congr. (Tokyo)*, **2**, 105-119, 8 fig.

CURRAY, J. R. (1975). – Marine sediments, geosynclines and orogeny. – *In :* FISCHER, A. G. & JUDSON, S. (ed.) : Petroleum and plate tectonics, 157-224. – Princeton Univ. Press, Princeton.

DALLMUS, K. F. (1958). – Mechanics of basin evolution and its relation to the habitat of oil. *In :* WEEKS, L. G. (ed.) : Habitat of oil, 883-931, 23 fig. – *Amer. Assoc. Petroleum Geol.*

DAVIS & LISTER (1974). – Fundamentals of ridge crest topography. – *Earth planet. Sci. Lett.*, **21**, 405-413.

DEBELMAS, J. (1976). – Théorie des plaques et genèse des chaînes de montagnes. Implication sur la notion de chaînes « géosynclinales ». – *Géologues*, **37**, 25-44.

DERCOURT, J. (1971). – L'expansion océanique actuelle et fossile, ses implications géotectoniques. – *Bull. Soc. géol. France* (7), **12** (1970), 2, 261 317, 17 fig.

DICKINSON, W. R. & YARBOROUGH, N. (1976). – Plate tectonics and hydrocarbon accumulation. – *Amer. Assoc. Petroleum Geol.*, Short course.

DIETZ, R. S. (1973). – Collapsing continental rises : an actualistic concept of geosynclines and mountain building. – *J. Geol.*, **71**, 314-333.

DIETZ, R. B. & HOLDEN, J. C. (1974). – Collapsing continental rises : actualistic concept of geosynclines, a review. – *Spec. Publ., Soc. econ. Paleont. Mineral*, **19**, 14-25.

DUNOYER DE SEGONZAC, G. (1969). – Les minéraux argileux dans la diagenèse, passage au métamorphisme. – *Mém. Serv. Carte géol. Als.-Lorr.*, **29**, 320 p.

DUNOYER DE SEGONZAC, G. (1970). – The transformation of clay minerals during diagenesis and low-grade metamorphisis : a review. – *Sedimentology*, **15**, 3/4, 281-346, 15 fig.

ELSBERG, J. N. VAN (1978). – A new approach to sediment diagenesis. – *Bull. canad. Petroleum Geol.*, **26**, 1, 57-86, 16 fig.

FISCHER, A. G. (1975). – Origin and growth of basins. – *In :* FISCHER, A. G. & JUDSON, S. (ed.) : Petroleum and plate tectonics, 47-82. Princeton Univ. Press. Princeton.

FAURE, H. (1972). – Paléodynamique du craton africain. – *Int. geol. Congr., 24th Sess., Canada (Montreal)*, **3**, 44-50, 1 fig.

HALBOUTY, M. T. (1970). – World's giant oil and gas fields, geological factors affecting their formation, and basin classification. – *Mem. amer. Assoc. Petroleum Geol.*, **14**, 502-555, 9 fig.

HEDBERG, H. D. (1974). – Relation of methane generation to undercompacted shales, shale diapirs and mud volcanoes. – *Bull. amer. Assoc. Petroleum Geol.*, **58**, 4, 661-673.

HELING, P. (1970). – Microfabrics of shales and their rearrangement by compaction. – *Sedimentology*, **15**, 3/4, 247-260, 6 fig.

HERITIER, F. & VILLEMIN, J. (1971). – Mise en évidence de la tectonique profonde du Bassin de Paris par l'exploration pétrolière. – *Bull. Bur. Rech. géol. min.,* (2) **3,** 11-30.

HOOD, A., GUTJAHR, C.C.M. & HEACOCK, R. L. (1975). – Organic metamorphism and the generation of petroleum. - *Bull. amer. Assoc. Petroleum Geol.,* **59,** 6, 986-996, 7 fig.

JANOSCHEK, R. (1958). – The Inner Alpine Vienna Basin. – *In :* WEEKS, L. G. (ed.) : Habitat of oil, 1134-1152, 10 fig. – *Amer. Assoc. Petroleum Geol.*

KLEMME, H. D. (1975). – Giant oil fields related to their geologic setting. – *Bull. canad. Petroleum Geol.,* **23,** 1, 30-66.

KÜBLER, B. (1967). – Anchimétamorphisme et schistosité. – *Bull. Centre Rech. Pau-SNPA,* **1,** 2, 259-278.

KÜBLER, B., BETRIX, M. A. & MONNIER, F. (1979). – Les premiers stades de la diagenèse organique et de la diagenèse minérale. – *Bull. Ver. schweiz. Petroleum-Geol. u.-Ing.,* **45,** 1-22.

LE PICHON, X., FRANCHETEAU, J. & BONNIN, J. (1973). – Plate tectonics. – Elsevier, Amsterdam; 300 p.

LE PICHON, X. & PAUTOT, G. (1976). – Le fond des océans. – Presses univ. France, Paris, 125 pp.

LONG, G. & NEGLIA, S. (1968). – Composition de l'eau interstitielle des argiles et diagenèse des minéraux argileux. – *Rev. Inst. franç. Pétrole,* **23,** 1, 53-69, 4 fig.

MAGARA, K. (1976). – Water expulsion from clastic sediments during compaction, directions and volumes. – *Bull. amer. Assoc. Petroleum Geol.,* **60,** 4, 543-553, 14 fig.

MAGARA, K. (1978). – Compaction and fluid migration. – Elsevier, Amsterdam; 319 p.

MORGAN, J. (1975). – Heat flow and vertical movements of the Crust. – *In :* FISCHER, A. G. & JUDSON, S. (ed.) : petroleum and Plate Tectonics, 23-46. Princeton Univ. Press, Princeton.

NATLAND, M. L. & KUENEN, P. H. (1951). – Sedimentary history of the Ventura basin, California, and the action of turbidity currents. – *In :* HOUGH, J. L. (ed.) : Turbidity currents and the transportation of coarse sediments into deep water. – *Spec. publ. Soc. econ. Paleont. Mineral.,* **2,** 76-107.

NORTH, F. K. (1971). – Characteristics of oil provinces : a study for students. – *Bull. canad. Petroleum Geol.,* **19,** 3, 601-658.

PERRODON, A. (1969). – Esquisse d'une géologie dynamique des bassins sédimentaires. – *Sci. Terre,* **14,** 4, 301-328, 20 fig.

PERRODON, A. (1971). – Essai de classification des bassins sédimentaires. – *Sci. Terre,* **16,** 2, 191-227, 28 fig.

PERRODON, A. (1972). – Réflexions sur la comparaison de quelques vitesses de phénomènes géologiques. – *C. R. Soc. géol. France,* **2,** 50-52.

PERRODON, A. (1976). – Logique des bassins sédimentaires. – *C. R. Acad. Sci.* (Paris), **D 283,** 1265-1268.

PITMAN, W. L. (1976). – Sea level changes as the cause of transgressive and regressive sequences of continental margins. – Abstract, Lamont.

POULET, M. *et al.* (1975). – Tectonique globale et évolution structurale des bassins sédimentaires. – *Proc. 9th World Petroleum Congr.* (Tokyo), **2,** 19-32, 8 fig.

PRICE, L. C. (1976). – Aqueous solubility of petroleum as applied to its origin and primary migration. – *Bull. amer. Assoc. Petroleum Geol.,* **60,** 2, 213-244, 19 fig., 9 tableaux.

PRUVOST, P. (1930). – Sédimentation et subsidence. – Livre jubilaire Soc. géol. France, 545-564.

PRUVOST, P. (1939). – Quelques observations sur le phénomène de plissement faites dans les bassins houillers. – *Bull. Soc. géol. France,* **19,** 6, 307-319.

RENZ, H. H. *et al.* (1958). – The Eastern Venezuelan Basin. – *In :* WEEKS, L. G. (ed.) : Habitat of oil, 551-600, 21 fig. – *Amer. Assoc. Petroleum. Geol.*

ROBERT, P. (1980). – Kerogen optical evolution applied to oil and gas exploration and geothermal history. – *In :* DURAND, B. (ed.) : Kerogen. Technip, Paris; 519 p.

RONA, P. A. (1973). – Relations between rates of sediment accumulation on continental shelves, sea floor spreading and eustacy. – *Bull. geol. Soc. Amer.,* **84,** 2851-2872, 16 fig.

RUDKEVICH, M. Y. (1976). – The history and dynamics of development of the West Siberian platform. – *Tectonophysics,* **36,** 1-3, 275-287, 9 fig.

SCHOLLE, P. A. & SCHLUGER, P. R. (1979). – Aspects of diagenesis. – *Soc. econ. Paleont. Mineral.,* spec. Publ., **29.**

SMALL, W. S. (1959). – Thrust faults and ruptured folds in Romanian oil fields. – *Bull. Amer. Assoc. Petroleum Geol.,* **43,** 2, 455-471, 7 fig.

SWEENEY, E. F. (1976). – Evolution of the Sverdrup basin, Arctic Canada. – *Tectonophysics,* **36,** 1-3, 181-196, 8 fig.

TANNER, W. F. (1978). – Future oil : what is the outlook ? – *World Oil,* Nov., 123-140.

TISSOT, B. & WELTE, D. H. (1978). – Petroleum formation and occurence. – Springer, Berlin, 538 p.

VIDAL, J., JOYES, R. & VEEN, J. Van. (1975). – L'exploration pétrolière au Gabon et au Congo. – *Proc. 9th World Petroleum Congr.* (Tokyo), **3,** 149-165, 7 fig.

WALCOTT, R. I. (1972). – Gravity, flexures and the growth of the sedimentary basins at a continental edge. – *Bull. geol. Soc. Amer.,* **83,** 1845-1848, 6 fig.

WATTS, A. B. & RYAN, W. B. (1976). – Flexure of the lithosphere and continental margin basins. – *Tectonophysics,* **36,** 1-3, 25-44, 8 fig.

WEEKS, L. G. (1952). – Factors of sedimentary basin development that control oil occurence. – *Bull. amer. Assoc. Petroleum Geol.,* **36,** 11, 2071-2124.

WEEKS, L. G. & HOPKINS, B. M. (1967). – Geology and exploration of three Bass Strait Basins (Australia). – *Bull. amer. Assoc. Petroleum Geol.,* **51**, 5, 742-760, 16 fig.

WEGMANN, E. (1950). – Le spectre des mouvements de l'écorce terrestre. – *Int. geol. Congr., 18th Sess., England* (London, 1948), **4**, 94-99.

WHITTEN, E. H. T. (1975). – Geodynamic significance of spasmodic Cretaceous rapid subsidence rates. – Continental shelf USA. – *Tectonophysics,* **36**, 1-3, 133-142, 3 fig.

WILCOX, R. E., HARDING, T.P. & SEELY, D.R. (1973). – Basic wrench tectonics. – *Bull. amer. Assoc. Petroleum Geol.,* **57**, 1, 74-96, 17 fig.

PART II

HYDROCARBON FIELDS

« ... *Les progrès réalisés dans les connaissances, dans les techniques, sont la source de perpétuels rebondissements et de constants rajeunissements de problèmes relatifs à une région donnée.*

... *La rapidité d'évolution de tout ce qui touche au pétrole rend cet état de choses inévitable : c'est une raison supplémentaire de se défier de tout ce qui est systématiquement dogmatique.* »

'... Advances made in knowledge and technology are the source of new developments and constant reshapings of problems concerning a given region.

... The speed of changes affecting everything relating to oil makes this state of affairs inevitable. This is one more reason for distrusting anything that is systematically dogmatic.'

R.G. Lévy.

1. – HYDROCARBONS

« ... *Au-delà des éléments, il y a le fait qu'ils forment un tout.* »

'... And beyond the elements lies the fact that they form a whole.'

Georges DUMEZIL.

1.1. CHARACTERISTICS OF CRUDE OILS AND NATURAL GASES

As opposed to kerogen, which is insoluble in organic solvents, oils and gases form the great family of bitumens (from the Latin *bitus,* resinous wood). The word asphalt — from the Greek *asphaltos* — corresponds more generally to thick or solid hydrocarbon substances. The word 'naphtha', of Mesopotamian origin, derives from one of the oldest known languages, from *napata,* to burn, more specifically applies to liquid petroleum products, and is found in the Arabic word '*al naft*'.

In technical language, bitumen is a derivative of petroleum. It exists in the natural state, in which case it is an alteration product of crude oil. It is also an industrial product, the heavy fraction produced by the distillation of certain crude oils. In the petroleum world, asphalt also designates a sedimentary rock impregnated with bitumen. The words *petroleum* and *Erdöl* are neologisms which mean stone oil, and which may appear inaccurate today in the light of our knowledge of the biological origin of crude oils. But the rocks sometimes have the same lineage.

Crude oils and natural gases are fluids which generally occur in the liquid state or the gaseous state in a given set of conditions. They consist essentially or chiefly of hydrocarbons, namely substances basically made up of carbon and hydrogen. One of the essential properties of these substances is to dissolve intimately in each other. In subsurface conditions, a crude oil may thus contain light alkanes that can be separated into a gas phase and heavy products which, if isolated, would normally occur in the solid state. However, these mutual solution possibilities are not unlimited, and vary in accordance with the hydrocarbon species, while benzene and toluene are the best solvents.

Conversely, resins and asphaltenes can 'precipitate' by the dissolution in water of the most soluble hydrocarbons, particularly the aromatics. This leads to the process known as crude oil 'deasphalting'. The same process can be triggered by the addition of methane, as the increase in the proportion of this gas causes asphaltenes to precipitate. Similarly, a drop in temperature and pressure can cause the precipitation of heavy paraffins in paraffinic crudes, as observed in Nigeria and Indonesia.

Due to these 'associative' properties, crude oils and natural gases are mixtures, in various proportions, of different hydrocarbon compounds and a number of other chemical compounds. They are complex substances which vary in accordance with the characteristics of the environment in which they are found, which are born, transformed, degraded, disappear, and consequently to some extent, live, like all natural objects. They represent a transition stage in the vast carbon cycle. Their physical and chemical properties provide an image of these mixtures, which are themselves consequences of the geological framework and history.

Nearly 500 chemical substances have been isolated in the different crude oils analyzed throughout the world, not including organometallic compounds. However, only 150 account for more than half of the components of crude oils. About 200 of these compounds are not hydrocarbons, but these are generally present in low concentrations. Exceptionally, certain heteroatomic 'nonhydrocarbon' compounds form the bulk of some heavy crudes, which have generally evolved very little if at all. Hence the Venezuelan 'Boscan', a Tertiary crude oil, known for its poor quality, only contains 35 to 38 % hydrocarbons.

As a rule, oils and gases can be grouped together under the term of hydrocarbons, which make up the bulk of their content.

1.1.1. Composition of hydrocarbons

1.1.1.1. *Chemical composition*

From the chemical standpoint, hydrocarbons are divided in practice into three major families :

— saturated hydrocarbons : alkanes or paraffins,

— unsaturated hydrocarbons : alkenes, naphtheno-aromatics and aromatics,

— resins and asphaltenes.

Saturated hydrocarbons

As a rule these are quantitatively the most important, with 50 to 60 % of all components. They are distributed in crude oils in three main families.

N-alkanes or n-paraffins (from the Latin *parum affinis,* little affinity) include saturated hydrocarbons with a linear chain and the general formula C_nH_{2n+2}.

They constitute the more or less long chains from C_1 to C_{40}, and the C_5 to C_7 group is generally most widespread. These hydrocarbons occur in three states :

— gas from C_1 to C_4,

— liquid from C_5 to C_{15},

— solid (natural paraffins) from C_{16}.

The n-alkanes with an odd number of carbon atoms, preferentially synthesized by living organisms, are true biological markers. The predominance of low molecular weight n-alkanes (C_{15}, C_{17}, C_{19}) is mainly observed in microscopic algae, whereas in the higher plants, n-alkanes with higher molecular weights (C_{21}, C_{23}, C_{25}, etc.) are encountered exclusively. On the other hand, light alkanes from C_2 to C_8, which are not present in living organisms, but abundant in crude oils, are the transformation products of organic matter after its deposition (Silverman, 1971).

The n-alkanes account for about 15 to 20 % of crude oils. They are especially important in crude oils derived from the organic matter of the higher plants. Their proportion tends to increase during diagenesis. Biodegradation reduces them, as certain bacteria feed preferentially on these substances.

Isoalkanes exhibit branches (methyl groups) in the C_6-C_8 families. Some compounds, alkanes with an isoprenoid chain such as pristane (C_{19}) and phytane (C_{70}), are derived from the molecules present in living organic matter (phytil chain of chlorophyll). Like some n-alkanes, they may be considered as true biochemical fossils.

Cycloalkanes or naphthenes are formed of cyclic hydrocarbons. Their general formula is C_nH_{2n}. Their proportion in crude oils ranges from 20 to 40 %, with an average of around 30 %.

Unsaturated hydrocarbons

This family essentially includes the aromatics, so-called because of their pleasant odor, and the naphtheno-aromatics. The latter are often combined with sulfur-bearing compounds characterized by very unpleasant odors. They account for an average of 20 to 45 % by weight of crude oils.

— True aromatics are molecules formed exclusively by aromatic rings — generally four or five : the simplest is benzene C_6H_6.

— Naphtheno-aromatics consist of one or more condensed aromatic rings, combined with naphthenic hydrocarbons and alkyl chains. They are extremely abundant in young and immature crudes, and also in oils that have undergone strong catagenesis. They are also often combined with sulfur-bearing compounds.

— Alkenes or unsaturated acyclic hydrocarbons are very unstable and therefore extremely rare, but not totally absent from natural crude oils.

Resins and asphaltenes

These are complex compounds with high molecular weight, relatively rich in N, S and O, and also nickel and vanadium. They form the heaviest fraction of crude oils. They appear to be formed chiefly of polyaromatic or naphtheno-aromatic nuclei and chains, partly heteroatomic, including compounds containing sulfur, nitrogen and oxygen. The resins have a molecular weight of 500 to 1200, the asphaltenes up to and above 50,000, and their diameter is about 40 to 50 Å on average.

Resins are relatively more soluble than asphaltenes in organic solvents and light alkanes. The former include acids and esters, and are generally slightly less aromatic than the latter. Resins are unstable in air and light and tend towards the structure of asphaltenes.

These substances account for 0 to 40 % of crude oils, with an average around 20 %. They are encountered abundantly in immature crudes and also altered and biodegraded crude oils, particularly tar sands. They account for 50 % or more of bitumens disseminated in sediments. Note that asphaltenes do not have any equivalent in living matter, and are a product of diagenesis/catagenesis.

1.1.1.2. *Isotopic composition*

The carbon in hydrocarbons is represented by two isotopes ^{12}C and ^{13}C, whose ratio $\delta^{13}C$, in relation to that of a standard sample, is given by the formula :

$$\delta^{13}C = \frac{(^{13}C/^{12}C) - (^{13}C/^{12}C)\ \text{standard}}{(^{13}C/^{12}C)\ \text{standard}} \times 1000$$

If $\delta > 0$, the sample is enriched in heavy isotope, and if $\delta < 0$, it is depleted. In terms of the value of δ, the accuracy of measurements of the isotopes of C is around $\pm$ 0.1 to 0.2 ‰, or 1 to 2 ppm of ^{13}C in relation to total C. In marine carbonates, the δ value oscillates some 3 ‰ about zero, and this is explained by the fact that the international standard is a belemnite (Figure 16). Non-marine carbonates are lighter, around — 5 $\pm$ 3 ‰/PDB.

Organic substances are always lighter than carbonates. The δ of marine organisms are statistically higher than those of continental organisms (by about 5 ‰). This difference is explained by the fact that marine plants use the carbonate and bicarbonate ions of sea-water by photosynthesis, while land plants utilize atmospheric CO_2 ($\delta \cong 7$ ‰/PDB).

This difference of 5 ‰ is also found in oils of marine origin and those of continental origin. The δ value for all oils and kerogens generally ranges from — 32 to — 2 ‰. For methane the range is — 85 to — 30 ‰, and in rare cases — 20 ‰. The carbonates are always heavier, and their δ value varies according to their origin.

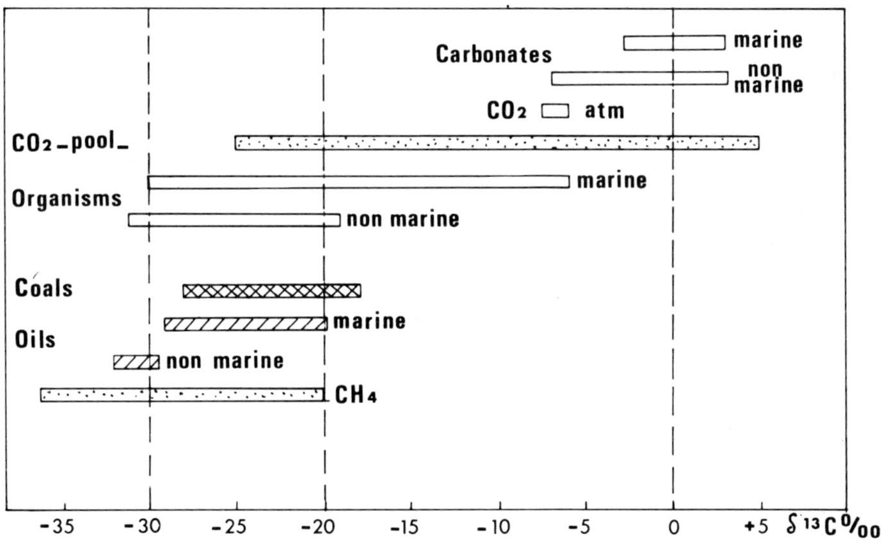

Fig. 16. — Variation in values of δ ^{13}C ‰ for different rocks.

1.1.2. **Principal hydrocarbon families**

In order of greater complexity, we shall examine the following in succession :
— natural gases,
— oils or liquid products,
— tar sands, oil shales and hydrates, or solid products.

These subdivisions are actually approximate and somewhat vague, since hydrocarbons form a continuous chain in nature. Pressure and temperature conditions prevailing deep in the earth superimpose the light oil and gas areas, and in these conditions the latter may be found in a liquid phase.

More accurately, it is customary to draw a distinction between dry gases and wet gases among the gaseous materials, and light oils (and condensates) and heavy oils among the liquid hydrocarbons. Dry gases consist essentially of methane — at least 97 % — and very small amounts of C_2, C_3, etc. Wet gases display high proportions of 'condensable' gases C_4, C_5, C_6, etc., and light liquid fractions (condensate).

Light oils consist chiefly of the light fractions of liquid hydrocarbons and gases, and heavy oils of the heavy fractions of liquid hydrocarbons, which keep the solid substances in solution. These heavy components may result from the absence of light products :

— by non-formation : immature oils,

— by elimination, alteration and biodegradation : degraded oils.

More or less solid materials are designated by the all-purpose term of bitumens or asphalts, while the term paraffin designates paraffinic hydrocarbons with C_{17} and higher, and clathrates are the hydrates of methane.

1.1.2.1. ***Natural gases***

The natural gases found in sedimentary basins often consist mainly of hydrocarbons, chiefly methane. They occur :

— in individual gas pools,

— associated with oil pools, saturated with gas and exhibiting a gas cap, or unsaturated and essentially in solution in the liquid phase.

Chemical composition

Methane is generally the essential constituent of natural gas, as much as 99 % or even 100 % in some pools. The remaining alkanes, ethane, propane, butane etc., account for decreasing proportions in the range of a few percent. Boiling points are — 165 °C for methane, — 42 °C for propane (dry gases), and 0.6 °C for butane for example (wet gas).

The isotopic composition of methane varies approximately from — 85 to — 20 ‰. These differences appear to be related essentially to the degree of maturation of the organic matter and hydrocarbons, the gas becoming increasingly 'heavier' as the diagenesis process is more complete (Figure 17). Hence methane of bacterial origin exhibits a δ of about — 90 to — 70 ‰, and gases resulting from weak diagenesis have a δ of — 70 to — 40 ‰. The gases formed concurrently with liquid hydrocarbons generally range from — 50 to — 40 ‰. Finally, the so-called diagenetic gases, or preferably catagenetic gases, usually exhibit values lower than 45 ‰, and possibly as low as — 30 to — 20 ‰.

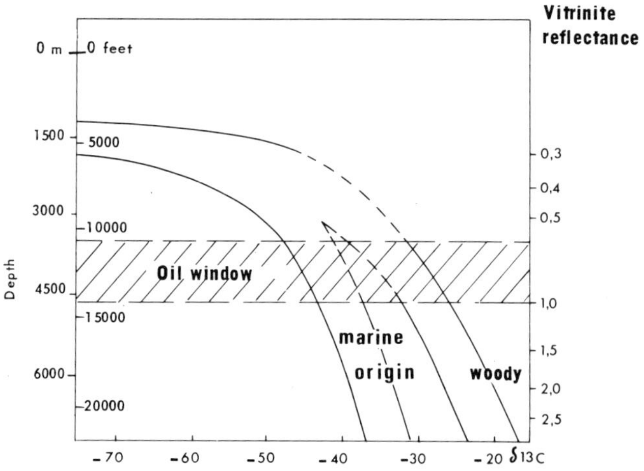

Fig. 17. — Diagram showing the variation in δ ^{13}C of methane versus vitrinite reflectance (Stahl, 1977).

Other constituents

Natural gases may contain variable amounts of other non-hydrocarbon gases, especially H_2S, CO_2 and N_2.

Together with carbon dioxide, hydrogen sulfide is a standard product of the diagenesis and catagenesis of organic matter, especially at great depths, by the action of methane on sulfates, and also due to magmatic activity. It is also produced by the bacterial reduction of sulfates, as observed in some lakes and in the Black Sea. It appears routinely in many provinces, especially in the Canadian Rockies. The proportion of H_2S is up to 15 % in the Lacq gas field, and as high as 88 % in a Devonian pool in Alberta. Gases containing H_2S, even in traces, are called 'sour' gases, as opposed to 'sweet' gases, which contain no H_2S.

Carbon dioxide appears initially at different stages in the sedimentation of organic matter, by oxidation, fermentation or decarboxylation of its constituents. It is also a product of metamorphism, especially of carbonates. Some regions, like the eastern part of West Germany, the Pannonian Basin, and New Mexico, are particularly rich in CO_2 ; some pools containing a gas consisting of 100 % CO_2. Carbon dioxide may also be formed by the oxidation of hydrocarbons by sulfates. In some cases, as in Hungary, the Gulf of Gabès, and in Sicily in the Ragusa Field, CO_2 may be associated with oil pools. The isotopic composition of CO_2 resulting from the oxidation of organic matter is close to that of the kerogens, around —25 to —15 ‰. That of carbon dioxide of volcanic origin is heavier, around —6 to —4 ‰. In many cases, the δ value is close to that of marine carbonates, + 3 to — 3 ‰.

Nitrogen, produced by diagenesis, or of crustal origin, is found in some fields, especially in North Germany and the Netherlands and in the adjacent part of the North Sea, in Oklahoma, and in Texas. The great gas fields of Slochteren in Holland and in the Texas Panhandle in the United States contain 14 and 11 % respectively. This percentage may be as high as 100 % in some pools. Nitrogen appears to exist more frequently in carbonate and evaporitic series than in shaly environments, where it is absorbed by the clays (Welte, Panel Discussion : Time and Temperature Relation, 9th World Petroleum Conference, 1975, **2**, 203). In the case of coal formations, it appears that the proportion of nitrogen in comparison with methane increases rapidly above an R_v of 3.2. In Holland, this value corresponds to a burial depth of about 5000 m (van Wijhe *et al.*, 1980).

All these incompatible gases generally exhibit more drawbacks than advantages, especially from the economic standpoint, if only because they supplant hydrocarbons in the traps. Whereas H_2S may offer a large source of sulfur, it is viciously poisonous and a cause of pollution and degradation, requiring its removal.

Helium and argon, often associated with nitrogen, are generally of crustal origin, and an increase in the percentage of the former is normally observed from the Tertiary to the Paleozoic, in a ratio of 1 to 2 and more. These two rare gases obviously have considerable economic value, even if they only occur in very small amounts.

Physical and economic characteristics

The physical properties of natural gases faithfully reflect their chemical compositions.

The heating value in particular, which accounts for practically their entire economic value, varies in accordance with the hydrocarbons in the composition and the other gas constituents. It ranges on the average from 38 to 40 MJ/kg or 9500 to 10,000 cal/g. Some examples are given by the gases from the following fields :

— Ripalta (Po Plain) (00 % C_1) : 8600 to 9500 cal/g or 34.4 to 38 MJ kg^{-1},
— Cortemaggiore (Po Plain) (92 % C_1) : 8150 to 9050 cal g^{-1} or 32.6 to 36.2 MJ kg^{-1},
— Frigg (North Sea) (95 % C_1, 4 % C_2, 0.8 % $N_2 + CO_2$) : 38 MJ kg^{-1},
— Hassi Rmel (Algerian Sahara) (75.6 % C_1, 9.3 % C_2) : 39 MJ kg^{-1},
— Groningen (Holland) (81 % C_1, 14 % N_2) : 8000 cal g^{-1} or 32 MJ kg^{-1},
— Lacq (France) (69 % C_1, 9.65 % CO_2, 15.3 % H_2S) : 38 MJ kg^{-1},
— Ekofisk (North Sea) (83 % C_1, 9 % C_2, 2.3 % $N_2 + CO_2$) : 11,500 cal g^{-1} or 46 MJ kg^{-1}.

The solubility of methane in water is relatively high, about 5 to 10 g dm^{-3} at 25 °C. It decreases with increasing temperature (up to 80 °C) and salinity, rises with increasing pressure to a value of around 70 bar, when the appearance of methane hydrate considerably decreases the dissolved volumes (Figure 18). Solubility decreases with an increase in the number of carbons (McAuliffe, 1979). The solubility of gaseous hydrocarbons in liquid hydrocarbons is much higher.

In relation to air, the density varies from 0.55, the density of methane, for gases consisting essentially of this hydrocarbon, to values of about 1 for wet gases rich in condensate or in H_2S and CO_2. The density of alkanes increases rapidly with atomic number :

$$C_1 = 0.55 \qquad C_2 = 1.04 \qquad C_3 = 1.52 \qquad C_4 = 2.00$$

The density in relation to water, or in g/cm^3, ranges from 0.00073 for methane at atmospheric pressure to 0.5 for a standard mixture at high pressure, between 350 and 700 bar, or 5000 to 10,000 psi (Schowalter, 1979).

Compressibility is another important property of natural gases, and in the conditions prevailing in the pool, their volume may be reduced by compression by a factor of 200 to 300.

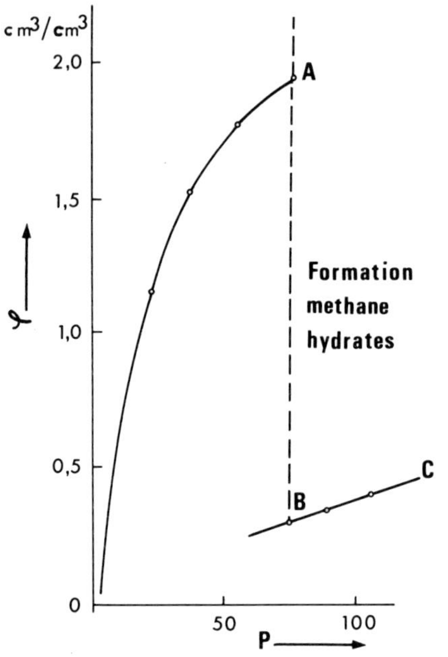

Fig. 18. — Diagram showing the variation in solubility of methane versus pressure (Makogon, Tsare & Cherski, 1971).

1.1.2.2. ***Crude oils***

Chemical composition

Crude oils generally consist mainly of hydrocarbons and polar compounds (resins, asphaltenes, etc.).

Hydrocarbons

Crude oils are formed mainly of liquid hydrocarbons, but the latter also contain gaseous and solid hydrocarbons in solution, in variable amounts. Light crude oils display a high proportion of the former, while heavy crudes contain large amounts of the latter. These different substances recover their individual identity in the distillation process. The isotopic composition of the oils generally ranges from — 30 to — 20 ‰, similar to the kerogens. Crude oils of continental origin, which are rich in waxes, are lighter, around — 32 to — 28 ‰, than the majority, which are of marine origin (Figure 19).

The δ value for oils is often closely comparable or slightly lighter, by 1 or 2 ‰, than that of the associated kerogens likely to be the parent formation and ranging from — 32 to — 21 ‰. As a rule, the isotopic composition becomes steadily lighter from the kerogens towards the saturates, and including the asphaltenes, resins and aromatics (Figure 19).

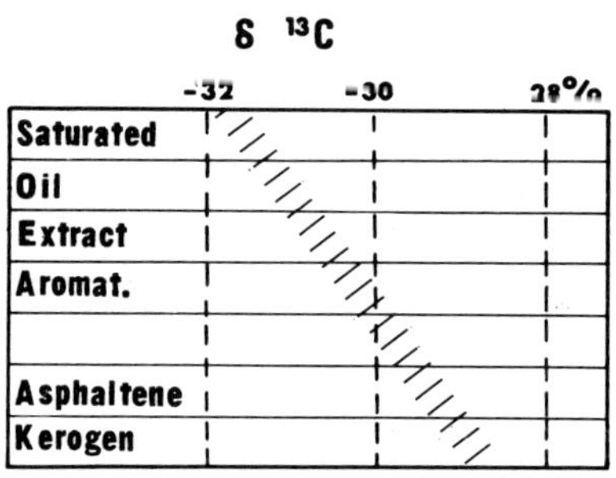

Fig. 19. — Variation in δ ^{13}C versus hydrocarbon type (Stahl, 1971).

Other constituents

Sulfur compounds are the most frequently found, especially in heavy crude oils, in which sulfur is the third most important element after C and H. It accounts for an average of 0.65 % by weight of the crude oils, and often more than 1 % (up to 5 to 6 %, for example, in the heavy oils of the Aquitaine Basin). Sulfur is generally present in the form of thiols or mercaptans, sulfides and thiophene derivatives, more rarely in the form of free sulfur. These thiophenic compounds are very abundant in crude oils which are rich in aromatics, resins and asphaltenes, where their proportion may be as high as 20 % in some cases. Sulfur compounds are often related to the depositional environments of evaporitic and carbonate source rocks. They show a tendency to decrease slightly with increasing depth. On the other hand, they increase with alterations, and the S content may be as high as 10 %, as in the Tertiary oils of the Camargue.

Nitrogen compounds are found less frequently and account for less than 0.1 % on the average. They are found in complex molecules with polyaromatic structures such as organometallic compounds, including the porphyrins which are also very rich in nickel and vanadium, and also in pyridine and quinoline.

Oxygenated compounds are chiefly associated with the acid function — COOH, particularly with the fatty acids and isoprenoid and naphthenic acids. They are found in low concentrations, mainly in immature crude oils.

Physical and economic characteristics

The chemical compositions of crude oils display certain correlations between constituents and condition their physical and economic properties. A close correlation appears between sulfur content and the proportion of aromatic constituents (thiophenic derivatives), resins and asphaltenes.

It is well known that the sulfur found in the constituents and in aromatic crude oils combines with the unsaturated hydrocarbons and facilitates the formation of cycles, particularly benzene rings (Tissot & Welte, 1978). Moreover, crude oils which are relatively rich in sulfur are generally heavy and viscous.

The chief economic property of crude oils is their heating value, for which they are used as an energy source. This heating value varies according to the density and chemical composition, ranging from :

47 to 44.5 MJ kg^{-1} (11,700 to 11,100 cal g^{-1}) for a crude of 0.7 to 0.8; 70 to 45 °API,
44.5 to 43 MJ kg^{-1} (11,100 to 10,675 cal g^{-1}) for a crude of 0.8 to 0.9; 45 to 25 °API,
43 to 42 MJ kg^{-1} (10,675 to 10,500 cal g^{-1}) for a crude of 0.9 to 0.95; 25 to 16 °API.

In comparison, a bituminous coal has a heating value of 22 MJ (5600 kcal) and 32 MJ (8000 kcal) per kg, yielding the approximate equivalence :

$$1.5 \text{ t coal} \cong 1 \text{ t oil} \cong 1000 \text{ m}^3 \text{ gas}$$

Crude oils also have the following uses :
— feedstock for the petrochemical industry,
— lubricant in oils and greases,
— road surfacing material with heavy fractions (for example, natural asphalt from the Val de Travers Mine in Switzerland).

All these uses depend closely on the chemical composition.

The chief remaining physical properties of crude oils are their density, viscosity and solubility. Added to these are secondary properties, which are often of great practical or theoretical value, such as fluorescence and the rotational effect on polarized light (see Perrodon, 1966).

The density of a crude oil reflects its chemical composition. It increases with the percentage of hydrocarbons and heavy products, including resins and asphaltenes. Hence heavy crude oils are often rich in sulfur. It decreases with the rise in temperature level at which the pool is found or has developed, namely with burial depth. Density is normally expressed in g/ml or degrees API, which varies inversely.

It ranges approximately :
— from 10 °API (1 g/ml or 1) for heavy crude oils (density equal to that of fresh water),
— to 50 °API (0.77) for light crude oils.

Density is a very representative indicator of the economic quality of the crudes and serves to set their prices.

Viscosity, the opposite of fluidity, represents the ability of a crude oil to flow. It depends on chemical composition, increasing with the percentage of heavy constituents and consequently with the density of the crude. However, some crudes which are relatively light but rich in paraffins of C_{30} and higher, display high viscosities (paraffinic crudes). Viscosity decreases sharply with an increase in dissolved gases and with rising temperature. It is expressed in centipoises and varies from less than 1 cP to tens of thousands of cP.

This property may be associated with the pour point of a crude oil, a property which is also very sensitive to the heavy paraffins content. Light and fluid crudes solidify at very low temperatures. High viscosity crudes, generally very paraffinic, may become pasty and solid at positive temperatures around 10 degrees Celsius (50 °F). Paraffinic crudes have a relatively high pour point. Bacterially modified heavy crudes, namely crudes which are very poor in paraffins, exhibit a relatively low pour point, which often suffices to identify them as a degraded product, considering their density.

The solubility of hydrocarbons is their ability to dissolve mutually in each other. Heavy fractions and solid phases are dissolved in the light hydrocarbons, gases dissolve the light liquid fractions and are dissolved in large quantities in the liquid fractions. This dissolution of the heavy fractions by the light hydrocarbons could facilitate their migration, as postulated for the Handil Field in East Kalimantan (Durand & Oudin, 1979).

Distillation is merely the separation of these gaseous, liquid and solid compounds gathered together in complex combinations in the crude oils. The most remarkable property of these mutual solubilities is that of the gas dissolved in the liquid phase, which can reach very high proportions. The ratio of gas dissolved in the oil, or the gas/oil ratio (GOR) may be as high as 1000 m^3 per ton (6000 $cfbl^{-1}$) for very light crude oils and condensates. This ratio varies inversely with the oil density, increases with rising pressure, in other words with burial depth, until it reaches the saturation pressure (the equivalent of the bubble point), above which a differentiated gas phase occurs and forms a gas cap. Conversely, it decreases with pressure and with the production of a field.

The gas/oil ratio is traditionally expressed in cubic meters of gas per ton ($m^3\,t^{-1}$) or cubic meter of oil ($m^3\,m^{-3}$), or in cubic feet per barrel ($ft^3\,bl^{-1}$) or standard cubic feet per barrel ($Scf\,bl^{-1}$), both expressions being related as follows :

$$1\ m^3\,m^{-3} = 5.65\ ft^3\,bl^{-1} \text{ or Scft per bl or}$$
$$100\ ft^3\,bl^{-1} \text{ or Scft per bl} = 17.7\ m^3\,m^{-3}$$

This dissolved gas increases the volume of the liquid phase by a quantity called the volume factor. The accurate knowledge of this factor is important in calculating reserves, since the volume of crude oil in a pool is in this case greater than the same quantity deprived of its gas in surface conditions. The volume shrinkage factor increases as the quantity of gas in solution is greater, i.e. as the crude oil is lighter, as shown by the following table.

volume factor	*GOR*	
	$m^3\,m^{-3}$	$ft^3\,bl^{-1}$
1.10	17.5	100
1.20	52.5	300
1.31	87.5	500
1.59	175.0	1000
1.85	260.0	1500
2.00	350.0	2000

The solubility of liquid hydrocarbons in water is generally very low and decreases sharply with rising molecular weight, but varies substantially in accordance with their chemical properties. Hence aromatics are far more soluble than alkanes with an equivalent number of C atoms. For instance hexane, cyclohexane and benzene exhibit solubilities of 9.5, 60 and 1750 $mg\,l^{-1}$ respectively. Benzene and toluene are by far the most soluble liquid hydrocarbons. Solubility in water increases with temperature at constant pressure, especially for aromatics. It declines with salinity and with pressure (McAuliffe, 1979).

The fluorescence of crude oils and more specifically of bitumens, under the action of ultraviolet radiation, varies from pale yellow to greenish blue. It is a characteristic property of aromatic hydrocarbons, and is associated with the presence of doublo bonds and aliphatic radicals. This property is extremely important for the discovery at the wellsite of small quantities of hydrocarbons, and also in the field. It generally makes it possible to distinguish them from coals.

Optical activity, or rotational effect on polarized light, is a general property of oils that testifies to their biological origin and particularly to their relations with lipids. It decreases gradually with geological time and is altered by the diagenesis/catagenesis process. Some authors claim that it is related to the production of organic compounds by microbes (Philippi, 1977).

Among the other physical properties are :
— resistivity, ranging from 10^{11} to 10^{18} ohms cm^{-1},
— dielectric constant, equal to 2, compared with 6 to 11 for rocks, 94 for ice,
— index of refraction, ranging from 1.7 to 1.48, which decreases slightly with increasing diagenesis.

Chief types of crude oil

These chemical properties, and especially the variations of the different components, give rise to the classification of crude oils, as made by Tissot & Welte (1978), according to the distribution of hydrocarbons that they contain.

Paraffinic crude oils, containing more than 50 % saturated hydrocarbons and over 40 % paraffinic hydrocarbons (iso- and n-paraffins), are light crude oils with density of about 0.85, sometimes highly viscous, containing less than 10 % resins and asphaltenes, and less than 1 % sulfur. This class includes the crude oils of the Paleozoic in the Sahara and the United States, the Cretaceous in Gabon and Congo, the Tertiary in Libya and Indonesia.

Naphtheno-paraffinic crude oils also contain more than 50 % saturated hydrocarbons, and less than 40 % paraffinic hydrocarbons (iso- and n-paraffins) and naphthenic hydrocarbons. They are also poor in sulfur, and may contain 5 to 15 % resins and asphaltenes and 25 to 40 % aromatics. This category includes a number of crudes from Alberta, the Paris Basin, North Aquitaine and the North Sea.

Naphthenic crude oils contain less than 50 % saturated hydrocarbons and over 40 % naphthenic hydrocarbons. This proportion sometimes results from the elimination of alkanes by the biodegradation of paraffinic or naphtheno-paraffinic oils. Naphthenic crude oils are found in particular in the Gulf Coast and the North Sea

Aromatic crude oils contain less than 50 % saturated hydrocarbons, over 50 % aromatics, resins and asphaltenes. Their sulfur content is often greater than 1 % and they frequently correspond to degraded products. These oils are heavy and viscous, often containing more than 25 % resins and asphaltenes.

According to the proportion of alkanes, a distinction is drawn between :

— asphaltic aromatic oils, containing less than 25 % naphthenes, and generally rich in sulfur; this category includes the Athabasca tar sands and they heavy oils of Aquitaine and Venezuela, constituting extremely large volumes of oil in place,

— naphthenic aromatics containing more than 25 % naphthenes, which are poorer in sulfur ($S < 1$ %); this group includes the Cretaceous oils of Gabon and Congo (Emeraude field).

Between the naphthenic and aromatic crudes, it is possible to identify a class of intermediate aromatics, characterized by 40 to 70 % aromatics, 10 to 30 % resins and asphaltenes. These crudes are generally rich in sulfur ($S > 1$ %) and rather heavy ($d > 0.85$). They include a large share of the Jurassic and Cretaceous oils of the Middle East, representing considerable reserves, oils of the Permian and Carboniferous of the Permian Basin of West Texas, and the Cretaceous/Jurassic in Aquitaine, Spain, and Sicily.

The large majority of crudes actually can be divided approximately into two main groups :

— intermediate aromatics corresponding to reducing marine environments,

— naphtheno-paraffinics and paraffinics, generally associated with deltaic or non-marine environments.

In each province, it is also necessary to consider the geothermal conditions and alterations which more or less obliterate the imprint of the depositional environment and leave a stamp of their own characteristics. Thus immature crudes are generally heavy and rich in nonhydrocarbon compounds and in asphaltenes.

Crude oils subjected to strong catagenesis are often relatively rich in paraffins due to the decrease in aromatics and heavy products. They become increasingly lighter with greater depth (or more precisely with rising temperature).

Alterations and biodegradations also modify the composition of crude oils :

— by favoring the relative concentration of complex heavy polar compounds, and especially resins and asphaltenes, by the degradation of alkanes and aromatics, to yield naphthenic and asphaltic aromatic crudes : this is the case in particular of 'oils' from tar sands,

— by eliminating a more or less broad range of paraffins by bacterial destruction, resulting in a pattern of naphthenic crudes.

Many of these effects often occur in combination, as in the case of shallow crude oils, which are immature and accordingly altered and degraded. These different effects are often difficult to identify individually. However, it appears that catagenesis cannot obliterate the marks of prior degradation.

Heavy crude oils

It is customary to apply this term to crude oils exhibiting a specific gravity between 0.93 and 1 or 20 to 10° API, and high viscosity in subsurface conditions, greater than 10,000 cP. Their sulfur content is often high, greater than 5 % and sometimes 10 %, as well as the proportion of metals (V, Ni) which may be as high as 600 ppm. The composition of the hydrocarbons is distinguished by a high proportion of C_{15}^+ and a very low proportion, if any, of light products. The share of these hydrocarbons in the composition of these crudes may be relatively small.

Heavy crude oils may be of various origins. They are often normal oils which have been degraded by the departure of light components, leaching by fresh water, oxidation, or microbial degradation. They may belong to varieties of crude oil which are especially rich in heavy compounds, such as paraffins and asphaltenes, giving them their very high viscosity.

Crudes of this type appear to be more specifically associated with carbonate source rocks. They may also be immature crudes whose composition, relatively poor in hydrocarbons, includes many polar and slightly evolved substances such as asphaltenes and resins. It is sometimes difficult to distinguish between different origins, as alteration affects mature as well as immature oils, especially since these two occurrences characterize shallow zones, together with the fact that maturation criteria are often obliterated by the destructive action of alterations.

Heavy crude oils represent considerable reserves in place, around 280 to 350 Gt, largely concentrated in the great tar sands accumulations in Canada and Venezuela.

1.1.2.3. *Solid hydrocarbons*

This category includes natural hydrocarbons which are relatively simple from the chemical standpoint, such as methane hydrates, or complex and often pasty, like the vast family of bitumens and asphalts, together with the sedimentary formations in which they are most often encountered, tar sands and oil shales. We shall deal with the following briefly :

— hydrates,
— the bitumen family.

Methane hydrates or clathrates

In certain low temperature and pressure conditions, methane can precipitate and crystallize in the form of methane hydrate or clathrate, with one methane molecule combined with six molecules of water. In these conditions, 170 m^3 of CH_4, measured in standard temperature and pressure conditions, condense to 1 m^3 of hydrate.

These compounds appear :

— at temperatures below 0 °C, if the pressures are less than 25 bar,
— for temperatures between 0 °C and 20 °C, at pressures of 25 to 250 bar (Figures 20 and 21).

In terms of burial depth, it is sufficient for the temperature to be less than + 15 °C at 1500 m, for example (Bily & Dick, 1974).

These conditions are perfectly satisfied in zones covered by a thick permafrost, and on the ocean floors. The Pleistocene glacial periods probably favored the development of hydrates. The latter appear to play an important role in the precipitation of methane dissolved in waters that are in contact with them, and hence in the formation of the gas accumulations of the Arctic zones (Katz *et al.,* 1959).

The bitumen family (tar sands, oil shales)

Bitumens can be defined from the producer's standpoint as natural viscous mixtures consisting chiefly of very heavy hydrocarbons, with subsidiary sulfur-bearing compounds. Their high natural viscosity (> 10,000 cP) and their density, lower than 10° API, prevent their economic recovery by traditional drilling operations. The semi-refined hydrocarbons extracted from bitumen are qualified as synthetic crudes.

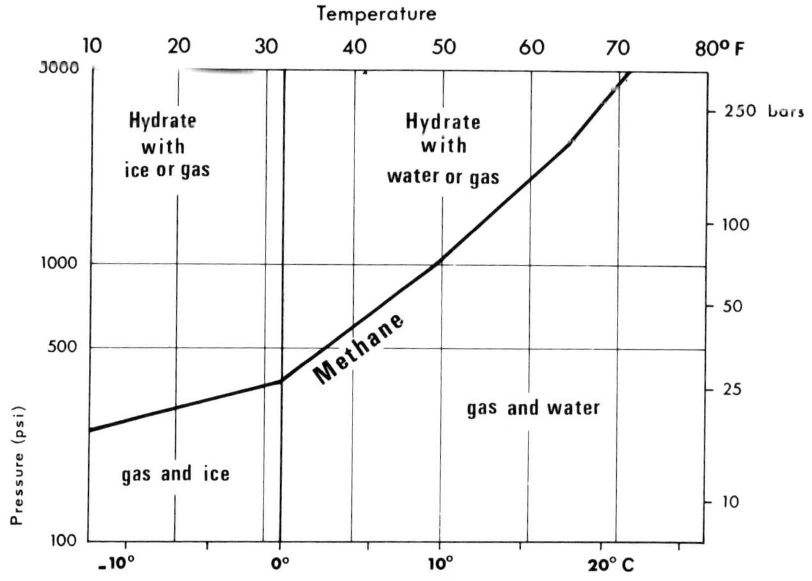

Fig. 20. — Formation conditions of methane hydrates (Katz *et al.,* 1959).

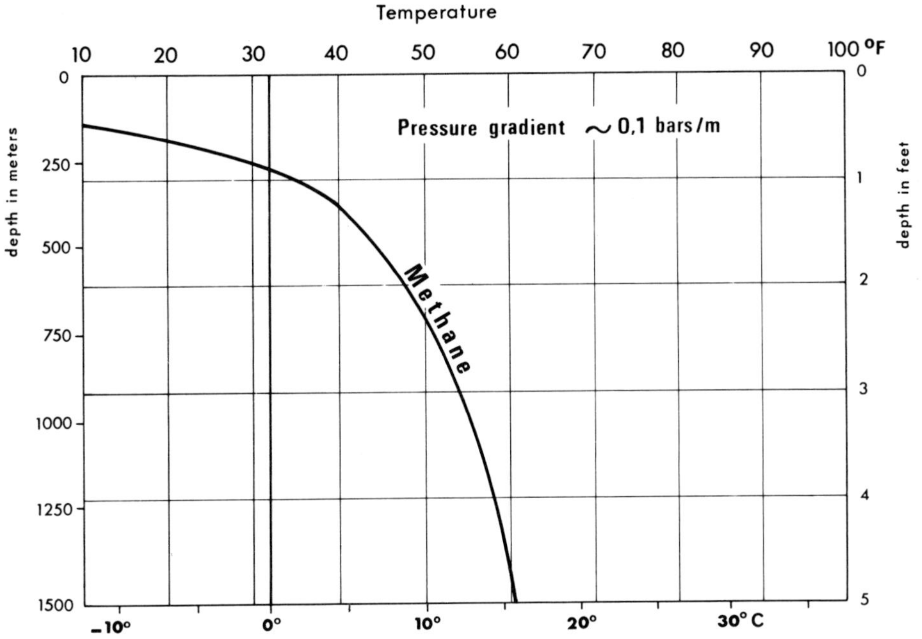

Fig. 21. — Depth/temperature curve for the formation of methane hydrates.

From the physicochemical standpoint, bitumens are compounds consisting cheifly of C, H, O and, to a lesser extent, N and S. They are complex mixtures of paraffinic, naphthenic and aromatic hydrocarbons with high molecular weight, and nonhydrocarbon heteroatomic substances, especially resins, asphaltenes and carbenes, whose differences of chemical composition are reflected by variable degrees of solubility in organic solvents. The molecular weights of bitumens range from 300 to 1500 for fractions comparable to crude oils and resins, soluble in n-pentane, from 600 to 10,000 for asphaltenes, insoluble in n-pentane and soluble in benzene, and probably more for carbenes, which are insoluble in benzene. Apart from sulfur and nitrogen-bearing compounds, vanadium and nickel are found in these heavy fractions and especially in the asphaltenes.

From the structural standpoint, bitumens appear to be complex polymers formed substantially of benzene rings connected by aliphatic chains (Wen *et al.,* 1978). The classification of bitumens, which is itself complex and debatable, is based on selective solubilities in different organic solvents. Solubility in carbon disulfide first distinguishes bitumens from pyrobitumens and kerogens, which are insoluble. Liquid bitumens soluble in pentane chiefly include oils.

Solid bitumens can be classified in three groups according to their solubility in different solvents :

— mineral waxes, consisting mainly of paraffinic chains, such as ozocerite, typical of the Uinta Basin (Utah) or mixtures of fatty acids and esters,

— asphalts, as encountered in the lakes of the same name in Venezuela and Trinidad, complex compounds with high molecular weight, containing 1 to 6 % S, and also containing vanadium,

— asphaltites, such as gilsonite and grahammite, important components of oil shales, polymers with aromatic rings rich in C, and also containing nitrogen (2 to 3 %) and porphyrins (Abraham, 1960, and Wen *et al.,* 1978).

Tar sands

Tar sands — also called oil sands — are sedimentary rocks, which may or may not be consolidated, containing bitumen and other heavy petroleum products, with very high viscosity, and which cannot be recovered by conventional methods. Depending on the burial depth, all the transition stages can be observed between these 'sands' and the heavy oils of certain accumulations (see Chapter 4 of Part 2). From the mineral standpoint, these sands are often coarse, well sorted, porous and permeable. They are wetted by water, which often accounts for less than 10 % of the voids. Hence the bitumen accounts for 18 to 20 % by weight of the impregnated rock.

From the geochemical standpoint, bitumens and heavy oils are characterized by a low percentage of saturates and light products, and by a predominance of asphaltenes and complex products rich in N, S and O.

These characteristics give rise to the postulation of at least three processes for the formation of these 'bitumens' :

— incomplete maturation of kerogen in case of shallow immature products relatively in place,

— physicochemical and biological alteration, especially by bacteria which destroy a number of paraffins, particularly in the case of seepages and in horizons that are relatively permeable to meteoric waters.
— hydrocarbon 'deasphalting' under the effect of the cracking of hydrocarbons during catagenesis and the formation of a more or less solid residue, in deeply buried series (Silverman, 1978).

Oil shales

These are generally argillaceous sedimentary rocks, but rather often carbonate rocks, very rich in organic matter (> 5 % kerogen), poor in bitumens (0.5 to 5 %), and capable of yielding hydrocarbons by pyrolysis at temperatures around 500 °C. The organic matter consists essentially of debris of lacustrine and marine algae. The chemical composition varies widely, but is always complex. Elemental analysis shows a high proportion of H, and a high H/C ratio, around 1.25 to 1.75. The O/C ratio is more variable, and ranges from 0.20 to 0.02. The molecular weight is about 1200. These rocks display low density, around 1, owing to the high organic matter content (Tissot & Welte, 1978) (see Chapter 4 of Part 2).

Summary and conclusion

Crude oils and natural gases are formed essentially of hydrocarbons, mostly saturated, of the paraffinic or ring type, including some 450 distinct chemical compounds. The diversity and the characteristics of these molecules make crude oils a highly valued raw material for the synthesis of a large number of chemicals. The very wide variety of combinations of the different hydrocarbons is reflected in surface conditions by their occurrence in three states, gas, liquid and, in rare cases, solid. At greater depths, the combined and complex action of pressure and temperature substantially alters the boundaries between these states.

The chief physical property of the hydrocarbons and their economic value is their high heating capacity, which itself depends on their chemical composition. This property is even more valuable because combined with the fluid state.

1.2. HYDROCARBON GENESIS

Hydrocarbons are formed in sedimentary formations, rich in organic matter, and subjected subsequently to a degree of diagenesis/catagenesis. These formations are qualified as source rocks. In certain conditions, and especially above a certain concentration, these products escape from the beds in which they were formed into porous and permeable rocks, called reservoirs, characterized in particular by lower internal pressure. This process is called 'primary migration'.

For a clearer understanding of the mechanisms that give rise to hydrocarbons and the characteristics of their source rocks, we shall examine the following in succession :
— the main constituents of living organic matter,
— the environments and mechanisms of deposition, in which the kerogen is formed,
— subsequent transformations during the diagenesis and catagenesis phases.

A sufficient number of facts, universally recognized, today allow us to assert the biological and sedimentary origin of petroleum, and inorganic theories are irrelevant except historically. In rare cases, however, methane may be of inorganic origin.

Throughout this chapter reference is frequently made to the recent book by Tissot & Welte, "Petroleum Formation and Occurrence" (1978), which offers the most complete review of the subject (see also Hunt, 1979).

1.2.1. The major constituents of living organic matter

All living organisms consist essentially today, as they did in the past, of a very small number of chemical substances, proteins, carbohydrates, lipids and lignin, and their proportions often vary substantially according to the groups and families of the animal and vegetable kingdoms.

— Proteins, which generally form a large share of organic matter, are polymers featuring a high degree of organization, formed of amino acids. They are present in the composition of animal tissues and enzymes. The amino acids are their monomer representatives, soluble in water.

— Carbohydrates include glucides and their polymers. They account for a large part of the tissues, especially with cellulose, chitin and sporopollenin. Cellulose is chiefly produced by higher plants and is a major constituent of wood. Chitin is mainly found in certain animals, and sporopollenin in spores and pollens.

— Carbohydrates are also a source of energy for living organisms.

— Lipids include all organic compounds which are insoluble in water and soluble in organic solvents (e.g. chloroform), and also unsaturated fatty acids, frequently found in vegetable oils. Lipids are abundant in algae, especially the botryococcus family, and in diatoms, in which they form up to 70 % of the dry weight, as well as pigments, grains and spores which are rich in C_{34} (Moldowan & Seifert, 1980), where they concentrate high energy potential.

— Lipids also contain fats, which are esters of fatty acids. Saturated fatty acids are represented by the simplified formula :

where each corner corresponds to CH_2. Unsaturated fatty acids are also frequently found. Lipids also include terpenic compounds consisting of isoprenoids, which can be polymerized into long chains or form cycles. Lipids are a constituent of a number of plant materials, such as gums, and also of sporopollenin, cutin, chorophyll and carotenoids.

— Lipids, which are insoluble in water and often more chemically resistant than the other components, play a predominant role in hydrocarbon genesis. Fine chemical analyses, and especially the $^{13}C/^{12}C$ isotopic ratio, show that crude oils are mainly derived from the lipid fractions of organic matter, while coals are produced from the remains of plants, and particularly from cellulose matter which is their major non-lipid constituent.

— Close chemical relationships have also been observed between hydrocarbons and lipids, especially through certain constituents called 'biochemical fossils' (Silverman, 1971). Lipids account for the bulk of the organic constituents capable of being converted into oil.

In addition to these basic materials, lignin and tannins deserve to be mentioned. Lignin is a polyphenol with high molecular weight aromatic structures that forms the skeleton of the tissue of the higher plants. Tannins are less condensed polyphenols that impregnate the cortex tissues.

Organic properties of some families making up the biomass

The biomass consists of a small number of large groups of living organisms : marine plankton, bacteria, algae and the higher plants.

— Marine and lacustrine plankton, especially microscopic algae, consist mainly of proteins for more than half of their mass, carbohydrates, and 25 to 5 % lipids. Certain microorganisms, such as diatoms, are especially rich in lipids, the latter consisting largely of fatty acids with 3 to 5 % hydrocarbons, particularly n-alkanes.

— Bacteria, among the earliest organisms, apart from water, consist mainly of proteins and may contain up to 10 % lipids, particularly hydrocarbons and mostly C_{10} to C_{30}. They feed largely on fatty acids.

— The higher land plants are formed mainly of cellulose and lignin, 30 to 50 % and 15 to 25 % respectively, although certain parts, such as seeds, pollens, waxes and resins contain significant fractions of lipids, especially n-alkanes with C_{27}, C_{29} and C_{31}.

The organic matter produced by plankton, algae and bacteria is generally deposited in place along the same vertical : this is called autochtonous. The organic matter from the higher land plants may be conveyed into sedimentary basins by air or river currents, and is qualified as allochtonous or inherited.

1.2.2. **Depositional environments and mechanisms - kerogen**

After the death of different living organisms, organic matter is found in their remains, disseminated or dissolved in water, and especially in combination with clay particles. To be able to settle, this organic matter must be degraded before deposition, and must be protected at the time of its deposition and subsequently, initially implying an aquatic environment. Protection before deposition requires an anaerobic environment, with water devoid of oxygen, i.e. perfectly calm, confined, and often stratified. As a rule, this preservation is facilitated by a slow time of fall for planktonic organic matter, which forms the bulk of marine organic sediments.

Most of the organic matter of planktonic origin is destroyed as it falls through the body of water at a speed of about 100 m per week. It is estimated that on the average :
— only 2 % reaches the shallow floors,
— about 0.02 % reaches the deep ocean floors.

Another part is destroyed by benthic organisms living on the sea bed. Hence the quantities that deposit finally depend on :
— the biomass produced, namely, the organic productivity of the environment,
— the physicochemical conditions of the depositional environment.

In other words, the characteristics of the basin directly condition the deposition of future source rocks (Curtis, 1980).

1.2.2.1. *Depositional environments*

Environments with high organic productivity, which generally create an oxygen shortage by their reactivity, are governed by the combination of an external energy, the sun, which furthers photosynthesis, and nutritive elements, such as phosphates and nitrates. These favorable and often exceptional environments are found at sea and on land.

Solar energy is limited to the surface layer of the oceans and land masses, but it extends to the entire surface of the Earth. It offers the driving force of the mechanism of photosynthesis, which has formed the basis of all biological activity for some two billion years. The nutritive elements, consisting mainly of nitrogen, phosphorus, silica and iron, are derived on land from the soil, and at sea from the continents by the drainage of rivers and waterways, or from deep waters, by upwellings which, especially in the trade wind zones, push the cold and more mineralized waters from the depths to the surface.

The most favorable environments, all of them aquatic, are thus found on the continental margins, shelves and slopes, in coastal waters, and also in lagoons, landlocked or silled seas, and lakes. Favorable environments may develop in the great deltaic units marked by high subsidence and sedimentation, like the deltaic zones and basins of the Niger, the Mississippi, the Ganges, and the Mahakam in Indonesia. They generally correspond to a warm, humid climate and high sedimentation, predominantly clay. They are rich in inherited organic matter, of the humic or lignitous type.

In the upwelling zones, such as those found off the coast of southwest Africa, the area of Walvis Bay, and the coasts of Chile and California in particular, the oyxgen deficiency created by the intense biological activity at the surface produces an anaerobic state in the underlying layers. The planktonic populations are also regularly poisoned by the proliferation of certain microalgae, causing water-blooming and the massive deposition of an organic mass. These environments may be considered as especially favorable to the production of source rocks. They often correspond to arid climates exhibiting low sedimentation and predominantly planktonic or sapropelic organic matter.

Similar mechanisms may also be found in lakes and lagoons, where certain algae proliferate, producing eutrophic conditions. It may also be considered that certain lacustrine basins, which are poor in detrital inflows, and display a high pH, constitute the best carbon traps. Certain lacustrine algae, like the botryococcus family, are particularly important constituents of source rocks. The wealth of the lacustrine basins of Reconcavo in Brazil, and Green River in Utah, provides an illustration of this (see Chapter 2 of Part 3). Some of the constituents of these algae, especially $C_{34}H_{70}$, have been found in the oils from Minas and Duri, Indonesia (Moldowan & Seifert, 1980).

Landlocked or semi-landlocked seas, like the Black, Baltic and Caspian Seas, the Gulf of California, and lagoons, like that of Maracaibo, also represent excellent models of environments for organic sedimentation. In these exceptional conditions, the percentage of organic matter in the sediments, normally less than 0.5 % (lower than 0.9 % in clays, 0.2 % in sandstones and carbonates) may reach values of 2 to 10 % and more. Hence the levels showing a high concentration of organic matter can be considered as the result of exceptional preservation conditions as well as the productivity of these conditions, and the alternations often found in these deposits correspond to variations in the sedimentation environment on the floor. It is these exceptional environments that form the cradle for the particular sediments that subsequently become oil source rocks.

The sedimentation rate plays a definite role in the dispersion and preservation of the organic matrix. Rapid sedimentation disseminates the organic matter in detrital materials. On the other hand, it guarantees rapid burial and effective protection of these delicate substances. By conveying these layers rapidly to greater depths and higher temperatures, it limits the period of bacterial destruction of organic matter. On the other hand, slow sedimentation favors the concentration of organic matter in certain beds. In this case preservation must be guaranteed by a rigorously confined environment. Thus the study of these environments is especially interesting from this point of view.

Clay minerals appear to play a more or less important role in the fixation and preservation of organic matter. They seem to absorb certain polar organic compounds and transform dissolved organic matter, reducing the risks of alteration in water (Dow, 1978). Two main types of depositional environment emerge from these considerations, corresponding to two clearly distinct biological families :

— a restricted marine or lacustrine environment, characterized by a predominance of planktonic or algal organic matter, qualified as sapropelic, or oilprone,
— an environment with a brackish tendency, and the rapid sedimentation of clay materials especially rich in ligneous or humic organic matter, also called gasprone.

This naturally includes a whole sequence of intermediate environments. We shall see below that the former provide the origin of oil provinces, often naphthenic, while the latter give rise to provinces which are rich in gas and paraffinic oils.

1.2.2.2. *Sedimentation mechanisms*

This section considers the initial and important transformations occurring in a freshly deposited sediment, especially in the mud. These are the first steps of the transformation of organic matter into hydrocarbons, corresponding to biological (or biochemical) diagenesis.

This initial stage of diagenesis begins immediately after deposition under the water/sediment interface, and continues while declining gradually but irregularly down to depths of a few hundred meters in cold zones, corresponding to temperatures below 60 to 70 °C (Philippi, 1977). During this phase, the sediments are subject to the energetic action of the microorganisms in the mud, and particularly the bacteria, in the first decameters of the deposit, especially in a shallow environment.

This biological activity is closely related to the situation of the depositional environment, and primarily its depth. Activity is intense in shallow environments, appears to decline fairly rapidly with the depth of the water, and is very low on the deep ocean floor (Pelet, 1974). During these reactions, the pH rises slightly and the environment becomes strongly reducing. Calcium carbonate and silica are partly dissolved. Simultaneously, CO_7, H_2S and CH_4 produced by the degradation of organic matter are evolved. The bacteria split the organic molecules to yield carbon dioxide and methane primarily, and a complex insoluble residue that forms part of the composition of kerogen. These transformations are also marked by the elimination of certain functions, and this results in a slight enrichment in the light ^{12}C isotopes.

The bacteria, particularly the sulfate-reducing strains, generate H_2S which is fixed by the clay minerals in the presence of iron to yield pyrite. In the case of carbonate source rocks, it is only fixed in small proportions and combines with the organic matter, passing subsequently into the hydrocarbons generated. In some conditions, as in fairly deep lakes without detrital inflows, such as Kivu Lake, the fermentation of deposited organic matter, chiefly cellulose, may yield considerable amounts of methane. This is one of the formation processes of what is called 'early gas'.

But the most important product of these early transformations is undeniably kerogen, the starting point of a long evolution which may culminate in hydrocarbons. More specifically, the true sources of the formation of hydrocarbons are represented by organic molecules, especially lipids, which resist the intense activity of the biological diagenesis of the upper layers of the deposit. This phase of biological diagenesis may be considered to correspond in fact to a stage of degradation of organic matter. It generates methane, apparently at the expense of liquid hydrocarbons.

Hence the rate of burial may play an important role : if rapid, it reduces the period of microbial action and degradation, halted by the rise in temperature, and allows the preservation of a larger fraction of the stock or organic matter; if slow, it will only allow the preservation of very little source rock materials (Coleman *et al.,* 1979).

Organic facies

In accordance with the characteristics of the organic matter, two main types of organic facies may be distinguished :

— humic or detrital facies,
— sapropelic or authigenic facies.

The former may be related to aquatic or eolian transport : plant tissues, spores, pollens, and particularly vitrinite (coal beds form the boundary of these two modes of transport); the latter are essentially aquatic, marine, brackish or lacustrine, and largely consist of algae and planktonic materials.

The sapropelic facies is generally represented by the presence of recognizable algae, and by the accumulation of undifferentiated algal materials exhibiting a fluffy appearance. These environments often appear to be devoid of any detrital inputs, especially humic, and particularly vitrinite. They are associated with evaporitic facies. In this case, they correspond to sub-desert climates. The humic material, largely formed of debris of higher plants, spores and pollens generally associated with active sand/shale sedimentation, usually corresponds to humid climates (Robert, unpublished document, 1978). These two facies are in some way merged with the chemical properties of the kerogens, defined in three categories I, II and III.

1.2.2.3. ***Kerogen***

The term kerogen is applied to the organic constituents of sedimentary rocks which are insoluble in organic or alkaline solvents. Kerogen is thus different from the organic constituents which are soluble in these solvents, and which are designated by the term bitumens. It is the most widespread form of organic matter on Earth, a thousand times more than coal and liquid petroleum, and fifty times more than the bitumens dispersed in non-reservoir rocks (Hunt, 1972; Tissot & Welte, 1978; Durand, 1980) (Figure 22).

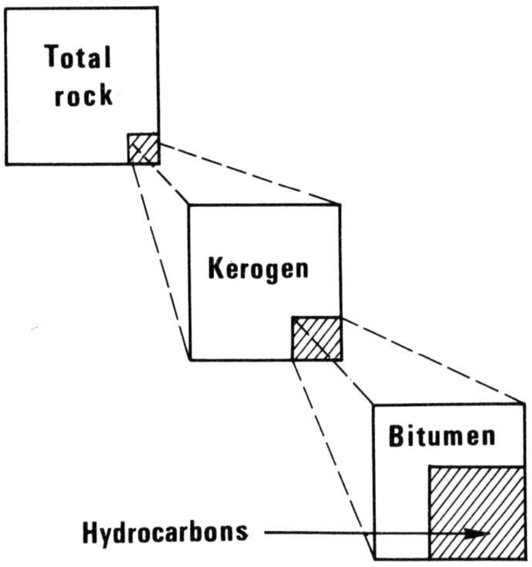

Fig. 22. — Ratios of hydrocarbons to bitumens, kerogen and total rock (Tissot & Welte, 1978).

From the chemical standpoint, kerogen is a macro-molecular complex composed of condensed benzene rings, connected by heteroatomic or aliphatic links. It may also contain lipids.

Its structure is disordered and only assumes a degree of organization during catagenesis. Under detailed observation, it appears to be composed of recognizable plant particles, of diffuse materials resulting from the degradation of previous constituents, and of secondary bituminous materials (the flaky amorphous material observed in palynological preparations is a concentrated form of diffuse materials, after the chemical destruction of the mineral support).

Certain specific organic compounds, synthesized by plants or 'animals', are also found in trace amounts in and outside the kerogen. These compounds are alkanes, fatty acids, terpenes, steroids, porphyrins and linear hydrocarbons which survive the ordeal of organic decomposition and diagenesis unscathed, and which are called geochemical fossils. These substances undoubtedly play a secondary role in the genesis of petroleum. On the other hand, they are of considerable importance in the identification of these mechanisms. Their structures reflect the architecture of organic substances present in the original organisms, and also the reactions to which these substances were subjected during their burial. Among these geochemical fossils is hopane, $C_{30}H_{52}$, of which many derivatives have been identified.

From the isotopic standpoint, the nature and the depositional environment of organic matter are reflected by certain variations in $\delta\ ^{13}C$. Kerogen, which is rich in aliphatic structures, often corresponds to more marine environments, and appears to be poorer in ^{13}C isotopes than the organic matter containing aromatic structures, often related to the higher plants and humic matter. The $\delta\ ^{13}C$ value appears to range from —23 to —30‰ in marine formations, and more often between —28 and —32‰ in organic matter and oils of non-marine origin (Figures 16 and 19). The isotopic composition also varies during burial :

— fairly rapidly, at the very outset of burial, the δ goes from —32 to —5‰ for the organic matter of living matter, to —32 to —20‰ for the organic matter of recent sediments,
— very slightly in the diagenesis and catagenesis stages,
— more sharply at the approach of metamorphism, when values of —10‰ are reached.

From the chronostratigraphic standpoint, a very slight enrichment in ^{13}C is observed with age. Hence Cenozoic oils exhibit a δ ^{13}C of around — 25 ‰, and those of the near Paleozoic a δ ^{13}C of — 30 ‰ (Stahl, 1977).

1.2.3. **Diagenetic and catagenetic transformations**

As a rule, organic substances are more fragile and more unstable than mineral compounds, and they undergo considerable transformations from the outset of the diagenesis stage. It is during these mutations subsequent to deposition that kerogen gives birth to oil and gas (Kübler, 1980). However, the presence of hydrocarbons closely comparable to crude oils in composition has been noted in recent sediments, as in the Santa Barbara Channel off the coast of California. This observation has led some investigators to postulate that the hydrocarbon generation process begins very early at the start of the sedimentary cycle (Copelin, 1979).

It nevertheless remains true that oil and, to a lesser extent, gas are the result of chemical transformations during burial. Hence the rapidity and intensity of burial, in other words, of subsidence, exert a major effect on the maturation of organic matter. These mutations and geneses fall within the general framework of diagenetic and catagenetic processes examined in Section 1.2. These effects, which lead to the formation of hydrocarbons, may be compared closely with coalification, which yields different categories of coal and methane, from organic materials derived from the higher plants (Robert, 1980).

Table I

Principal parameters of the diagenetic transformation of hydrocarbons and coals.

Stage	Vitrinite reflectance %	LOM	TAI	Hydrocarbons	Coal	% V.m
Diagenesis	0.30 0.5	2- 4- 6- 7	2.5	Biogenic and early diagenesis gas (Immature zone)	Peat Lignite Sub-bituminous	
Catagenesis	0.5 1 1.5	7 8- 10- 11 12-	2.5 3 3.5 4	Oil (Oil window) Wet gas (Condensate) Fluorescence	Bituminous: high volatile medium vol. low vol.	45 40 30 20
Metagenesis	2 2.5 4	14- 16- 18-	5	Methane ?	semi-anthrac. Anthracite	10 5
Metamorph.	4	20-		?	meta-anthrac.	

In both cases, these rearrangements tend towards a higher degree of order and a more stable product, as the most disordered structures, particularly the functional groups, are gradually eliminated by the breakage of the weakest bonds. The separated radicals thus yield CO_2, H_2S and water. From this standpoint, the genesis of hydrocarbons is only one aspect of a general redox reaction scheme, or of the disproportionation of organic matter, simultaneously reflected by the genesis of a reduced substance enriched in H, and an oxidized compound depleted of H and assuming an aromatic structure, according to the following pattern :

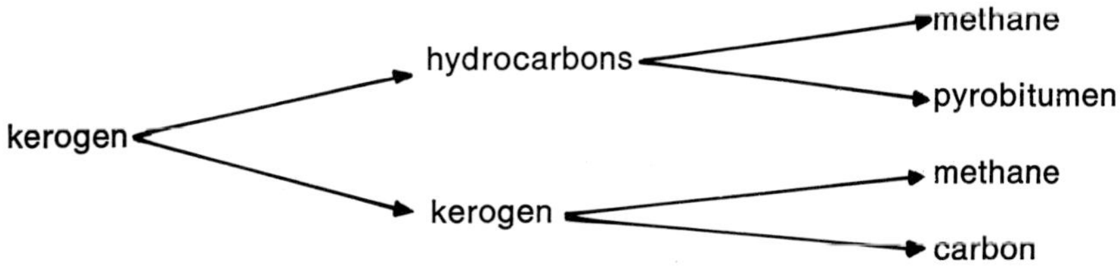

The hydrocarbons thus appear as a necessary step in the conversion of organic matter under the action of temperature, reaching towards a higher degree of order and greater stability.

The chemical properties of the early kerogens are grouped under three types defined by the 'evolutionary paths' and qualified I, II, III.

— Type I is characterized by a high H/C ratio (1.5 or more) and by a low O/C ratio (less than 1). It consists largely of lipids, particularly aliphatic chains, resulting from the accumulation of algal materials, for example of the botryococcus family, or from the elimination of the other constituents by biodegradation. Type I is especially prolific in oil, or oilprone. It is found in large quantities in the Green River oil shales It may be compared with the alginites (van Krevelen, 1961) and the sapropelites (Combaz, 1975 and 1980).

— Type II is frequent, especially in the marine sediments of restricted environments. It is characterized by relatively high H/C ratios and fairly low O/C ratios. It is fairly rich in aliphatic chains and naphthenic and benzene rings, and often in sulfur. Type II may be compared with the exinite and liptobiolit of the previous authors. It is also called oilprone.

— Type III, derived chiefly from the constituents of the higher continental plants, but also from marine planktonic organsims, mainly contains condensed and oxygenated polyaromatic groups, with a few aliphatic chains. The H/C ratio is low, but the O/C ratio is relatively high. The petroleum potential of this type of kerogen has a low yield, about five times less productive than that of Type II (Tissot, 1973). It only produces mediocre source rocks, chiefly containing gas, as well as carbon dioxide and water and is gasprone. It is comparable to the vitrinite of coals.

This classification does not cover all the organic matter existing in sedimentary rocks. Organic matter also exists with an H/C ratio lower than that of Type III kerogen, including oxidized organic matter, called inertinite, consisting of 'dead carbon', which is practically unaffected by rising temperature, and is consequently incapable of generating gas.

These different types are related to the characteristics of the depositional environment and the materials that have deposited. Hence types I and II appear to be predominant in carbonate facies, while type III can be found in a marine sand/shale environment.

Type III kerogen is opposed to both bituminous materials and sapropelic coals (bogheads and cannel-coals). It corresponds to the humic part (vitrinite) of normal coals, which include two other groups of materials, inertinite (fusinite) and liptinite (exinite).

***

Following the same plan as above, we shall now analyze the great phases of diagenesis and catagenesis.

1.2.3.1. ***Diagenesis***

The transformations of organic matter and hydrocarbons can be observed in detail in the fine sediments where they have good chances of being produced, and where errors related to migration can be discarded (Hunt, 1977).

These transformations are the result of biological action (bacteria) and thermo-catalytic action. Bacterial action appears to be limited to the first hundred meters of sediments in the total sulfate-reducing zone. As the pressure and temperature increase, their activity declines due to the development of toxic materials or the depletion of the stock of organic matter. Hence they diminish in number, but very irregularly, with depth. However, in certain low temperature conditions, below 30 °C, microorganism activity could extend down to 1000 or 2000 meters (Claypool & Kaplan, 1974; Rice & Claypool, 1981). The bacteria degrade the organic matter and sometimes certain categories more specifically, such as carbonaceous materials, and possibly sapropelic kerogens. This could be the cause of a large proportion of the early gases, but also of the alteration and destruction of potential source rocks. These biogenic gases could account for about 20 % of the total gas reserves (Rice & Claypool, 1981).

In this initial stage corresponding to an R_v lower than 0.5 % and to average temperatures below 50 to 60 °C, the first chemical reactions, undoubtedly favored by the catalytic action of certain minerals, including smectite and dolomite, cause the breakage of the weakest links of the organic structure, in other words, chiefly the splitting of the acid and ether functions (decarboxylation). This is reflected on a van Krevelen diagram by a rapid decrease in the O/C ratio for type III, and to a lesser extent for type II, and a smaller decrease in the H/C ratio (Figure 23). Especially for kerogen type III, this results in the formation of H_2O, CO_2, CH_4 and heavy heteroatomic compounds, such as resins and asphaltenes. Hence it is partly from organic matter qualified as immature that the stage of genesis occurs of methane and small amounts of ethane, which constitute the 'early diagenetic gases' (Weber & Maximov, 1976).

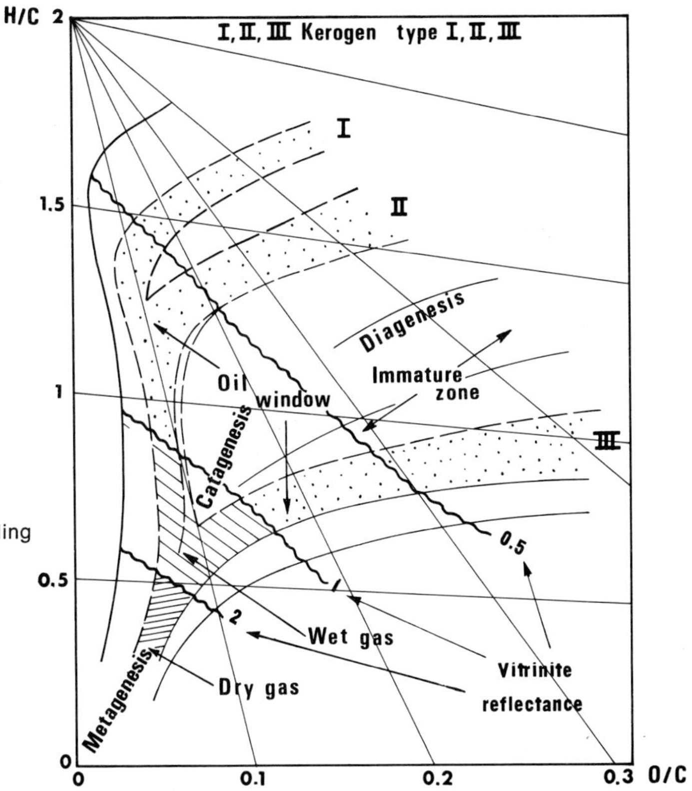

Fig. 23. — General scheme of kerogen evolution according to van Krevelen diagram (Tissot & Welte, 1978).

These early gases are differentiated on the isotopic level from the later gases by a $\delta\ ^{13}C$ less than — 60 ‰ (— 60 to — 95 ‰), close to the value of biogenic CO_2, and the remaining gases generally display values above — 50 ‰ (Rice & Claypool, 1981). Also encountered in this stage are 'geochemical fossils', in other words, molecules synthesized by living organisms at the time of deposition and preserved without any notable change. Their value is more theoretical than practical, because they never represent more than a small part of the hydrocarbons formed.

1.2.3.2. ***Catagenesis***

It is above temperatures of around 50 to 60 °C, but variable depending on the characteristics of the kerogen, that the organic matter 'matures' and 'cooks', to give birth to oil and gas. Two main phases can be distinguished depending on temperature : the first corresponds to the principal phase of the genesis of liquid hydrocarbons, and is often qualified as the 'oil window', while the second is characterized chiefly by the formation of wet gases.

Weak catagenesis stage or 'oil window'

It is in the temperature range from 50 to 120/150 °C, and mainly above around 100 °C, that the transformations of kerogen into hydrocarbons, and particularly the C_{15}, C_{40} fractions, appear to be the most extensive (Hunt, 1977) (Figure 24).

This zone of formation and preservation of liquid hydrocarbons, called an 'oil window', is characterized by the following organic markers (see Chapter 2 of Part 1, p. 42), while time and the composition of the kerogen cause these limits to vary slightly within the following ranges :

— R_v vitrinite 0.5 to 1.0/1.2
— TAI 2.5 to about 3.7 to 4
— LOM 7 to about 11 to 12.

For an average geothermal gradient, this corresponds to approximate burial depths of 1.5 to 4 km, or for Tertiary formations to maximum paleotemperatures of 60 to 130 °C. These temperatures may be lowered for pre-Tertiary series and for type I and type II kerogens (or sapropelic organic matter). From the practical standpoint, note that the TAI is more sensitive in areas of weak diagenesis, while the R_v is sensitive above the value 0.5.

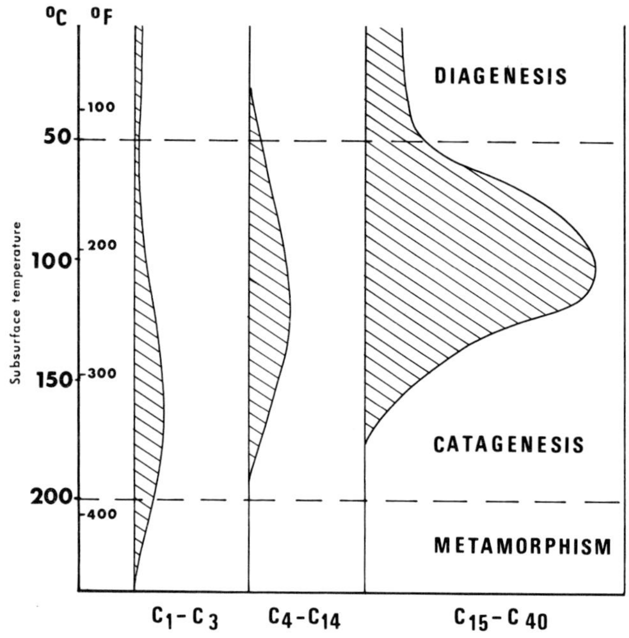

Fig. 24. — Diagram of variation in quantities of hydrocarbons contained in fine-grained sediments versus temperature (Hunt, 1977).

The 'paths' of this transformation have been clarified by van Krevelen on a diagram, according to the composition of the kerogen, and are often designated by his name (Figure 23).

From the chemical standpoint, most of these transformations are reflected by a faster drop in the H/C ratio than the O/C ratio. This pattern varies substantially from one type of kerogen to another, and is distinguished by the decrease in :

— the O/C ratio, more rapid in the case of wet kerogen, which chiefly yields gas,

— the H/C ratio from 1.25 to about 0.5 in the case of a type II kerogen.

If the chemical analysis of the hydrocarbons formed proceeds further, as catagenesis becomes more intense, a shortening of the chains is observed as well as a reduction in molecular weight, an increase in the content of n-alkanes in comparison with iso-alkanes and cyclanes, and a progressive increase in the benzene ring/saturated ring ratio. In the n-alkane family, the disappearance of the predominance of individuals with an odd number of C is usually observed simultaneously (although parity is found in some

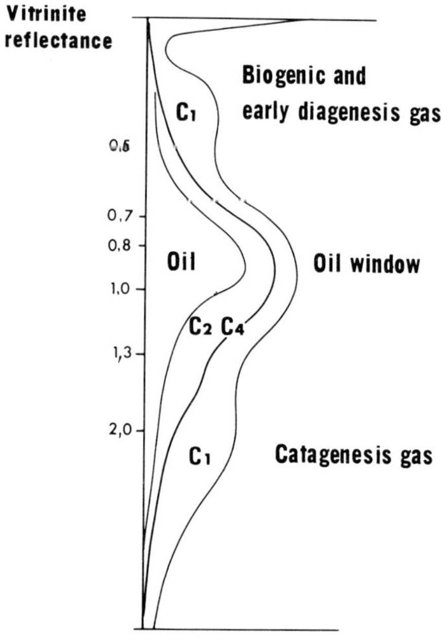

Fig. 25. — Diagram showing the intensity of hydrocarbon formation as a function of diagenesis/catagenesis (Chilingarian & Yen, 1978).

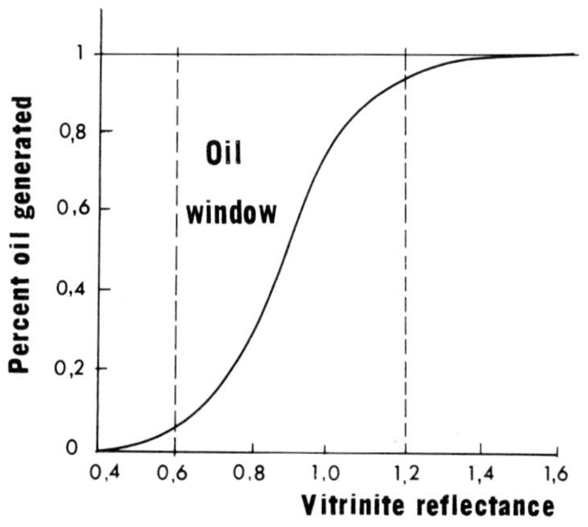

Fig. 26. — Variation in percentage of oil generated versus vitrinite reflectance (Waples, 1979).

sediments and organisms, and even sometimes the predominance of an even number of C) (Connan & van der Weide, 1978).

The rate of these transformations does not appear to be constant as a function of temperature, and accelerates exponentially (Philippi, 1965). This is shown by the curve depicting the transformation of kerogen or percentage of oil generated as a function of the increase in R_v, for example for a medium-grade kerogen, consisting of 50 % alginite and liptinite (exinite). In this case, the graph shows that half of the oil was formed when the R_v of this genesis occurred in the reflectance range between 0.6 and 1.2 %, or the 'oil window' (Waples, 1979) (Figures 25 and 26).

The presence of aromatics, more stable and more resistant to thermal cracking, appears to consolidate resistance to evolution. These characteristics seem to be offset by the inherent features of the carbonate source rocks, especially their high content of lipids, which finally guarantees a high yield in oil genesis.

From the mechanical standpoint, this birth of hydrocarbons is reflected by an increase in volume, and consequently, if the rock is only slightly permeable or impermeable, by a rise in pressure. It also introduces certain changes in the properties of the source rock, including increased resistivity, as observed in the Bakken formation in the Williston Basin (Meissner, 1978) (Figure 36).

Advanced catagenesis stage

Above temperatures in the neighborhood of 110 to 120 °C and up to around 200 °C, in the geological series corresponding on the average to a vitrinite reflectance of 2 %, new splits occur in the C-C bonds of the remaining kerogen and the hydrocarbons already formed. The latter become increasingly lighter and are eventually only represented by gases, and finally methane. The H/C ratio falls steadily in the kerogen and reaches zero when only insoluble organic matter is left. It increases in the hydrocarbons formed to a value of 4 when only methane remains.

From the isotopic standpoint, the δ ^{13}C of these gases, which rises during the maturation process, generally ranges from —60 to —30 ‰. The value for kerogen also increases to —10 ‰ at the limit of metamorphism boundary (Figure 27). In these conditions, the carbonates form CO_2, which can be added to the hydrocarbons, and may even supplant them. Large volumes of H_2S may also be formed from the sulfur remaining in the kerogen, by the reduction of sulfates by organic compounds, and of N_2 from benzene rings (Tissot & Bessereau, 1982). These gases may play an important role by increasing the permeability of the carbonates and hence facilitating hydrocarbon migration (Erdman, 1975).

Changes in fluorescence

At the outset, the fluorescence of the organic matter varies according to the type of compound involved — it is associated with the existence of the C=C bond (or benzene rings in the hydrocarbons) — and it also varies with the degree of catagenesis.

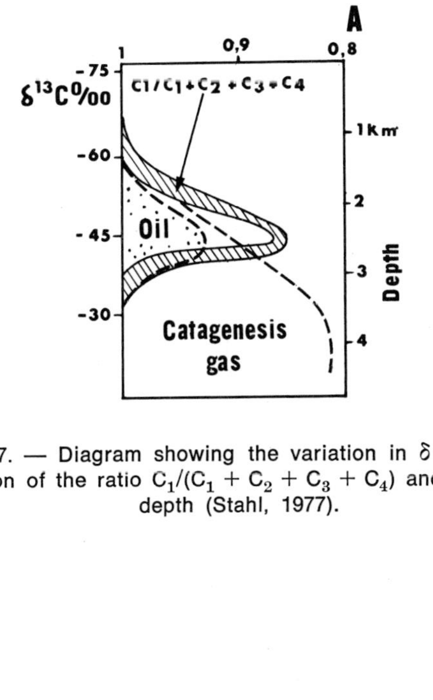

Fig. 27. — Diagram showing the variation in δ ^{13}C as a function of the ratio $C_1/(C_1 + C_2 + C_3 + C_4)$ and of burial depth (Stahl, 1977).

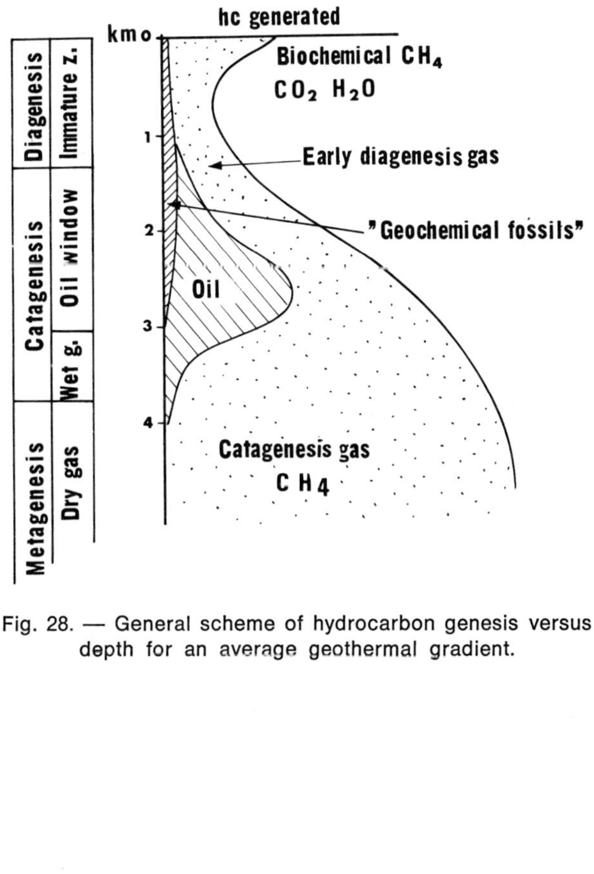

Fig. 28. — General scheme of hydrocarbon genesis versus depth for an average geothermal gradient.

Lacustrine algae, particularly the botryococcus family, display the following changes in their fluorescence colors :

— green in immature zones ($R_v < 0.5$ ‰),
— yellow in the zones at the beginning of the oil window,
— orange in the zones at the end of the oil window,
— reddish in zones with R_v 1 to 1.2 ‰

(P. Robert, 1978, unpublished document).

1.2.3.3. ***Evolutionary pattern***

This series of transformations with increasing pressure, and especially temperature, may be summarized by a graph showing the burial depths on the *y* axis, and assuming an average geothermal gradient, with the quantities of hydrocarbons formed on the *x* axis. This diagram shows the following (Figure 28) :

— at shallow depth, an immature zone characterized by a vitrinite reflectance lower than 0.6, the area of the formation of H_2O, CO_2 and some early hydrocarbons, including CH_4, sometimes with the presence of the same quantities of natural gasoline (Connan *et al.,* 1978),
— below, an oil genesis zone, initially heavy and rich in polar compounds, then in naphthenic resins and finally paraffinic, up to an R_v of 1 to 1.3, corresponding to a temperature of about 150 °C, a zone of wet gas genesis up to an R_v of 2,
— deeper still, a zone of dry gas genesis (methane) beyond an R_v of 2, but with a lower R_v for type III kerogens; the destruction of the benzene rings at temperatures above 150 °C may be accompanied by the formation of H_2S and N_2.

These transformations appear to depend on the composition of the organic matter. Type III kerogens give liquid hydrocarbons at higher temperatures than a type II kerogen. In these conditions, the oil window concept must be interpreted according to the type of kerogen, especially if type III predominates.

A vitrinite reflectance greater than 2 ‰ marks the entry into the metagenesis stage, distinguished by the exclusive presence of methane, and a stable residue consisting chiefly of stacks of aromatic layers. As a rule, the methane remains stable at high temperatures, above 550 °C. However, the presence of water, a permanent subsurface constituent, causes a degree of dissociation. At depths around 9000 m, depending on the geothermal gradient, 5 to 40 ‰ of the methane is destroyed. The presence of sulfur or sulfur compounds significantly decreases its stability (Barker & Kemp, 1979).

In the case of carbonaceous source rocks, the formation of nitrogen predominates above an R_v of 3.2 (about 5000 m), and some authors use this figure to define a 'gas window' (R_v between 1.2 and 3.2) (van Wijhe *et al.,* 1980). The analysis of the hydrocarbons imprisoned in the fine-grained source rocks, which may be considered as a first approximation to be in place, confirms this statistical scheme and shows that 82 ‰ of the gas, 91 ‰ of the liquids and 60 ‰ of the asphaltic compounds are found in the catagenetic conditions, corresponding to a temperature interval lying between 50 and 200 °C (Hunt, 1977).

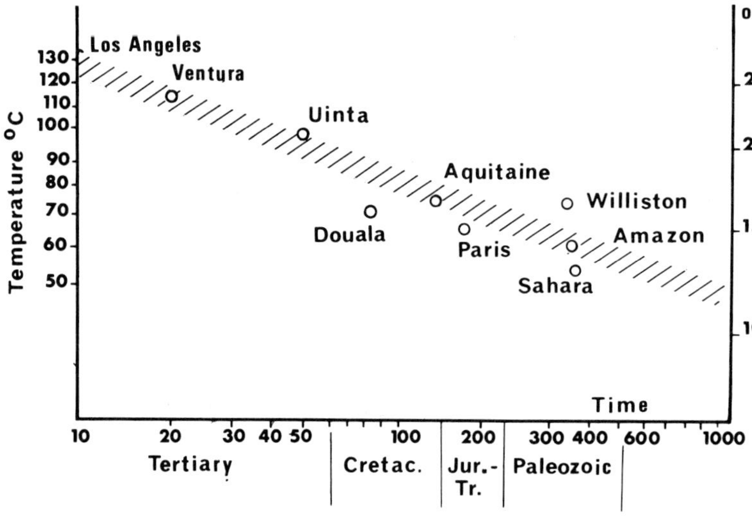

Fig. 29. — Variation of oil genesis temperature versus time.

Importance of the time factor

Time plays a major role in hydrocarbon genesis, which is known to require even higher temperatures for more recent source formations (see Chapter 2 of Part 1). Simultaneously, laboratory experiments have shown that temperatures around 300 °C must be obtained to 'degrade' organic matter into hydrocarbons in the space of weeks or months. In this respect, the investigation not only introduced an acceleration of the time factor, but altered the quality of the reactions, and is not a faithful reflection of the transformations occurring in nature with the passage of geological time. The study of different basins provides the following approximate 'critical' temperatures (Figure 29) :

— 50 to 60 °C for the Silurian/Devonian in the Sahara,
— 60 to 70 °C for the Toarcian in the Paris Basin,
— 100 °C for the Eocene in the Utah Basin,
— 115 °C for the Mio-Pliocene in the Los Angeles and Ventura Basins.

This genesis may extend over a more or less long period, possibly several million years, in very subsident basins, and over a hundred million years in the case of platform provinces. This more or less long 'manufacture' probably passes through intermittent phases in case of accelerated subsidence or the increase of heat flows. Everything occurs as if this transformation takes place from a sort of threshold concentration, as Zieglar & Spotts (1978) pointed out for the Great Valley of California (Figure 30).

These graphs, which are evidently based on a number of assumptions, show the difference in historical behavior between the northwest and southeast zones of the basin. In the northwest, where subsidence is relatively weak, the diagram suggests that hydrocarbon genesis proceeded during less than five million years, and that only the pre-Miocene formations reached the 'oil window'. In the southeast part, which is more subsident, the Miocene has been brought to the 'cooking' depth required for more than five to six

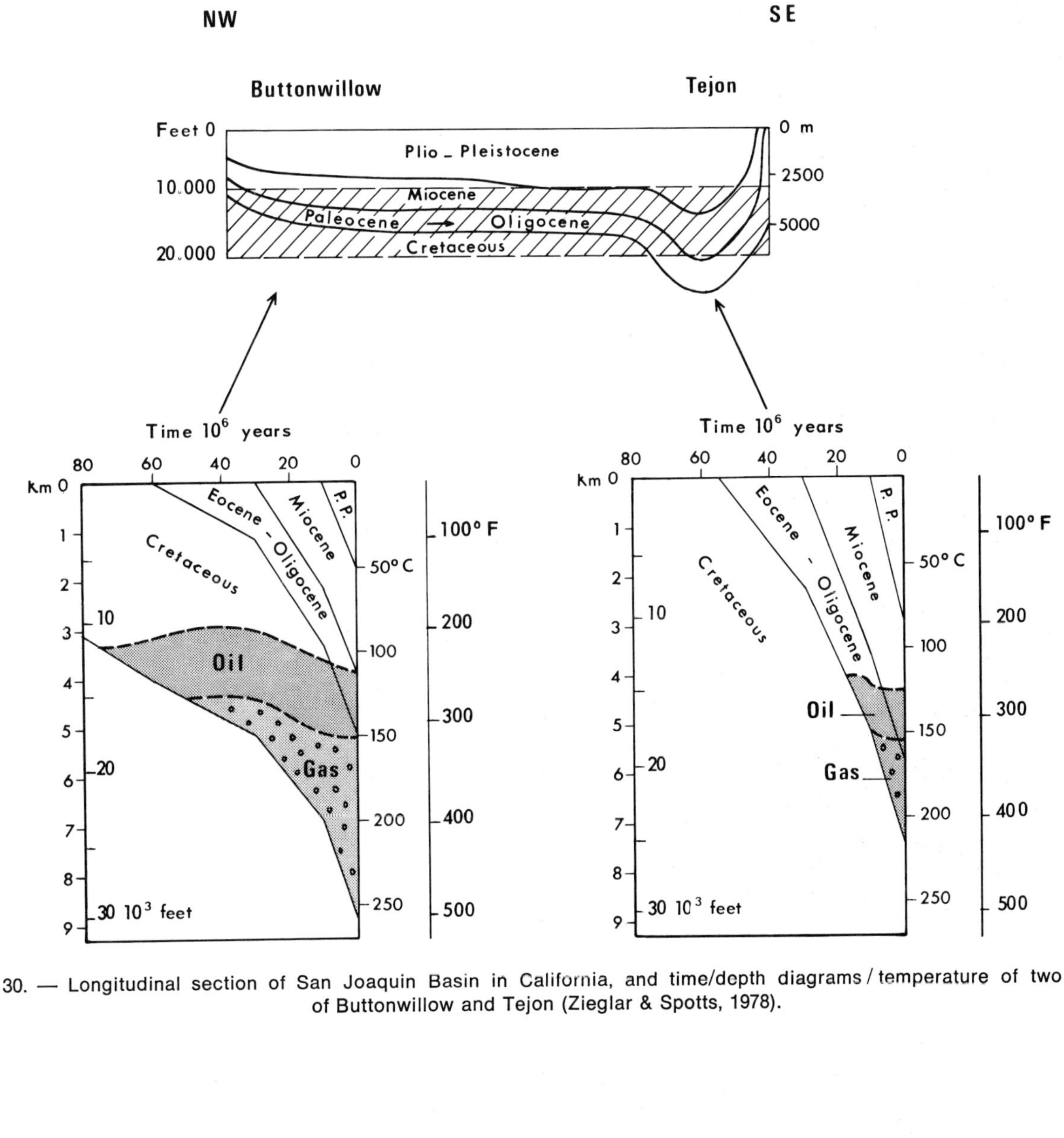

Fig. 30. — Longitudinal section of San Joaquin Basin in California, and time/depth diagrams / temperature of two zones of Buttonwillow and Tejon (Zieglar & Spotts, 1978).

million years, and the older formations for more than ten million years. This southern region turned out to be richer, with a cumulative oil output of more than 600 Mm^3 (see Chapter 3 of Part 3). The process is more complex and somewhat different for gases, and gas formed in the Plio-Pleistocene, in other words less than one million years ago, is found in the Po Plain and in the Gulf of Mexico.

In all cases, the thermal history must be reconstructed, particularly by the analysis of vitrinite reflectance values, of the basin or the zone concerned. Like Lopatin (1971), Waples (1980) recommends the consideration of a time/temperature maturity index.

Evolution of coals and oil shales

This evolution of bitumens occurs simultaneously with that of coals during their 'coalification' :

— the immature phase of the organic matter corresponds approximately to peats, and to soft and dull lignites,

— the oil window corresponds to high-volatile bituminous coals and gas coals,

— the gas zone of catagenesis corresponds to semibituminous and nongaseous coals and to anthracites.

This 'coalification' also corresponds to a progressive concentration of carbon and a concomitant increase in the reflectance of vitrinite, the main constituent. Volatile compounds are liberated throughout this catagenesis, carbon dioxide and water. Liquid hydrocarbons may also be generated from coal, but normally in very small amounts, possibly due to the absence of drains for their migration. No commercial oil accumulation appears to have originated in this way (Tissot & Welte, 1978).

Oil shales form another large family of carbonaceous rocks, characterized by high kerogen content, and especially type I kerogen. These rocks, particularly tasmanites, correspond to lagoonal or epicontinental sea deposits, and the matrix is argillaceous or carbonate. The amorphous organic matter is essentially of algal origin, and the *Botryococcus* family plays an important role. The H/C ratio is often higher than 1.5 %. They contain more than 2.5 % organic matter and often 5 to 10 %. Two-thirds of the organic compounds are susceptible to conversion to bitumen, with proportions reaching 50 to 100 and even 200 liters of oil per ton of rock (25 to 50 US gallons per short ton).

An oil shale subjected to sufficient catagenesis can generate liquid hydrocarbons.

***

The evolution of kerogen with depth and temperature, and its transformation into hydrocarbons, may be reflected by the diagram in Figure 28, which shows that gases, especially methane, originate more clearly (apart from the eventuality of gas of deeper origin) :

— at the time of deposition, gas of bacterial origin,

— in the first phases of diagenesis, partly early gas,

— in the advanced catagenesis phases, so-called catagenesis gas, produced from both sapropelic and humic organic matter, as well as coals.

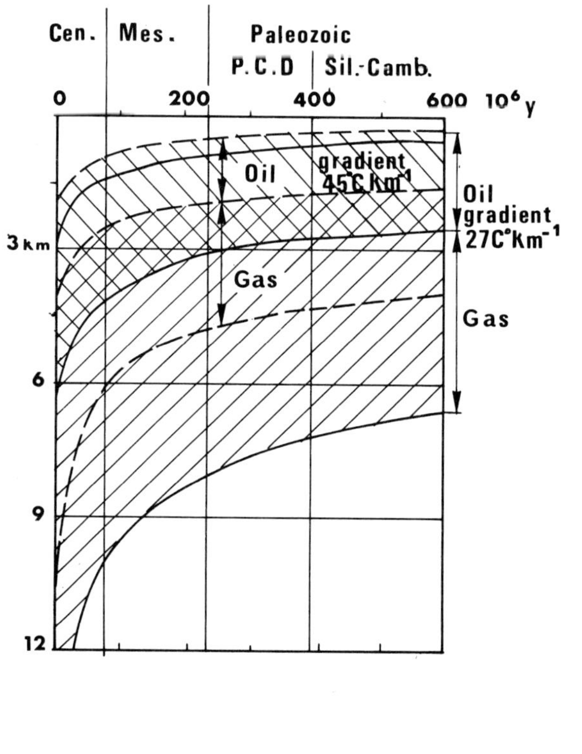

Fig. 31. — Graph showing the depth of oil and gas genesis zones versus time, with respective geothermal gradients of 45 °C/km at the center and 27 °C/km on the right (Magara, 1978).

Gas, particularly methane, which is more resistant and hence more omnipresent, is thus the culmination of the different evolutionary pathways.

The isotopic composition of the carbon in the methane offers direct help in identifying the origin of a natural gas. Hence gases in immature zones, up to depths around 1100 m, exhibit a $\delta\ ^{13}C$ ranging from — 95 to — 55 ‰ on average. In the intermediate zone, around 1000 to 1700 m, the $\delta\ ^{13}C$ values are roughly between — 55 and — 65 ‰. Below approximately 1700 m, in the catagenesis zone, the $\delta\ ^{13}C$ average is between — 58 and — 36 ‰. At extreme depths, below 6000 to 7000 m, $\delta\ ^{13}C$ reaches values around — 10 ‰ (Alekseyev, 1974).

In the central area, liquid hydrocarbons occur essentially in the approximate temperature interval of 60 to 120 °C, called the 'oil window'. As shown above, the depths of this interval vary according to time and geothermal gradients. They appear to be shallower in older series and with higher temperatures, as shown by Figure 31.

Roughly it may be stated, as by Tissot *et al.,* (1975), that periods of hydrocarbon genesis correspond substantially to subsidence eras, in other words the downwarping periods of the basins.

1.2.4. **Relationships between crude oils and depositional environments**

From the sedimentological standpoint, the different depositional environments favorable to the formation of source rocks or kerogen rich rocks may be related to two factors : sand/shale sedimentation of the paralic type, and carbonate/shale sedimentation with an evaporitic tendency. However, some continental environments of the lacustrine type may exhibit remarkable abundance and quality. These environments often strongly condition the characteristics of crude oils, especially if the action of catagenesis is weaker.

1.2.4.1. ***Paralic sand / shale environments***

These environments are characterized by thick pyritic black shale series, often with coal seams and sand intervals, in variable proportions, often lens-shaped, and of lacustrine, brackish or marine facies. They correspond to high subsidence areas, with intense alluvial activity and rapid burial, generally situated in the neighborhood of the mouths of the great rivers. The organic matter, of continental origin, consists of debris of the higher plants, rich in cellulose and lignin, conveyed directly to the lakes, lagoons and the sea by currents, with the detrital materials, sands and clays. Rapid burial or a deep environment pose obstacles to reworking on the floor.

In areas of slower sedimentation, these materials are attacked in the mud by microorganisms. Cellulose and part of the lignin are attacked, and very little remains but the waxes and long chains of n-alkanes. The kerogen resulting from these transformations, usually type II, yields paraffinic crudes, and rarely naphtheno-paraffinic oils. Saturated hydrocarbons account for 60 to 90 % of these crudes, aromatics 10 to 30 % of total hydrocarbons, and resins and asphaltenes generally less than 10 %. Sulfur is normally absent or in proportions less than 0.5 %.

Some of these crudes of non-marine origin, which are remarkable for their high proportion of waxes, display high viscosities. They are encountered in particular in rift type basins, especially in the pre-drift Cretaceous coastal basins of Brazil, Magellan and Gabon, as well as the Tertiary Basins of the Orinoco, Indonesia and the Rocky Mountains (Uinta Basin in particular) (Tissot & Welte, 1978).

Another closely studied example of this type of crude is offered by the oils of the Handil and Bekapai Fields in the Mahakam Delta (East Kalimantan) (Combaz & de Matharel, 1978) (see Chapter 2 of Part 3). These crudes, most probably originating in a type III kerogen, are highly paraffinic, particularly rich in high molecular weight paraffins, and poor in S (around 0.06 %). Their specific gravity oscillates between 0.82 and 0.87 — 41 and 31 °API —, decreasing with depth. Their pour point is + 30 °C for Handil crude.

The crude oil of the Gamba Field in Gabon, which also has a high paraffin content (13 %), and a specific gravity of 0.86, has a viscosity at 25 °C of 165 cP and a pour point of + 23 °C (see Chapter 2 of Part 3).

1.2.4.2. ***Carbonate and evaporitic environments***

The organic matter in purely marine environments is rich in protéins, carbohydrates and lipids derived from plankton and algae. It deposits to yield a type II kerogen that evolves normally towards naphtheno-paraffinic or intermediate aromatic crude oils. Saturated hydrocarbons account for 40 to 70 % and aromatics 25 to 60 % of total hydrocarbons. Resins and asphaltenes may be relatively abundant, together with sulfur, especially in poorly evolved crudes. Good examples of these oils are found in the Jurassic and Cretaceous formations of the Middle East, the Devonian formations of Alberta, and the Upper Cretaceous formations of Equatorial Africa.

Environments with an evaporitic tendency are generally represented by black shales and dark limestones, often bituminous, phosphatic and radioactive, and sometimes siliceous. They frequently correspond to older continental margins or marginal depressions in 'upwelling' and eutrophic zones in a generally arid climate. Detrital deposits are meager, stratification is very finely bedded, and the environment often highly saline. Sedimentation may be relatively slow and the proportion of organic matter high. Benthic organisms are generally absent and planktonic forms predominate, characteristic of a eutrophic environment. The sedimentation is formed of very fine beds, very thin alternations of various types, denoting the absence of currents and organisms on the floor.

The kerogen of carbonate source rocks has the reputation of being of excellent quality. It normally consists of amorphous organic matter, presumably of algal or sapropelic origin. The absence of ligneous or reworked components goes hand in hand with the limited clastic deposits, and these factors contribute to the development of the high quality of this kerogen. Its H/C ratio is high, giving it high evolutive potential (Elf Aquitaine, 1977).

While the carbonate muds are rapidly lithified, it appears that they subsequently develop very dense fissures due to their lack of plasticity. These microfractures are probably the reason for the effective expulsion of the hydrocarbons formed, reaching a clearly higher level than in shale formations. The crudes formed in these environments are usually heavy, because they are rich in polar compounds, resins and asphaltenes, and in sulfur, which may reach 5 % and even 10 % in the altered aromatic base crudes. They are rich in trace metals, Cu, Ni, Co, Mn and U. The branched chains of these hydrocarbons are more resistant to alterations and produce most of the shows (Weeks, 1968).

Crude oils of this type are found in the Middle East, especially in Syria and Iraq, and also in Aquitaine, Venezuela and Mexico. The Wyoming Basin in the Rocky Mountains offers an interesting example of the superposition of two hydrocarbon families. Sulfur-rich naphtheno-aromatic crudes are found in the carbonate and evaporitic formations of the Paleozoic, while gasoline-rich paraffinic and naphtheno-paraffinic crudes are found in the clastic shale series of the Cretaceous. In these conditions, the deepest crudes are the heaviest.

Reef environments are very prolific, but the organic matter they contain is largely degraded and mineralized. It can only be preserved and accumulate in the sheltered depressions which are often formed behind the reef barriers. Highly saline environments, often displaying a very high organic matter content, derived from a few rare species of living organisms, slight elastic deposition, and evident preservation characteristics, can provide excellent source rocks (Kirkland & Evans, 1981).

1.2.4.3. *Lacustrine environments*

While most source rocks correspond to marine environments, this is not indispensable, and in different basins, gas and oil clearly appear to be derived from lacustrine formations. Among these are the following :

- Tertiary of the Cook Inlet Basin, the Middle Magdalena (Colombia), and the Pakistan and Chinese basins,
- Carboniferous in the British Midlands,
- above all, the oil shales and oils of the Eocene Green River formations in the Green River and Uinta Basins (Wyoming, Utah, Colorado).

Lakes and some lagoons may offer especially favorable eutrophic conditions for the development of algae, particularly the botrycoccus and cyanophyceae algae as in Lake Maracaibo, which are capable of using atmospheric nitrogen directly. In these environments, the density of plankton and other organisms rapidly stops the penetration of light, creating a contrast between the surface waters which are very rich in organic matter and the deeper levels which are stratified, stagnant and anaerobic. Thus life develops intensely at the surface, and the organic matter of dead organisms finds ideal conditions on the floor for putrefaction and preservation. These eutrophic environments thus constitute true 'kerogen factories' (du Rouchet), on highly variable scales, ranging from the simple pond to inland seas, like the Black Sea.

Summary and conclusion

Hydrocarbon genesis is a fundamental chapter in the geochemistry of organic matter. From the moment when protids, lipids and carbohydrates, which make up the bulk of the planktonic biomass, together with humic materials inherited from the continent, are buried with the sediments in anaerobic conditions, the action of pressure and especially temperature, exerted during subsidence, necessarily transforms this insoluble organic matter or kerogen into hydrocarbons and coal.

This birth occurs during burial between certain depths, called the 'oil kitchen', and corresponds to a degree of catagenesis bounded by vitrinite reflectance values between 0.5 and 1 to 1.2 %, while the characteristics of the organic matter and the age of the deposits slightly alter these limits. Deeper down, the kerogen and hydrocarbons formed are transformed into gas, chiefly methane, and into coal. Catagenesis thus gradually obliterates the characteristics associated with the conditions of sedimentation.

1.3. PRIMARY MIGRATIONS AND THE SOURCE ROCK CONCEPT

The study of hydrocarbon genesis cannot be separated from that of their expulsion from the source rock towards the reservoirs. A close relationship is generally observed between the oil genesis peak and the onset of primary migration (Jones, 1981), and one may well ask whether the 'manufacture' of oil or gas in the source rock is not facilitated by the more or less simultaneous departure of the resulting products. Moreover, hydrocarbon expulsion mechanisms appear to exert an influence on the very nature of these bodies.

It is customary, and logical, to distinguish between primary and secondary migrations, not for reasons of mechanisms, but due to the difference in characteristics of the pore volumes. The first occurs immediately after hydrocarbon genesis, from fine-grained rocks to a reservoir, and only involves short or very short distances, from around a meter to about ten meters; the second occurs within the permeable rocks and may continue and extend over distances and periods which are much longer (Roberts & Corpel, 1980). Primary migration processes may be divided into two main phases :
- expulsion of hydrocarbons from the kerogen and the source rock,
- transfer of these hydrocarbons through the source rock to its boundaries, in contact with more porous and permeable series.

1.3.1. Different modes of primary migration

It is first necessary to specify the dimensions of the objects to be moved and to compare them with the void configuration of the solid medium through which they are to pass. Migrating hydrocarbons include methane, a gas whose molecules are about 4 Å in diameter (the water molecule has a diameter of 3.2 A), and heavy, solid compounds of the asphaltene type, with molecular diameters up to 50 to 100 A. As for the solid, the pore or micropore diameters of shales range from 50 to 100 Å at a depth of some 2000 m, and are less than 50 Å deeper down.

Theoretically, hydrocarbons may be considered to migrate in the following forms :
- true molecular solutions,
- colloidal and micellar solutions,
- individualized oil or gas phases.

Transport may be considered in the form of a homogeneous flow or in the form of diffusions through a concentrated 'solution'. It is also important to consider whether the water and hydrocarbons transported form a single- or multi-phase mixture, or whether the solid medium is preferentially wetted by oil or by water. The water/hydrocarbon relationship must also be examined, as these hydrocarbons migrate through a porous medium which is always invaded by water.

Migrations by molecular solutions

An important role may be played by interstitial water and water liberated by diagenesis. At shallow depth in particular, large volumes are expelled, and some authors, including Baker (1962), Cordell (1972, 1973), and Rumeau & Sourisse (1975) have postulated that the hydrocarbons entrained in this flow could possibly continue their maturation and conversion into hydrocarbons within the reservoirs themselves. However, it is worth noting that this protopetroleum and its precursors, representing an intermediate stage in the formation of hydrocarbons, have never yet been identified.

We have seen above (Section 1.1 of Part 2) that hydrocarbons display very slight and variable solubility in water, depending on their chemical composition, and that the most soluble compounds are benzene and toluene. Hobson & Tiratsoo (1975), Dickey (1975), Tissot & Welte (1978), and Jones (1981) showed that in these conditions the water volumes expelled would have to be far greater than the quantities normally produced by compaction, in view of the solubility measured, or the hydrocarbon solubility would have to be increased to values of 10,000 ppm to ensure the accumulation of the known oil reserves of a basin.

The first observation is the great discordance existing between the distribution of the different hydrocarbon species in a crude oil and the distribution that would result in accordance with their solubilities in water. The most soluble substances, such as benzene, always occur in very small proportions. Crude oils are especially rich in saturated hydrocarbons, that are among the least soluble and, conversely, they only contain traces of the more soluble hydrocarbons.

Fine hydrocarbon analyses carried out in a source rock, in the adjacent reservoir that it appears to have irrigated, and in the transition zone between them, reveal a depletion of the source in soluble materials (by organic solvents) and a concomitant relative enrichment in heteropolar compounds. On the other hand, the reservoir crudes are enriched in saturated hydrocarbons and depleted of polar compounds containing N, S and O. These observations show the importance of chromatographic absorption mechanisms throughout migration (Tissot and Welte, 1978).

To compensate for the very low solubility of hydrocarbons, many authors, including Baker (1962) and Cordell (1973), suggested the presence of solubilizing agents in the migration waters, including surfactants, which have the property of substantially raising hydrocarbon solubility. These solubilizers occur in the form of microemulsions of micelles with diameters from 100 to 2000 Å. This theory comes up against the following objections :

- the lack of any observation of these solubilizers, which should occur in significant quantities in order to play an active role,
- the dimensions of the micelles, which would prevent them from circulating in fine-grained rocks
- their negative electrical charge identical to that of the mineral surfaces (McAuliffe, 1979; Jones, 1981).

Migration by colloidal or micellar solution

Considering the very low solubility of hydrocarbons, it is tempting to postulate their dispersion in the form of colloids or micelles. However, the dimensions involved would be of the same order of magnitude and normally larger than the pore spaces, about 60 Å on the average, lending little credibility to this hypothesis. In addition, the electrical charge opposition between micelles and surfaces of argillaceous minerals further reduces the likelihood of this process. At best it may be imagined in very slightly compacted series which have not yet reached the critical burial depth of 2000 m.

Transport in the form of bubbles and droplets involves the respective diameters of the objects transported, constrictions between pores, and capillary pressure mechanisms. Movement is only possible if the droplet diameter is smaller than the constrictions, or capillary forces would have to be overcome for the droplets to be deformed and to pass through. For some authors, the rise in pressure in the liquid phase (water and/or hydrocarbons) during compaction associated with fracturing (geotectonic or hydraulic) would enable this process to occur.

Primary migration in aqueous form is undoubtedly not entirely impossible, but it appears to be limited to a very small number of cases, chiefly in the sand/shale zone. This could be the source, for example, of the small neogene oil and gas accumulations of the Nagaoka Plain in Japan, where the particularly high geothermal gradient could have resulted in rapid maturation of the hydrocarbons (Magara, 1968).

Gases are more soluble and, especially at great depth and in deltaic provinces, dissolved gases could well convey the liquid hydrocarbons that they keep in solution (Hedberg, 1979; Price, 1976, and Jones, 1981).

But water is also the vehicle for oxidizing agents and, in this case, it is an important factor in the transformations and biodegradation of oils. Its composition, or more specifically its salinity, plays a major role in this respect. Water is very probably a transport agent whose mechanical or physicochemical action may be considerable, positive or negative. However, its effectiveness must decrease fairly rapidly with depth.

Migration in individualized oil or gas phase

It is logical to believe that, in the initial phase, as long as the macromolecules making up the kerogen are relatively unaltered by link-splitting due to defunctionalization or to breakages of C-C bonds, expulsion by geostatic pressure is hardly credible. But the transformation of 10 to 20 % of the kerogen mass into hydrocarbons (which appears to occur in the 'oil window') must substantially alter the situation : the chemically-disintegrated micromolecules must be transformed into a complex of smaller molecules, mutually free from each other. Subjected to a sufficient stress field, the kerogen thus broken up and softened is plastically deformed. The most mobile molecules probably move in the direction of the weakest main stress (du Rouchet, 1978, internal report). This liberation property could be related to the distribution of organic matter in the rock. This could explain the very low yield of the oil shales, and, conversely, the high yields of carbonate and phosphate rocks (van der Weide, 1978, internal report).

When the oil saturation reaches 30 to 50 % in a source rock which has reached its mature stage, the interstitial water is structured and fixed to the pore walls, if these are not oil wet. Under the action of the pressure gradient, these conditions allow the movement of the oil in a continuous phase (Dickey, 1975; Tissot & Welte, 1978; McAuliffe, 1979; Jones, 1981; du Rouchet, 1981). Analysis by fluorescence microscopy at high magnification (500 ×) reveals oil veins about a micron in size in a mature source rock. This observation, added to the fluorescence of 'exudatinites' in the oil window, confirms the theory of the direct liberation of hydrocarbons from kerogen (Robert, unpublished report) (Plate 1). From the chemical standpoint, a drop in the H/C ratio, corresponding to the expulsion of the product richest in H from the kerogen, is observed in the source rock at these levels. This observation is confirmed by an increase in resistivity on the electric logs at this level. From the isotopic standpoint, an enrichment in ^{13}C occurs during migration.

Kerogen, which is generally hydrophobic, more or less retains the mobile materials in the matrix of its complex structure. Thus its structural organization directly governs the expulsion of hydrocarbons, which is facilitated by the presence of microfissures. In a second stage, the hydrocarbons formed force their way across the barrier posed by the fine-grained sediments of the source rock, by opening or re-opening microfissures above the level at which the thrust of the fluids can overcome the confining pressure. This 'breakthrough' is partly facilitated by the increase in pressure caused by the expansion of the aquifers under the effect of temperature, and the individualization of methane microbubbles when the saturation pressure is reached (Hedberg, 1979). It also appears to depend on the quality of the source rocks. Geomechanical confinement is also significantly weakened by tectonic movements (Segor, 1965; du Rouchet, 1978, 1981).

Magara (1978 a, 1978 b, 1980) stresses the role of the oil-permeability of the shales during compaction, while the water-permeability decreases with compaction. This oil-related permeability increases with greater oil saturation, namely following the expulsion of the water with compaction, at least up to a certain point. Thus at a certain stage of compaction, when the bulk of the water is already expelled, but where the relative permeability is still substantial, permeability and oil-related permeability display values offering especially favorable migration conditions for the continuous oil phases (Dickey, 1975; Magara, 1978 a) (Figure 32).

In these conditions, the expulsion of oil in a continuous phase appears to be the most likely hypothesis in a large number of cases, at least when the source rock is sufficiently rich and concentrated. This is acknowledged today by most investigators, including McAuliffe (1979), Baker (1962), Chiarelli and du Rouchet (1977), Dickey (1975), du Rouchet (1978 c) and Jones (1981). This would explain in particular the formation of fields with dense and compact source rocks, like those of the Cretaceous limestones of La Luna in Venezuela, the Devonian limestones of the Swan Hills in Canada, the Devono-Mississippian 'Bakken Shales' of the Williston Basin (Meissner, 1978), and the Silurian shales of the Algerian Sahara (Tissot & Welte, 1978).

1.3.2. **The geological framework of primary migration**

In overall terms, hydrocarbons begin to take shape in the depths of the sedimentary basins when most of the water initially present is expelled. In fact, the genesis of liquid hydrocarbons occurs in a temperature interval of about 65 to 150 °C (150 to 300 °F), generally corresponding to an advanced degree of compaction, when the bulk of the interstitial water has already been expelled (around 88 % at 500 m, 95 % at 1500 m, 98 % at 2500 m). Primary migration appears to start only above a certain level of kerogen transformation, when the kerogen begins to react to the stresses no longer as a solid, but as a liquid. It continues at depths around 3000 m and beyond in recent series (Young *et al.,* 1977).

Dickey (1975) postulates that a shale buried between 1500 and 4500 m, substantially corresponding to the 'oil window', loses some 11 % of its porosity, mostly because of the loss of water. Magara (1978) states that the water could continue to migrate very slowly but for a very long time in compacted shales with very low porosity. Everything else remaining equal, the volume of water passing through shales of 10 md permeability during one year is equivalent to the quantity passing through a shale of 10^{-5} md during one million years.

An attempt may be made to determine the most favorable period for this expulsion. As noted by Magara (1978 a), the maximum liquid hydrocarbons present marks the start of their transformation into gas, and consequently the beginning of the decline or oil genesis. Consequently the peak phase of oil genesis is situated above, namely before the oil accumulation peak. And the main migration phase could well correspond to this first oil generation peak (Figure 33).

Some authors, including Magara (1974), also claim that osmotic flows, caused both by compaction and by salinity differences between the shale and sandstone waters, add their effects and stimulate primary migration. By raising the salinity in a closed reservoir, this electro-filtration mechanism could also contribute to the concentration of hydrocarbons, by 'precipitating' those that are present in solution (Hobson & Tiratsoo, 1975; Cordell, 1972, 1973 and 1976). Many authors, notably Neglia (1979) and du Rouchet (1979), place the

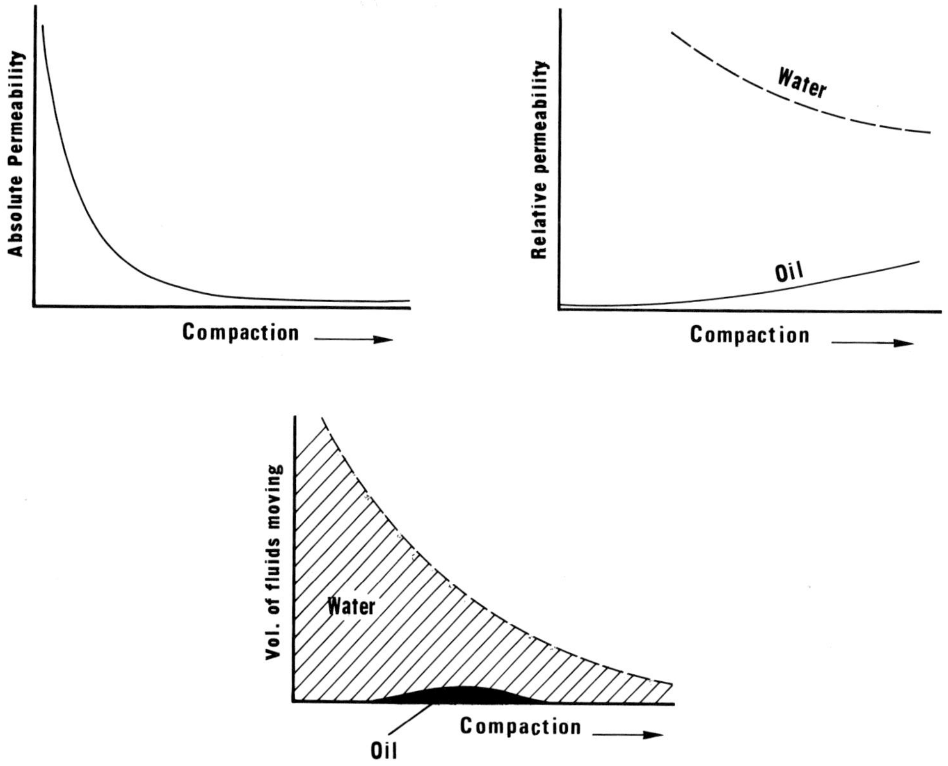

Fig. 32. — Diagrams showing assumed relationship between absolute and relative permeability and oil and water movements as a function of compaction (Magara, 1978).

PLATE 1

Visualization of a source rock by fluorescence microscopy

Essises No. 1 well (France) : Hettangian

Figures 1 and 2 : 2812 m Figure 3 : 2812.75 m

Catagenesis level : 1 % vitrinite reflectance, end of the ‘oil window’

Observations on massive polished rock, without other treatment

Fig. 1-2. — Images of a stratified source rock

1 : dry front 2 : oil immersion

Shaly rock consisting of alternating

- lamination 20 to 60 μ thick of a fluorescent matrix (yellowish) : this is the productive part of the source rock,
- non fluorescent shale (black), devoid of oil-generating properties.

The fluorescent matrix is sapropelic (its scattered organic fraction is of algal origin) ; it is flecked with small fragments of orange-colored algae.

The exuded oil appears bright green to bright yellow in the fracture fillings (1-2) and in the shale pores (lower quarter Figure 2).

Fig. 3. — Oil identification

Dry-front observation with low magnification.

The low magnification prevents observance of the rock fluorescence, which is less intense, but it shows the geometry of the network of oil-filled fractures (green) over a broad field.

The left-hand part is normal. At the right, the same subject, after cleaning with a solvent pad : the surface oil film has disappeared and the fractures are largely empty.

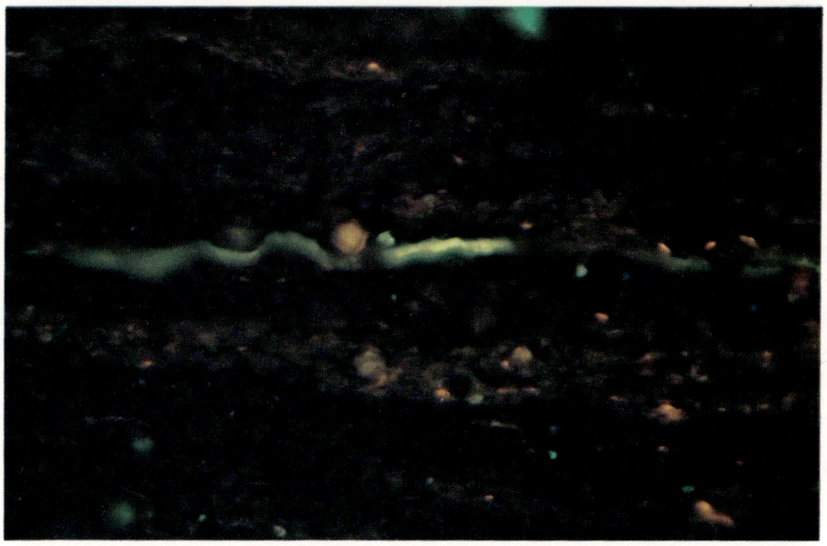

1 *Objectif à sec* 100 μ

2 *Objectif à immersion d' huile* 100 μ

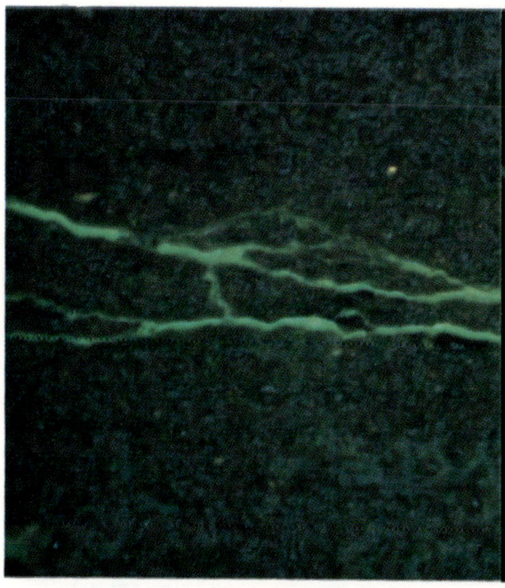

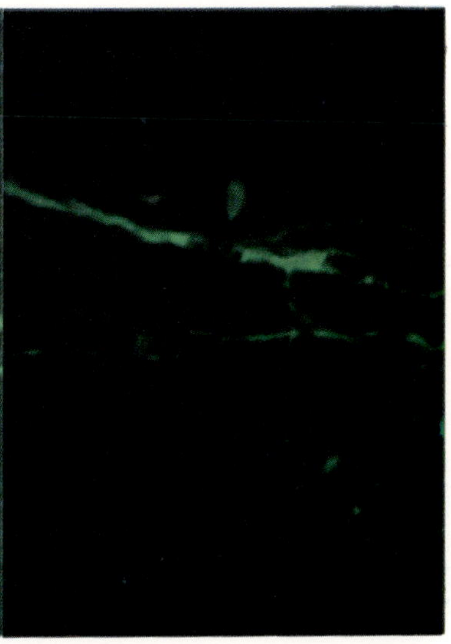

3a *Objectif à sec* 3b 100 μ

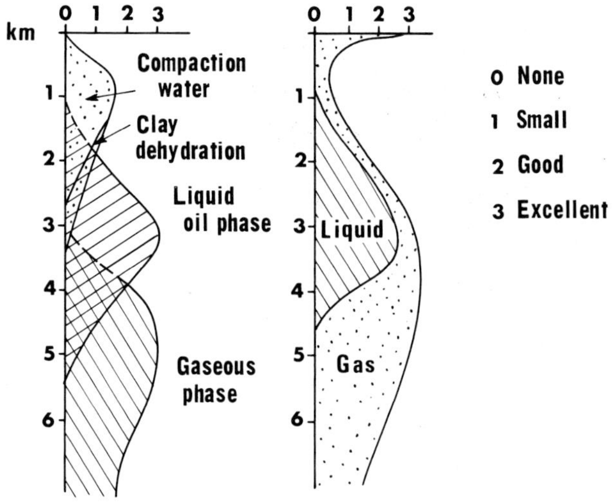

Fig. 33. — Diagrams showing the following versus depth ·
— (left) areas of action of main migration modes;
— (right) oil and gas genesis areas.
(Tissot & Welte, 1978).

accent on the importance of microfractures affecting the compacted shales and carbonates at these depths, partly due to the rise in pressure caused by the increase in fluid volumes triggered by rising temperatures and hydrocarbon genesis. It also appears that these fractures develop more easily with tighter geomechanical tensile patterns.

While the boundaries of this 'oil kitchen' are fairly well defined especially with respect to the geothermal gradient, considerable uncertainty surrounds the quantity of hydrocarbons expelled. Magara (1978 b) attempted to illustrate this unknown parameter by diagrams.

In the large majority of cases, the yield of this primary migration appears to be low, around 5 to 10 %, and very rarely more, as in the Middle East (Momper, 1978). These movements seem to occur over short distances, around 1 to 10 m. They obviously largely depend on the petrophysical characteristics of the rocks, and are encouraged by the existence of natural migration paths, permeable detrital intervals offering drains for lateral transfers, microfissures and fractures for vertical movements.

It appears that these hydrocarbon expulsion flows occur intermittently throughout geological history, depending on the heat flows and periods of intensified subsidence. In these particular conditions of advanced compaction, in which primary migration appears to occur, special attention is paid to the water produced by catagenetic transformations and to the undercompacted formations.

The role of 'catagenetic water'

In the absence of a good correlation between the expulsion of compaction waters and hydrocarbon genesis, the accent has been placed on the waters expelled during the late compaction phases, and especially the dehydration water of certain minerals like smectite, during genesis and catagenesis (Burst, 1959 and 1969; Weaver, 1959; Powers, 1967; Dunoyer de Segonzac, 1969 and 1970). Magara (1975 and 1978 a) showed that this special release of catagenetic water was a relatively continuous occurrence, and was added to the compaction water, providing a relay in time and guaranteeing a degree of continuity in the water expulsion process. Many researchers have thus stressed the role of smectite, whose transformation into illite in the oil kitchen zone releases water in the proportion of nearly 50 % of its volume (Magara, 1978). A correlation has also been observed between oil pools and smectite.

The case of undercompacted formations

Zones of undercompacted shale, and, in particular, the transition zones above them, play an important role in oil genesis. In fact, these are fairly often beds of sapropelic marine shales with a favorable source rock facies, and in many countries are qualified as 'bituminous zones'. The delay in compaction, which is reflected by abnormally high porosity for the burial depth at which they are found, facilitates the mobilization of the hydrocarbons despite a certain lag in diagenesis/catagenesis. Undercompacted formations thus offer the advantage of delaying the expulsion of large quantities of water at depths where the temperatures are higher and close to those corresponding to hydrocarbon genesis. If these formations are or were more widespread than one is led to believe, this would offer a favorable pattern for certain primary migrations (Chapman, 1972).

The transition zones surrounding them appear to be especially favorable for the expulsion of the hydrocarbons formed, due to their pressure gradient. In the Mackenzie Delta, for example, 90 % of the oil and gas pools lie within this depositional interval (Evans *et al.*, 1975) (Figures 34 and 165). Similar observations have

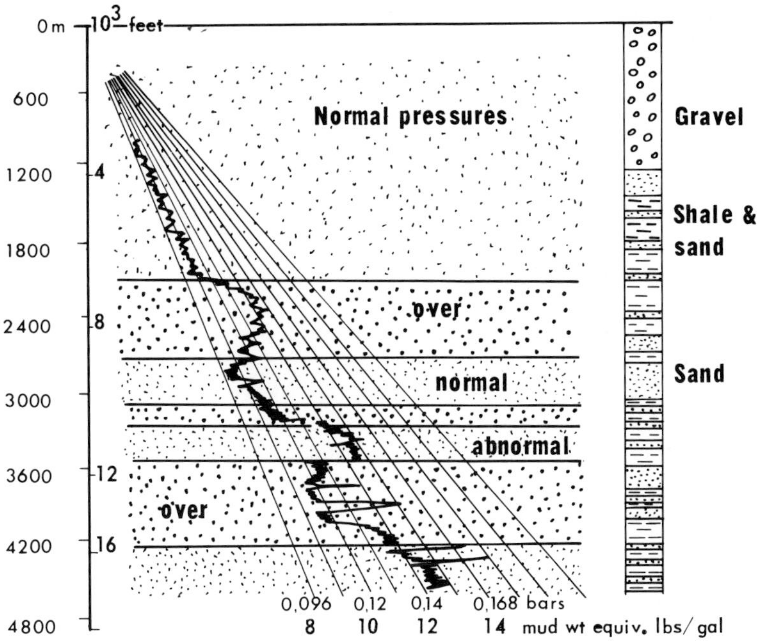

Fig. 34. — Pressure/depth profile in Taglu Field, Mackenzie Delta (Bruce & Parker, 1975).

been reported in the Handil oil zone in East Kalimantan (Durand & Oudin, 1979) (see Chapter 2 of Part 3, page 231).

The migration of gas

Gas migration very probably obeys rules which are different to those of oil, and dissolution in water seems to be a decisive factor.

As Hedberg (1964) states, it may be considered that "overlapping probably occurs between the methane formation depths by biochemical and thermochemical processes, one supplanting the other progressively with depth", the first generally characterized by a higher percentage of isotope ^{12}C than methane of thermochemical origin. The solubility of the methane rises rapidly with pressure, but decreases with higher water salinity. For a fresh water, it increases from 0.7 $m^3\,m^{-3}$ at 35 bar pressure to 5 $m^3\,m^{-3}$ at 700 bar (Magara, 1978).

At great depths, the quantities of gaseous hydrocarbons dissolved in the aquifers may reach high values, around 5 % by weight, the same order of magnitude as the solubility in oil (Jones, 1978). Considerable quantities are thus accumulated. For instance, it has been estimated that the deep high-pressure aquifers in the onshore areas of Louisiana and Texas could contain more than $6700 \cdot 10^{12}$ m^3, which may be compared with the $80 \cdot 10^{12}$ m^3 of gas discovered to this day in all gas fields worldwide. Some investigators postulate that these high-pressure aquifers release part of their dissolved gas if they undergo pressure drops, as when moving upward under the action of rising gradients. The gases thus released could replenish the traps located in their paths (Hobson, 1978).

Gases could also migrate by diffusion, in particular by a substantial difference between the gas composition in the shales, which contain around 50 % C^1, and those migrating into the neighboring sandstones, where C^1 accounts for 80 to 90 % (van der Weide, 1977; Hinch, 1978). The gas is found to be enriched in ^{13}C during its migration (Stahl, 1977).

The very high molecular diffusion of the gas — especially methane — in water tends to imply that this process is an important mechanism of primary gaseous hydrocarbon migration through the water-impregnated shale series (Leythaeuser *et al.,* 1982). Measurements taken on models by these authors show that this process is compatible with the formation of commercial accumulations in a few million years. However, concentration in pools also provides preferential sources of gas diffusion through the seal.

One may thus draw the conclusion that no universal primary migration mechanism exists, but rather several complementary patterns, the relative importance of which depends on the hydrocarbon composition and the burial depth. At shallow depths (1500 m) methane could migrate easily in solution in the water, possibly together with some heavier hydrocarbons. Below 1500 m, in optimal maturation zones, and chiefly in the case of concentrated source rock, liquid hydrocarbons, initially heavy, and then lighter at greater depths, probably migrate mainly in a continuous phase.

In provinces with relatively dispersed source rock and a low geothermal gradient, like the deltaic basins, migration could occur by less effective processes, especially associated with the aquifer phase. Deeper down, where gas and condensates predominate, mechanisms involving gas as a vector of liquid hydrocarbons have been considered. However, "it is doubtful that gas contains the heavier hydrocarbons and N-S-O compounds in high molecular weight crude oil fractions" (McAuliffe, 1979). Moreover, the departure of very large quantities of gas would have to be assumed. Most important is the existence of fairly strong subsidence which, with the burial of several thousand meters of maturing organic matter, guarantees its transformation into hydrocarbons and their expulsion from the rock in which they originated. Despite the name, what is referred to as 'morts terrains' or overburden, plays an active and fundamental role in pool formation.

Once they reach a continuous reservoir horizon, the hydrocarbons continue their travels : this stage is called secondary migration. If a permeability barrier or trap is found in the neighborhood of the source rock, secondary migration does not occur — or only over a very short distance — and the pool may be considered 'in place'. This is referred to as a primary trap or 'trap-reservoir'. This is often the case of reefs, as in Alberta and West Texas, and also in the Jurassic fields of the North Sea, Cormorant, Brent and Magnus, as well as the great Permian (Khuff) gas fields of Iran, Kangan, Nar and Agar.

1.3.3. **Characteristics of source rocks**

A source rock is a sediment containing a certain quantity of organic matter and having generated appreciable quantities of oil or gas. By extension, this term is applied to facies identical to those of a source rock, but placed in different diagenetic conditions, namely a rock which has produced or could produce hydrocarbons. In this case, one speaks of an exhausted or potential source rock. These rocks seem to be characterized by :

— a percentage of organic matter between 0.5 to 1 and 5 to 6, depending on composition,
— the presence of extractable products (EOM) * denoting at least the possibility of hydrocarbon genesis, with EOM/OM ratios greater than 3, higher for carbonate source rocks than for shales,
— some dispersion of organic matter in a shale phase representing at least 30 % of the rock,
— the existence of laminations, layers and diastems, featuring beds with high concentrations of organic matter and horizontal and/or vertical drainage surfaces.

The composition of the constituents of the organic matter is of vital importance for the petroleum generating potential of a source rock, and hence for the quality of the basins (Rogers, 1979). A relationship has been established between the following two aspects :

— the potential of the Gippsland Basin, southeast of Australia, where some 340 Mt of liquid hydrocarbons have been discovered, and the composition of the shales of the Latrobe/Valley Group, considered to be the main source rock (84 % vitrinite, 12 % exinite and 4 % inertinite) (see Chapter 2 of Part 3, page 228).
— the poverty of the Cooper Basin, which presently contains only 14 Mt of oil and 100 Gm^3 of gas, and the composition of the Permian source rock (Gidgealpa Formation) : 43 % vitrinite, 4 % exinite, 53 % inertinite (from Shibaoka *et al.*, 1978).

The presence of clay minerals in relatively large quantities allows the buildup of high internal pressures, probably causing microfissures, and thus accounting for the expulsion of the fluids formed and especially the hydrocarbons. Carbonate and phosphate source rocks exist, possibly in the form of fine shaly laminations. On the average they contain nearly ten times less organic matter than shales, but substantially the same amount of extractable products, as the organic compounds are generally of the sapropelic type rich in lipids. They also exhibit a more open matrix which probably facilitates transformation into hydrocarbons.

Source rocks exhibit tremendous diversity depending on the composition and evolution of the organic matter (Claret *et al.,* 1981). The hydrocarbons formed do not appear to be extractable except for methane, and large quantities of extractable products remain imprisoned. A source rock is identified by the correlation between the hydrocarbons it is presumed to have contained and the traces that have remained in it. This operation is often difficult due to the differential migrations of the different hydrocarbon species, and the variations recorded by them, especially by chromatographic effect, during their transfer (Figure 35).

(*) EOM : extractable organic matter.

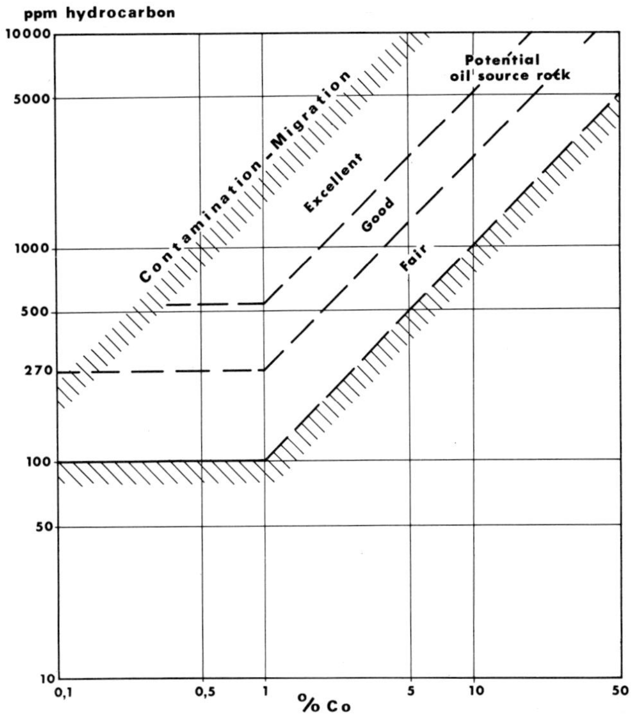

Fig. 35. — Diagram showing percentages of organic C and characteristic hydrocarbon quantities of potential source rocks.

A source rock is characterized by its content of non-migrated extracts, and by the ratio EOM/TOC (*) :

	EOM	EOM/TOC
— very good grade	1500 ppm	5 to 10
— good grade	150 to 1500 ppm	2 to 5
— mediocre or poor	50 to 150 ppm	2

Higher values of the EOM/TOC ratio correspond to carbonate rocks, lower values to shales.

To remove any doubt, it is usually necessary to make correlations based on very fine analyses and reliable markers, the geochemical fossils. Among these are the four- and five-ring saturated hydrocarbons, polyaromatics, isoprenoids and n-alkanes with C_{15}^+, as well as the relationship between the isotopic carbon ratios. All such correlations must be based on several independent parameters which are correctly located in the geological context (Welte *et al.*, 1975). The 'geochemical fossils' that may be found in the source rock and in the crude oils are a very important correlation factor today. The isotopic correlation of kerogen and oils appears to be an increasingly important and reliable criterion for correlation between different crude oils, extending beyond their possible alterations, and also between crude oil and kerogen.

Thus in the Parentis Basin, the oils of the main fields, Cazaux (Albian), Lugos and Parentis (Neocomian), and Lavergne la Teste, display very similar $\delta\,^{13}C$ values around — 25 ‰. On the other hand, the deep Parentis and Le Teich 2 crudes appear to be somewhat lighter and seem to belong to another family (Sourisse, Esquevin & Menendez, 1979, unpublished).

Hydorcarbon genesis is observed in the source rock :

— by the existence of sorbed gases, chiefly alkanes from C_1 to C_5 trapped in the mineral matrix, and the C_2/C_1 and C_3^+/C_1 ratios in particular will be examined,

— by the observation of oil microdroplets in microfluorimetry.

Fluorescence shows an intensity peak in immature source rocks close to the oil genesis stage. It declines gradually with advancing maturation. Correlations between the oil of a field and a presumed source rock are always difficult to make and generally require very detailed analyses.

It is first necessary to make sure that the extracts of the probable source rock are correctly in place, namely produced from the organic matter of this source rock and not migrated from another source. The fact that a rock is not a reservoir does not certify that it is not invaded. It should also be noted that the composition of the oils changes during their migration. They grow richer in saturated hydrocarbons, less

(*) TOC : total organic carbon.

in aromatics, and lose oxygen, nitrogen and sulfur compounds. In many cases, particularly when the reservoir is found in an immature zone, the hydrocarbons may originate alternatively in :

— older and deeper series, by means of vertical migrations,
— contemporary series buried deeper in neighboring synclines, with lateral migration.

An example of a source rock : the Bakken Formation of the Williston Basin

(Meissner, 1978)

The Bakken Formation is a thin unit that has the special feature of being carbonate as well as rich in organic matter during a major cycle of onlap-offlap sedimentation at the Devonian/Mississippian transition zone, or at the extreme base of the Mississippian, on the Devonian formations. It belongs to a vast sequence of black shales of the same age which covered the North American continent at the time (Figures 36, 165 and 192).

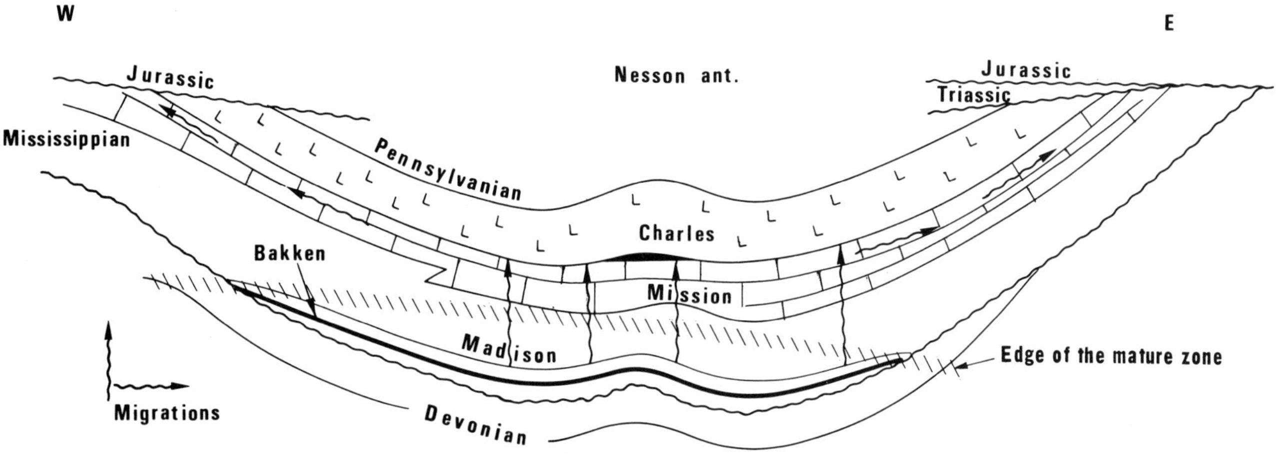

Fig. 36. — Section of Williston Basin showing extension of Bakken source rock and maturation zone (Meissner, 1978).

From the petrographic standpoint, this formation consists of a black dolomitic siltstone, sometimes laminated, bounded by two strongly radioactive more shaly levels, slow on the sonic log and displaying very high or very low resistivity. The central part contains 0.6 to 10 % organic carbon, with a mean of 3.8. When buried deeper than 6300 ft, it functioned as a source rock and produced some 1.4 Gt — 10 billion barrels of oil in the deep part of the basin. The very low speed on the sonic log can be explained by a high percentage of organic matter, and the wide variations in resistivity by the presence or absence of hydrocarbons in the formation, depending on which side of the maturation limit one is placed. This boundary lies between 6200 ft and 8200 ft, in the neighborhood of 70 °C or 160 °F, a critical temperature for a carboniferous source rock. One may in passing note the application of electric logs in detecting the maturation of a particularly rich carbonate source rock, an application which is fairly exceptional but probably offers tremendous possibilities.

Also worth noting is the close correlation between the maturation zones and the zones in which this formation exhibits high internal overpressure, which would imply a direct relationship between hydrocarbon genesis and the presence of high pressures, also observed in the Uinta and Powder River Basins. In these cases the pressure anomalies may be the direct result of the genesis of hydrocarbons, which appear to be the only fluids in excess in this confined environment.

The Bakken Formation is also oil-producing in five fields of the Williston Basin, and especially in the Antelope Field. This pool has produced some 1.4 Mt — 10 million barrels of oil — and about 800 Mm^3 — 29 billion cubic feet — of gas from 44 wells, practically without water inflows. This implies that the hydrocarbons are the only liquids capable of moving in this ‘reservoir’. In these overpressure zones, the oil moves through vertical microfissures formed and opened. This offers a visible example of migration (and production) in a fracture system, caused by a high pressure fluid. Wider vertical fractures are very probably the source of vertical migrations on a larger scale, conveying the oil to the main carbonate reservoir of the basin, the Mission Canyon, which is well sealed by the evaporites of the Charles Formation.

Summary and conclusion

Death and transfiguration appear to symbolize the destiny of the organic matter deposited throughout burial. From a complex, decomposing substance, in a few million years, in the depths of the sedimentary basins, temperature creates products with a simpler structure, endowed with powerful energy potential.

Like seeds, organic matter gives way to a new product with great possibilities.

We have seen that this genesis is closely related to the expulsion of the hydrocarbons formed from the rock which gave birth to them, and that these products have excellent chances of leaving in a continuous phase, in a certain sense fully armed. Water probably plays an important role, but it is not quite clear whether its positive effects are greater than its negative effects, which are better known. On the other hand, like the facies distribution, the geochemical context plays an essential and undeniable role.

Throughout these transformations, we have also noted the imprint, initially powerful, of the depositional environment become gradually obliterated during burial under the effect of temperature. Due to their sensitivity and their structural diversity, hydrocarbons thus illustrate perfectly the many aspects assumed by sediments all along their history.

2. – RESERVOIRS AND SEALS

« Plus la science progresse et plus s'impose l'idée d'une évolution quasi permanente des roches. »

'The further science advances, the more we realize that rocks are in a state of virtually permanent evolution.'

Marcel Roubault, 1965.

Once the hydrocarbons are formed from their source rocks, they flow through the porous and permeable subsurface horizons that constitute the reservoirs, bounded by impermeable series that play the role of seals. This reservoir/seal unit is permanently occupied by aquifers within which the oil and gas are concentrated.

After having examined aquifers in Part 1, this chapter deals with the two components of this container, or more precisely the reservoir/seal combination.

2.1. RESERVOIRS

Reservoirs are rocks containing void spaces, pores or fractures, interconnected and allowing the circulation and accumulation of fluids within them. Their petrochemical characteristics are basically expressed in terms of porosity and permeability, of which the relationships are often complex. They are the result of the entire geological history of these deposits, and in particular of the sedimentation conditions and the diagenetic mechanisms that follow. Some examples of the chief types of sandstone and carbonate reservoirs illustrate these general statements.

In a more precise manner than in other chapters of petroleum geology, the knowledge of the mechanisms that determine the reservoir characteristics is of fundamental importance, because it serves to predict certain variations and properties before drilling. This is important information when assessing prospects.

2.1.1. Petrophysical characteristics (*)

The pore volume of a rock is generally very heterogeneous on the microscopic scale. Intergranular voids (which assume an increasingly flatter shape with compaction) display irregular forms and are interconnected by gulleys and tiny, narrow, twisting and complex channels. In other words, the petrophysical characteristics depend closely on the rock/fluid relationships, and especially on the interfaces between the container and its contents.

2.1.1.1. *Adsorption and interface tension mechanisms*

Mineral surfaces exert forces of attraction on fluids, or Van der Waals forces, which are reflected by adsorption processes, or interfacial tension at the rock/fluid contact as well as wettability.

Interfacial tension may be defined as the work required per unit area to broaden the interface between two immiscible fluids such as oil and water. The interfacial tension of an oil/water combination varies according

(*) The reader can refer to the work by Serra (1979) for a thorough study of this subject, and to Perrodon (1966) for general information to the chemical composition of the oil, the presence or absence of surfactants, of gas in solution, the pH of the water, as well as pressure and temperature. Its values range from 5 to 35 dyn/cm^2 in subsurface conditions (Schowalter, 1979).

The interfacial tension is opposed to the displacement pressure exerted by an external fluid. This displacement pressure may be defined as the minimum pressure required to force a stream of oil or gas, in a water/hydrocarbon system, through the widest pores and channels. Hence for a given fluid, the interfacial tension resists pressures that rise in proportion to the interfacial tension itself. A rock displaying a certain interfacial tension thus forms a barrier to a given oil or gas pressure. In these conditions, the interfacial tension determines the possibilities of migration, or, in other words, the characteristics of the seals and closures, as we shall see below.

Physical adsorption may retain several monomolecular layers on a solid surface, while chemical adsorption only retains a single layer. Several monomolecular layers of water may be adsorbed physically on the shale surfaces. The relationships between the two elements are not only physical, but also chemical : the hydrogen atoms form a bond between the oxygen atoms of the mineral and especially silica, and the water molecules. This water, bound to the surface of a clay mineral, has a higher degree of order and constitutes the irreducible water. In most reservoirs, and in particular in silicious rocks, the water thus preferentially wets the rock. Small proportions of shale may substantially increase the quantities of interstitial water.

In this complex space with narrow pores connected by fine constrictions, interfacial tension mechanisms thus play a fundamental role, and the wetting liquid, generally water, has a tendency to rise by capillarity.

Wettability

Wettability corresponds to the work required to separate a fluid from the solid that it wets. It is normally expressed by the angle of contact of the water/oil interface against the rock or the pore wall. As a rule, water is a fluid that wets rocks, while oil is a fluid that does not. However, especially in carbonate reservoirs, oil sometimes at least partly wets the pore walls in some cases. Heteropolar compounds appear to be highly attracted to mineral surfaces. Fine layers of organic matter are also liable to cover the rocks and thus alter their wetting properties. In the case of shales rich in dispersed organic matter, for example, the polar radicals are adsorbed on the surface of the layers, and in the case of smectite, they are adsorbed between the layers. Hence, rocks with lower porosities, exhibiting internal voids of less than 50 A, could be oil wet. These heteropolar compounds also lower the water/oil and even water/gas interfacial tensions considerably.

2.1.1.2. ***Porosity***

The porosity (Φ) of a rock is its property of exhibiting voids, pores and fissures. It is expressed quantitatively by the percentage of the pore volume related to the total volume of the rock. In addition to this total or absolute porosity, an 'effective' porosity is defined (Φe), the porosity corresponding to the interconnected voids or the volume of voids susceptible to occupation by fluids. It corresponds to the total porosity less the volume occupied by irreducible water, according to the equation :

$$\Phi e = \Phi - Sw$$

where Sw is the water saturation of the rock.

One also speaks in this case of 'connected' porosity, a concept that introduces the idea of permeability. The effective porosity is generally lower by 20 to 25 %, sometimes by 50 %, than the absolute porosity. This coefficient rises as the granulometry of the rock decreases, thus accentuating the action of capillary mechanisms. In the case of very narrow pores and interstices, the consequence of a very fine granulometry, interstitial water remains adsorbed irreducibly on the grains under the action of capillary forces. This water saturation may be caused by upward flows due to capillarity, varying according to the porosity characteristics. The surface separating the layers impregnated with oil and water respectively thus obeys the characteristics of the reservoir and may not be spread along a horizontal surface.

As opposed to the matrix porosity, which is the result of intergranular voids, there is fissure porosity. The latter usually has lower values, but may supplement matrix porosity. In detrital rocks exhibiting uniform particle size distribution, porosity is independent of grain size, but not of shape. It decreases with increasing burial depth, or more precisely with increasing temperature, as we shall see below.

The porous rock, which contains one or more fluids saturating it, displays a conductivity, or its opposite, resistivity. The ratio of the resistivity of a porous rock to that of the water impregnating it is designated by the name of formation resistivity factor or formation factor (FF). Thus the porosity corresponds to the storage capacity for fluids and particularly hydrocarbons, in other words, to reserves in place. Hence it can be expressed in terms of capital. To determine the pore volume of a reservoir, the effective porosity is multiplied by the corresponding height (h), and the product often expressed by the factor Φh. This product gives the quantities of hydrocarbon in place in the pool per unit area. This concept of reserves per square kilometer or per acre is an important factor in appraising a prospect or a field.

2.1.1.3. ***Permeability***

A porous medium only allows the displacement of fluids to the extent that its pores are interconnected. In this case it is stated to be permeable. Thus permeability represents the ease with which a formation allows a fluid with a given viscosity to pass through it. If the fluid is homogeneous, and has no significant effect on the enclosing medium, the permeability is called absolute. It is derived from Darcy's Law and is expressed in darcys (d) or more often in millidarcys (md). The darcy can thus be defined, in the SI, as equivalent to one square micrometer.

If Q is the flow rate of the fluid passing through the rock, S is the cross-sectional area, dp/dx the pressure gradient, and η the fluid viscosity, the permeability K appears as the coefficient of 'ease of passage' :

$$Q = K \frac{S}{\eta} \frac{dp}{dx}$$

Its dimension is the square of a length.

It depends on the grain size and their sorting. It is multiplied by a factor of 4, for example, for each increase in mesh size of the Wentworth sieve (Taylor, 1978). Note that this equation is only valid if :

— the compressibility of the rock is greater than that of the fluid which it contains, or the grains making it up,

— forces of inertia are negligible in comparison with the viscosity forces (case of laminar flows),

— the porous medium is only crossed by a single fluid.

In actual fact, a petroleum reservoir generally contains two or sometimes three fluids (water, oil and/or gas). This implies the need for concepts of two-phase flow and relative permeability. If a reservoir contains two fluids, their flow rates interfere and the effective permeabilities for each of the fluids are less than the absolute permeability. The effective permeability for a fluid expresses the property of a reservoir of being traversed by this fluid in the presence of one or more other fluids. Hence it depends not only on the rock itself, but also on the respective proportions of the different fluids present.

The relative permeabilities correspond to the ratios of the effective permeabilities to the absolute permeability. The relative permeability of a given fluid varies directly with the saturation of this fluid in the rock and is expressed as a percentage displacement of one fluid in relation to the other. In a water/oil system, for example, the oil permeability is maximum and approaches 1 when the oil saturation is a maximum, i.e. around 70 to 80 %. It is a minimum and the water permeability a maximum when the oil saturation is very low (Figure 37).

The oil permeability decreases rapidly as the oil saturation decreases, and the water permeability first remains nil or negligible. Starting from a water saturation around 45 to 50 %, a water permeability appears

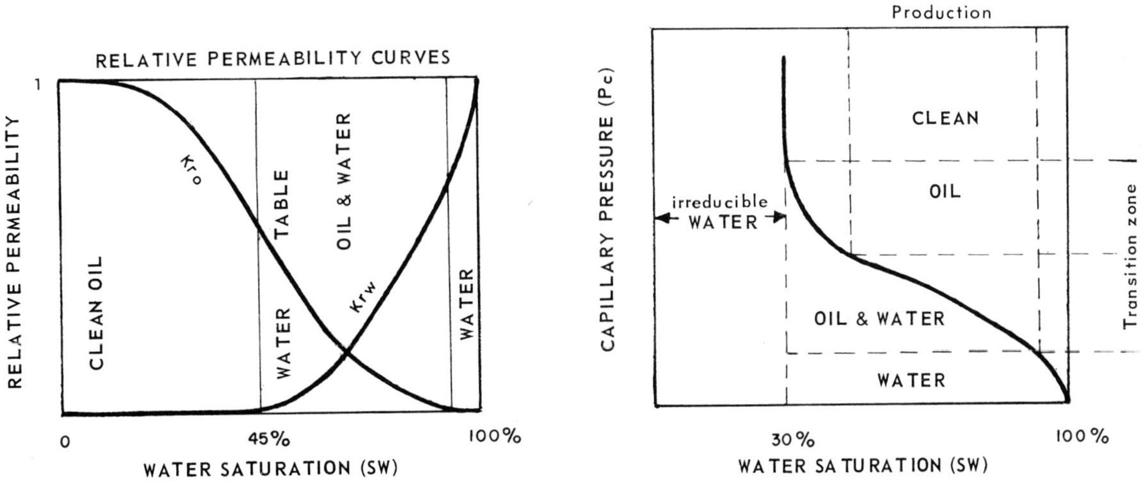

Fig. 37. — Diagrams showing relationships between water/oil saturation and relative permeability : (above) saturations and capillary pressure, (below) different production zones of a reservoir (Arps, 1964).

and rapidly grows. For Sw values of 80 to 90 %, the oil permeability becomes practically nil, and that of water approaches 100 %.

In terms of production, this diagram indicates that anhydrous oil can be extracted from a well at oil saturations above 50 to 55 %. At lower values, an inflow of a water/oil mixture will occur, with the proportion of water increasing with Sw. At values above 80 to 90 %, only water would be obtained. Permeability determines the start of a well and its production capacity, and thus corresponds to a notion of income. In a similar manner to porosity, the production capacity of a well or pool can be characterized by the product of the permeability multiplied by the corresponding height h, and is expressed by the product Kh.

The performance of a reservoir is frequently indicated by the determination of its productivity index (PI). This is the quantity of oil produced per day per unit of pressure differential between the reservoir pressure and the downhole pressure. In other words, it provides the measure of a reservoir's capacity to supply oil at the reservoir horizon. The productivity index generally decreases with time, with the decline in oil permeability and increase in oil viscosity and its flow resistance after the expulsion of dissolved gases.

2.1.1.4. *Porosity / permeability relationship*

Although a certain relationship exists in clastic rocks between porosity and permeability, no direct quantitative relationship is generally observed between these two fundamental characteristics of a reservoir. This is the case of shales, which are generally porous, but whose pores and extremely fine channels reinforce the effects associated with interfacial tensions, which tend to restrict the possibilities of fluid movement. The opposite case is that of fissured rocks which are often compacted, particularly carbonates, generally exhibiting high permeability but low porosities.

2.1.1.5. *Reservoirs and recovery*

The trapping of a non-wetting phase depends strongly on the characteristics of the pore system, and particularly on the type and distribution of non-random heterogeneities. Conversely, the recovery of hydrocarbons depends directly on the pore geometry, and especially their degree of heterogeneity. The latter is generally lower in sandstones than in carbonates. Hence final recovery does not appear to show good correlations with pore geometry in the case of carbonate reservoirs. In other cases, the existence of good correlations between mercury injection and porosity may constitute a correlation factor with recovery (Wardlaw, 1976 and 1980).

A few figures will serve to emphasize the difficulty of determining the characteristics of a reservoir on the scale of a field. The development borehole network remains relatively loose, with wells located several hundred meters to one kilometer from each other. The cored and recovered height normally only represents a small fraction of the reservoir. On the Hassi Messaoud Field, at the end of the first development grid, 16,200 samples were obtained representing only 0.10 m^3, for an impregnated volume estimated at $60 \cdot 10^9$ m^3. In other words, the percentage of the reservoir analyzed and measured in this initial stage only amounted to $0.10/60\ 10^{-9}$.

2.1.2. **Action of geological factors**

The present characteristics of a reservoir, like those of a rock in more general terms, are the result of the geological history of the basin, and more specifically of the conditions of deposition and transformation by diagenesis and catagenesis. To a certain degree, the present situation is merely a 'snapshot' whose explanation lies in geological history.

2.1.2.1. *Influence of the depositional environment*

The characteristics of the depositional environment closely condition the shape, extension, structure and nature of a reservoir. As a rule, rocks exhibiting good porosity and permeability characteristics correspond to deposits in high energy environments. If detrital components have been deposited, sands normally occur, and rarely calcarenites. An agitated environment, devoid of detritus, often favors the development of carbonates, especially reef structures and their debris, offering good matrix porosity characteristics. One exception, chalk, which may offer a good reservoir on the regional level, is deposited in a calm environment.

From the geometric standpoint, two main depositional categories may be distinguished :

— great detrital spreads of the blanket type, exhibiting fairly uniform thicknesses and characteristics over vast areas,

— in a far greater number, sand and carbonate bodies, essentially intermittent in space, which are defined by their geometry of deposition, in terms of form, size and orientation, with the understanding that the subsequent geological history can make sweeping changes in this original morphology.

Observation tends to show that the continuity of a reservoir very often varies inversely with its porosity/permeability characteristics. From the geological standpoint, the following will be dealt with in succession :
— detrital bodies,
— built-up units.
These account for the very large majority of reservoirs. Chalk reservoirs are dealt with elsewhere (see Section 2.1.3.2 of Part 2).

Detrital bodies

Depending on the paleogeographic and paleooceanographic conditions, deposits of sand or carbonate particles may correspond to :
— well individualized bodies, eolian dunes or offshore bars on the continental shelf, fluviatile or submarine channels, deltas or turbidites,
— great blankets with fairly constant thicknesses and characteristics, sometimes made up of vast turbidite spreads.

Examples of these detrital bodies are provided in Section 2.3.

Environments with relatively high energy, generally shallow, favor the deposition of 'clean' sands, while the finer fractions, especially clay, are deposited in calmer regions. These sands or more generally detrital bodies often find their well characterized depositional forms made gradually uniform by compaction and diagenetic mechanisms. In a section, the characteristic forms of a levee or a channel fill assume a lens shape as burial proceeds (Figure 38). As a rule, it is always difficult to identify the true morphology and the inherent characteristics of the sedimentary unit from thickness data alone, both for a sand body and a vast sedimentary unit. Isopachs must always be checked by knowledge of the depositional environment (Rittenhouse, 1961) (Figure 39).

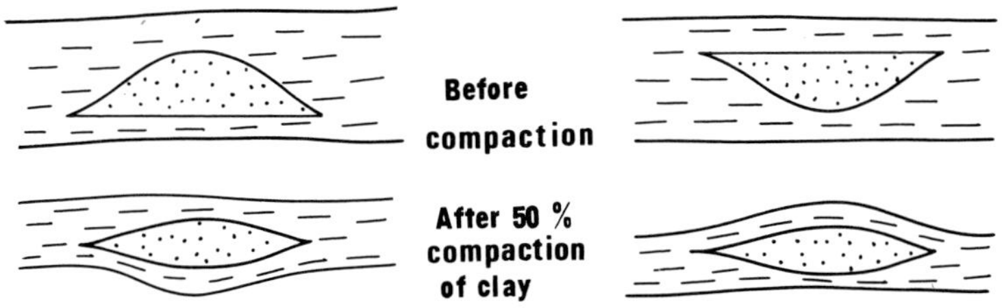

Fig. 38. — Diagrams showing the effects of 50 % compaction of shales on two types of sand body : sand bar and channel (Rittenhouse, 1961).

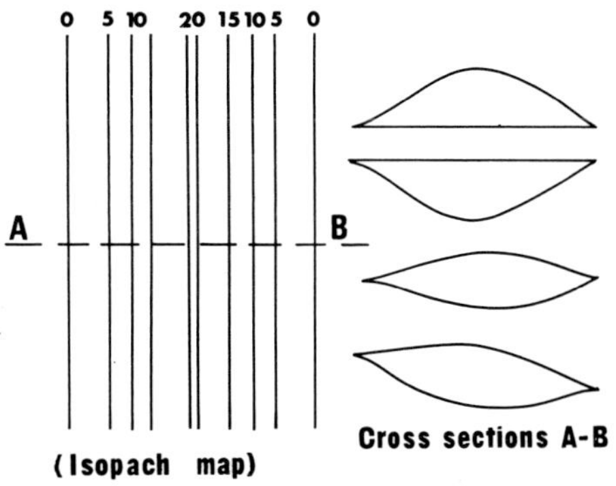

Fig. 39. — Part of isopach map of an elongated sedimentary body and different interpretations of a cross-section (Rittenhouse, 1961).

Built-up units

These units, which are generally carbonates, are associated with two main categories, both represented by biogenic structures, barriers and atolls or pinnacles :
— barriers correspond to structures of a linear reef character which may stretch over several hundred kilometers,

— reef formations, of the 'individual' or circular type, display dimensions which may in rare cases be as large as several dozen kilometers, but are normally not more than one kilometer; as in the case of the present oceans, they assume the form of atolls, cusps and pinnacles, and may rise 100 meters and more; they often develop in front of a barrier and their insertion in impervious, shaly or evaporitic series provides them with an effective seal.

These types of reservoirs and traps are illustrated by the reef formations of the Devonian in Alberta, the Cretaceous in Libya, and the Miocene in the Salawati Basin in West Irian.

2.1.2.2. *Influence of diagenesis / catagenesis*

Cementation and pore destruction

Like all other components of sedimentary basins, reservoirs are never static for a long time. Their 'fixedness' is merely the expression of an instantaneous observation on the geological timescale. They, and more specifically their cement, the true living part of the rock, are the culmination of a series of transformations by additions and subtractions, caused chiefly by variations in temperature and the circulation of aquifers.

The characteristics of the depositional environment directly influence diagenetic transformations. Limestones deposited in shallow water exhibit variable and complex facies. They contain unstable minerals (aragonite, Mg rich calcite) which make them highly sensitive to post-deposition alterations. Small differences in burial or uplift may give rise to zones of porosity protection or destruction.

The depositional conditions, especially of a carbonate, may closely condition its development during diagenesis. Moore & Druckman (1981) have shown that in the Smackover Formation, the freshwater facies displayed very early stabilization of the minerals that protected porosity during burial, whereas the lateral marine facies, which exhibited early, local and surface diagenesis, were rapidly cemented during burial. As a rule, early diagenesis tends to fossilize a certain porosity which remains partly protected during subsequent diageneses.

On the other hand, limestones in a deep environment generally consist of far more stable minerals, such as low Mg calcite, and preserve more homogeneous facies. Their porosity disappears with increasing burial depth. The rate of this loss of porosity, or pore destruction, appears to rise with higher temperature and pressure, and with the tendency of interstitial solutions to allow greater ion exchange. Pressure and temperature are usually conditioned by burial depth. The presence of complex silicates, especially clay minerals and feldspars, but also carbonates, tends to stimulate ion exchange.

As we have seen above (Section 2.2. or Part 1), different stages of compaction and diagenesis/catagenesis are observed during burial, and these largely overlap. Initially the compaction mechanism appears to be the predominant factor. The various rocks react to this in different ways depending on their petrographic characteristics. In a second stage, chemical processes play an essential role, which expands particularly with rising temperature, by altering the chemism of certain minerals.

Under the effect of burial, the pressures exerted at the contact points of the grains or at textural irregularities (for example, yielding stylolites in carbonate formations), add their effects to those of temperature and facilitate the dissolution of materials concentrated in the pore spaces in the form of cement, while part of the interstitial water escapes. These chemical reactions are intensified by rising temperature. Dissolution by pressure becomes effective from 100 m in carbonates, and from about 1000 m in sands and sandstones. It increases fairly steadily with burial depth, up to maximum values of 20 to 25 %.

In Abu Dhabi, for example, the porosities of the aquifer carbonate reservoirs of the Lower Cretaceous (Thamama) decrease from 35 to 30 % in the 0 to 1200 m interval to zero below 3600 m (Clarke, 1975).

Pore destruction in sandstones generally develops beyond a thermal diagenesis threshold depending on the petrographic characteristics, and also because a sandstone tends to show greater resistance to a loss of porosity if it is 'cleaner'. This thermal threshold is expressed by the vitrinite reflectance (R_v), which may vary widely from 0.6 to 1.75 for different sandstones in the coastal basins of Equatorial Africa, for example (Cassan, 1978). Since paleotemperatures vary independently of maximum burial depth, and in accordance with geothermal gradients, good reservoir characteristics may be maintained at considerable depths, especially in 'cold' zones and for 'clean' sandstones facies.

Measurements have shown that the porosity of the Cretaceous and Tertiary formations of the Gulf Coast declined on the average by 0.38 % per 100 m (Atwater & Miller, 1965), and 0.46 to 0.56 % per 100 m in those of the San Joaquin Basin in California, where the temperatures are higher (Zieglar & Spotts, 1978). This decrease is accentuated with higher geothermal gradients and with less 'clean' rocks (Figure 40). From an analysis of over 8000 samples, Maxwell (1964) observed a gradient declining by 3.28 % per kilometer for Oligocene sands exhibiting a geothermal gradient of 23 °C per kilometer, going to a decrease of 8.4 % per

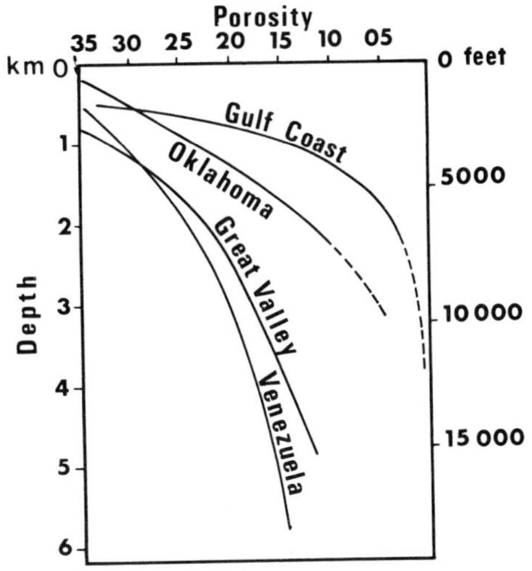

Fig. 40. — Variation in porosity versus depth in different basins.

kilometer for a temperature gradient of 32.8 °C per kilometer. These data relate to present temperatures, and can be compared with the peak temperatures for basins in a stage of continuous subsidence, like that of the Gulf Coast, where these measurements were taken.

These considerations do not apply exactly for basins like that of the North Sea, which has a more complex geological history. In this region, decreases of 9.8 % per kilometer have been observed in Tertiary sands, 6.5 % in Jurassic sandstones, and 5 % in Permo-Triassic sandstones (Selley, 1977). In short, the porosity and permeability of a reservoir decrease as it becomes less 'clean' and is buried at higher temperatures and pressures.

The fossilization of porosity

This loss of porosity may be stopped if the formation water is replaced by an 'inert' non-reactive fluid, like a hydrocarbon. In fact, it has been observed that impregnated reservoirs (with hydrocarbons) very often exhibit better characteristics than the same neighboring water bearing reservoir. This factor must be considered in assessing the reservoirs penetrated by dry wells (generally aquifers). In the chalky reservoirs of the Danian in the North Sea, for example, the porosity of the oil impregnated horizons amounts to 30 to 40 %, while it is no more than a few per cent in the adjacent aquifer zones (Scholle, 1977 and 1978).

In extreme cases, the limit of the oil level may be strongly cemented and may become impermeable, with the accumulation more or less sealed. This type of 'diagenetic' pool is discussed in Section 4.3. A comparison of the porosities of the contiguous oil and aquifer parts of a single reservoir also make it possible to calculate the burial depth at which the hydrocarbons arrived in place, as the 'preserved' or 'fossilized' porosity approximately represents the burial porosity at the time that the oil arrived.

But other chemical mechanisms, more or less complex, may alter the 'normal' loss of porosity with burial. The dissolution of carbonate or quartz particles, for example, may be blocked if the environment is closed to the circulation of water, when enclosed between evaporite beds for instance. In such a case, the values reached rarely exceed 4 to 6 %. The cementation of a porous rock turns it into a closed environment, in which the different fissures are cemented progressively (Hortenbach, 1977).

In some cases, as in the Paleocene limestone reservoir of Zelten, the porosity, which is 40 % at 1500 to 1600 m depth, appears to be partly preserved in the 'grainstones' by the presence of an early cement which seems to have stopped compaction, and in the 'packstones' and 'wackestones' by a porous carbonate mud. The reservoir characteristics were then improved by leaching during periods of erosion (Bebout, 1979). In the case of the Smackover Formation, which is a significant producer in the Jay Field in the Eastern Gulf Coast, the porosity appears to be protected by an early cementation in a freshwater environment, and subsequently improved by leaching processes (Stoudt, 1979). Early diagenesis, especially of carbonates, thus very often plays an important role in determining reservoir characteristics.

Secondary porosity

By investigating the Tertiary sediments of the Rhine Graben, Heling (1970) showed that the rocks buried deeper than 1000 m underwent mineralogical and chemical transformations, such as dolomitization, and the dissolution of phosphates, some of which could help to increase the porosity. Hydraulic fracturation mechanisms can also lead to some improvement in reservoir characteristics.

In the case of sandstone reservoirs, from the petrographic standpoint, this 'secondary porosity' chiefly results from the dissolution of feldspars or carbonates, which occurs later and which requires rather deep burial, and which can increase the porosity by 7 to 8 %. More generally, as a rule, compaction at certain depths causes overpressure in fluids that may lead to the fissuration of the rock.

2.1.2.3. ***The influence of tectonics***

Tectonic deformations (mainly tension and shear mechanisms, but also folding) can cause the fracturation of compact rocks and give them certain porosity and, especially, permeability characteristics. The action of fractures, of whatever origin, through the succession of deposits, tends to open new paths for the circulation of water, which is liable to penetrate into the deep horizons. A positive or uplift movement also gives rise to :

— a decrease in pressure and temperature, by the erosion of the overlying series, which may be reflected by chemical transformations of the reverse type,

— the intrusion and circulation of meteoric waters and the revival of the movement, mainly along permeable beds, of the aquifers which act by subtraction or by addition depending on their characteristics and those of the environment.

In the latter case and contrary to what may first be surmised, the tectonic fracturation of a limestone reservoir, which is perfectly homogeneous and shows no facies variations, is liable to decrease its porosity and its permeability. In fact, certain anticlines reveal a decrease in reservoir properties with an increase in flexuration and dip. In the case of massive beds in particular, the decrease is caused by pressure/dissolution and stylolitization before the start of folding. The pores are then filled by precipitation in the pores of calcite dissolved along the stylolitic joints (Trémolières & Reulet, 1978).

As a rule, fissuration may be considered to create fracture permeability in the initial stage. The solutions then deposit a cement into these voids, clogging them progressively. In other words, recent fractures produce permeability; old fractures are most likely to be cemented and impermeable.

*
**

Porosity, one of the fundamental characteristics of a reservoir, thus appears as the result of a whole sequence of geological events (Figure 41) :

— at a sedimentological level, it increases with the energy of the depositional environment and the 'cleanliness' (or absence of cement) of the deposit,

— at the level of geological history, it decreases with increasing burial, the extent and duration of subsidence, the temperature reached and the age of the tectonic deformations.

The vicissitudes of post-depositional history may largely obliterate the characteristics of the initially porous and permeable sediment, but these are two-way transformations, and a highly compacted reservoir may subsequently regain effective properties of porosity and especially permeability.

2.1.3. **Geological characteristics. Examples of certain reservoir types**

Reservoir rocks are divided into two main families, detrital reservoirs, mainly siliceous (sands and sandstones) and carbonate reservoirs, including limestones and dolomites. Considering the world's giant fields (reserves over 70 Mt — 500 10^6 bbl — of oil or gas equivalent), sandstone reservoirs account for nearly 60 % of oil reserves, and carbonates, which represent only 16 % of the material of the sedimentary basins, account for about 40 %. This proportion rises to 2/3-1/3 if one only considers the giants of the United States, while the Middle East has a large part of its reserves in carbonate formations (Halbouty *et al.,* 1970). The share of detrital reservoirs is even larger where gas is concerned. Reservoirs belonging to other lithologic categories (cherts, argillite, non-sedimentary rocks) only represent 1 % of the total.

2.1.3.1. ***Sandstone reservoirs***

Detrital reservoirs, generally consisting of stable components, exhibit fairly uniform porosity and permeability, of the intergranular type, conditioned by grain size and the degree of cementation. The relationships between porosity and permeability are generally good, and decreases in one or the other with burial are fairly uniform as a function of temperature.

A statistical analysis prepared in 1960 covering 7241 sandstone reservoirs in the United States, ranging from the Cambrian to the Pleistocene, shows that most of these formations cover a limited area, generally

Fig. 41. — Variation in porosity/depth relationships of sandstone reservoir in accordance with different geological processes (Zieglar & Spotts, 1978).

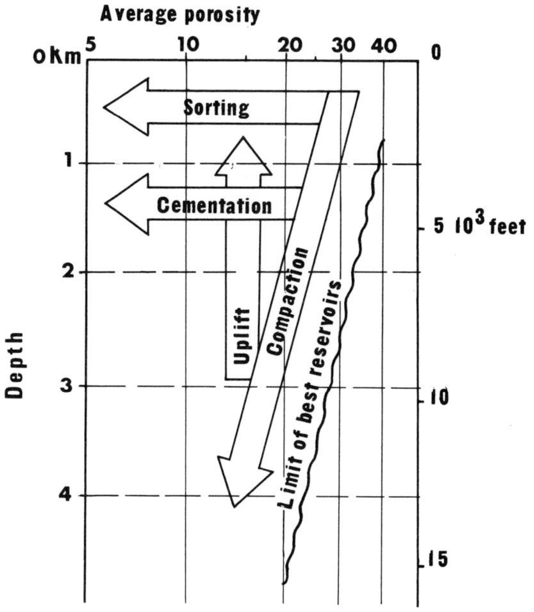

less than 260 km², and exhibit an average thickness of 12 m. Whatever the type of trap involved, sandstone reservoirs with a limited local extension clearly predominate over the continuous detrital reservoirs that extend over vast areas (Curtis *et al.,* 1961).

Most of these sand bodies have an average length of about 5 km, and a length/width ratio of 3 to 1. They belong to various depositional environments : eolian, desert, alluvial, deltaic, intertidal, beach, barrier, platform and deep sands.

Eolian sands

Accumulations of eolian desert sands are probably the most favorable of all detrital reservoirs. They may be as thick as 200 m, as in the Leman gas field. These Rotliegendes eolian formations extend over a belt of 1000 by 300 km, from the British Isles to Poland. Under the North Sea, the Netherlands and Germany, the reservoirs formed by these continental sandstones contain some 2600 Gm³ of gas. They exhibit average porosities of 25 % and permeabilities of 1 d (Taylor, 1978).

The reservoir of the Groningen Field, the largest in Western Europe, varies in thickness from 80 m in the south to 200 m at the northern limit. It consists of reddish sandstones and heterogeneous conglomerates. Average porosity lies between 15 and 20 %, and permeability varies from 0.1 to 1 d. These deposits, whose material originates in the south, correspond to fluviatile and eolian environments. The average well production varies from 0.7 to 1.2 Mm³ per day (Figure 70) (from Stauble & Milius, 1970) (see Chapter 4 of Part 2).

Fluviatile sands

Deposits of this type correspond chiefly to meanders or braided streams of the alluvial plain. They are represented by sand bodies 250 m to 20 km long (average 4 km), with thicknesses of 6 to 75 m (average 30 m). Their porosity ranges as a rule from 10 to 25 %, and their permeability from 60 to 2000 md (Figure 42). The Permo-Trias Sadlerochit Formation in Prudhoe Bay Field is one of the richest illustrations of this type of sand body, containing some 2.8 Gt out of the 3.3 Gt reserves in place in the overall field (Conybeare, 1976; Jamison *et al.,* 1980) (see Chapter 4 of Part 2 and Figure 56).

In the Denver Basin, sand bodies corresponding to channel fills are particularly characterized by :

— a general elongated and often sinuous shape,

— sudden lateral transition from sand to shales,

— a certain irregularity in porosity and permeability, and the presence of pebbles.

They generally yield low oil columns, less than 15 m, but allow active hydraulic drainage. The traps usually consist of the combination of a nose and a transverse fault. The enclosed volumes are small and reserves very slight, 1.5 Mt in place on the average (Exums & Harms, 1968).

Alluvial fan deposits

These deposits show extreme variations in geometry and structure, and are often very sensitive to diagenetic effects. This type of environment includes the alluvial sands of the Lower Cretaceous in the Sirte

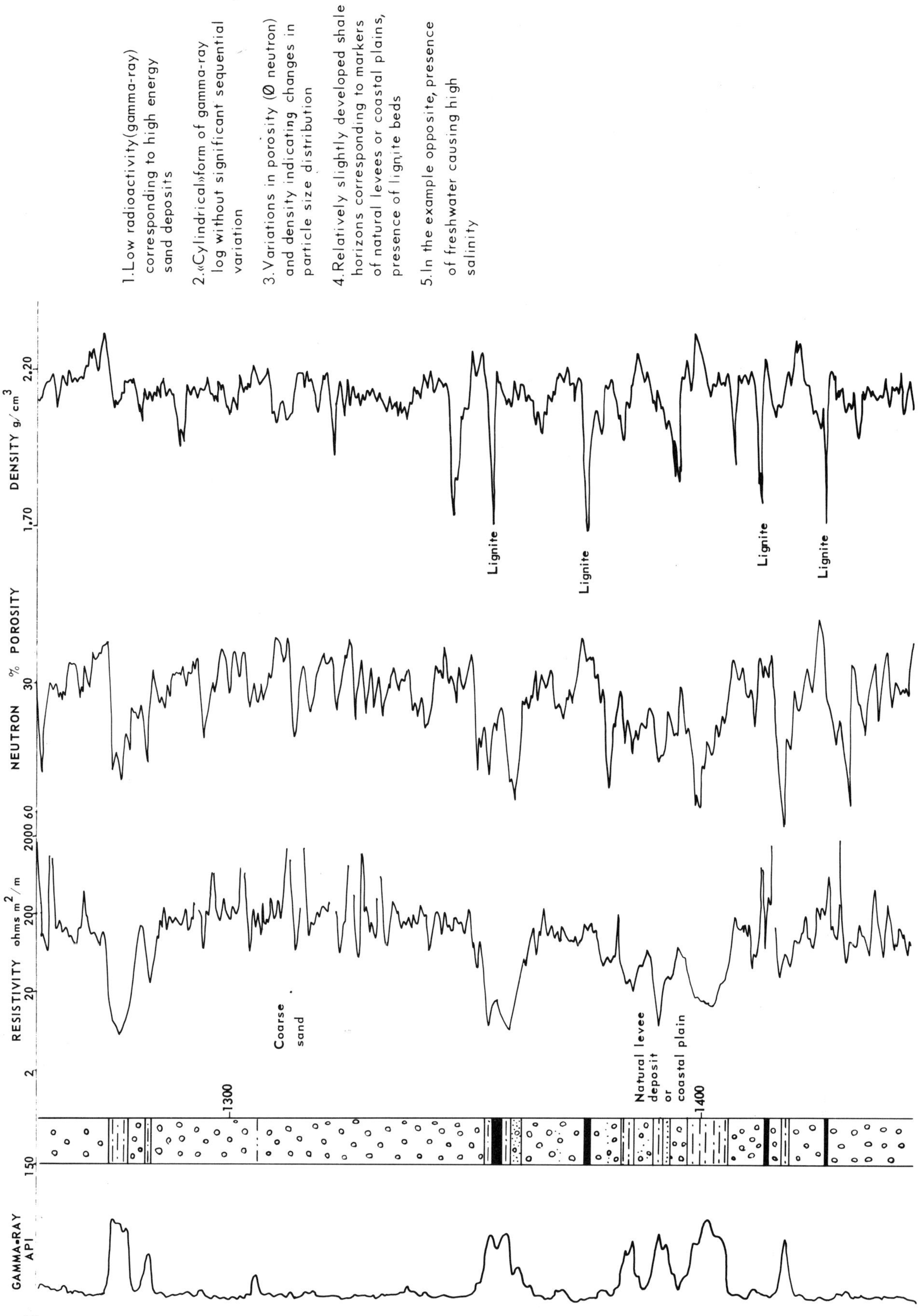

Fig. 42. — Example of fluviatile sands by logs.

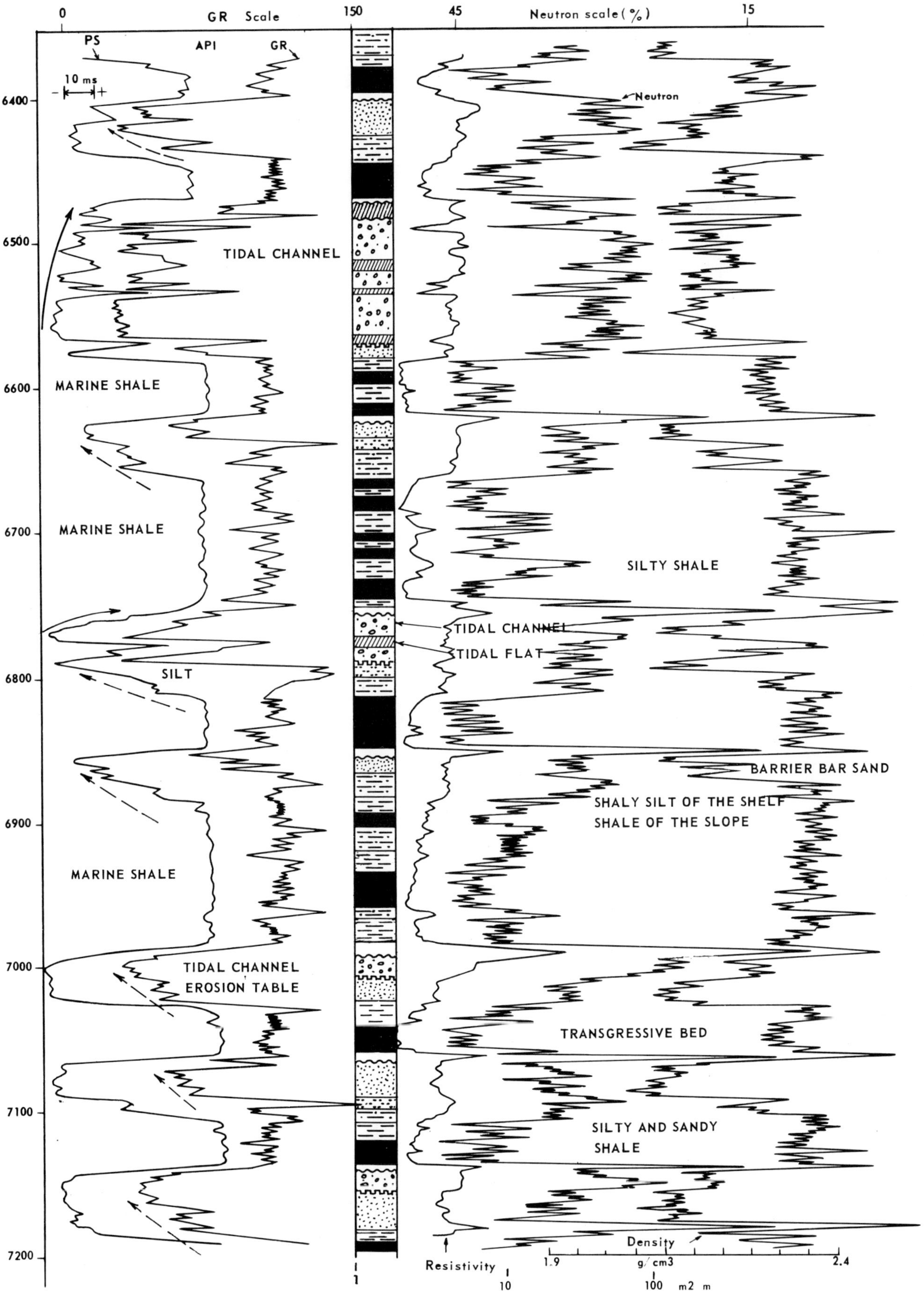

Fig. 43. — Representation of littoral and delta sands by logs.

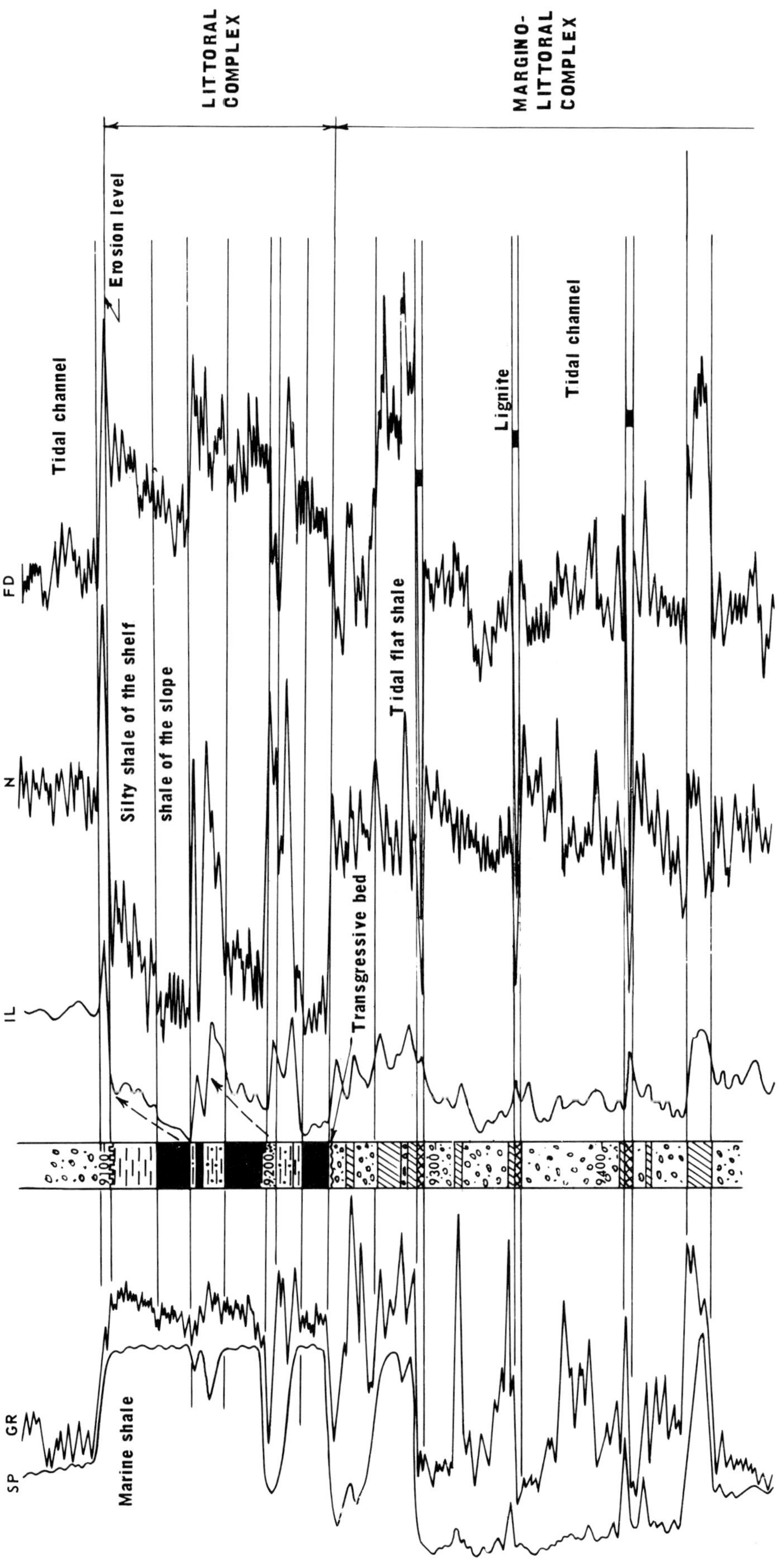

Fig. 44. — Representation of littoral, with erosion channel, and margino-littoral complexes by logs.

Basin in Libya, supplied by a granitic basement and effectively sealed by the Senonian shales. They form the reservoir of the great fields of Sarir and Messia in particular (Taylor, 1978) (see Figure 206).

Beach and barrier sands

These formations, aligned parallel to the shorelines, represent a major group of potential reservoirs. A statistical study dealing with eighteen North American fields shows that these sands exhibit an average length of 40 km, width of 4 km, and thickness of 20 m. The porosities range from 15 to 20 %, and permeabilities from 50 to 2000 md (Conybeare, 1976). Figures 43 and 44 show the log illustration of littoral and margino-littoral sands, intersected by erosion channels.

The Jurassic reservoir of the Piper Field, in the North Sea, with reserves of 85 Mt, consists of a group of sand bodies of the beach and barrier deposit type, with an imbricate structure. It varies in thickness from 35 to 100 m, in porosity from 22 to 27 %, and in permeability from 0.15 to 10 d (Conner *et al.,* 1975).

Marine platform sands

Sand bodies corresponding to offshore bars with a littoral marine facies are characterized by :

— usual elliptical shape,

— good quality reservoir, often containing glauconite,

— progressive decrease in shale percentage upward.

The traps whose reservoirs belong to this type are often stratigraphic. In the Denver Basin, for example, oil columns may be as much as 30 m, and reserves in place some 5 Mt per accumulation (Exum & Harms, 1968).

These sand bodies correspond to transgressive or regressive deposits. The former are generally more extensive, but thinner and characterized by a smaller particle size distribution and transition to shales towards the top of the series. The best sands of this type exhibit porosities of 15 to 28 % and permeabilities of 80 to 3000 md, with thicknesses of 6 to 90 m. The degradation of the reservoir characteristics towards the top is a disadvantage, and deposits of this type appear to contain smaller reserves than those of the regressive series which, on the contrary, display a coarser particle size distribution but sometimes raise problems of sealing (Mackenzie, 1972).

Deep detrital deposits

Turbidites and deep sea fans form vast relatively homogeneous spreads over tens or hundreds of square kilometers. They are deposited in the depressed and often deep zones of the depositional basins, whose morphology conditions their geometry. These deep sands, generally deposited in periods of regression, exhibit a succession of detrital deposits, graded, separated and enclosed by shale beds which may play the role of a seal, often characterized by microfauna indicating relatively high water depths, while the redeposited beds initially correspond to shallow environments.

This sedimentation is discontinuous in time with brief intervals of very high accumulation rates. The base of a sequence marks a sudden contrast with the underlying beds, which are sometimes eroded. The upper portion may change into more pelagic deposits (Figure 45). The proximal facies, corresponding to channel fills excavated in the slope, generally exhibit a complex distribution scheme. The distal facies, which result from gravity deposits, usually appear in the form of vast and fairly uniform onlaps.

One often tends to believe that the interpretation of deep sea deposits is largely a question of fashion. While a better knowledge of these present deep deposits, and in particular of turbidites and deep sea fans, accounts for a fair share of our interpretation of older deposits, reservoirs of this type are not new. In his memorandum "Geometry of Sandstone Bodies", published in 1960, Sullwold includes in this class the productive horizons of the Permian in the Delaware Basin, the Atoka sandstones, the Pennsylvanian of Oklahoma, the Ordovician and Devonian sandstones of the Appalachians, different Tertiary sands of the most 'offshore' part of the Gulf Coast, and, above all, the rich Pliocene sands of the Los Angeles Basin.

The Lower Pliocene Repetto sands of this basin, forming one of its chief reservoirs, occupy the entire central portion with thicknesses up to 700 m in the most depressed zone. The superposition of this excellent reservoir on a deeper source rock series is one of the reasons for the outstanding potential of this small but fertile province (Figures 46 and 161) (Sullwold, 1961).

In the Parentis Basin, the Albian sands of the Cazaux Field clearly illustrate another type of reservoir consisting of turbidites (Figure 47). The latter occur in the form of elongated bodies, superimposed and offset in time, some 50 m thick and 1 to 2 km wide. Each one exhibits a positive sequence starting with conglomerates and ending with clays, and they are all interpreted as submarine channel fills (from Bessaquet & Martin, 1977).

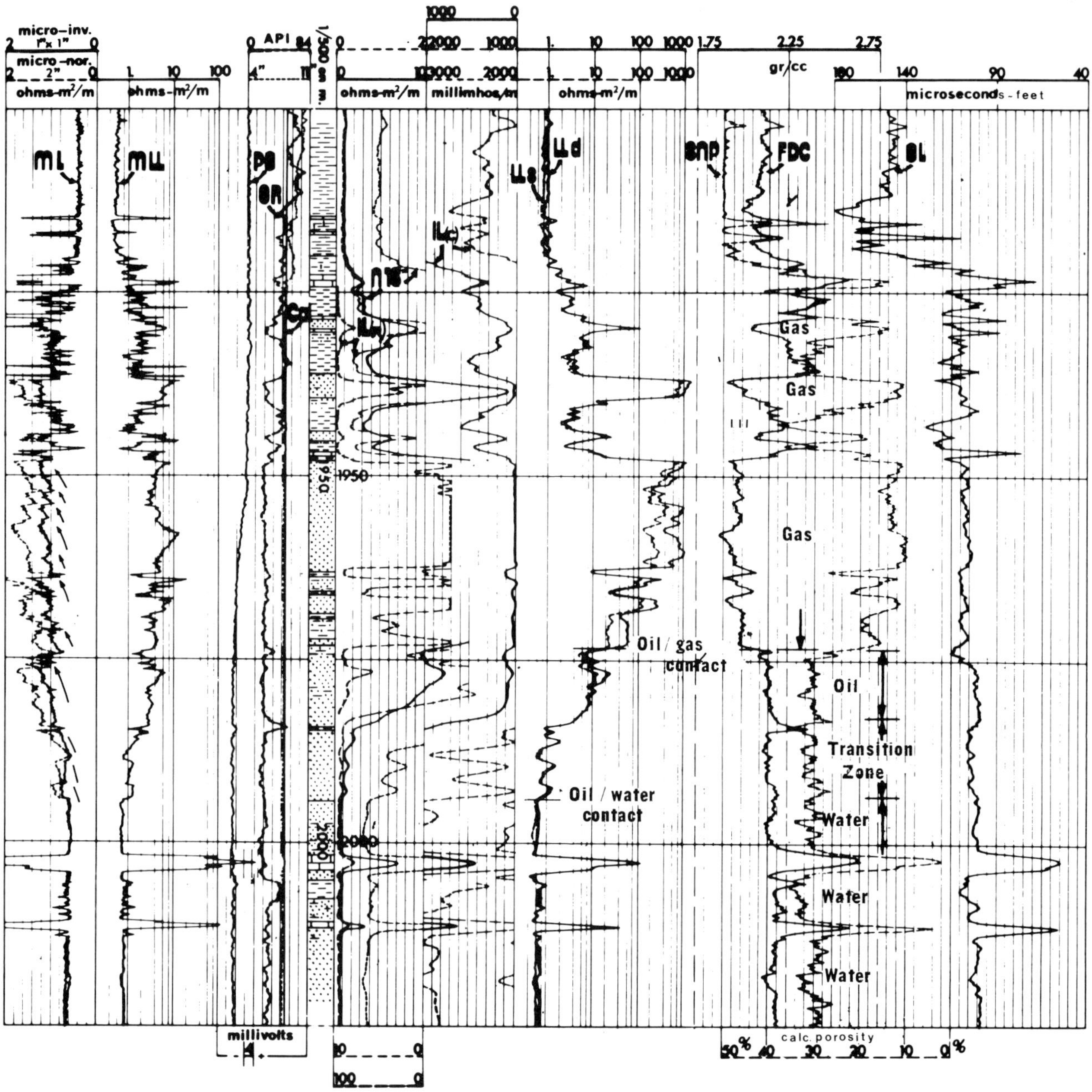

Fig. 45. — Representation of deep sea fans by logs.

The Paleocene sands of the Viking Graben in the North Sea offer another illustration of these deep deposits. They display a complete series of transitions, from the major fan in the Lower Paleocene, to a slope edge situation in the Upper Paleocene. These sands originate in an emerged region of the Scottish coast and the Shetlands, probably related to an uplift accompanying the North Atlantic opening, and accumulate in eastward progradating series, in a deepening sea (Parker, 1978) (Figure 193). The Frigg gas field, which is productive in these formations, and its satellites, clearly depict one of these fans, so that the structure earned the name of 'Chinese Butterfly' from the seismic survey (Figure 48) (Heritier *et al.,* 1980).

The Chicontepec Basin in Mexico, southwest of the Faja de Oro, which could have considerable potential, also corresponds to the filling of canyons excavated in Paleocene and Cretaceous, locally Jurassic formations, by Paleocene/Eocene turbidite deposits. They are up to 500 to 900 m thick in the deepest parts. The porosity and permeability characteristics are mediocre and the well production rates low.

Trapping is chiefly of the stratigraphic type, and the accumulations are widely disseminated. The main field is President Aleman, in a pinchout position (Busch & Gouela, 1978).

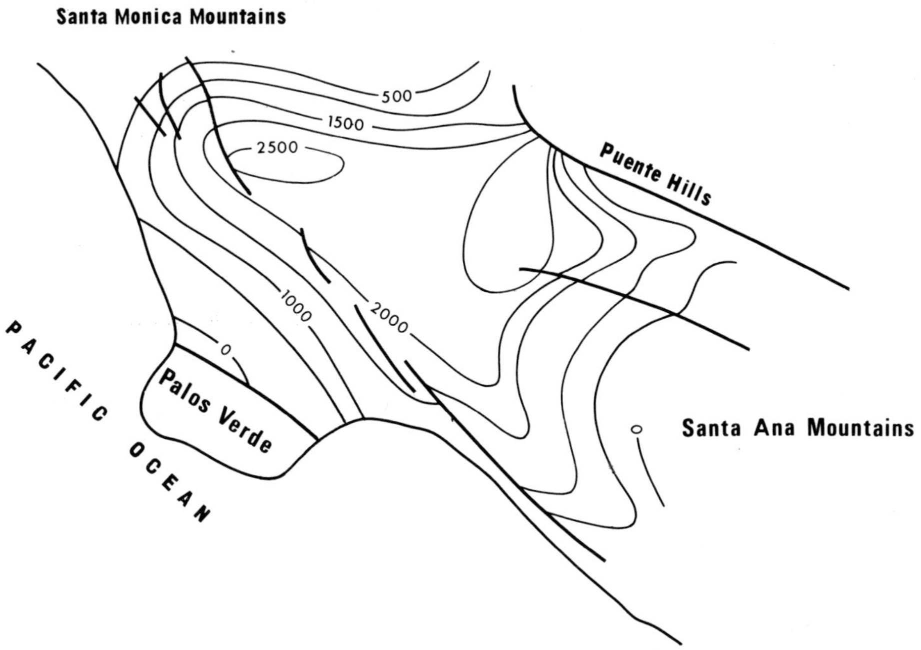

Fig. 46. — Isopach map of Repetto Formation (Lower Pliocene) in Los Angeles Basin (in feet) (Conrey, 1959 and Sullwold, 1961).

Deltaic provinces

Deltaic environments, which may be considered somewhat as vast sand traps, often associate these different types of deposit, including the juxtaposition and superposition of fluviatile sands and offshore bars by a transgressive movement, in a generally favorable environment for both petroleum generation and trapping. Deep sea sands may also be encountered at the front of the deltaic unit. In the Gulf Coast Basin, for example, the Frio and Vicksburg sands, and the Wilcox sands (Eocene), well known in their deltaic facies, extend southward as a turbidite facies, where they are gas producers (Berg, 1979). The Mahakam Delta (Kalimantan) offers an excellent example of the present deltaic system (Allen *et al.*, 1979).

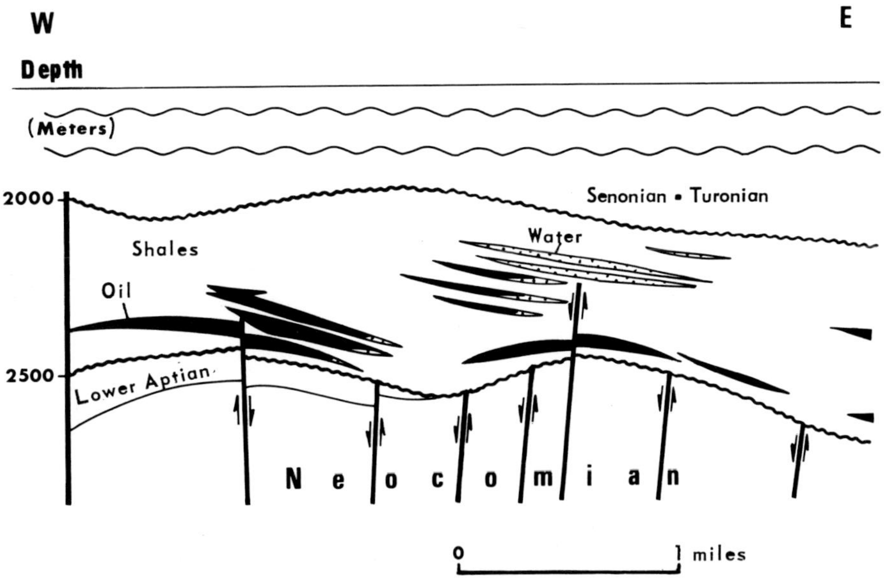

Fig. 47. — Structural section across Cazaux anticline axis (Bessaguet & Martin, 1977).

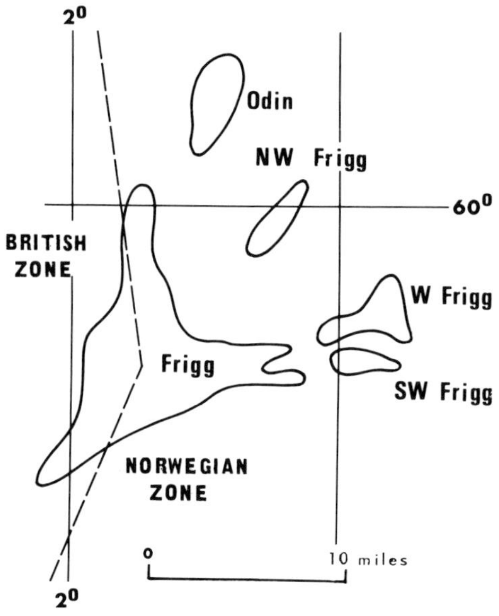

Fig. 48. — Map of Frigg Field and satellites.

2.1.3.2. ***Carbonate reservoirs***

Carbonate reservoirs are characterized by a very broad range of porosities and permeabilities. They often occur in massive form, with large pores and fractures in the case of reefs, finely stratified and laminated in back-reef and shoal deposits. The primary porosity is mainly of the intergranular type. It varies widely due to the diversity of size and shape of the components, some of which may add their own inherent or intergranular porosity. It varies closely with increased energy of the depositional environment, especially in the case of reefs which generally emerge from the sea. Porosity is highest on the flanks of the reef structures, where organism debris accumulate.

Added to the heterogeneities caused by the varieties of deposition environment is that caused by diagenetic action or secondary porosity. In particular, dolomitization is reflected by an increase in the pore volume in the case of a compact limestone, and by a simple modification of the architecture of the pore volume in the case of a reef limestone. A statistical survey made in western Canada showed porosities and permeabilities of 8 % and 68 md in limestone reservoirs, 9 % and 800 md in dolomite reservoirs, revealing a more significant increase in permeability than in porosity. Sometimes, however, dolomitization may have a negative effect by creating a dense network of crystals.

The constrictions between grains are very broadly and very irregularly affected by diagenesis, resulting in a wide dispersion of the diameter of these 'bottlenecks', and hence of permeability, as recrystallization mechanisms often lead to the building of large calcite crystals. This means greater porosity, but often lower permeability (Jardiné *et al.,* 1977).

The form of the pores and their relationships are also of considerable interest, especially for the production engineer. For example, Wardlaw (1976) showed that in the dolomites, the connections between pores are generally flattened and not tubular. Porosity declines with dolomite crystal growth, and the initially polyhedral pores become tetrahedral and finally lamellar. Considering simpler geometric forms, the size ratio of the pores to the constrictions separating them increases, implying that the constrictions become smaller.

Fracturation can create permeability in rocks which are initially impermeable. It favors the circulation of water, which exerts leaching or cementing action. It may also cause aquifers to rise in producing hydrocarbon reservoirs. The uniformity of the sedimentary bodies is often obscured by secondary mechanisms which may be superimposed on them, such as emersion and the formation of karsts, compaction, dolomitization and fracturation. While the characteristics of a carbonate reservoir primarily depend on the conditions of deposition, they are also determined by subsequent processes, more so than in the case of sandstone reservoirs.

Reef reservoirs

Reef structures and their destruction products which surround them may offer outstanding reservoirs by their porosity and permeability. A number of reefs of Alberta and West Texas have displayed reservoir thicknesses of 170 to 180 m. Through the action of recrystallization, dissolution and sometimes fracturation, the well production rates may be as high as 10,000 bopd to 1400 tons per day and even more. Different examples of this of reservoir and their geological context are discussed below (Section 4.2.2.2. of Part 2, Section 2.1 of Part 3, and also Sections 3.4, 3.5 and 3.7.3 of Part 3 including the monographs on the provinces of western Canada, West Texas and the Sirte Basin).

Chalk : an exceptional reservoir

Chalk is a very finely detrital rock composed essentially of the microscopic skeletons of coccoliths. It is a stratified marine sediment, deposited at shallow depth and also as deep as 1000 m, at sedimentation rates ranging from a few meters to some 50 m per million years. The coccoliths, whose skeletons are 1 to 20 μ in size, consist of calcite that is very pure and consequently very stable chemically. From the petrophysical standpoint, chalk is generally a very porous and relatively impermeable rock. Its porosity at the time of deposit reaches 70 to 80 % and then decreases steadily as a function of burial depth, with average values of 30 % between 1000 and 2000 m, and less than 20 % deeper down. The pore volume exhibits a dense network of very fine pores no more than 1 to 2 μ in diameter. The permeability rarely exceeds a few millidarcys, and is often less than 1.

Chalk nevertheless constitutes petroleum reservoirs generally of average quality, in the Cretaceous of the Gulf Coast (Austin chalk) and the Netherlands. In Texas, for example, the porosity is only 10 % at a depth of about 2000 m, and permeability is always less than 0.5 md. The production of wells in these reservoirs falls sharply from about 40 to a few tons per day. However, in a few regions of the Middle East (the Zakum and Bu Hasa Fields) and in some Gulf Coast and North Sea fields, especially the Ekofisk zone, chalk is an outstanding reservoir. The Danian reservoirs of the Ekofisk fields, in the central part of the North Sea, exhibit a thickness of 400 to 1400 m at a depth of 2000 to 3000 m. They have preserved a mean primary porosity of 30 %, with peaks over 40 % normally corresponding to a burial depth of 1000 m, and a matrix permeability of 1 to 2 md, but effective permeability measured on tests is 10 to 12 md. The wells yield exceptional output around 10,000 bopd to 1400 td^{-1} (Figures 184 and 186).

These remarkable characteristics are explained by the combination of a number of specific mechanisms. The high porosity, higher in the oil zone than in the aquifer, is chiefly due to the high pressures (500 bar or 7100 psi at 3050 m) related to the Paleocene undercompacted shales overlying the reservoirs, and to a growth of cement in the water bearing zones. Also noteworthy are :

— the presence of fractures related to anticline folding under the probable thrust of a salt dome,
— high oil saturation of 70 to 90 %,
— very low oil viscosity, 10.24 cP in downhole conditions, itself associated with a high GOR (265 to 350 m^3 per m^3, 1000 to 2500 cf per barrel).

The 56 km^2 Ekofisk Field is formed by an anticline. But the oil column which is greater than the structural closure may be explained by the cementation of the water bearing part of the reservoir. At 3120 m the temperature is 130 °C, corresponding to a geothermal gradient of 33 °C per km. Recoverable reserves, for a recovery factor of 20 %, amount to 134 Mt and 112 Gm^3 of gas. The oil, which is undersaturated, has a density of 32 °API (Masson, 1977; Scholle, 1977; Bark & Thomas, 1980).

2.1.3.3. ***Fissured reservoirs***

Fractures sometimes create high permeability, but the porosity always remains low to very low. In other words, fractures may substantially improve the productivity of reservoirs containing hydrocarbons within a matrix porosity. They do not in themselves imply a good reservoir. The permeability that they generate rapidly reaches very high values. Hence the permeability of a 0.25 mm wide fracture could be higher than 5000 darcys, and a single 1 mm fracture could guarantee production of 1000 to 1500 td^{-1} (quoted by Russel, 1960).

The fissures affect reservoirs formed of compact rocks, such as sandstone, limestone, dolomite, chalk, cherts, indurated shales, and sometimes also volcanic or eruptive rocks. The fractures develop particularly in the vicinity of faults, especially in the external zone (above the neutral point) of anticline folds. The more recent fractures are more open and permeable, as the circulation of water containing salts progressively cements them. They may also be formed by compaction and diagenetic processes, especially during the formation of stylolites, or by the overpressure of the fluid contained in the rocks. The most striking fissured reservoirs are represented by the folded limestones of the Asmari Formation in Iran.

2.1.3.4. ***Reserves per square kilometer***

For a given oil flowing normally in downhole conditions, the reserves in place per km^2 of a field essentially depend on the reservoir characteristics, its effective thickness and its porosity. Theoretically, a reservoir which exhibits an effective porosity of 10 % with a thickness of 10 m is equivalent to a continuous oil column of 1 m and yields reserves in place of 1 Mm^3 per km^2. Assuming 25 % recovery the recoverable reserves amount to 0.25 $Mm^3\ km^{-2}$. In similar conditions, a thickness of 100 m would mean in place and recoverable reserves of 10 and 2.5 $Mm^3\ km^{-2}$ respectively.

From the statistical standpoint, the oil volume of a sandstone reservoir increases approximately according to the square of its mean thickness, the reservoir/source rock relationships being taken into account (see Chapter 1 of Part 3) (Figure 49). As for gas, the thickness of the reservoir must be multiplied by a power of

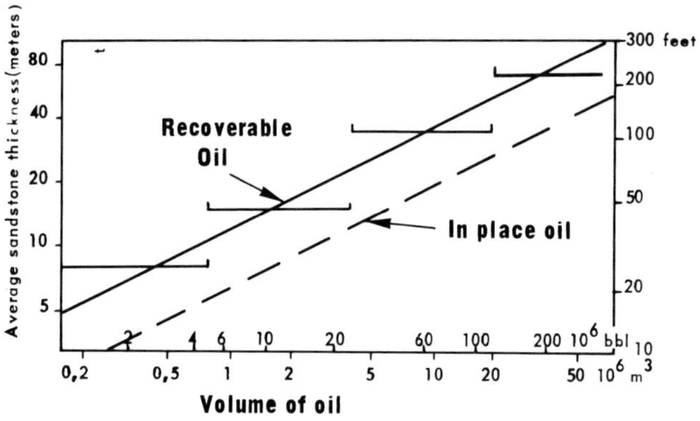

Fig. 49. — Graph showing relationships between average sandstone thickness and volume of recoverable and in place oil in pools (Smith *et al.*, 1975).

2 to 3. The statistical analysis also shows good correlations between the thickest sands and the largest accumulations. Their porosities and permeabilities are generally better than those of thinner horizons. A thick sandstone tends to exhibit a broader extension and is a better reservoir than its relative thickness would suggest (Curtis *et al.*, 1961).

Here are some examples of rich accumulations :

— the Long Beach Field in California, in the Pliocene Repetto sands, which form 50 % of the series, contains 10 $Mm^3\ km^{-2}$, and the Ventura Field in the basin of the same name contains 15 $Mm^3\ km^{-2}$ (375,000 barrels per acre), a record for the United States (*),

— the fields in the Baku region of the USSR, 18 $Mm^3\ km^{-2}$,

— the Salem Field in Illinois, in Mississippian sandstones and limestones, 12 $Mm^3\ km^{-2}$ (3000 barrels per acre),

— the Ekofisk Field in the North Sea, 3 $Mt\ km^{-2}$ in Danian chalk,

— the Grondin Field in Gabon, 1.4 $Mt\ km^{-2}$ in Cretaceous sands.

A good comparison of the potential of the different facies of a reservoir of the same age is given by Riche & Prestat (1979) for the Mid-Cretaceous of the Middle East :

— sands of the Burgan zone, 15 to 2 Mm^3 oil per km^2,

— bioclastic limestones of southeastern Iraq, around 1 Mm^3,

— fractured limestones of the folded zone of the Zagros range, less than 0.125 Mm^3.

The gas problem is more complex, considering the high value of its compressibility factor. Hence the volume of a gas with specific gravity 0.7 situated in pressure and temperature conditions :

— of 300 bar at 60 °C is compressed by a factor of 290,

— of 500 bar at 100 °C is compressed by a factor of 320,

— of 700 bar at 140 °C is compressed by a factor of 350.

In these conditions, 1 Gm^3 of gas located at depth in the above conditions yields gas volumes of 290, 320 and 350 Gm^3 respectively at the surface. Some examples may be given of reserves per km^2 :

— Kangan 330 Gm^3 for an area 440 km^2 at 2200 m, mean porosity 1 to 6 % : 750 $Mm^3\ km^{-2}$,

— Hewet 80 Gm^3 for an area of 100 km^2 at 2700 m, mean porosity 20 % : 800 $Mm^3\ km^{-2}$,

— Lacq 170 Gm^3 for an area of 110 km^2 at 300 m, mean porosity 3 % : 1500 $Mm^3\ km^{-2}$,

— Frigg 200 Gm^3 for an area of 110 km^2 at 1800 m, mean porosity 25 % : 1800 $Mm^3\ km^{-2}$.

Reservoir/fluid relationships

Good quality continuous reservoirs are generally associated with normal pressures, and often with brackish aquifers. On the contrary, heterogeneous and discontinuous reservoirs are related to sudden and high pressure and salinity gradients. Other relationships, resulting from this observation, have been noted between reservoir quality and crude oil characteristics.

(*) Equivalence : 1 barrel per acre is approximately equivalent to 50 $m^3\ km^{-2}$.

Homogeneous and permeable reservoirs are often preferential flow paths for aquifers, leading to two consequences :

— from the chemical standpoint, these aggressive waters are liable to alter the hydrocarbons and to be associated with heavy crude oils,
— from the mechanical standpoint, as we shall see below, these easy flow paths require very well sealed traps to hold an accumulation.

On the contrary, irregularly porous and permeable beds are often content with more mediocre closures, as stratigraphic factors may enhance the structural features, and the crude oils in this case are often light and rich in gas. Good, well supplied reservoirs can offer major fields if the traps have sufficient volume, and in such a case, the habitat may be concentrated. Irregular reservoirs are normally the source of dispersed habitats.

2.2. **SEALS**

Seals are beds that prevent the movement of hydrocarbons perpendicular to the strata. Hence they are opposed to the reservoirs, but not all non-reservoir rocks provide effective barriers. Some of them may even act as source rocks. They are not necessarily impermeable to water.

2.2.1. **Petrophysical and geologic characteristics**

These characteristics are essentially impermeability and plasticity. Impermeability is determined chiefly by injection pressure, which, in all cases, is higher than that of neighboring beds. This high injection pressure is associated with :

— very fine pores giving rise to very high capillary forces,
— often very high interstitial fluid pressures,
— stress fields.

Plasticity enables the rock to preserve its sealing properties after deformations, especially structural deformations. These properties are associated with a number of mineralogical, hydrogeological and tectonic characteristics.

— The mineralogical characteristics primarily condition the nature and arrangement of the minerals or crystals making up the cap rock. Smectite, for example, thanks to its platy layers, offers better characteristics than kaolinite, which has a more polyhedral structure. In addition, diagenesis plays an important role in the sealing properties of shales. In particular, the porosity of these rocks decreases up to a burial depth of 1200 m, below which only extremely fine porosity subsists. Deeper down, and chiefly below 2000 m, their impermeability is altered by microfissuration effects.

— Internal overpressures are often related to hydrogeological mechanisms, and especially to the presence of compaction or formation waters which remain imprisoned in the rock. This is the case of undercompacted shale formations, in which the pressure approaches the geostatic pressure. This is also the result of the action of the molecular water of some minerals, especially smectite and gypsum, at the time of their transformation into more anhydrous materials. Overpressures, which are not necessarily associated with impermeable rocks, but sometimes with reservoir horizons, play a role of great importance. Through this process, the capacity of a seal may be strengthened.

— The tectonic framework also affects the properties of cap rocks, especially if the latter are less plastic. Hence the extension phases reduce the quality of the seals, while periods and zones of compression and tectonic confinement tend to preserve and to reinforce them.

The paleogeographic situation of a seal is important for its action on the scale of the basin. An overlapping position on the margins guarantees very good protection of the underlying series. This is the case for example of the evaporitic series of the Sahara Triassic and of the Jurassic (Hith Formation) of the Arabian platform, whose oil potential is well known. This overlapping arrangement of series at the end of a sedimentary cycle is relatively exceptional and corresponds to a particular extension phase of the basin tectonics.

2.2.2. **Main cap rocks**

Two rock families mainly perform the role of cap rocks :

— claysand shales, the most widespread,

— evaporites, the most effective, impermeable to both gas and water.

In addition, permafrost and the clathrates generally associated with them can also offer an excellent seal.

Clays and shales

These rocks possess two major sealing qualities of high injection pressure and plasticity. The high injection pressure is due to the fine particle size of the clay minerals, and the pores that they contain. It depends on their characteristics and especially their dimensions. This high injection pressure may be reinforced by the establishment of internal overpressure. Shale permeability first decreases with increasing burial depth, until the appearance of microfractures, and then increases, particularly if the sediment is subjected to tectonic stresses. Plasticity is the result of the arrangement of mineral layers.

Evaporites

These sediments are characterized by a crystalline structure, and the crystals are often in chemical equilibrium with their brine. In many cases, diagenesis causes the expulsion of water which increases the internal pressure. Due to their structure, they generally have no effective porosity or permeability. They also possess very high plasticity, especially halite, allowing the reclosing of fault and fracture zones. However, this extreme mobility, giving rise to intumescences and diapirs, could cause discontinuities through salt migration, going as far as a sieve pattern, as conceived for the Aptian salt of the Congo (Figure 58). Anhydrite, which is more widespread than halite, is an excellent and relatively frequently encountered cap rock. Halite is also an excellent conductor of heat.

Miscellaneous seals

In rare cases, certain carbonates or siltstones may perform a sealing function, especially in basins which are not or only slightly tectonized. In the case of carbonates, their properties are enhanced by the presence of stylolites, affected and sometimes nullified by the formation of microfissures.

A good example is a seal formed of a compact limestone in the small Sunoco-Felda Field in Florida. The reservoir, impregnated to a thickness of about 10 m, consists of a limestone containing algae, gastropods and debris, exhibiting a mean porosity of 18 % and permeability of 60 md. It is probably supplied by underlying calcareous black shales about 10 m thick, and often by an impermeable brown limestone only 3 m thick. This seal is surmonted by a rudist limestone reservoir aquifer, substantially of the same thickness, and the overall unit is covered by anhydrite beds, which probably form the main cap rock. This small limestone seal thus creates an apparent anomaly in the fluid distribution, by placing the oil level below an aquifer. It is important to recall these examples when monitoring the drilling of a well (Tyler & Erwin, 1976).

Also noteworthy is the sealing function played by permanently frozen ground, especially in the Arctic zones. This permafrost, as well as the presence of solid clathrates, appears to be the originating factor of the huge gas resources of western Siberia.

Although they are relatively porous, water bearing horizons at a higher pressure than deeper beds ('mixed compaction facies') provide an effective seal. As observed particularly in the Beaufort Sea Basin (Evans *et al.,* 1975), these high pressure horizons can exert indirect action on the underlying seal formations, which only play the role of a cap rock when they are surmounted by the high pressure bed which thus performs the role of the principal or ultimate seal (Figure 50).

Conversely, certain non-reservoir formations do not necessarily constitute seals. They allow fluids to filter through and are referred to as false cap rocks. This concept is especially important for the definition of a structural closure (Figure 66).

Cap rocks must therefore be considered from the relative and dynamic standpoint. In fact, the barrier properties of a seal vary in accordance with :

— the composition of the hydrocarbons, gas being more mobile than oil,

— the pressure exerted on the barrier, which gives way naturally if the fluid pressure is greater than the injection pressure,

— the tectonic state, extension or compression of the basin.

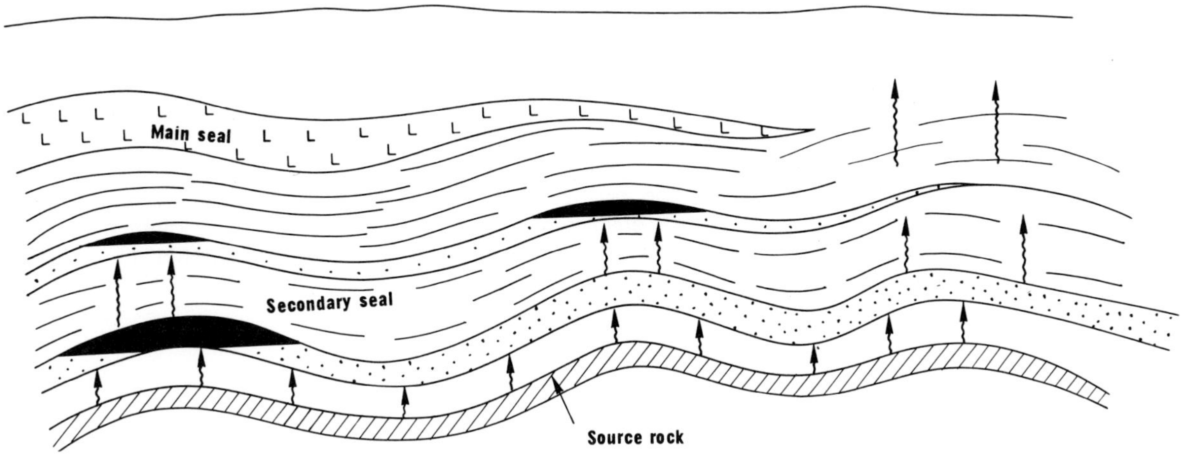

Fig. 50. — « Induced » seal.

Neglia (1979) showed that below a burial depth corresponding to the bubble point pressure of an oil/gas mixture, a shale cap rock is liable to allow gas microbubbles to pass through and only retain the oil. These properties give rise to differential traps. Thus seals may be considered as valves, which effectively prevent the passage of hydrocarbons in certain conditions, while they allow certain quantities to pass through, especially when the pressure is too high, which may be the case when the height of the hydrocarbons accumulated updip becomes too great.

These aspects show that the properties of cap rocks very often condition the characteristics of a province and, in particular, of the hydrocarbons. A very effective seal, halite, for example, may pose a perfect barrier and prevent the passage of any hydrocarbons. This is the case of the basins of the northeast Sahara and Arabia, which are completely devoid of seepage, with the exception of Bahrain Island. Very good or thick cap rocks often retain gas or light oils, as illustrated by these two provinces, whereas mediocre or thin seals often only retain heavier products.

*
**

Close correlations are observed between the characteristics of reservoirs, seals, aquifers and hydrocarbons. Roughly speaking, it has been observed that homogeneous, continuous and permeable reservoirs often contain waters of low salinity with uniform pressure gradients. On the contrary, discontinuous, lens-shaped reservoirs generally contain aquifers with variable salinity and pressure, often of the chloride rich type. These reservoirs retain hydrocarbons more easily than continuous reservoirs.

Generally speaking, homogeneous continuous reservoirs with good permeability often contain low salinity water with stable pressure gradients. On the other hand, discontinuous lenticular reservoirs generally have water of variable salinity and pressure gradients, often chloride rich, and these retain hydrocarbons more easily than continuous reservoirs.

Effective and continuous cap rocks generally retain chloride rich water with high salinity, exhibit pressure regimes often higher than hydrostatic, and well conserved hydrocarbons. On the contrary, mediocre or discontinuous cap rocks often cover low salinity aquifers, or variable pressure — frequently hydrostatic — and relatively altered, heavy hydrocarbons, low in gas content.

Brackish aquifers often containing some carbonates and sulfates are preferentially related to heavy, more or less altered crude oils, whereas chloride waters, which contain more salts, are associated with light and often paraffinic crude oils, which may be rich in gas. No preferential seal apparently exists for gas, and the thickness of a seal does not appear to play a significant role, since perfectly sealed beds a dozen meters thick can effectively shelter high pressure pools, as in the Permian gas fields of Iran. On the other hand, their extension often closely conditions the distribution of the accumulations.

Summary and conclusion

The reservoir/cap rock unit constitutes one of the indispensable elements for the existence of a pool. The reservoir is characterized by its porosity and permeability properties, enabling the hydrocarbons to flow and accumulate in it, while the seal performs a barrier role, preventing their escape from the reservoir. From the physical standpoint, the difference between the two resides in the values of the injection pressure: if low, it allows the fluids to pass through; if high, it opposes the lower pressure fluid. Hence all the transitions exist between reservoirs and cap rocks, and many rocks are neither one nor the other.

These two entities are determined both by deposition and burial conditions. Apart from chalk, reservoirs are generally the result of high energy depositional environments, whereas cap rocks, like source rocks, result from low energy environments. As a rule, the former are degraded during burial, while the latter appear to be favored to a certain degree. The combination of these two types of rock consequently requires rapid variations in the characteristics of the depositional environment, and these differences are often accentuated during geological history. Here again, factors which are favorable to one of the components, like fractures, are detrimental to the other.

3. – HYDROCARBON FLOWS, SECONDARY MIGRATION, TRAPPING AND SEEPAGE

« A égale distance entre le cristal et la fumée, j'ai vu vivre la surface de l'Ecorce terrestre. »

'Equidistant between crystal and smoke, I saw the surface of the Earth's crust come to life.'

Georges Millot, 1978.

The migrations of oil and gas are a particularly important yet still poorly understood link in the formation of hydrocarbon fields. Unlike primary migration, secondary migration occurs essentially in permeable rocks and fractured zones. These two schemes are actually often combined, and culminate in the formation of an accumulation, which may be modified or destroyed by tectonic movements, causing new hydrocarbon migrations.

3.1. DRIVING FORCE AND MECHANISMS OF SECONDARY MIGRATION

3.1.1. Physical dynamics

Although secondary migration is better understood than primary migration, the drainage of producing fields offering a good field of investigation, the mechanisms of these transfers are still partly hypothetical. As a rule, it is accepted that fluids in movement, hydrocarbons in particular, proceed from high pressure zones towards lower pressure zones.

Firstly, oil migrates independently of the aquifers in which it advances essentially under the action of buoyancy. The densities of the two fluids are quite different, the specific gravity of oil being less than 1 and often less than 0.9, while that of water is equal to or greater than 1. In these conditions, it is perfectly reasonable to assume that the oil moves in a continuous phase within carrier beds which are usually water wet. The speed and degree of separation are obviously directly proportional to the difference in densities.

Secondly, the slow movement of the aquifers conditions hydrocarbon migrations. The hydrodynamic gradients facilitate or work against trappings and seepage from the simple mechanical standpoint, according to whether they oppose or enhance the buoyancy. Furthermore, the waters may, in certain cases, transport liquid hydrocarbon elements in solution, possibly in the form of micelles. More probably, they convey gaseous hydrocarbons in solution, and the latter may in certain cases themselves contain liquid hydrocarbons. The movement of hydrocarbons in an individualized phase is increasingly sensitive to higher hydrodynamic gradient and increasing oil density (King Hubbert's Law).

Thirdly, migrations are related to the petrophysical characteristics of the reservoir rock, to its permeability, or, more precisely, the value of the injection pressures in the rock or the capillary pressure. To cross a space between pores, if its diameter is greater than that of the space to be traversed, a hydrocarbon element or droplet must be deformed to a degree that varies directly with the difference in diameter. The interfacial tensions oppose this deformation and hence the movement, with greater force as the constriction becomes narrower. This opposing force is the essential feature of the impermeability of fine-grained rocks. Capillary forces thus tend to oppose the two above driving forces, buoyancy and hydrodynamism (Figure 51).

Migrations are controlled by the framework and structure of the sedimentary basin : dip and continuity of the carrier bed, faults and fractures. They follow the deformations resulting from geological history, and especially any accentuation in dip, the increase or decrease in pressure accompanying subsidence and uplift movements, and they obey any new tectonic stress.

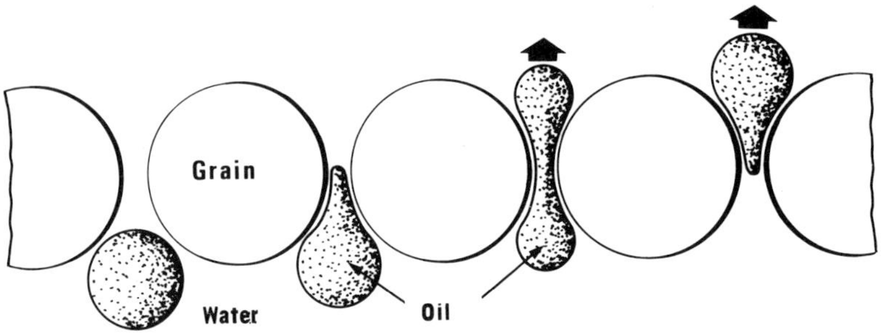

Fig. 51. — Penetration of an oil droplet in a water wet detrital rock (Berg, 1975).

Finally, hydrocarbon flows are closely related to the genesis and expulsion of new hydrocarbon sources, chiefly in case of subsidence. Movements appear to be easier if the continuous oil phase is larger, and the oil flows probably tend to combine as their paths converge (Chiarelli, 1973). These impregnations are probably the source of many shows observed in drilling.

3.1.2. **Chemical dynamics**

By selective retention, hydrocarbon migrations through relatively non-porous and impermeable rocks cause physicochemical variations which are often significant. As a rule, an increase in the percentage of non-polar hydrocarbons is observed along the migration path, due to the adsorption of polar compounds of the oil on the oil/mineral or oil/water interfaces, which can lead to the 'substitution of oil for water as the wetting fluid', a decrease in the content of asphaltene aromatics and resins, and the 'selective extraction of light compounds of the oil by the circulation of gas' (du Rouchet, 1978 a). This process is reinforced by the greater water solubility of these compounds, and the statistical result is a substantial increase of about 4 ‰ of isotope ^{12}C (Stahl, 1977).

These mechanisms could explain the observation according to which the 'effects of the concentration of thiophenic sulfur or trace metals are more marked as the cap rock quality decreases' (du Rouchet, 1978 a). A geochemical study of the fields of the Handil group in East Kalimantan (see Chapter 2 of Part 3) showed that a crude oil could be enriched in light products during its migration, even yielding an 'abnormal' distribution of hydrocarbons on the scale of the field, with the lighter components located in the higher horizons (Durand & Oudin, 1979). The action of secondary migrations on gas chemistry is less well known.

In brief, and in simple terms, migrations, especially through seals, can lead to heavier crude oils accompanied by an increase in the contents of sulfur, trace metals, and aromatics in particular.

3.2. **CHIEF MODES OF MIGRATION AND TRAPPING**

3.2.1. **Migrations in the basin framework**

Hydrocarbon migrations are conditioned by the major sedimentary, paleogeographic, structural and hydrogeological features of the basin. They are qualified as lateral or vertical, according to whether they pass along the carrier beds, or through the fractures which are not parallel to the strata.

Sedimentary and paleogeographic criteria

Transgressive series, beginning with impermeable horizons, limit the displacement of the underlying fluids towards the atmosphere. On the other hand, overlapping surfaces resting on detrital base series offer excellent roads for migration and seepage. Positive sedimentary sequences, especially when they terminate in evaporites, provide an effective barrier to vertical migration. The combination of positive sequences and an overlapping scheme on the basin margins offers the best seal on the regional scale. Conversely, regressive series and negative sequences generally offer few possibilities of sealing and facilitate dysmigrations and seepage.

Structural criteria

Large marginal faults may form effective barriers to migration, although the tightness of the faults is a complex question and varies in space and time in accordance with the pressures, facies and diatrophism. On the other hand, zones in a distension régime, characterized by open fractures and fissures, are the most natural vertical migration paths.

Hydrogeological criteria

Young sedimentary basins, or those which are still in the subsidence stage, with series undergoing compaction, are characterized by centrifugal gradients. The highest water pressures are situated in the central and deep areas. Tectonic basins, subjected to structural deformations, uplifts and folding, especially on their margins, generally exhibit centripetal hydraulic gradients. The central areas are characterized by relatively low water pressures. Hydraulic pressure gradients may affect the action of migrations and determine the most favorable accumulation zones (Chiarelli, 1970 and 1973; Coustau, 1975).

3.2.2. **Lateral migrations**

Lateral migrations flow at the top of continuous and permeable horizons, in contact with the impermeable seals, and the flows proceed with greater ease if the contact surface is flat and continuous. They probably play an important role in the concentration of hydrocarbons and in the formation of accumulations at the top or on the flanks of high regional zones, from source rocks which are more deeply buried within the neighboring synclines. These migrations seem to be able to continue over long distances, ranging from 10 to 100 km in the most favorable cases, especially in platform basins and on stable margins.

If the reservoirs are discontinuous and bounded updip by a permeability barrier, the scheme is a typical 'through cap rock' migration pattern, in which the lightest hydrocarbons occupy the upper levels, similar to the vertical migration scheme (Figure 52).

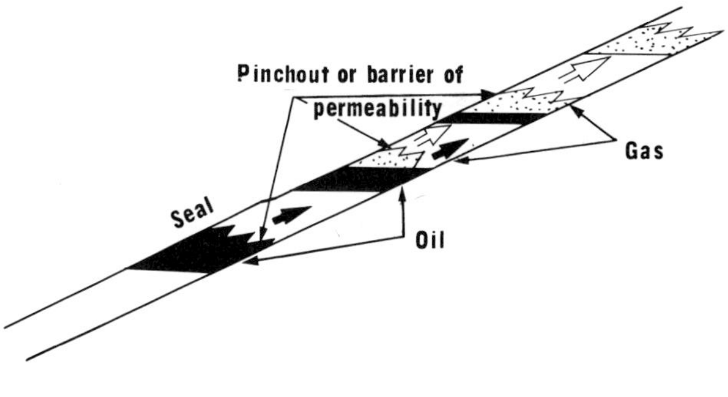

Fig. 52. — Differential trapping in stratigraphic pinchout system.

Continuous reservoirs, and the lateral migrations which they stimulate, are often the source of considerable dysmigration and seeps, which appear with aquifer horizons and are sometimes likely to constitute large impregnations, especially of tar sands like those of Athabasca, the Orinoco, and Melekess in the USSR (Demaison, 1977). On the contrary, it has often been observed that discontinuous sand horizons have more chances of being productive than continuous horizons of the 'blanket sand' type. The latter, subjected to active hydrodynamism, very often only offer productive reservoirs if they are perfectly sealed by pronounced folds or by particularly tight faults. The presence of overlapping surfaces, starting in particular with detrital horizons, also offers a preferential flow path. The Gamba sands in Gabon offer a good illustration of this. Nevertheless, these large migration paths may play a major role in the formation of vast hydrocarbon concentrations, if the hydrocarbon supply is itself prolific. This appears to be the case of the fields of Arabia, the northeast Sahara, and the Rocky Mountains. In these cases, especially in West Canada, in the Mid-Continent, on the east flank of the Denver Basin, lateral migrations may continue over long distances, up to 100 km according to some authors (Clayton *et al.,* 1980; Price, 1980). On the other hand, a barrier in the migration path could result in certain sterile zones (Price, 1980).

In case of a succession of anticline folds or reefs, as shown in the Michigan Basin for example (Figures 53, 146 and 147) (Gill, 1979), lateral migrations give rise to differential trapping, which was described more or less simultaneously by Gussow (1954) in the United States, and by Maximov in the USSR. This pattern is familiar. Starting from a deep source rock, the flow occurs along a carrier bed. The closed anticlines nearest

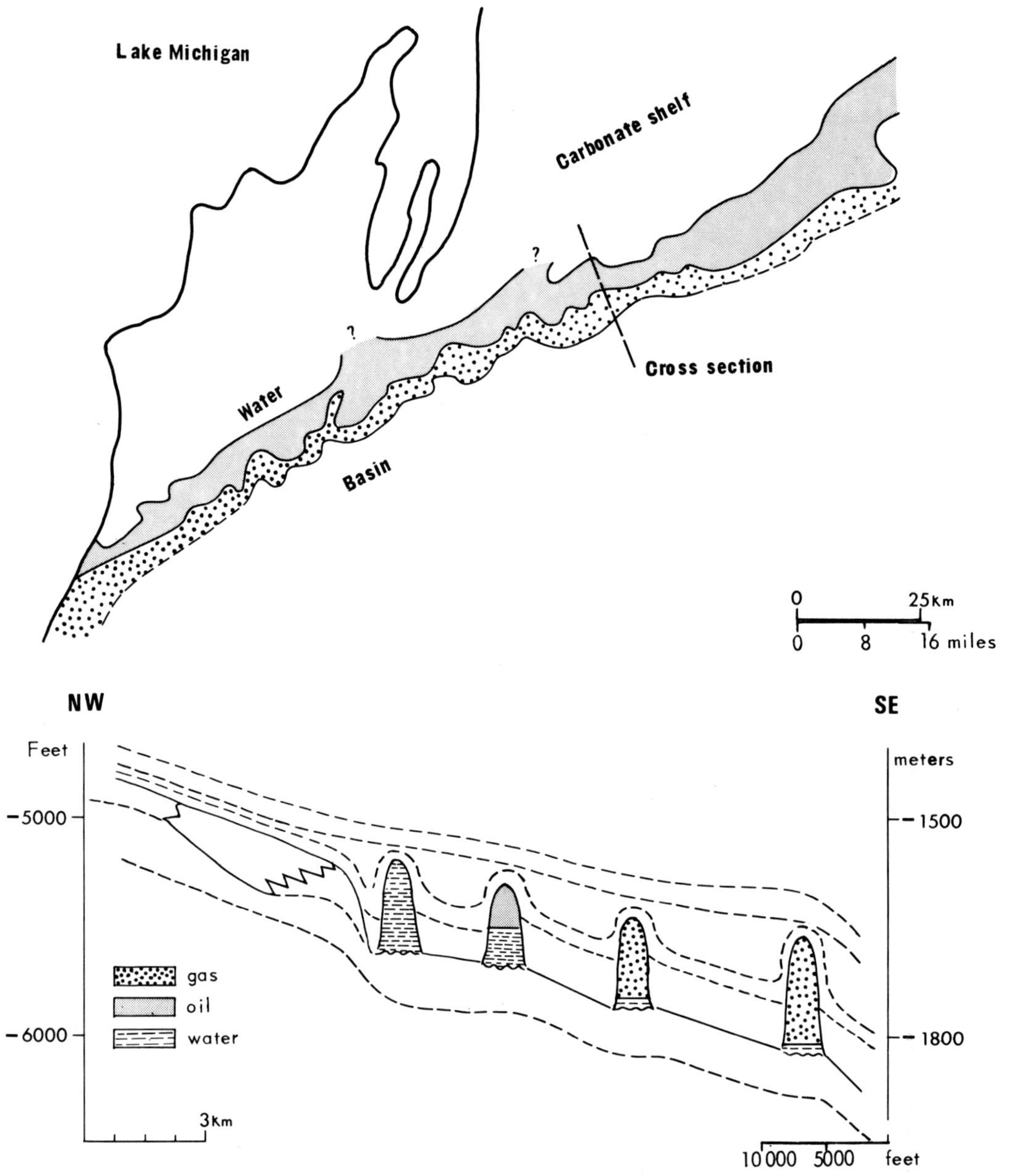

Fig. 53. — Map and section of Silurian reef trend in northwest Michigan (Gill, 1979).

to the source, namely the deepest, are filled first. The gas occupies the upper part of the closed volume and spreads gradually by expelling the oil. When the oil reaches the closure boundary, the spill point, it escapes along the same reservoir horizon towards the neighboring structure, which is higher, and so on. The final distribution is as follows (Figure 54) :

— gas in the deepest folds and in the central part of the basin,
— oil in the structures located above and in the boundary zones of the basin.

The achievement of this pattern presumes :

— perfect continuity of the reservoirs and seals,
— a succession of folds,
— the absence of faults or any source of leakage towards the seals,
— a sufficient flow of hydrocarbons.

The flow may be fed by the genesis of new quantities of oil and gas, which occurs particularly in subsident zones. But deformations of the horizons may also cause the hydrocarbons to be set in motion. Tilting may diminish a geometric closure and cause a new flow to begin (Figure 54). An uplift, resulting in erosion, decreases the bed pressure, increases the volume of the gas cap, and can allow part of the oil to escape, again without an additional inflow of hydrocarbons (Hobson & Tiratsoo, 1975).

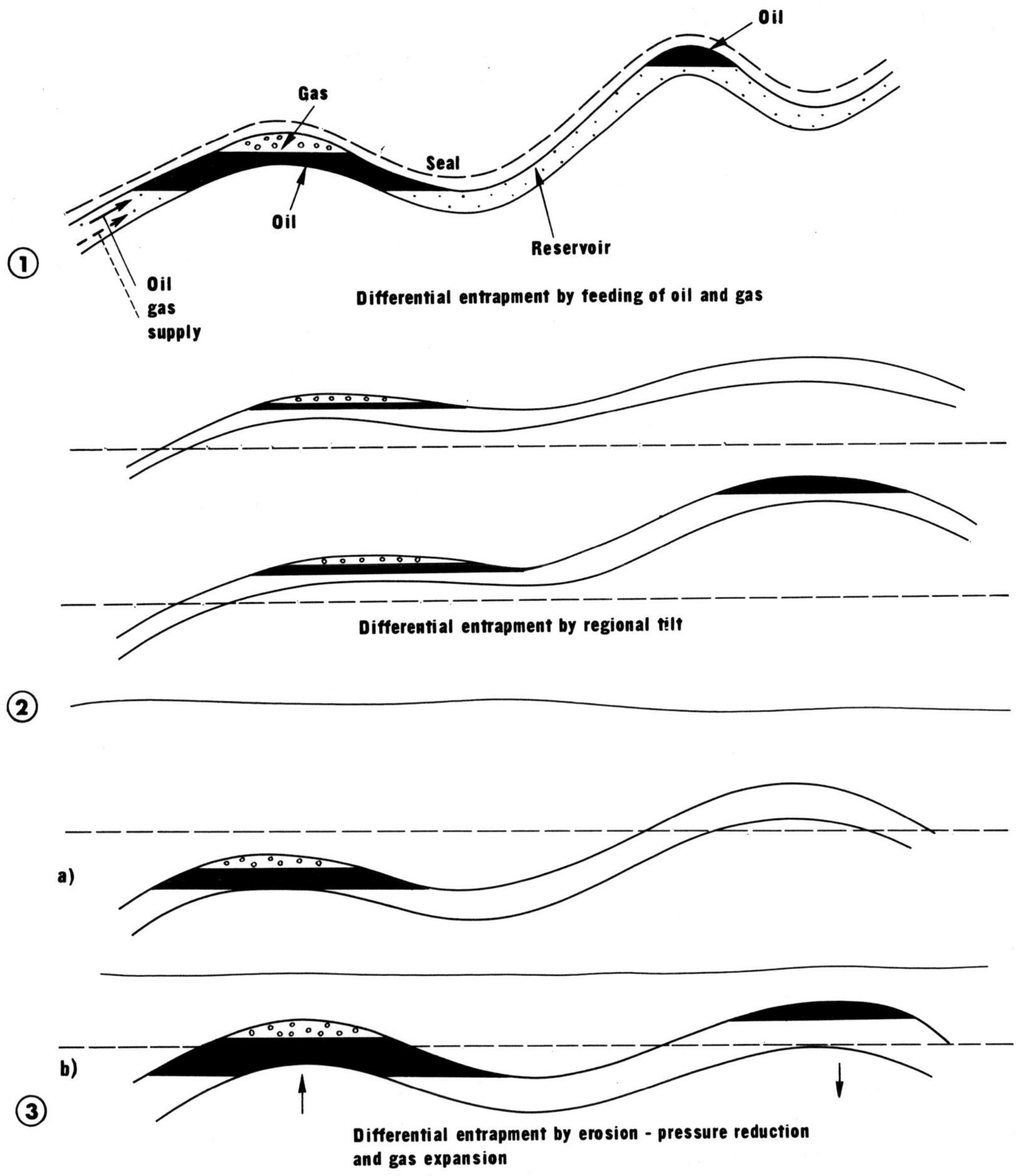

Fig. 54. — Various differential entrapment patterns (Gussow & Hobson 1954 ; Tiratsoo, 1975).

3.2.3. **Vertical migrations**

The mechanism of vertical migration consists of a sort of direct or indirect impregnation of the reservoirs located below or above the source rocks, or the passage of the seals and other impervious rocks above them. It is comparable to the primary migration process. These vertical migrations can only occur when the pore pressure of the hydrocarbons is sufficiently high to overcome the confining stress and this is rendered possible by an increase in the fluid pressure, especially under the action of tectonic forces, and, in particular, at the core of anticlines (du Rouchet, 1981; Magara, 1981).

Downward or per descensum migrations

Chapman (1972) postulated that as a rule, the hydrocarbons formed in transgressive series often show a tendency to migrate downwards, generally over very short distances, and to supply stratigraphic traps, especially the reef traps which occur frequently and in a majority in these stable provinces in the absence of large domes. Apart from Canada, he offers the examples of Mexico and Libya.

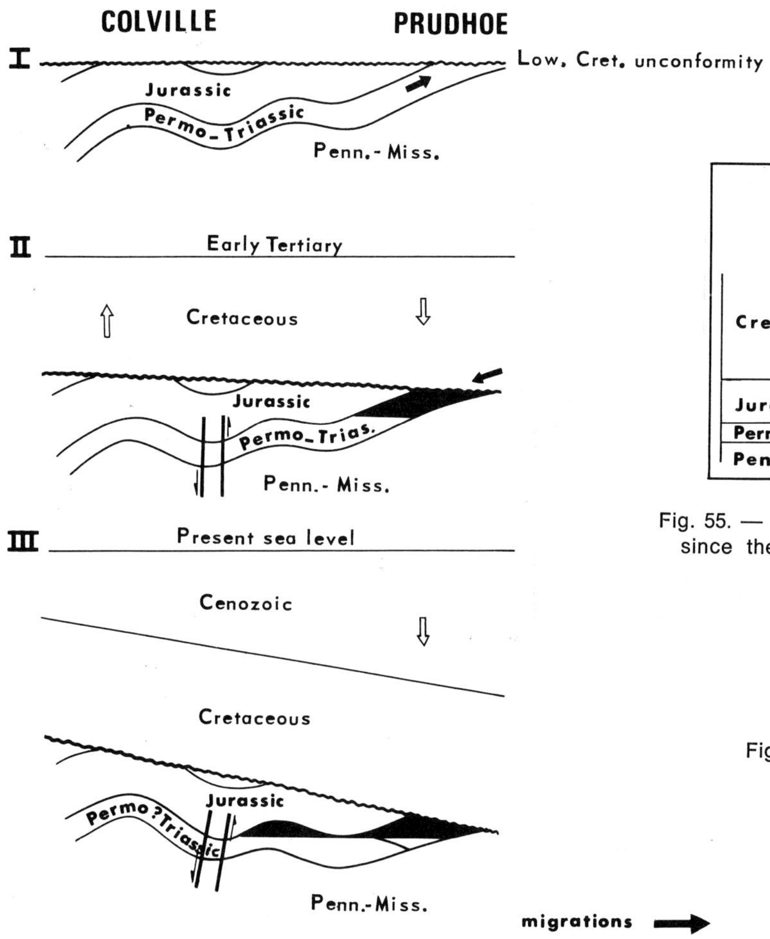

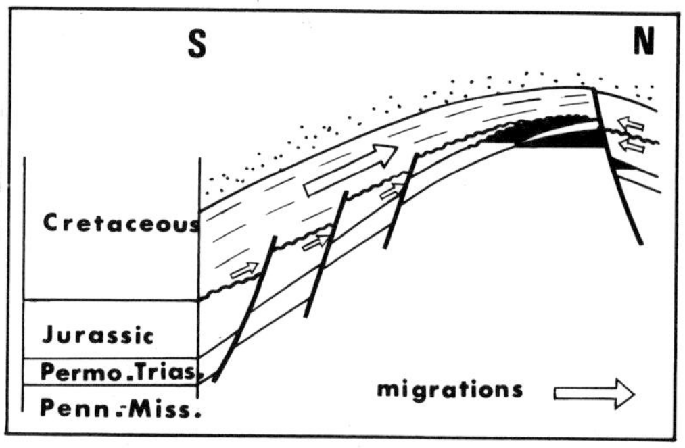

Fig. 55. — Migrations that supplied the Prudhoe Bay Field since the Early Cretaceous (Jones & Speers, 1972).

Fig. 56. — Section of Prudhoe Bay Field (Jones & Speers, 1972).

Example of migration from a transgressive source rock : the Prudhoe Bay Field

Geochemical correlations imply that the oil from the Carboniferous and Permo-Trias reservoirs of the Prudhoe structure derive mainly from the Lower Cretaceous shales which overlap the different eroded components ranging from the Carboniferous to the Lower Cretaceous (Figures 55 and 56).

However, more thorough geochemical analyses, based in particular on the analysis of steranes and terpanes in the different source rocks and in the crudes from various reservoirs, have helped to identify more complex relationships. Apart from the transgressive Neocomian Formations, the shales of the Triassic (Shublik) and the Jurassic (Kingak) thus appear as effective source rocks which supplied the adjacent reservoirs (Seifert *et al.,* 1979).

The impregnated reservoirs are in contact with the source rock, either directly or through faults. The field is found on a regional high peaking at — 2800 m, and may be supplied from the neighboring syncline areas, at depths reaching 5000 m. The density of the crude oil increases with depth : 25 to 30 °API at 2650 to 2700 m; 15 °API at 2750 m (Jones & Speers, 1976). The same pattern seems to be illustrated in particular by the Upper Cretaceous of the Sirte Basin, Northwest Rankin (Australia) and West Texas.

This migration scheme seems to prevail when the overlapping source rock series simultaneously plays the role of a cap rock, in other words when it has a relatively high injection pressure. Moreover, downward migration is facilitated by the existence of an altered surface with good porosity at this level.

Upward or per ascensum migrations

These migrations occur through stratified seals along fractures and fissures. In this scheme, it is mainly the faults, and the fractured zones associated with them, that play the role of drains. We are familiar with the often ambiguous role of faults, sometimes an escape route, especially when they have been in recent movement, and sometimes as a tight overlay, when they are plugged. They are often the source of dys-migrations and seeps, sometimes extensive. They may also become effective barriers to migration, allowing oil and, more rarely, gas to accumulate (Section 4.1.2. of Part 2).

The classic example of vertical migration, and one of the first investigated, is the foothills province of Iraq and Iran, where the oil of the giant Asmari fields is most certainly of Cretaceous origin (Dunnington,

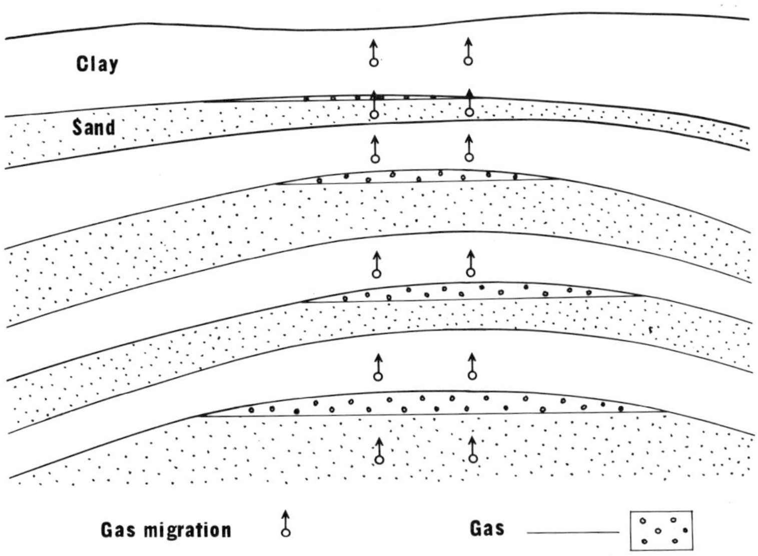

Fig. 57. — Vertical migration through shale seals (Neglia, 1979).

1958; Young *et al.,* 1977) (Section 3.1. of Part 3). This migration is frequent in the portion of the anticlines located above the neutral point of the stresses, which is the case in this zone. The Tertiary basin of the Niger Delta is another remarkable illustration, in a deltaic province of this type of migration and trapping (Evamy *et al.,* 1978) (Section 4.1.2. of Part 2). This also applies, at least partly, to Louisiana (Young *et al.,* 1977) (Chapter 1 of Part 3).

In the absence of faults, weak zones of the cap rock are the source of seepages if their injection pressure is lower than the pressure exerted by the fluid, which is notably the case at the top of the accumulations. 'Paradoxically', as du Rouchet reports (1977), "closed high zones are potentially preferential zones of vertical migration". Hydrocarbons migrate upwards more easily if :

— they are lighter (this type of migration is especially extensive and frequent for gas),

— the injection pressure of the rocks forming the barrier is lower.

In all cases, the petrophysical characteristics of the cap rocks play a major role in the distribution of hydrocarbons. Since their barrier quality depends on the degree of compaction, it is observed, for example, that their retention capacity increases with depth. In these conditions, the shallowest horizons only hold small amounts of oil and gas, as may be seen in the Pliocene and Pleistocene series of the Po Plain. This often goes hand in hand with a weaker structural closure (Neglia, 1979) (Figure 57).

In this case, good chances exist of finding light hydrocarbons and gases in the upper beds, and oils in the deeper zones. This means that, apart from the early gases, the vertical arrangement prevailing is the reverse of the classical diagenesis pattern. Often, impregnated horizons are found preferentially in the lower reservoirs, the ones closest to the source rocks, if they are well sealed. The upper reservoirs, which are not supplied, are aquifers in this case. Vertical migrations appear to be more frequent and more extensive than generally believed before the acquisition of precise knowledge about maturation zones. In the Denver Basin, for example, the height of these movements is as much as 700 m (2500 ft) (Clayton *et al.,* 1980).

It appears that lateral and vertical migrations most often occur together to produce a combined system. The first accumulations occur by a lateral migration pattern. Leaks are then liable to affect the seals of some of these anticlinal pools, leading to vertical migrations and differential trapping of this type. This is the case in particular in the Illizi Basin (Algeria), where the oil migrated laterally in the Devonian series, and then vertically along faults from the Upper Devonian to the Carboniferous. This also seems to apply to the Paleocene and Danian fields of the North Sea, supplied by Jurassic source rocks (Figure 193). At Frigg, for example, which produces gas in the Eocene, the deeper horizons are aquifers, and virtually empty, whereas in the neighboring Statfjord, Cormorant and Brent Fields, the hydrocarbons have remained trapped in the Cretaceous and Jurassic horizons, and the upper reservoirs are aquifers.

In the Saliferous basin of the Congo, the following complementary pattern appears to exist :

— vertical migrations through the Lower Cretaceous, especially through the discontinuities in the Aptian Saliferous seal, from the origin in Pre-Saliferous source rocks,

— lateral migrations under the shales of the Upper Cretaceous in the direction of the boundary of the basin, where significant outcrops of tar sands occur (Figure 58).

A comparable process appears to be at the source of the Rotliegendes gas fields in the southern part of the North Sea (Leman, Indefatigable).

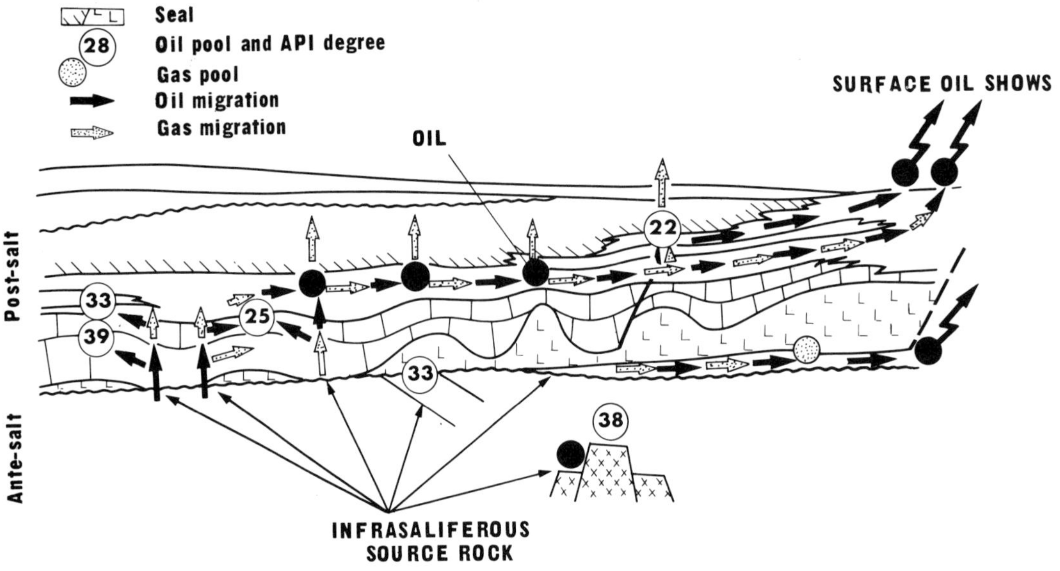

Fig. 58. — Section of Congo Coastal Basin showing probable migration pattern (Augier, Chiarelli & du Rouchet, 1977).

The origin of the oil is often complex and diverse for different reservoirs in the same zone, and even in the same field. The fields of the Maracaibo Basin offer outstanding examples of this. In this province, oil is mainly encountered in the Cretaceous, Eocene and Miocene. Detailed surveys have shown that part of these hydrocarbons originates in each of these three units. However, it appears that the bulk of the Miocene oils originated in Eocene source rocks and escaped through faults, while the reservoirs of this age were probably supplied in many cases by carbonate source rocks of the Cretaceous (Young *et al.,* 1977) (see also Chapter 3 of Part 3).

In addition to leaks through faults and fractures in the seal, losses occur through diffusion, and this is especially true of gaseous hydrocarbons. Smith *et al* (1971) calculated that an average field of some 3 Gm^3 of reserves loses 21 Mm^3 of methane and 0.6 Mm^3 of propane per year, not counting losses through any fractures. These considerations corroborate the need for more or less continuous supply upstream to preserve a relatively large gas field.

3.3. DYSMIGRATIONS, ALTERATIONS AND SEEPS

Hydrocarbon dysmigrations which reach the surface give rise in a way to outcrops of oil and gas, also called seeps. These seeps can be analyzed from two complementary standpoints, from the chemical standpoint, with the alteration mechanisms, and from the geological standpoint, by attempting to interpret them from the point of view of exploration.

3.3.1. Alteration and degradation

Hydrocarbons in an oxidizing environment are considerably altered, and often destroyed. The light and volatile fractions, and especially the gases, are lost to the atmosphere. Emanations of natural gas are familiar in certain regions, especially when they burn naturally to form fen fires or *ignis fatuus,* as in Mesopotamia. The observation of these spills is easier in an aquatic environment, and some of them can be detected in the high definition seismic profiles (Plate 2). Liquid products are altered by oxidation and polymerization to yield heavy, pasty or solid products, black in color, which are capable of withstanding meteoric agents for a certain time. The most important destructive action seems to be caused by microorganisms, and especially by bacteria.

The results obtained by laboratory experiments partly agree with the observations made in the field and in the oil-bearing beds. Bacteria cause a genuine metamorphosis of the saturated hydrocarbon fraction : complete elimination of n-alkanes, acyclic hydrocarbons with isoprenoid chains (pristane, phytane, C_{18}, C_{16}, C_{15}), intense destruction of the homologous series of iso- and antiiso-alkanes, probable metabolism of certain

1

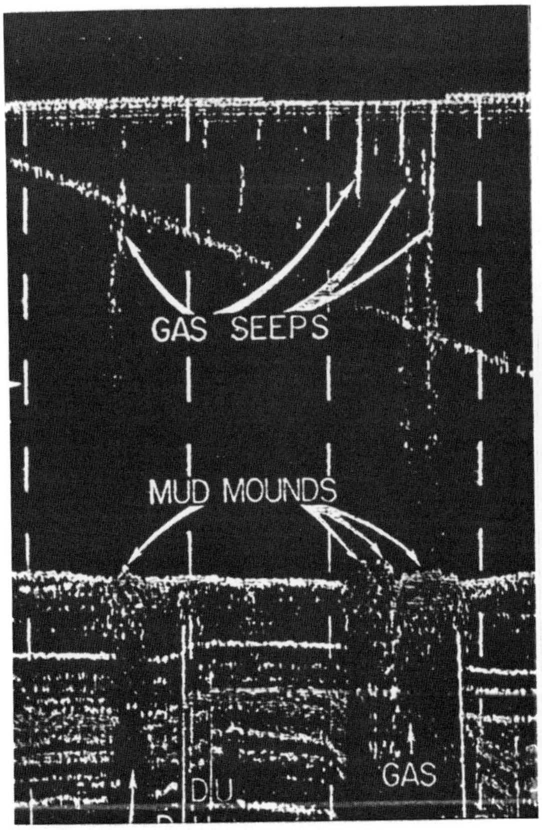

2

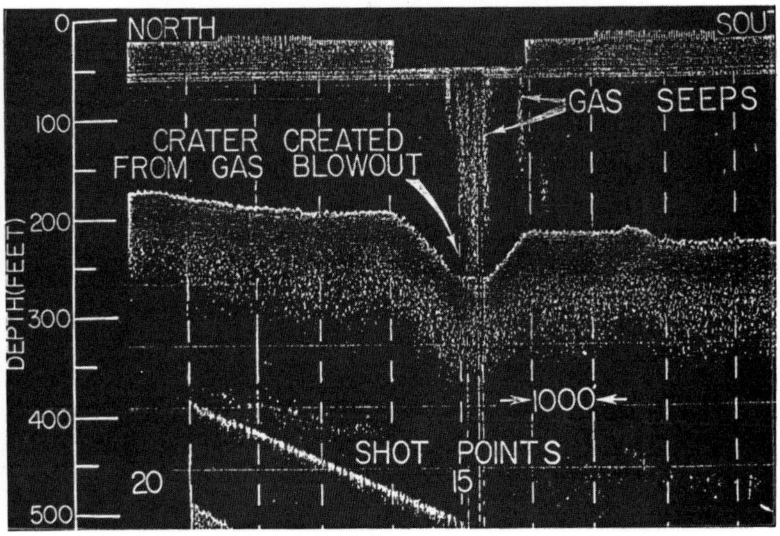

3

PLATE 2

The migration of hydrocarbons
Oil production in a well — seepage of gas at the seabed

Fig. 1. — Trickle of oil, filmed in open hole while flowing from the Emeraude reservoir, Congo.

Fig. 2. — Seepage of gas at the seabed, seen in a seismic profile.

cyclo-alkanes with low molecular weight. This alteration has the effect of concentrating polycyclic biological markers, steranes and triterpanes (Connan, unpublished report). It contributes to transform the crude oils into naphthenic crudes at temperatures below 6 °C (150 °F) (Philippi, 1977). In nature, however, unlike *in vitro* tests, aromatics are also liable to be considerably degraded (Connan, 1981). Biodegradation also destroys part of the petroleum-generating potential of these products, even under the effect of subsequent catagenesis, making them approach low yield products, such as humic kerogens and coals (Connan & van der Weide, 1978).

Bacteria are capable of finally obliterating the main specific characteristics of alkanes (Figure 59). Hence crude oils with very different alkane compositions can, after bacterial attack, yield products consisting mainly of a mixture of comparable iso- and cyclo-paraffins. In other words, it becomes very difficult to characterize the origin and the history of these crude oils, and especially to distinguish between the results of maturation and of alteration.

3.3.2. **Geological significance of seeps**

It is well known that the geological context in which seeps occur is of vital importance for their interpretation. Seeps supplied by lateral migration along a monocline are of little value in comparison with vertical migration products, occurring along faults and fissures, especially in an anticline zone, which are capable of revealing an underlying accumulation (Link, 1952; Perrodon, 1966). The finding of living, in other

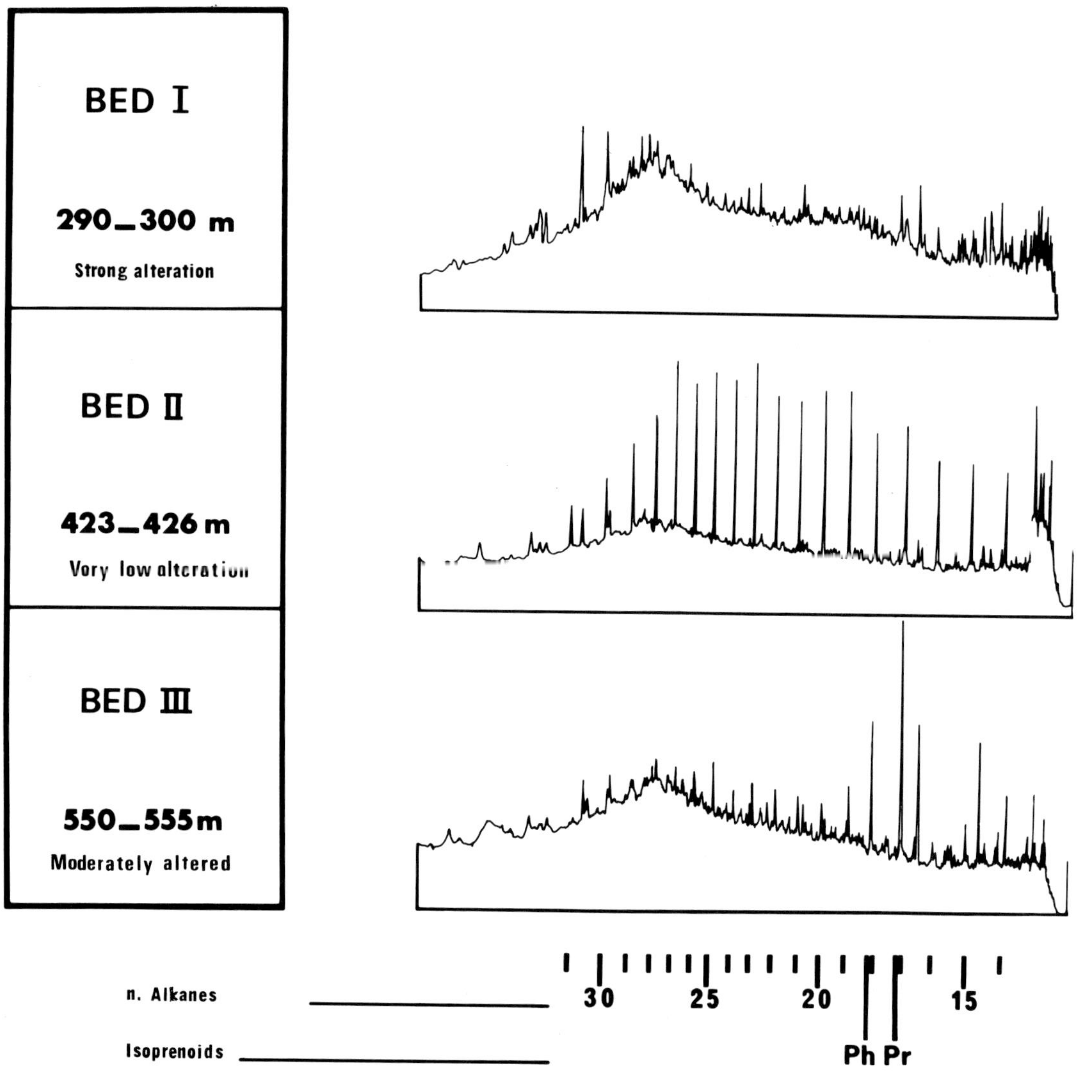

Fig. 59. — Chromatograms of saturated hydrocarbons showing the disappearance of paraffins by biodegradation on Emeraude Field, Congo (Claret *et al.*, 1977).

words constantly replenished seeps in a borehole may be interpreted as passing through an accumulation, or as the passage of migrations.

Beyond this analysis, the deeper meaning of these shows, even living ones, is still fundamentally hypothetical, and advances in the analysis of geochemical prospecting cannot remove this uncertainty. The presence of active shows is undeniably the proof of the genesis of hydrocarbons in the basin or in a zone. But it cannot indicate whether they represent the vanguard of great battalions that have remained in the earth, or merely the last survivors of a decimated rearguard.

3.4. TRAPPING AND ACCUMULATION

Secondary migration transfers the hydrocarbons to zones of lower pressure and lower temperature than those of the source zones. This offers the possibility of physical and/or chemical transformations, some of which may assist in the concentration and trapping of the hydrocarbons. This is the case of the gas in solution in aquifers which encounter methane hydrates at relatively low temperatures.

The solubility of the gas in water decreases sharply in the presence of methane hydrate (or clathrates). For instance, at a temperature of 10 °C, this solubility increases steadily from 0 to 2 $cm^3\ cm^{-3}$ as the

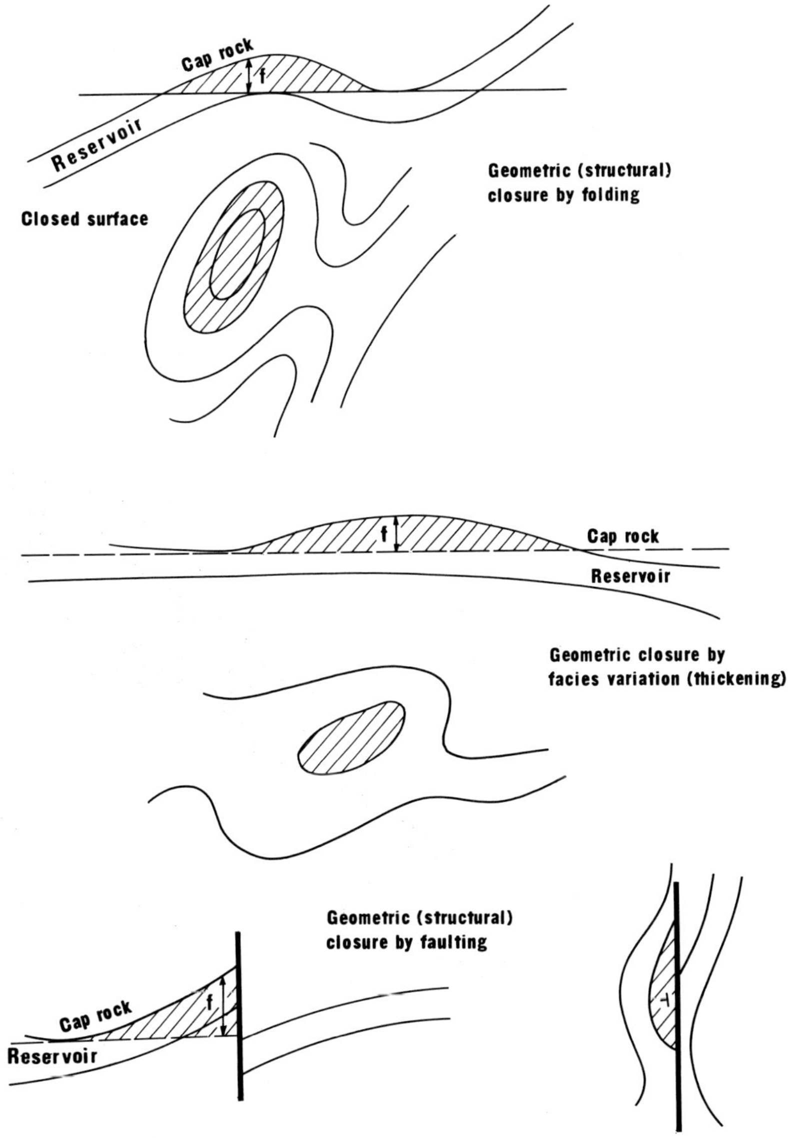

Fig. 60. — Examples of geometric closures.

pressure goes from 0 to 75 bar (Figure 18). At this stage, which corresponds to the hydrate formation pressure in these conditions, the equilibrium concentration falls to 0.5 $cm^3\ cm^{-3}$ only. This drop upon contact with hydrates could also constitute a significant mechanism of the liberation of gas in solution in aquifers, and could be the source of very large accumulations of dry gas in the Siberian regions (Makogon et al., 1971; du Rouchet, 1978) (Section 1.1 of Part 2). But the chief factors in trapping are of a geometric and dynamic nature.

3.4.1. **Closure and trapping concept**

A zone is called 'closed' and provides a trap when it is bounded by a barrier that forces the hydrocarbon flow to accumulate upstream. The barrier is provided by the concave basal surface of a seal or formations with a high injection pressure. To a certain extent, closure, on a local scale, is comparable to a cap rock on the regional scale. A trap thus allows the accumulation of oil or gas in its upper part and thus forms a pool. The base of the hydrocarbon accumulation is a more or less flat surface separating the hydrocarbons from the aquifer, and is called the water table. The closure concept must be considered from geometric and dynamic standpoints.

Geometric aspect

In this case, the closure consists of the concave basal surface of the bed forming the seal. In general, different facies are superimposed between the reservoir horizon and the cap rock series. From the geological standpoint, this pattern corresponds to the following :

— in the most frequent case, the surface of a normal stratigraphic series, folded into a dome or anticline (structural trap by folding) (Figure 60),
— the non-synchronous warped surface of a facies variation corresponding to particular deposits, constituting a sedimentation anomaly, such as a reef structure or a sand robe (Frigg) (lithologic or stratigraphic trap),
— a combination of both : in this case, the real closure of the field may be higher than the one measured by seismic reflection, which has a tendency to follow a synchronous deposition surface.

In the case of faulted anticlines, the fault may or may not participate in the closure, depending on its tightness (Figure 61).

In the case of a fault trap, part of the reservoir closure is provided by the contact with an impermeable formation. In this case, minor adjacent faults often play a significant role in the distribution of closed volumes and fluids. A rather complex example is offered by the Jordanton Field in Texas, at the level of the productive limestones of the Albian Edwards Formation. The structural map of this horizon (Figure 62) shows the distribution of gas and oil zones in accordance with secondary features, above the closed surface (Cook, 1979).

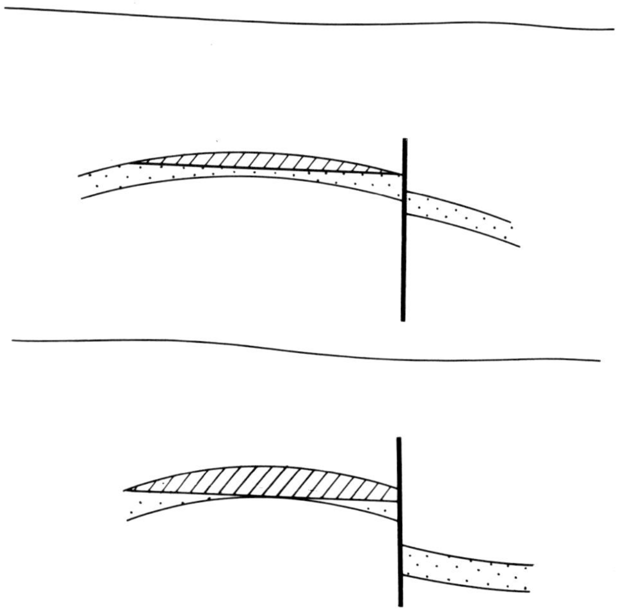

Fig. 61. — Examples of closures :
— simple anticline with fault, probably not sealed (above),
— on sealed fault (below).

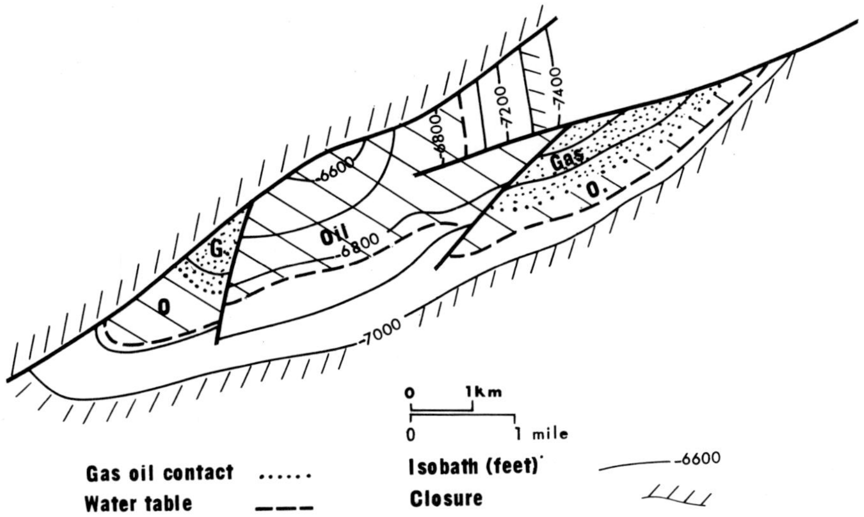

Fig. 62. — Structural map of Jordanton Field, Texas, example of trap against fault, showing areas impregnated with gas, oil and closed (Cook, 1979).

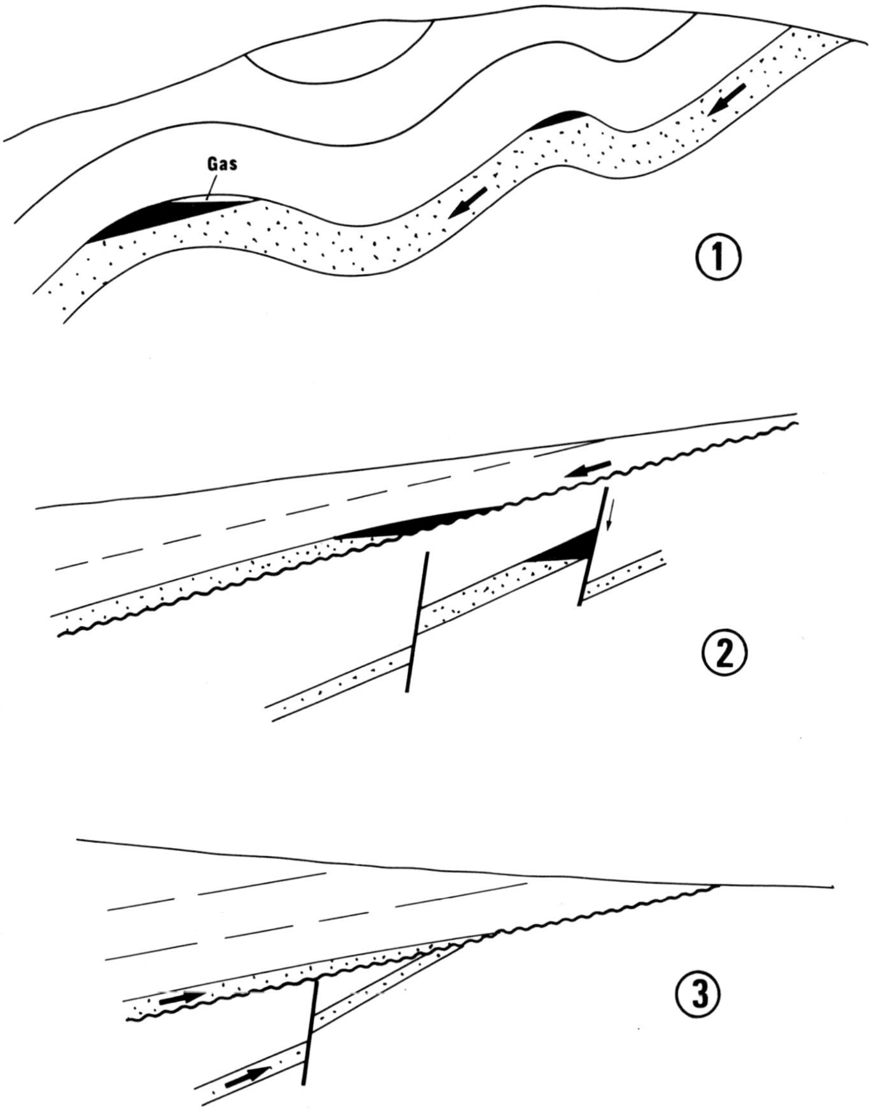

Fig. 63. — Influence of hydrodynamic gradients :
— in folded zone (1),
— on pinchouts and favorable (2) and unfavorable (3) fault blocks.

Dynamic aspect

From the dynamic standpoint, the closure must be considered as a barrier to the migration of hydrocarbons, in other words, the passage from a porous and permeable facies to a fine-grained sediment exhibiting an injection pressure which is greater than the outward thrust exerted by the fluids. Thus the properties of this barrier are related to the fluid characteristics, and especially the dimensions of the molecules and the pressure exerted. From this standpoint, two major types of closure may be distinguished schematically :

— closures related to a flow perpendicular to folded beds in an anticline,

— closures related to a flow parallel to the beds, with which we may associate fault closures, corresponding to a flow not parallel to the beds.

The former correspond to concave closures of anticlinal traps, and from the dynamic standpoint, are identified by the characteristics of the cap rocks. These are more numerous and more effective, because if the first bed does not display high interfacial tension, there is a good possibility that another level possesses a sufficiently high internal pressure. In this case, it is the bed exhibiting the highest interfacial tension that prevails.

The latter, characteristic of non-anticlinal traps, stratigraphic traps, fault traps etc, are more subtle and closely dependent on the hydrodynamic or geomechanical context. In this case, the barrier is essentially created by the injection pressure of the stratigraphic levels or fissures prolonging the reservoir rock, and just

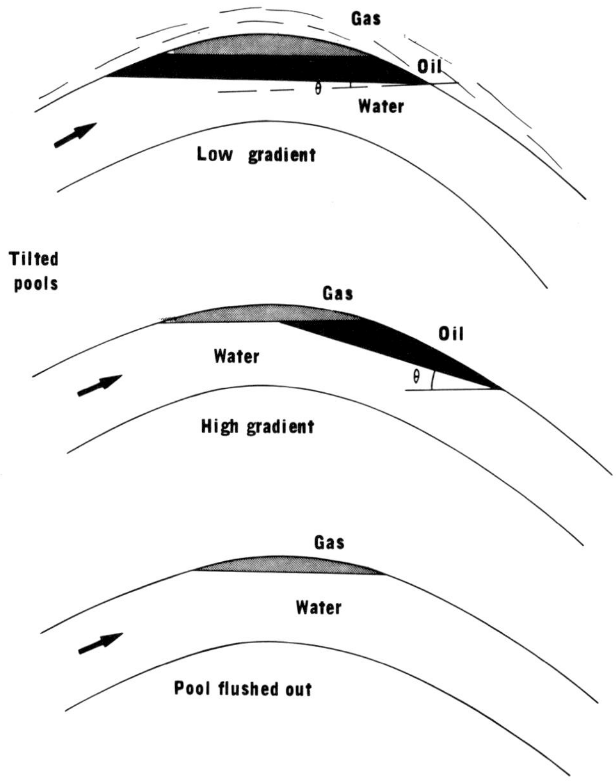

Fig. 64. — Examples of hydrodynamic trapping.

(A)

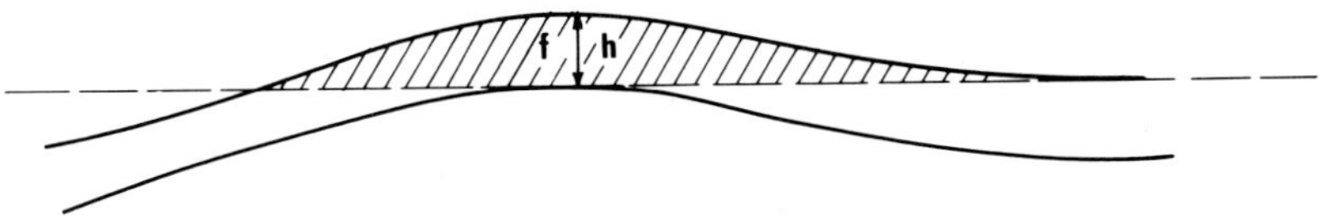

The oil or gas column "h" is the same as the structural closure "f". Filling ratio equal to 1.

(B)

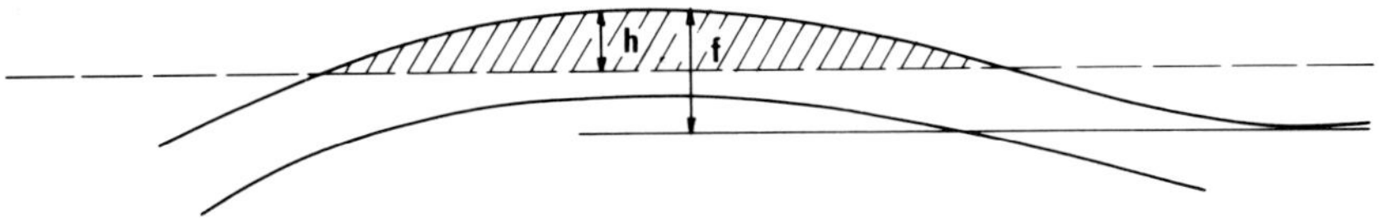

The oil or gas column is less than the structural closure "f". Filling ratio smaller than 1

(C)

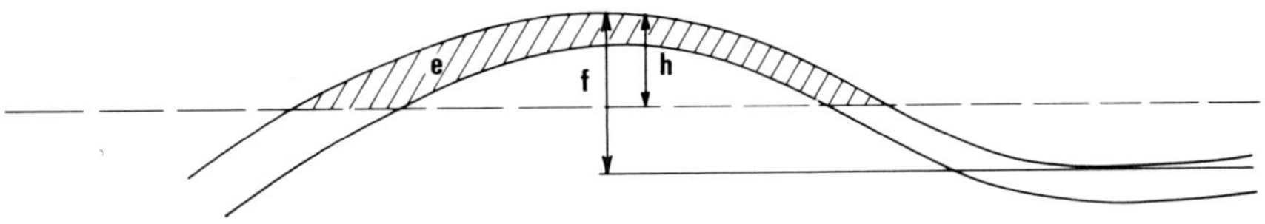

The oil or gas column is less than the structural closure "f" and the thickness of the reservoir is less than the impregnated height.

(D)

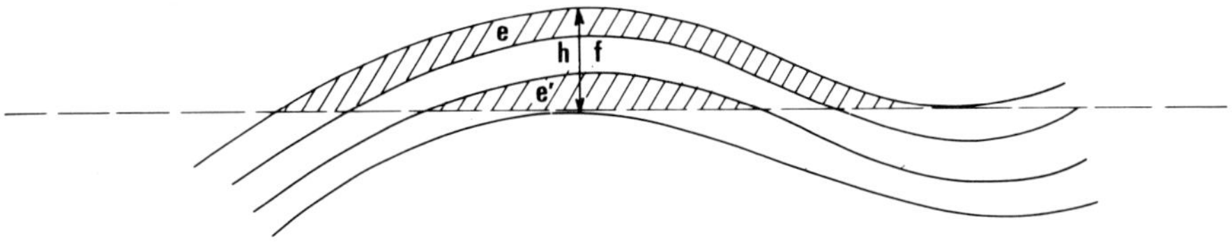

Multiple layer traps with single oil / water contact, the column is equal to the structural closure.

(E)

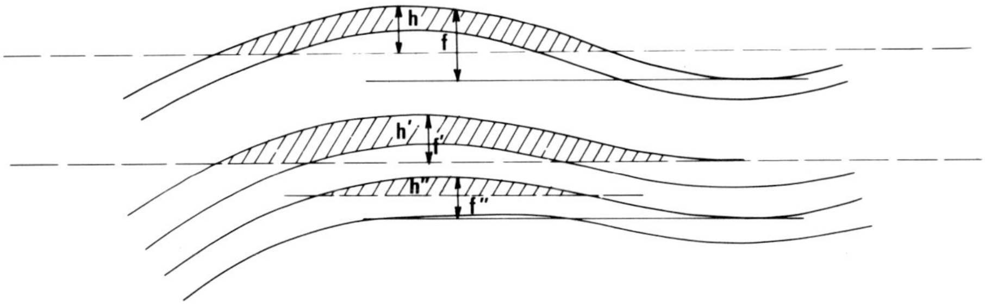

Multiple layer traps with separate oil / water contacts, the total oil or gas columns (h + h' + h") is greater than the largest structural closure (f).

Fig. 65. — Examples of different types of closure.

as a pinhole suffices to deflate a balloon, a single more detrital bed or a single fissure exhibiting a lower injection pressure can allow the oil or gas to migrate. In these conditions, any external factor, decreasing hydrodynamic gradient or geodynamic pattern in compression, can play a fundamental role.

Like the cap rocks, closures thus appear to be effective with certain fluids and not with others, depending on pressure and temperature. This implies a sequence of hydrocarbon inflows and outflows, and in most cases a trap may be considered to be filled to the extent that it is supplied from downdip at a rate substantially equal to that of its updip losses. In these conditions, it is clear that the situation of the trap in the basin, and the characteristics of the basin, condition its filling or drainage. It may be noted in particular that the zones that generate hydrocarbons, chiefly areas undergoing subsidence, offer greater possibilities of containing full traps than recently uplifted areas. In this pattern of fluid dynamics, the aquifers (and the hydrodynamism) play a particular role.

Closures and hydrodynamism

In cases of traps which are not formed by folds, the hydrodynamic gradient parallel to the carrier beds or to the fault planes is also parallel to the preferential drainage paths of the hydrocarbons, and exerts considerable action on the effectiveness of the closure. If its direction is opposed to the oil flow, it has a tendency to prevent migration and reinforces the closure. If not, it tends to diminish and even nullify it (Figure 63). Without attempting a complete explanation, it is worth noting the gas accumulations in the low porosity/permeability reservoirs of the synclinal areas of some basins (San Juan, Alberta), where the centripetal hydrodynamic gradients could contribute to trapping in an open pinchout context (Section 4.3. of Part 2).

On a far smaller scale, the anomalous trappings characterized by the presence of gases, downdip from an oil accumulation, within stratigraphic traps, have been described in the Pennsylvanian sands of Kansas. This situation could be explained by the application of differential trapping to the stratigraphic traps (Mannhard & Busch, 1974). In the case of fold trapping (Figure 64), the gradient parallel to the beds has the effect of shifting the oil accumulation sideways by tilting its base surface or water table, in accordance with Hubbert's Law (1953) :

$$\tan \theta = \frac{dh}{dx} \, \frac{dw}{dw - dh}$$

where :

θ = angle made by the water/oil table with the horizontal,
dh/dx = hydraulic slope of the aquifer,
dw = density of water,
dh = density of the hydrocarbons.

This formula first shows that the tilt of the water table is greater with a higher hydrodynamic gradient, and it is perfectly reasonable to believe that, in many cases, this gradient expels the oil 'bubble' outside the trap. In this case it is said that the pool has been flushed. In actual fact, the accumulation, tilted or shifted on the limb of an anticline, appears as an intermediate solution, a sort of equilibrium, between the hydrostatic gradient and the oil flushing. Hubbert's rule also shows that tilting is more pronounced when the oil and water densities are comparable and closer to specific gravity 1. In the extreme case, this would imply the absence of any separation between the two fluids if their densities tend towards the same value, because buoyancy forces would have no effect in this case. The application of this formula also implies that the environment is uniformly permeable (tilts could be caused in hydrostatic régimes by porosity variations), and that the oil remains fluid.

Practical closure concepts

A distinction must be drawn between the theoretical closure and the practical closure or impregnated column height (Figure 65). The theoretical closure is that which can be measured from geological data, generally structural data about the trap, provided by seismics and/or boreholes. It offers a first assessment of the order of magnitude of the reserves which may be expected to be found in the trap. The practical closure is "the pore space between the underlying water table and the overlying roof rock" (Levorsen, 1956) or the height really impregnated in the closed volume of the trap; it is related to the concept of a pool. The impregnated column height may be different from the geometric closure. The cumulative column height of several impregnated reservoir beds, separated by seals and possessing distinct water tables, may be greater than the structural closure. If the water table is a common one, it remains lower or at most equal.

In most cases, and whenever only one impregnated bed is present, the oil or gas column height is equal to or less than the structural closure. In practice, the definition of a structural closure is often delicate. To begin with, it is difficult to identify exactly the horizon which constitutes the true cap rock in the field conditions. Not all rocks that are not reservoirs perform the function of seals, and many beds absorb hydrocarbons without either stopping them or allowing them to flow : these come under the category of false seals (Figure 66). The closure concept is especially vague in the case of fault trapping, partly because the tracing of the latter is generally imprecise, and chiefly because the tightness of a fault is extremely variable according to the subsurface conditions, and especially the formation pressure. Very often, in the case of faulted anticlines, the true closure is limited at the fault, especially in the case of transverse or recent faults (Figure 61).

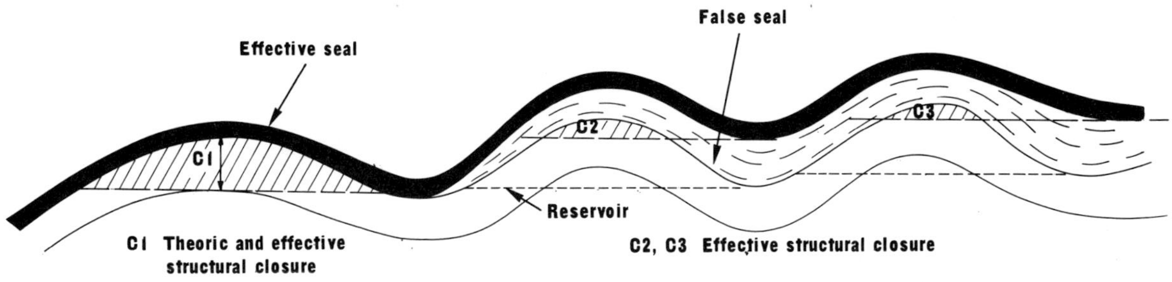

C 1 Theoretical and effective structural closure

C 2, C 3 Effective structural closure

Fig. 66. — Examples of real structural closures in the case of effective and false seals.

Filling factor

The ratio of the impregnated column height to the structural closure defines the filling factor. It is equal to 1 if both terms are equal, and generally lower than 1 if not. In rare cases it may be greater than 1, when stratigraphic mechanisms are added to the structural data.

This factor is generally high in rich zones, low in poor zones. It depends on the hydrocarbon flow, which supplies the trap and which may also drain through the seal. The inflow may be limited by the volume and saturation of the neighboring source rocks, and by the drainage radius of the trap. The reefs of the Rainbow-Zama region, in northwestern Alberta, may appear to offer a good example of a filling factor determined by the ratio of the volume of the traps to the inflow. The volume of the traps varies according to the type of reef structure, while the volume of hydrocarbons available to replenish the trap is relatively

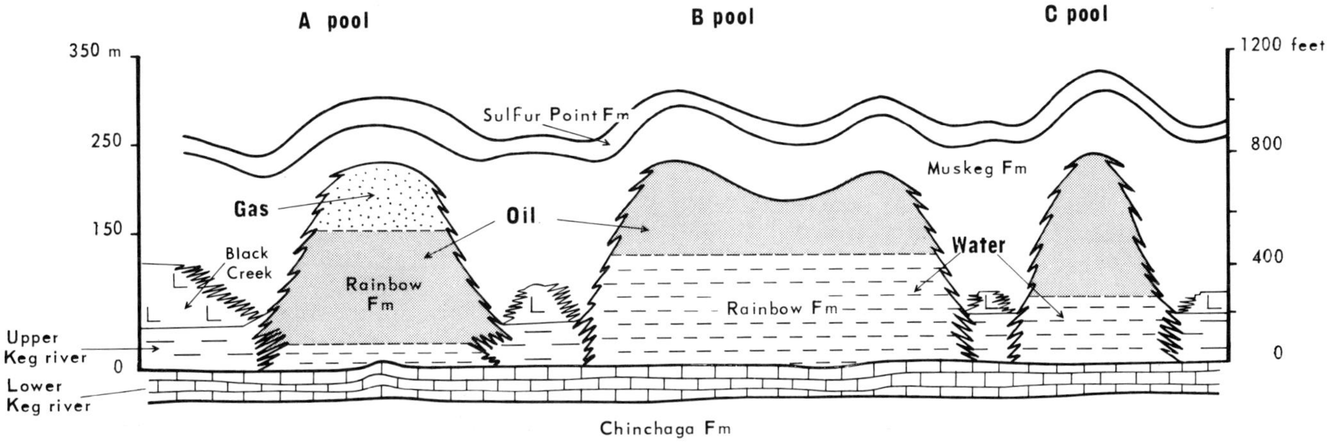

Fig. 67. — Section of pools A, B and C of Rainbow showing irregularities in gas/oil and oil/water table distribution (Langton & Chin, 1968).

constant. This gives rise to variable impregnated column heights and hence filling factors (Figure 67). Other authors believe that this pattern is the result of different drainage surfaces (Dumestre, 1969).

Seepage through the seals favors the escape of light hydrocarbons and gases, and the fluids retained tend to become progressively heavier. The filling factor is also altered by variations in temperature and pressure, often related to uplift and subsidence movements, which affect the fluids. Hence subsidence movements tend to increase the volume of dissolved gas, while positive movements may create and expand a gas cap, and increase the volume and filling factor of the field (Figure 68).

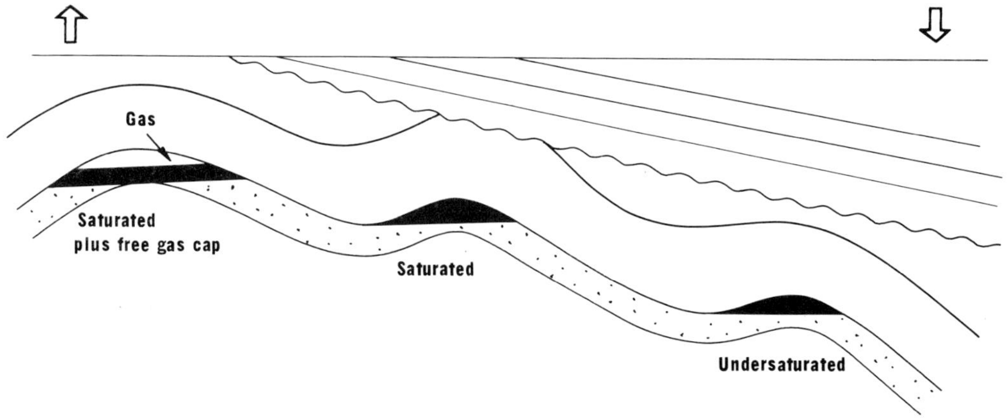

Fig. 68. — Distribution of a given mixture of oil and gas conveyed to different pressures and depths (Levorsen, 1954).

3.4.2. **Pool and reserve concept**

The concept of trap and closed volume introduces the concept of a pool, which is a trap whose closed volume is partly or totally occupied by hydrocarbons. This volume of trapped oil and/or gas constitutes the pool's reserves in place. These reserves are calculated, in a first approximation, from the volume of voids within the trap occupied by hydrocarbons. If V is the volume of impregnated rock, Φ the porosity of the reservoir, and Sw the saturation of irreducible water, the reserves are given by the formula :

$$R \text{ oil in place} = V \times \Phi (1 - Sw)$$

Before determining the impregnated volume, the capacity of the trap (or the reserves which it is likely to shelter) can be estimated by the same formula. Depending on the petroleum context, assumptions are made concerning the filling factor. This approach is necessary to assess the economic value of a prospect. Reserves identified by wells are proved reserves in place, also called 'accumulations'.

The production of a pool only allows partial recovery of the oil in place. The percentage recovery varies for oil from a few percent (average 25 % worldwide) to 50 % and more in the best instances, depending on the characteristics of the reservoir (porosity, permeability, pressure) and the fluids (hydrocarbon fluidity, aquifer régime). The gas recovery factor is around 80 %.

These recovered or producible quantities are the recoverable reserves. They are calculated from reserves in place by the formula :

$$R \text{ recoverable} = R \text{ in place} \times RF \times FVF$$

where RF is the recovery factor, and FVF the formation volume factor, itself dependent on the GOR (Chapter 1 of Part 2). A field corresponds to a pool, or to a group of pools belonging to the same geological unit (different horizons of an anticline, flank pinchouts etc.).

Summary and conclusion

Hydrocarbons, which generally occur in the fluid state, move within permeable water-bearing rocks, chiefly under the effect of buoyancy. Secondary migration, which appears to occur mainly in the form of oil streams, in other words in a continuous phase, follows the reservoir horizons or fissures, and may or may not be parallel to the beds. If the oil or gas fails to encounter a barrier opposing its movement, it reaches the surface and is lost, after having produced seeps. If not, it accumulates at depth, where it is concentrated in the form of pools.

4. – THE MAJOR TYPES OF TRAP AND POOL

« Pour eux (les naturalistes), la notion fondamentale est celle de la réalité du fait observé ; peu importe que l'on en connaisse ou non l'explication, que l'on comprenne ou ne comprenne pas, le fait d'abord, le fait prime tout. »

'For them (naturalists), the fundamental concept is the reality of the observed fact. Explanation and understanding are inessential. It is the fact that counts, the fact above all.'

Marcel Roubault, 1972.

According to the argument so far, the different types of trap and pool can be explored by starting with the simplest, the primary traps, directly associated with primary migrations, and then dealing with traps supplied by secondary migrations. This leads to a distinction between :

— lithologic traps, in which the reservoir is simultaneously the trap, and which may be called trap-reservoirs, like reefs and sand mounds,
— stratigraphic traps, formed chiefly by sedimentary and paleogeographic mechanisms, such as pinch-outs and unconformities,
— structural traps associated with synsedimentary or postsedimentary formation deformations, producing folds and fault blocks.

In this chapter we nevertheless make the classic distinction between the two major types of trap and pool qualified as :

— structural,
— stratigraphic,
— we must also add the miscellaneous and unconventional traps, which do not fall easily under the above two headings, and which involve large hydrocarbon resources often qualified as unconventional.

In actual fact, these different structural, stratigraphic and other criteria are very often closely associated, if not combined, producing complex traps, qualified as combination traps when the predominating factor cannot be identified.

This chapter consists chiefly of short monographs covering pools which are fairly well known, and which may to a certain extent be considered as exemplary. These descriptions often highlight the limits and the difficulties of classification. Hence the sometimes apparently repetitive aspect of certain pages. But then reality itself is often deceptively redundant.

Different statistics describe the distribution of pools. Percentages sometimes vary substantially according to the populations analyzed, while structural traps, and especially anticlines, largely predominate. Two statistics covering the 45 giant fields (over 70 Mt — 500 million barrels — of oil or gas equivalent) of North America (NA) and the 306 major fields (over 14 Mt — 100 10^6 bl — or equivalent) of the United States (US) yield the following results in percentage of the number of pools (Halbouty, 1976).

	NA (giants)	US (major)
structural	49 %	78 %
stratigraphic	31 %	11 %
combination	20 %	11 %

The difference between these two groups stems largely from the weight of the Canadian reef reservoirs.

If one considers the statistics based on the 266 giant fields of the world, a considerable advantage emerges on the side of structural traps, which account for 89 % of the total number of giants, compared with 10 % for stratigraphic traps. The Middle East, where the formations are nearly exclusively anticlines, weighs heavily in favor of structural traps (Halbouty, 1970).

4.1. STRUCTURAL POOLS

The closure of this category of pool is provided essentially by structural criteria, folds and faults, creating two major types of trap :

— by folding, in which the closure is provided by the concave surface of the reservoir / cap rock interface,
— by fault, in which the fault itself directly or indirectly creates the permeability barrier.

It is customary to identify a *third group,* related to salt and shale intumescences, and including folds, closures against faults and against domes.

4.1.1. Fold traps or dome and anticline pools

These traps are by far the most numerous and the largest, and they form the large majority of the giant fields. They are also the easiest to identify, both by the surface geology and by seismics. This has contributed to the formulation of the anticline rule, which still remains, if not the rule, at least one basic criterion for exploration.

Anticline traps exhibit widely varying forms, from the simplest, produced by a barely pronounced arch, to the more complex, often faulted, injected with salt or shale. The causes of the formation of these domes and anticlines are themselves extremely varied, from the simple draping of a horst to the most complex deformations of tangential tectonics. Sedimentary and stratigraphic processes, especially unconformities, often related to the deformations that caused the fold, in which transcurrent movements often play an important role, can complicate these structures even further.

In the following pages, we shall describe a number of anticline fields going from the simplest to the most complex :

— broad domes,
— "Appalachian" anticlines,
— disharmonic anticlines,
— unconformity anticlines,
— faulted anticlines,
— overthrust anticlines.

4.1.1.1. *Broad domes*

The simplest domes and anticlines, which are also the most numerous, consist of arches with a wide radius of curvature. In these cases, variations in thickness of the different beds of the sedimentary series may give rise to movements which offset the position of the closed zones at depth. The detection of these traps requires very fine seismics as well as a thorough knowledge of the productive horizon. Among illustrations of this type of trap are the following fields :

— Forties, in the British zone of the North Sea,
— Ekofisk in the Norwegian zone, very probably related to a deeper salt dome and to more pronounced dips (Figure 186),
— Romashkino (Figure 176) and the large neighboring fields of Tuimazy and Arlan (Figure 177) in the Volga Urals basin of the USSR,
— Samotlor, Urengoi and the other giants of Western Siberia (USSR),
— Ghawar, Burgan (Figure 170) and the other giant fields of the Arabian platform.

Certain anticlinal areas correspond to regional high zones very often affected by shear faults, and thus display a certain stratigraphic and structural complexity, although the dips generally remain very low, around 1 degree or less. This is the case of a large number of giant fields, and especially the largest :

— Ghawar (2300 km^2 closed area),
— Samotlor (2000 km^2),
— Romashkino (3000 km^2),
— Hassi Messaoud (1300 km^2),
— Hassi R'mel (2600 km^2).

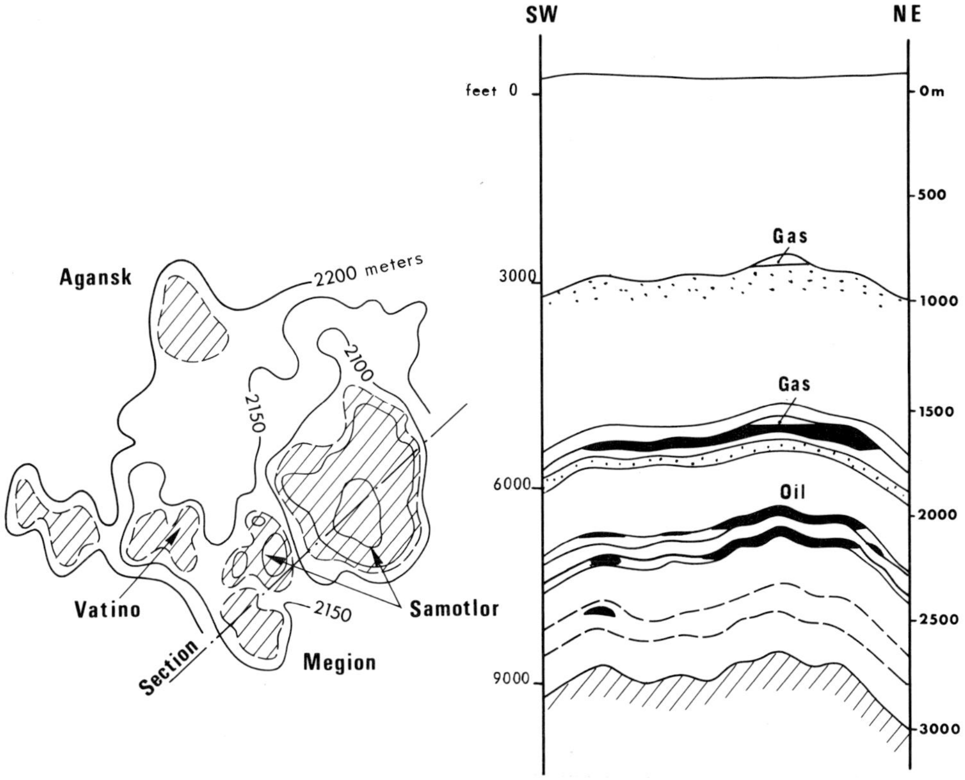

Fig. 69. — Map and section of fields in the Samotlor zone, Western Siberia (Rovnin *et al.,* 1975).

The Samotlor Field, the USSR's largest, is located in the central part of the vast sedimentary basin of Western Siberia. It is situated on a broad dome where the crystalline basement rises to 3000 m. The trap is created by an anticline with a large radius (Figure 69). The reservoirs are represented by Cretaceous sandstones, with seven levels within a section of more than 600 m at a depth between 1500 and 2100 m. They correspond to shallow marine platform deposits. Permeability is up to 500 md, the oil column is 50 m, and cap rock is provided by 40 m of shales. The productive area of the main pool covers 334 km^2. The formation pressure is 245 bar at 2000 m, and the temperature is 88 °C. Average output per well is 114 t day^{-1}, with values up to 1000 t day^{-1}. The field produced 150 Mt in 1980, and reserves have been estimated between 1.5 and 2.2 Gt (1 Gt of cumulative production at the end of 1980).

The Groningen Gas Field, the largest in Western Europe, was discovered in 1959 (Figure 70). The trap consists of a regional high zone with an area of 775 km^2, plunging steadily towards the north, and cut by many faults and horizontal strike slips dating from the Carboniferous, post-Carboniferous and post-Permian. It appears that an initial shoal developed in the end of the Carboniferous, and more clearly in the Uppermost Jurassic. The stratigraphic series is interrupted by three large unconformities. The reservoir is formed by Rotliegendes sand formations of the dune and fluviatile type, overlapping an eroded Carboniferous. The reservoir, which thickens from 30 m at the south to 75 m at the north, exhibits porosities of 15 to 20 % and permeabilities of 0.1 to 1 darcy. It is surmounted by shales and by a thick series of Upper Permian evaporites, 600 to 1500 m thick, forming the seal. The Lower Cretaceous, which onlaps on to the Trias, marks the return of the sea after a hiatus and a period of considerable erosion. The overall Cretaceous exhibits a thickness of 400 to 1000 m. The Tertiary, which succeeds a new period of faulting accompanied by uplifts and erosion, and the Quaternary, are 400 to 1000 m thick.

The source rock of the gas appears to be the Upper Carboniferous coal formations, and possibly the Zechstein bituminous Kupferschiefer formations. The gas mainly appears to originate in the syncline zones located to the east, buried deeper and more mature than the beds situated vertically below the field (Bartenstein *et al.,* 1974, Stahl, 1977). The gas has the following composition :

81 % methane,

4 % ethane and higher alkanes,

14 % nitrogen,

1 % carbon dioxide.

The initial pressure of the pool was 350 bar at 3000 m depth and the temperature was 107 °C. Initial reserves were estimated at 1830 Gm^3 (Bungener, 1969; Stauble & Milius, 1970). About 800 Gm^3 were extracted

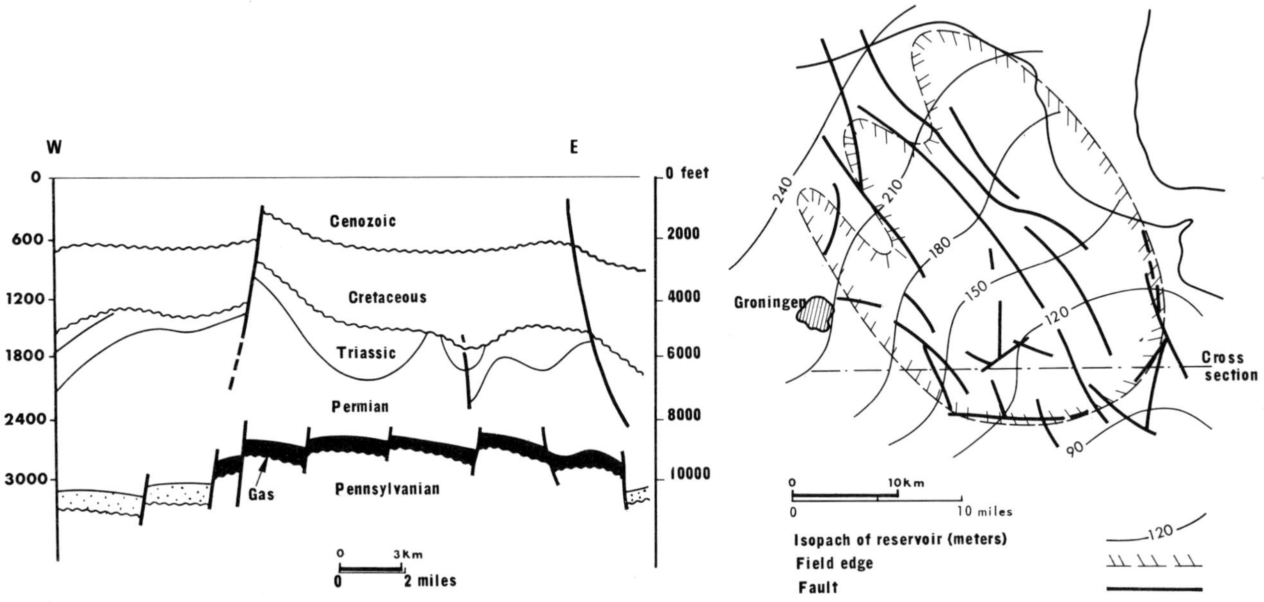

Fig. 70. — Map and section of Groningen Field (Stauble & Milios, 1970).

at the end of 1981. The size and potential of this field is related not only to the dimensions of the closed volume of the trap, but to its high filling factor, reflecting a copious supply of hydrocarbons as well as effective sealing. This difference is illustrated by a comparison between sections of this accumulation (Figure 70) and the offshore zone, which only contains relatively minor pools (Figure 71).

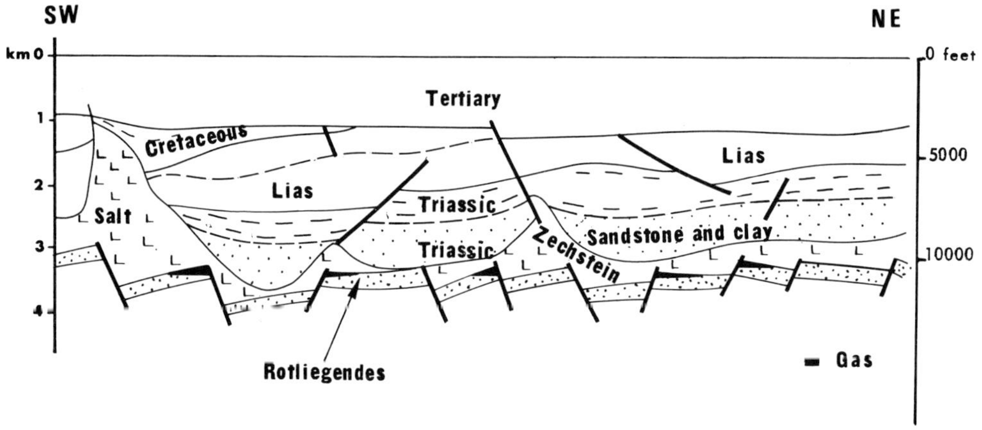

Fig. 71. — Section showing features of Rotliegendes gas pools on Netherlands continental shelf : only the upper part of the fault blocks is impregnated.

The Conroe Field (Texas), which initially contained 100 Mt of oil, is another example of a slightly oval dome intersected by transverse faults, probably associated with a deep salt dome. The same group includes the Grondin pool (Gabon), discovered in 1971 at a distance of 40 km offshore (Figure 72). The trap is formed by a faulted anticline, cut in its central part by a small graben. A deep salt dome lies at the origin of the structure. The reservoirs are formed by Campanian/Lower Maastrichtian sandstones (Batanga Formation) that display an effective thickness of 25 to 80 m, mean porosity of 25 %, and permeabilities up to several hundred millidarcys. They correspond to marine deposits of subaquatic detrital fans. Well output has reached 700 t day^{-1}. Sealing is provided by the Upper Maastrichtian/Paleogene shales, and the source rocks consist probably of Turonian shales rich in organic matter. Initial reserves are estimated at 28 Mt of oil with density 28 °API (Vidal, 1980 and unpublished SNEA (P) documents).

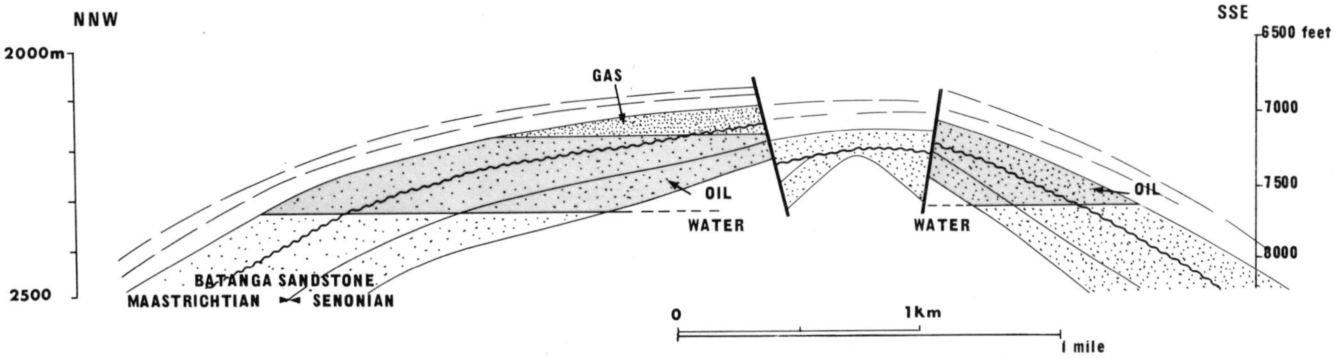

Fig. 72. — Section of Grondin Field, Gabon (Vidal, 1979).

4.1.1.2. "*Appalachian*" *anticlines or* "*whalebacks*"

Anticlines with pronounced dips also form very important traps. Outstanding examples are available in the foothills of Iraq and Iran, in the Rocky Mountains, and in California. In this type of trap, the age of the tectonic process is very important. A single and recent folding phase affecting a competent sedimentary series yields simple folds, often well reflected by the topography.

Examples are the anticlines of the southeast part of the Zagros foothills zone in Iran. The sedimentary series, over 7000 m thick, extending from the Permian to the Pliocene, is continuous. The lithologic and mechanical characteristics are reflected by folds exhibiting a wavelength of 12 to 20 km with amplitudes of 2 to 3 km (Kent & Warman, 1972). The youth of these folds, formed since the Miocene until the present day, is reflected by close conformity of the structures and the topography, and the nearly complete absence of erosion. This yields a relatively simple framework of high and broad anticlinal arches, separated by narrow and more complex synclines. Conversely, deeper erosion would have brought into evidence narrower and faulted anticlines between very gentle synclines (Figure 73) (Colman-Sadd, 1978).

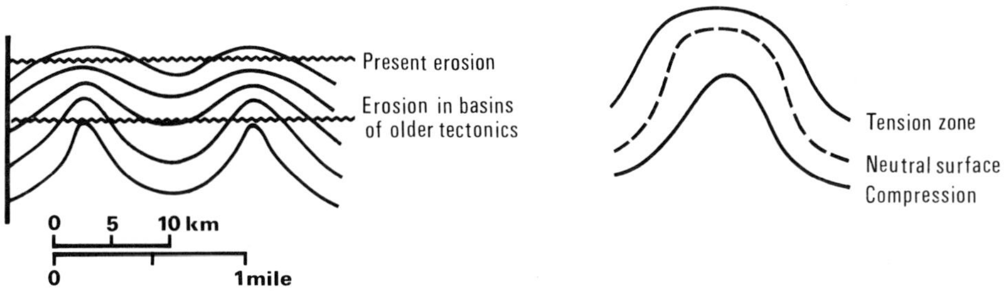

Fig. 73. — Competent folded series of the Zagros foothills type showing :
— (left) features of anticlines and synclines according to degree of erosion ;
— (right) distribution of extension and compression zones (Colman-Sadd, 1978).

The Kangan gas field and its neighbors (Nar, Agar, etc) are magnificent examples of the first type (Figure 74). These pools consist of vast anticlines which the Asmari limestones mould on the surface, long structural arches 8 to 12 km wide and 1.5 to 3 km high (Plate 3). Breaking this imposing regularity, distension mechanisms cause the formation of grabens in the axis of many of these anticlines. In addition, transverse faults displace these folds and affect the pool closures. The gas is found in the Permian carbonates (Khuff Formation), effectively sealed by anhydrite evaporitic series with high internal pressure (Sudair Formation) (Reyre, 1975). Reserves are estimated at more than 330 Gm^3.

4.1.1.3. *Disharmonic anticline*

It is rare when a folded saliferous series does not give rise to deep disharmonies between the upper series and the underlying beds, in which productive reservoirs sealed by evaporites are often found. For example, the Saliferous series of the Fars in the foothills of Iraq and Iran introduces a mechanical discontinuity which is reflected by very deep disharmonies between the deep productive series and the surface beds,

PLATE 3

Anticline architecture and movements. Recording of deformations

Fig. 1. — Althar whaleback anticline. Asmari limestones. Zagros foothills, Iran.

Fig. 2. — Subvertical striae in Daryan limestone. Zagros foothills, Iran.

Fig. 3. — Surmeh anticline. Asmari limestones. Zagros foothills, Iran.

Fig. 4. — Tensile joints in the flysch formations of the Côte Basque (France).

PLATE 4

Mud volcanoes due to seepage of gaseous hydrocarbons

Fig. 1 to 4. — Mud volcanoes on the Berca anticline (Romania).
Surface manifestations of seepage of gaseous hydrocarbons from deep layers, carrying along clays and lumps of rock.
(Photos A. Perrodon)

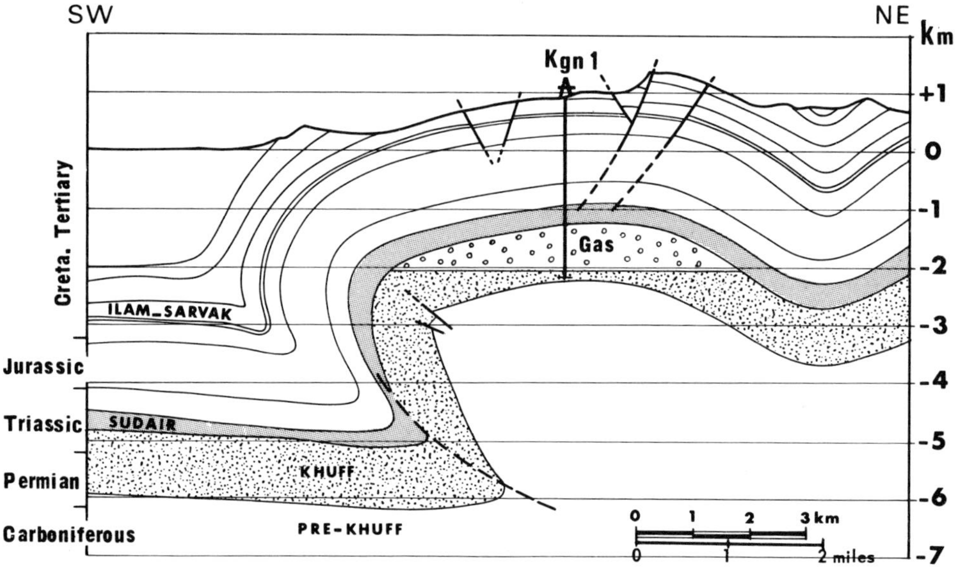

Fig. 74. — Cross-section of Kangan pool (Iran) (Reyre, 1975).

yielding the characteristic asymmetrical structure of this region. While these structural complexities have made the geologist's task more difficult, they in no way affect the potential of the fields in the region, far to the contrary. In these anticlines, the neutral surface separating the distension zones from the compression zones is located relatively deep, approximately around the Jurassic. The distensive forces improve the reservoir characteristics and facilitate the migration of hydrocarbons, which are nevertheless perfectly trapped by the presence of excellent evaporite and shale seals.

As is well known, this region of the Middle East contains some of the world's biggest and most productive fields, including those of :

Kirkuk, 2100 Mt,
Gash Saran, 2000 Mt (Figure 75),
Marum, 1500 Mt,
Aga Jari, 1900 Mt,
Ahwaz, 2400 Mt,
Pazanun, 500 Mt and 1400 Gm^3 of gas.

These disharmonies may extend as far as complete structural inversions, in which the deep anticline is surmounted by a syncline zone, as illustrated by the Permian fields of the Anglo-Dutch basin, in the southern part of the North Sea, and especially by the West Sole and Indefatigable Fields (Figure 76) (Chapter 3 of Part 3).

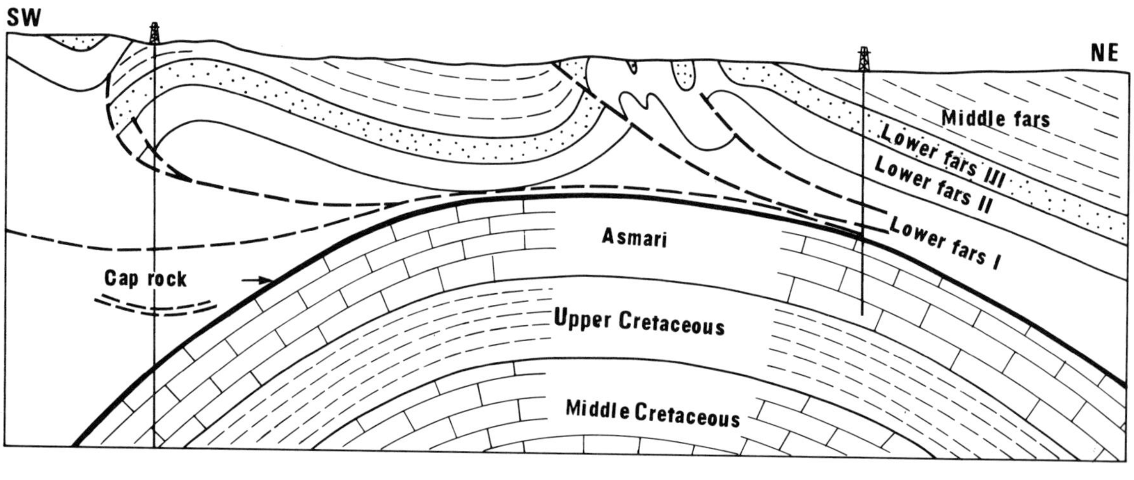

Fig. 75. — Section of Gash Saran pool (Iran) (British Pretroleum, 1956).

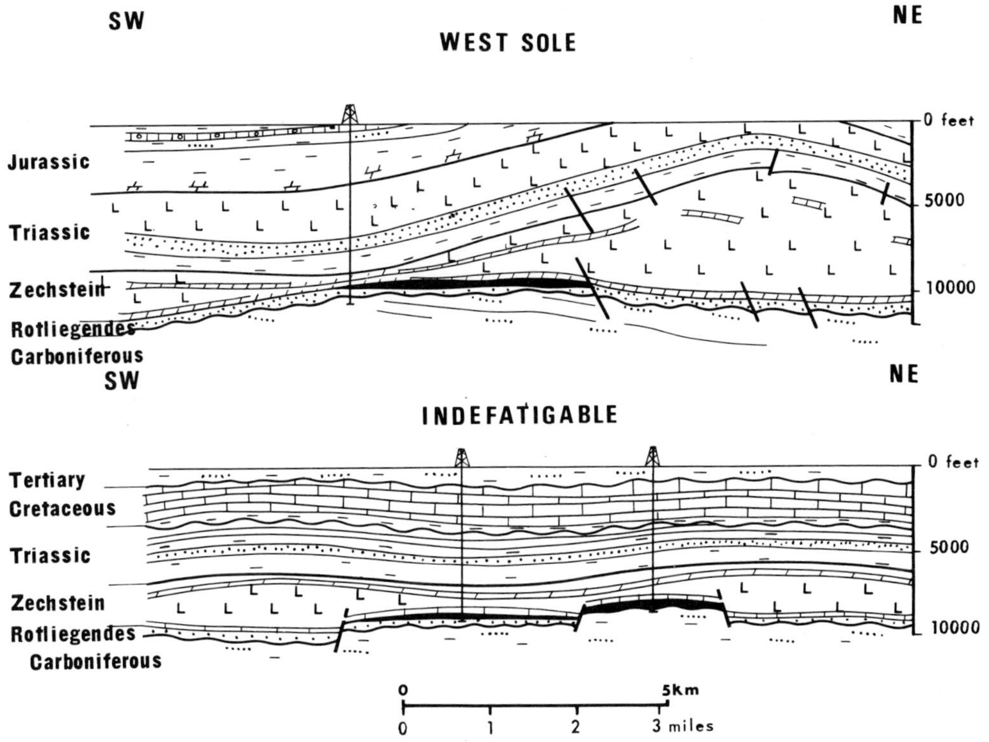

Fig. 76. — Section of West Sole and Indefatigable Fields (Kent & Walmsley, 1970).

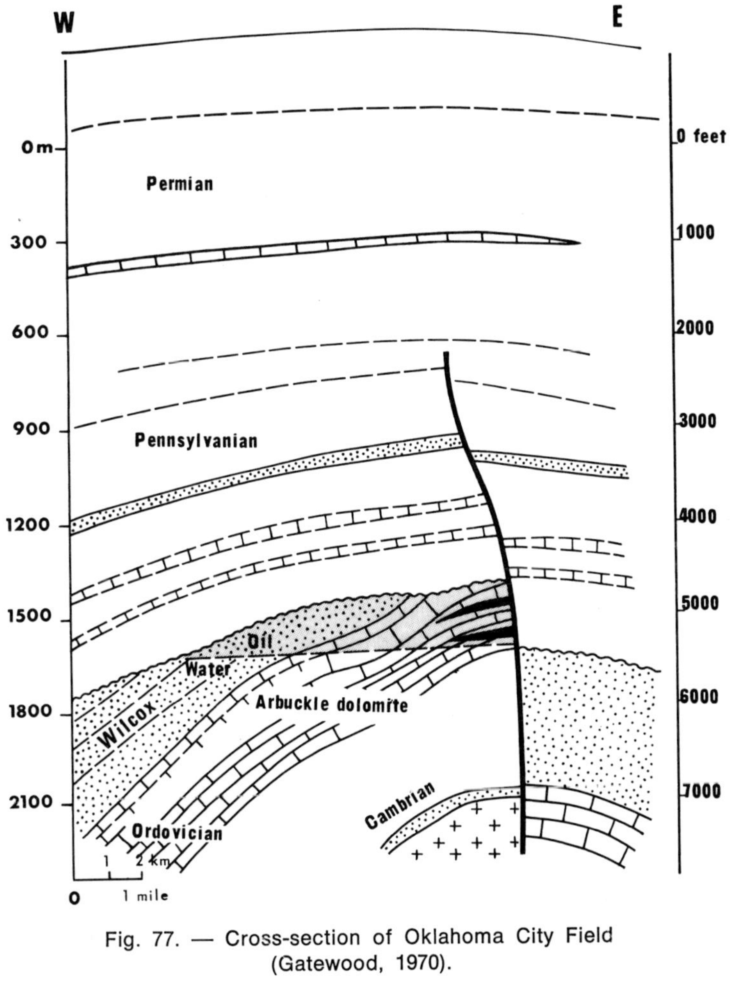

Fig. 77. — Cross-section of Oklahoma City Field (Gatewood, 1970).

4.1.1.4. *Unconformity anticlines*

Unconformities record the coming and going of the sea, or, in other words, the pulsations of an unstable zone. It is perfectly normal for a number of high zones, and particularly domes with a large radius of curvature, which have been uplifted early during sedimentation, to be characterized by local unconformities. To the extent that fairly diversified series are deposited to form the reservoirs and seals, thick enough to guarantee effective sealing, anticline zones bounded by unconformities, or simply marked by sedimentary thinning, but marking a former high position, generally represent favorable zones and traps. On the other hand, excessively long gaps, resulting in bald anticlines, are a negative criterion. In this case, pinchout accumulations may nevertheless be found on the flanks, such as the Cut Bank Field on the edge of the Sweet Grass Arch (Montana) (Figure 106) (see page 161).

The Oklahoma City field in the Mid-Continent province is a remarkable illustration of this type of anticlinal trap (Figures 77 and 190). This field, discovered in 1928 on a gentle surface anticline, is located on the Nemaha Ridge in Oklahoma. The structure is faulted at depth and marked by a major unconformity of the Pennsylvanian on the Permian. The productive area covers 83 km^2 (32 square miles) with an impregnated height of 300 m. Structural analysis shows that the anticline began to take form in the Cambrian, and was completed by a major folding phase following the Ordovician and some movements following the Permian. The reservoirs consist principally of :

- — Arbuckle dolomite (Ordovician) productive at the core of the structure, with certain wells with high output (record 6000 t day^{-1} — 43,000 bopd) and even very high production (170,000 t — 1.223.868 bbl — of cumulative production for the best well).
- — Wilcox sandstones (Ordovician) above, with the highest potential, which produced more than 2.4 Mt km^{-2} (600 bbl/acre foot) in 1968, for a mean thickness of 30 m (225 ft) with porosity of 22 % : initial recoverable reserves are estimated at about 105 Mt (750 10^6 bbl) of which slightly more than 100 (737 10^6 bbl) have been extracted (1981) (Gatewood, 1970).

Another example of a faulted anticline below an unconformity is illustrated by the large Sarir field (Libya), whose initial reserves are estimated at 1100 Mt (Figures 78 and 206). The Hassi Messaoud field in Algeria, which is productive in the eroded Cambrian sealed by the Triassic onlap, illustrates another type of dome below a plane of unconformity (Figures 79 and 204). The light oil (specific gravity 0.8) was generated in the shale series rich in organic matter of the neighboring Silurian synclinal zones, and expelled under the Mesozoic sedimentary overburden whose thickness is sometimes more than 3000 m (Perrodon, 1966, Balducchi *et al.*, 1970). Reserves are estimated at about 800 Mt. Examples of fields more directly associated with unconformities will be discussed later.

4.1.1.5. *Faulted anticlines*

It is rare for a fairly pronounced anticline to be unaccompanied by a faulted arch. Assigning to the fault traps in the true sense those which are sealed by the fault plane, we shall examine here only those traps in which faults play a minor role. We can place, under this heading, many anticlines of the roll-over type, bordered by a synsedimentary fault, characteristic of deltaic provinces.

The fields of the Vicksburg trend, in the coastal plain of Texas, offer a remarkable illustration of traps associated both with a basin margin synsedimentary fault system and a regional facies distribution. This sequence of pools extends over more than 500 km, a short distance from the Coast, and contains over 420 Mt of 30·10^6 bbl of oil and some 560 Gm^3 — 20 tcf — of gas in Oligocene (Frio) and Miocene sands.

The traps of this unit, all located at the base of faults, are formed by roll overs, and less often by a fault, in particular by small antithetic faults accompanying the main fault and isolating a minigraben. In this case the productive zone is found in a relatively high position (Figure 80). The synsedimentary faults were active mainly during the Oligocene and Upper Miocene, and were completely extinct in more recent horizons. The Oligocene formations (Vicksburg) in particular pass from a thickness of 180 m in the upthrown block to 1200 m on the downthrown side, testifying to intense subsidence in this period at the base of the fault. The largest field of this trend, the Tom O'Connor, discovered in 1934, contains more than 110 Mt — 800·10^6 bbl — of oil and 30 Gm^3 — 1 tcf — of gas, distributed among about fifteen sand horizons. The trap is formed chiefly by a phenomenon of downbending (roll over and reserve drag) (Mills, 1970; Stanley, 1970).

The large fields of the California basins, and especially those of Ventura and Wilmington, illustrate another type of faulted anticline. The Wilmington Field, discovered in 1936 on the coastal zone of the Los Angeles Basin, is a broad asymmetrical anticline cut by a series of transverse normal faults which subdivide the productive beds into separate pools. It is productive in a Miocene and Lower Pliocene sand series bedded

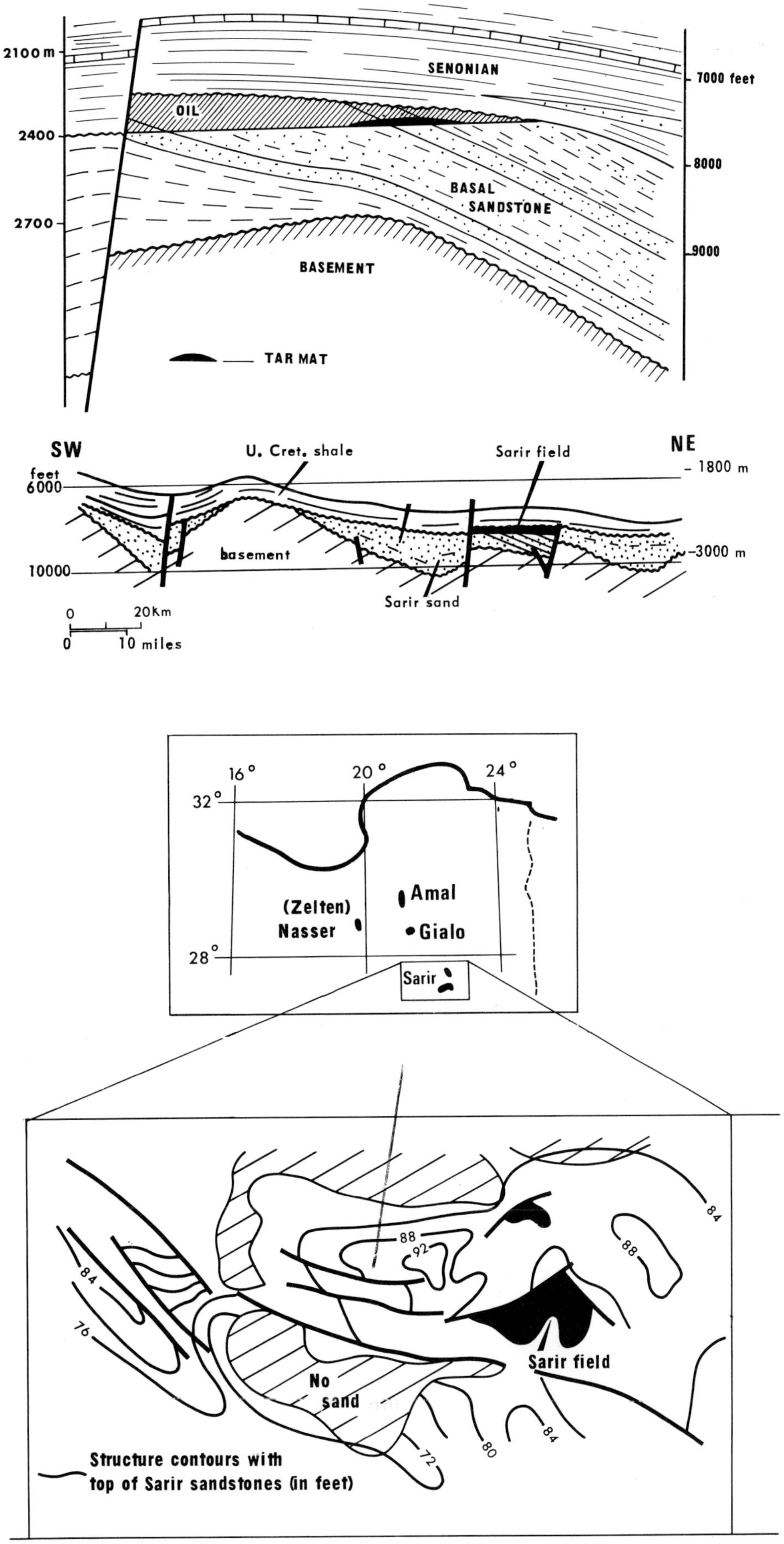

Fig. 78. — Map and section of Sarir Field, Libya (Gillespie, 1967 & Sanford, 1970).

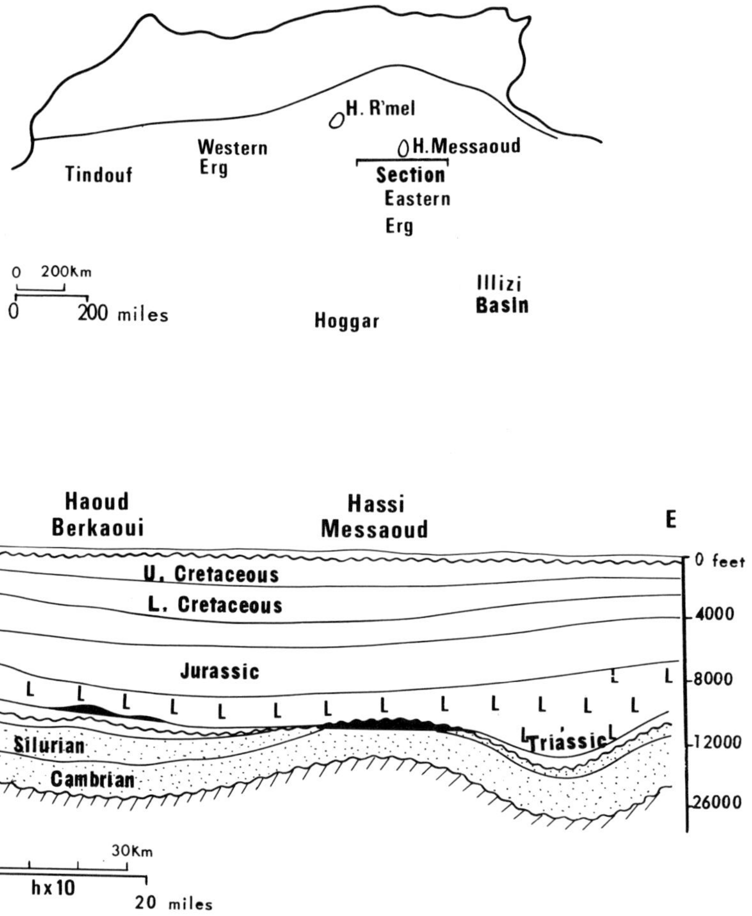

Fig. 79. — Section across Haoud Berkaoui and Hassi Messaoud Fields, Algeria (Balducchi & Pommier, 1970).

with shales (Repetto Formation), intersected by marls forming seals and probably, at least partly, the source rock. This unit is overlaid unconformably by the Upper Pliocene (Pico Formation), which provides an overall cap rock (Figures 82, 161 and 162). The percentage of sands in the productive series varies from 25 to 70 %. The faults provide effective barriers that subdivide the field into distinct pools containing hydrocarbons with different characteristics. The Wilmington Field has recoverable reserves of more than 700 Mt — 3 B bbl — of which slightly more than 330 Mt — 2400 10^6 bbl — have been produced.

The Shebelinka Gas Field, discovered in 1950 in the Dnepr-Donets depression, corresponds to a faulted anticline with a Devonian salt core, productive in Permian fractured dolomites. The faults are sealed and plugged by Upper Permian evaporite formations (Figure 83). This field, the largest in the Ukraine, covers an area of 450 km^2, and its reserves are estimated at 500 Gm^3.

4.1.1.6. *Diapiric anticlines*

The combination of faults and undercompacted shales or deep salt yields brachy anticlines with a diapiric core, assuming the form of mud volcanoes in the former case. A number of Romanian fields in the Ploesti district display these diapiric structural characteristics. In this intensely folded and well-known basin, all forms of diapiric structure are encountered, as the extreme mobility of the salt engenders a very wide variety of forms, which represent the extreme case of traps associated with salt domes (Figures 84 and 163) (see Chapter 2 of Part 3).

The pools of the Baku region are characteristic brachy anticlines, with a core of diapiric shales. The Bibi-Ebat field, located on the southern coast of the Apsheron Peninsula, perfectly illustrate this type. This field, discovered in 1882, is formed by an anticline enclosing seventeen productive sand horizons, in which sands form 70 to 75 % of the series. Its cumulative production is over 150 Mt (Figures 85 and 179) (see Chapter 3 of Part 3).

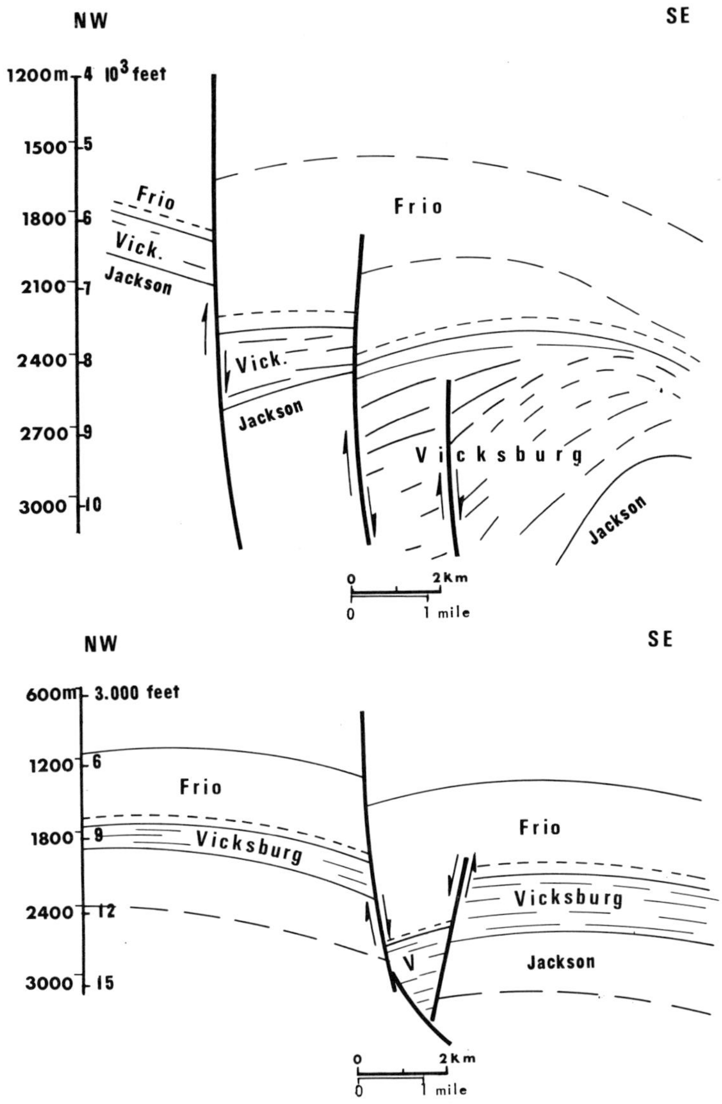

Fig. 80. — Sections of Vicksburg Fault (Texas) showing :
— thickenings and anticlines at fault base ;
— closures against antithetic fault (Stanley, 1970).

4.1.1.7. ***Overthrust anticlines***

The Waterton gas field (Alberta), discovered by seismics in 1957, offers a good illustration of pools of the anticlinal family in an area of tangential tectonics. It is located just at the boundary of the overthrust of the Canadian Rockies, 200 km south of Calgary, near the United States border. The trap is formed by an overthrust fold directed northeast and plunging about 30 degrees towards the southwest, resulting in an elliptical anticline measuring 40 × 8 km in the plan view (Figures 86 and 187). The reservoirs are formed by massive carbonates of the Devonian and Mississippian, partly dolomitized, with an effective thickness of about 50 m at 300 m depth. The mean porosity is 6 % and matrix permeability is 1 md, but many fractures with an average opening of 1 mm allow unit flows of 3 Mm^3 day^{-1} with a Δp of 2 bar (35 psi).

The sour gas has the following composition :

— C_1 .... 67 %
— C_2 .... 4 %
— H_2S .... 16 %
— CO_2 .... 4 %

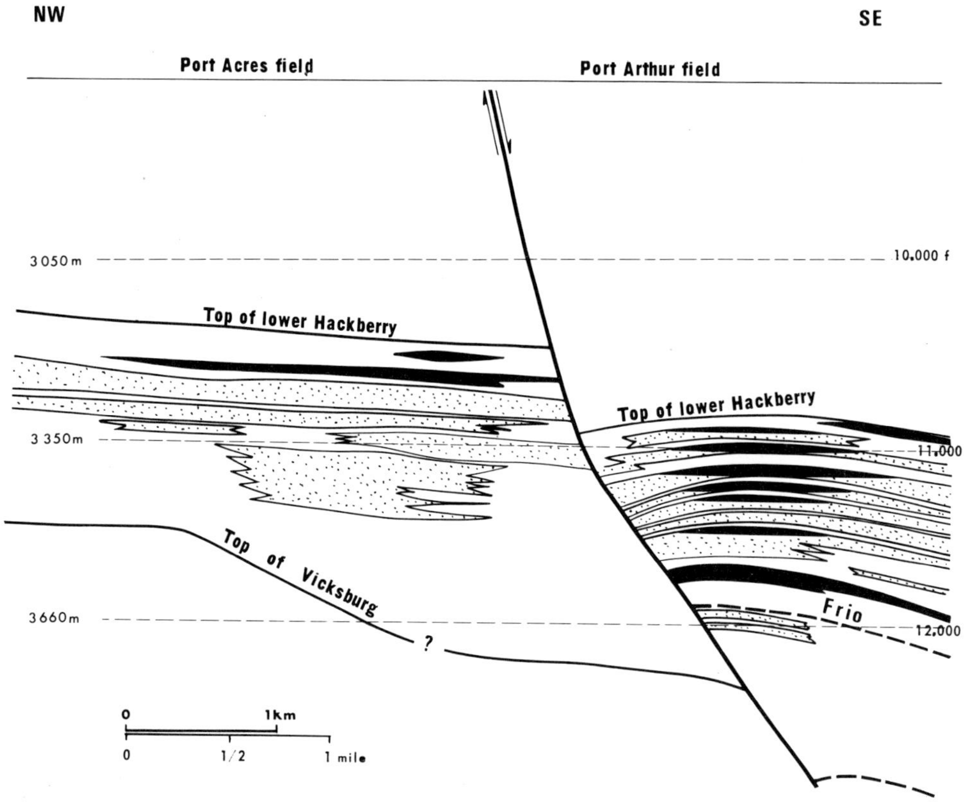

Fig. 81. — Section across Port Acres and Port Arthur Fields (Texas) (Halbouty & Barber, 1972) (see situation map, Figure 195).

With reserves of 100 Gm^3, this field is the largest in the Canadian foothills. It offers proof that seals effective for gas can even exist in an area of complex tectonics (Guidebook of Excursion A25-C25 of the 24th International Geological Congress, Montreal, 1972).

The Santa Maria a Mare field (Italy), discovered by seismics in 1974, on the Adriatic shore south of Ancona, is formed by a faulted anticline surmounted by an overthrust fold to the east (Figure 87). The reservoir is represented by Upper Cretaceous/Eocene calcarenite and limestone breccia, exhibiting an effective thickness of about 100 m at various horizons. The porosity is about 15 %, permeability is 10 md, but is improved by fractures. The oil is heavy (specific gravity 0.92) of the naphthenic type, and the GOR is 65 $m^3\ t^{-1}$.

4.1.2. **Fault traps**

In certain conditions, faults that accompany vertical displacements as well as strike slip movements may form barriers to the migration of hydrocarbons and trap them. Very often, however, they facilitate migrations and dysmigrations. The tightness of a fault varies in accordance with the geological, hydrodynamic and pretroleum context. As a rule, young faults or those which become active again, while under tension, are open and permeable. With the passage of time and a certain stability, the brines circulating therein 'mineralize' the fault plane and cement it.

But the most important factor is probably the lithologic nature of the beds on opposite sides of the fault. Reservoirs facing shales are often closed by the shales. Thus statistically, data such as the sand/shale percentage provide a useful indication about the sealing probabilities of a fault system. Plastic shales and other materials often tend to be injected into the fault plane and to plug it. This occurs in particular in Tertiary basins with a thick and plastic shale series. The pressure régime also plays a vital role in the behavior of faults. At a certain depth, where the pressures are high, especially in the vicinity of the high pressure zones, fluids tend to force themselves through the fault planes. In the neighborhood of reservoirs at lower pressure, the fluid pressure decreases and may be insufficient to overcome the injection pressure at the fault plane. The latter then becomes impermeable and may serve as a barrier.

Hydrodynamic gradients, depending on whether they enhance or oppose the migration of hydrocarbons, facilitate seepage or contribute to trapping of the oil. Finally, the tightness of a fault depends on the

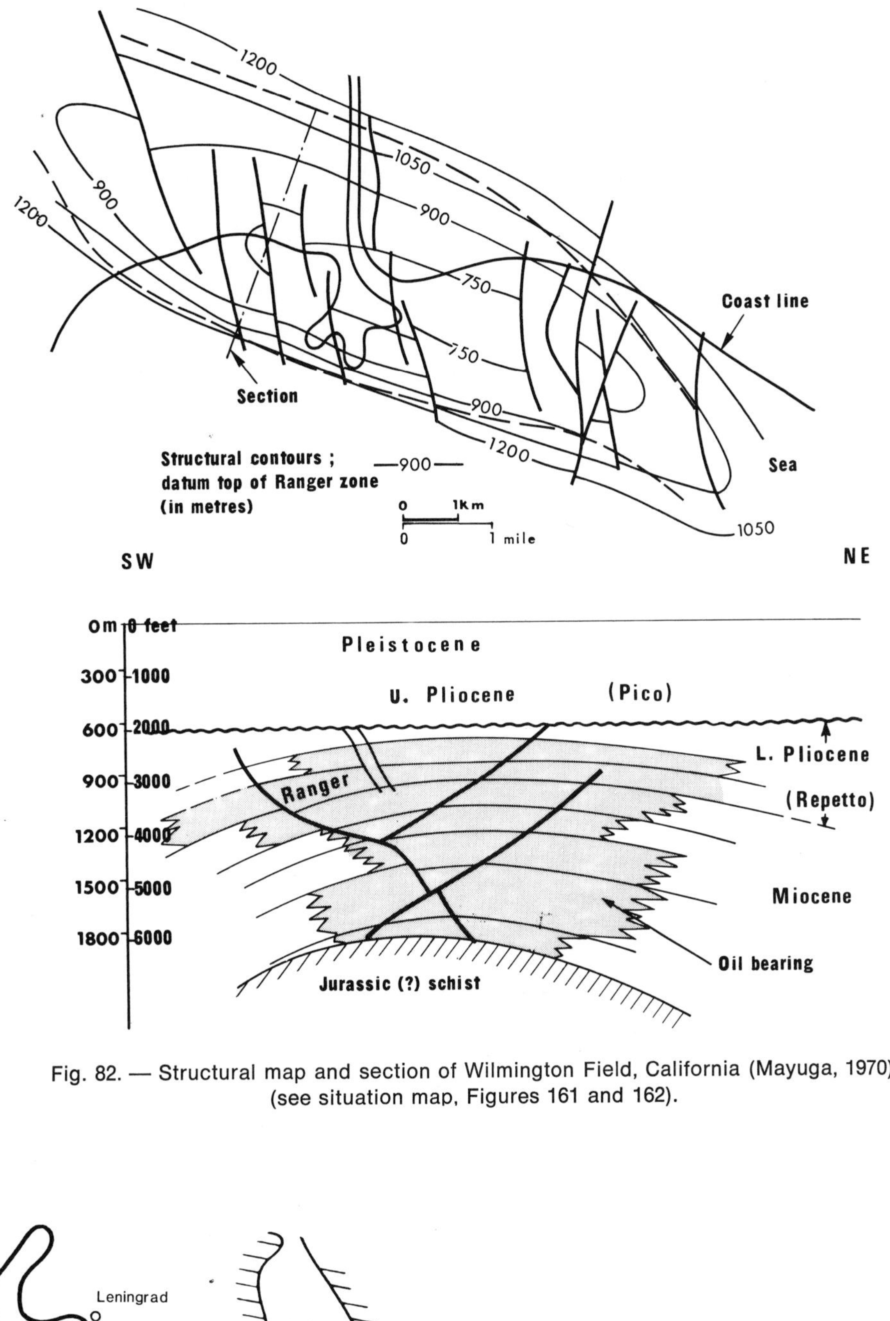

Fig. 82. — Structural map and section of Wilmington Field, California (Mayuga, 1970) (see situation map, Figures 161 and 162).

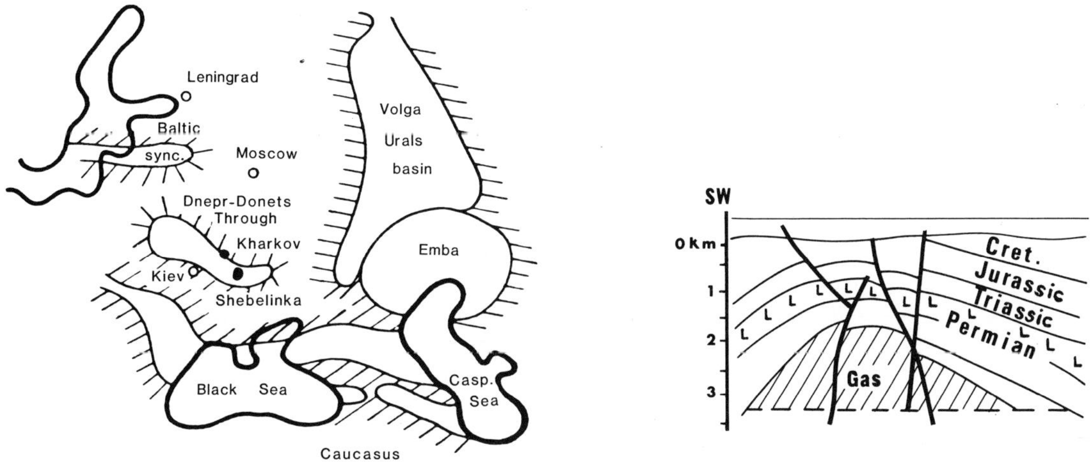

Fig. 83. — Section of Shebelinka gas field.

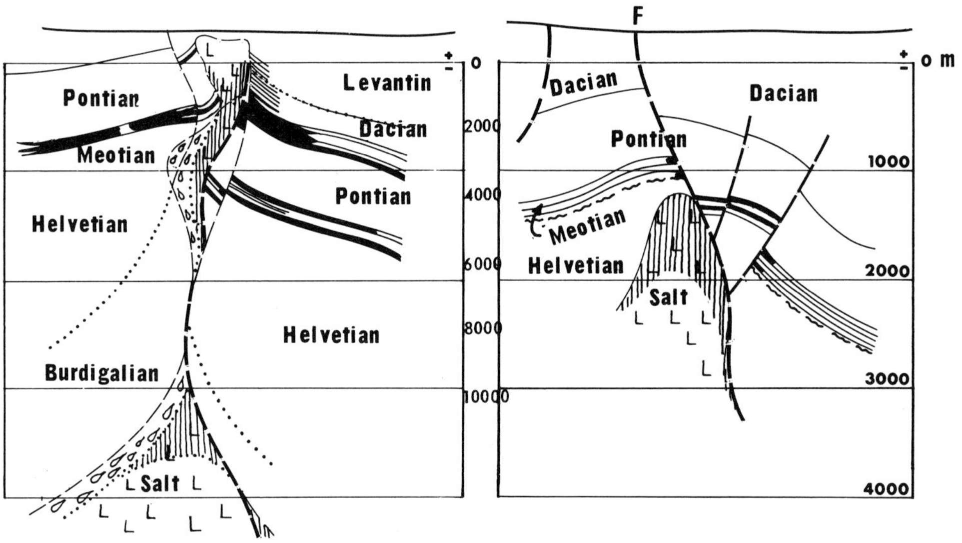

Fig. 84. — Cross-sections of Moreni/Gura Ocnitei Field (Romania) (Paraschiv & Olteanu, 1970).

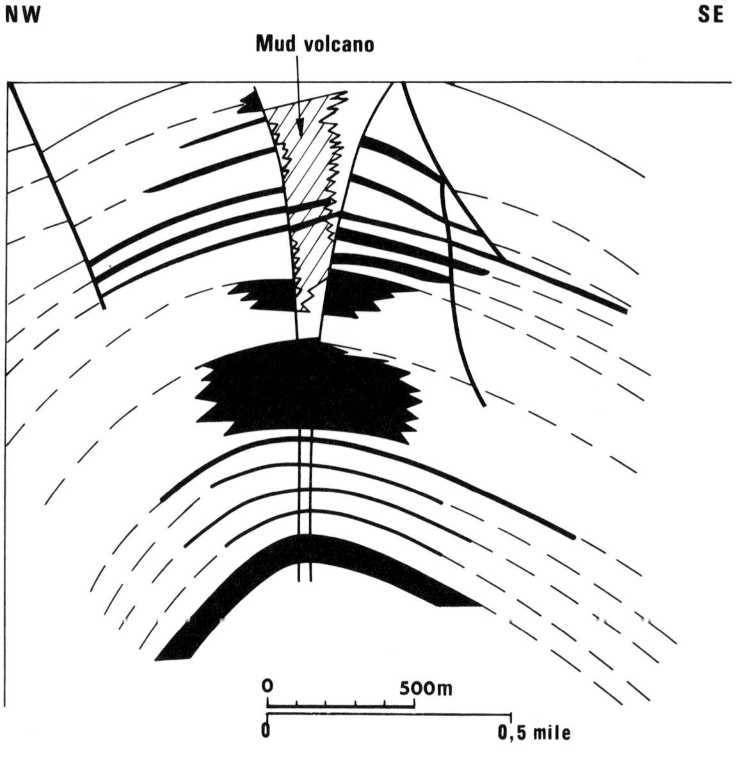

Fig. 85. — Section of Bibi-Ebat Field (Baku region, USSR).

composition, and in particular the fluidity, of the hydrocarbons. In Nigeria, for example, it has been observed that faults are often permeable to gas but form a barrier to the migration of oil.

The hierarchy of a fault system also plays an important role, as may be observed in the large deltaic provinces in Nigeria and the Gulf Coast in particular. In the latter basin, the major growth faults generally appear as migration paths, while secondary faults normally form barriers. In this case, the productive zone is located in a relatively high position, as shown by the fields of the Vicksburg trend (Figure 80) and those of the Edwards limestones in the Albian.

Fault traps are encountered in particular :

— in paralic basins with thick plastic sand/shale series characterized by synsedimentary faults,
— in more stable provinces, which may or may not be folded, and where the features are generally the result of fracture tectonics.

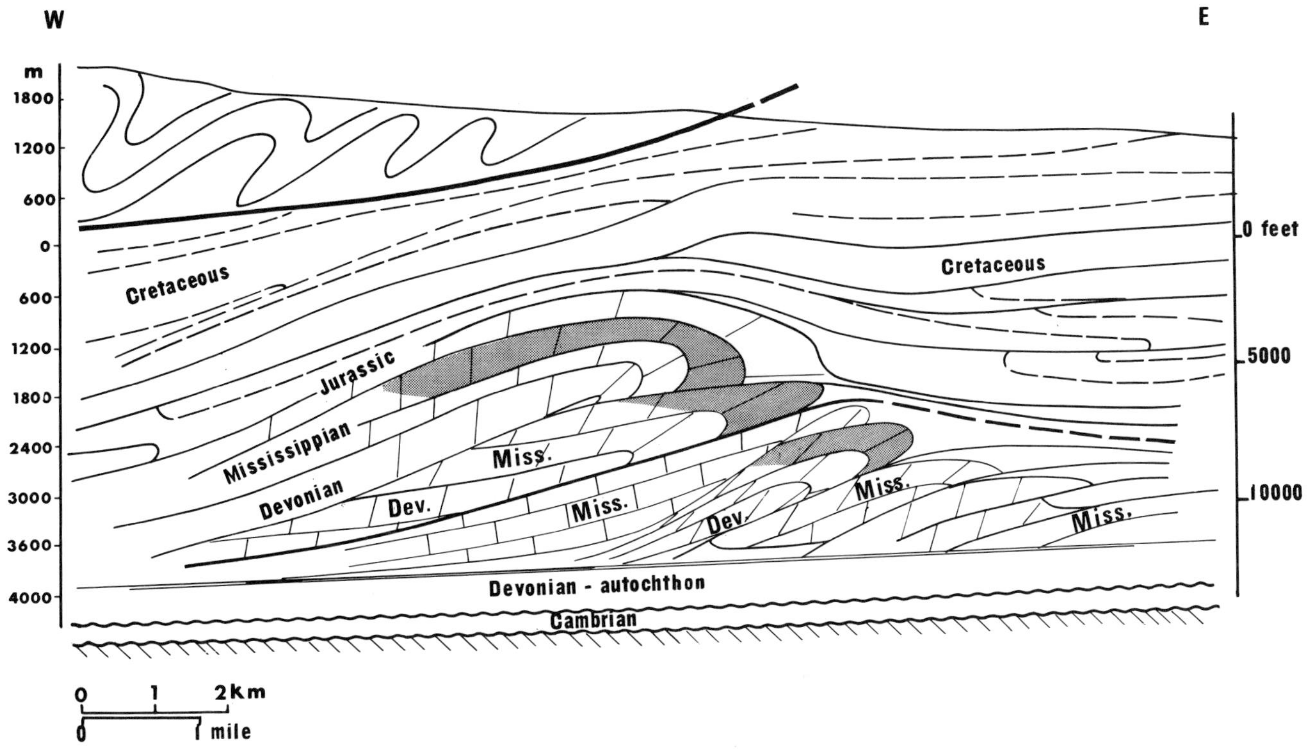

Fi.g 86. — Section of Waterton Field (Alberta) under overthrust front of the Canadian Rockies (Document 24th Int. Geol. Cong. 1972).

However, it is often in the nature of a fault to be active repeatedly, and all possible transitions may be found between these two groups. Very schematically speaking, we shall now examine traps associated with synsedimentary faults and those related to mainly post-sedimentary traps.

4.1.2.1. ***Synsedimentary fault traps***

Faults of this type, also called listric faults, appear to form barriers more often than seepage zones, at least on the whole. A much debated question is that of the respective value of the closed blocks in the upper and lower positions. Accumulations of oil and gas are known in the downthrown blocks of normal and reverse faults. This is the case for the former in the paralic basins with growth faults as observed especially in Nigeria and the Gulf Coast (Figures 88 and 89). But it appears that in both provinces, the oil only accumulates in the downthrown block if it is directly in contact, through the fault, with the source rock which thus supplies it directly (Webber & Daukoru, 1975).

The closed volume is generally small, although it may give rise to a number of giant fields, including Hibernia (150 Mt) offshore eastern Canada. This type of trap is found very frequently in Tertiary basins with thick sand/shale series. It is illustrated by many fields, particularly in deltaic provinces, like the Gulf Coast, Nigeria, and the Mackenzie Delta Basin. Examples given below are the fields of Port Arthur (Texas) and Taglu (North Canada).

— The Port Arthur Field (Texas), discovered in 1954, near the stratigraphic field of Port Acres, is located in the trend of the Frio sands, in a slightly positive zone. The trap is formed by the downthrown block, and the roll over associated with a listric fault (Figures 81 and 195). The reservoirs are represented by a dozen sand horizons in the Frio and Hackberry deltaic formations (Oligocene). The productive area covers 7 km^2, and reserves are estimated at 11 Mm^3 of gas and 3 Mm^3 of condensate (Halbouty & Barber, 1972).

— The Taglu Field, discovered in 1971, is located in the Beaufort Basin, below the Mackenzie Delta (North Canada). This deltaic basin is filled with Tertiary sand/shale formations up to 9000 m thick lying on Cretaceous, Jurassic and Paleozoic series (Figures 90 and 165).

The field is positioned near the outer fringe of the delta, and the trap is formed by the upthrown block of a normal fault that places the productive sands in contact with brackish shales of substantially the same

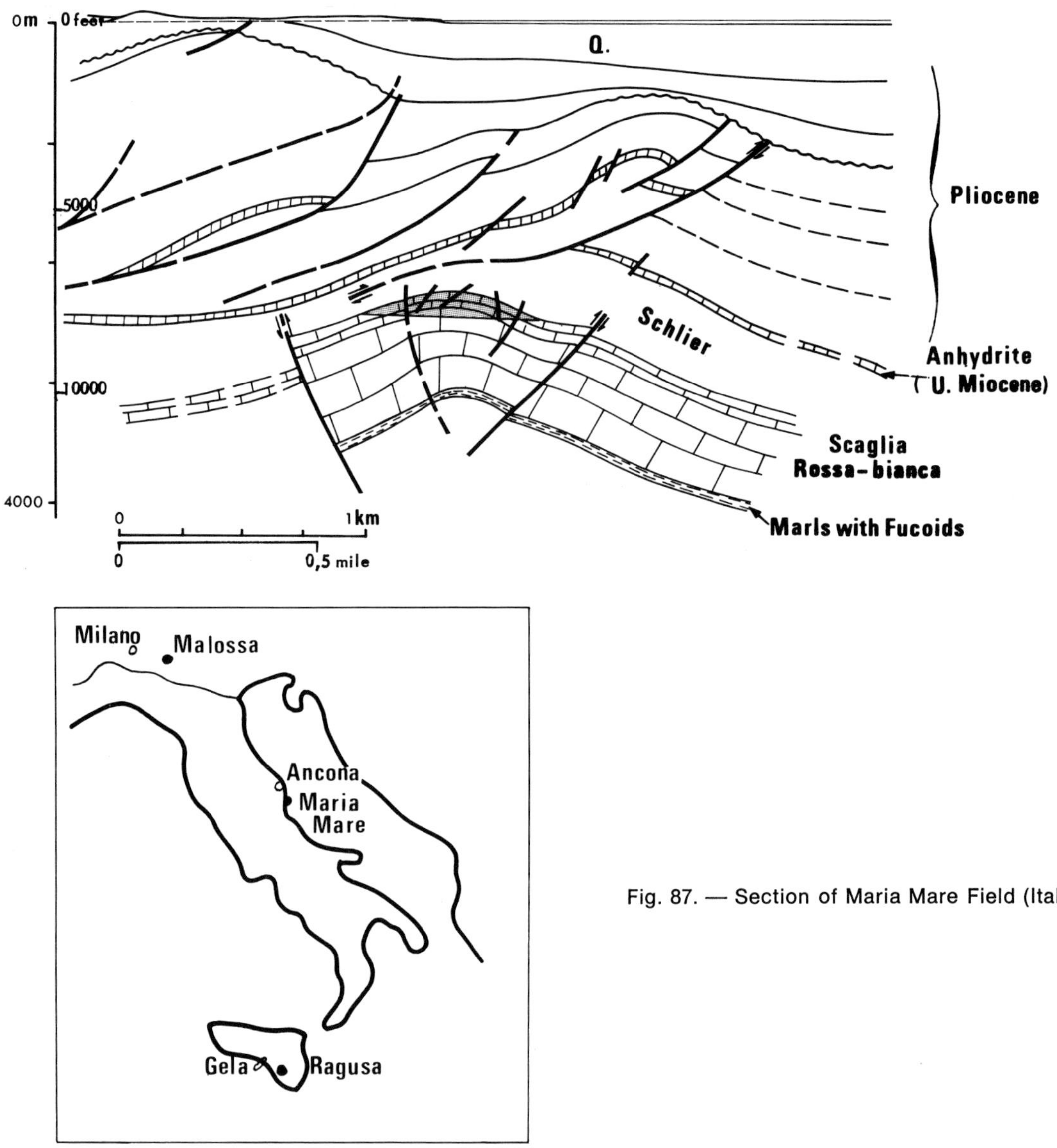

Fig. 87. — Section of Maria Mare Field (Italy).

age. It produces gas from the Paleogene sands at a depth of about 3000 m. These sands correspond to delta front deposits, mainly beaches and offshore bars, with the following characteristics :

porosity :	12 to 22 %
permeability :	2 to 260 md
water saturation :	25 to 55 %
gas column :	400 and 500 m.

The formation water, containing 8 g/l of salt, exhibits the special feature of containing predominantly sodium bicarbonate. The productivity of both levels is 0.2 and 0.25 Mm^3 day^{-1} respectively. The gas contains 94 % C_1 and condensates with a high proportion of aromatics. The hydrocarbons originate in shales of non-marine brackish facies, immature and intercalated between beach and stream mouth bar sandstones. However, migrations may occur from deeper horizons. Wet-gas reserves are estimated to be in excess of 3 Tcf - 90 Gm^3 (Hawkings *et al.,* 1976).

A special case of a synsedimentary fault and associated trap consists of the faults caused by collapses above a salt dome. One example of this type of trap is offered by the Hastings Field (Texas), located in the southern suburbs of Houston (Figures 91 and 195). The trap, which is of a general rounded shape in plan view, mainly consists of the upper block of a normal fault, and also a faulted anticline on the downthrown flank, containing a small gas cap. The fault has a throw of 300 m. The Frio sands making up the reservoir contained initial recoverable reserves of 108 Mt (Bleakley, 1968).

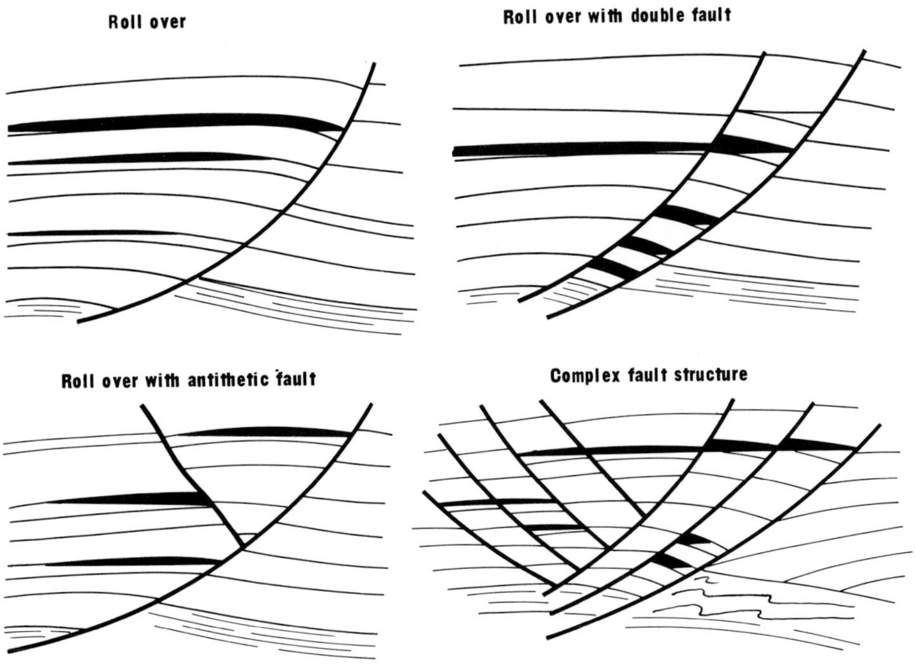

Fig. 88. — Different examples of roll overs in paralic area, Nigeria type.

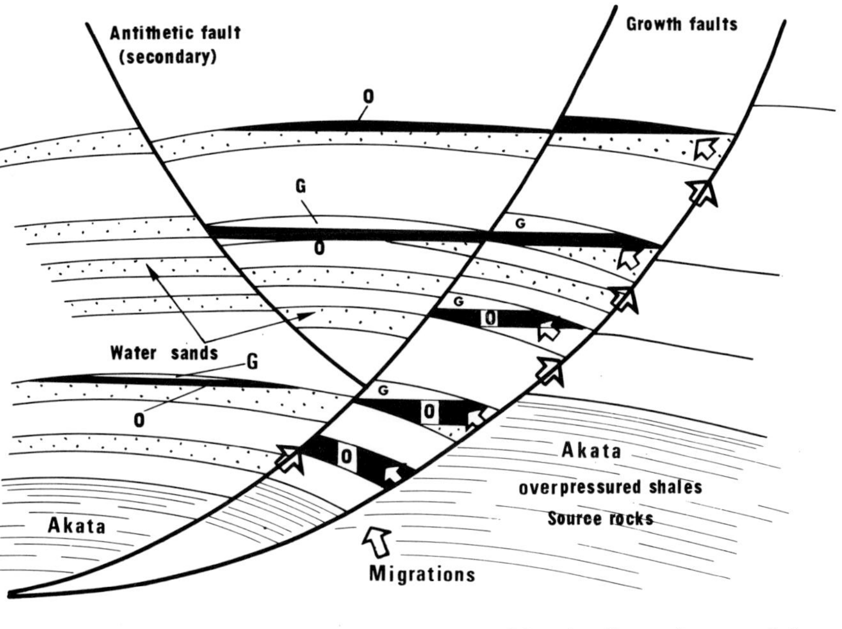

Fig. 89. — Section of a Nigerian field showing different possible migration and accumulation paths (Precious in Weber *et al.*, 1975).

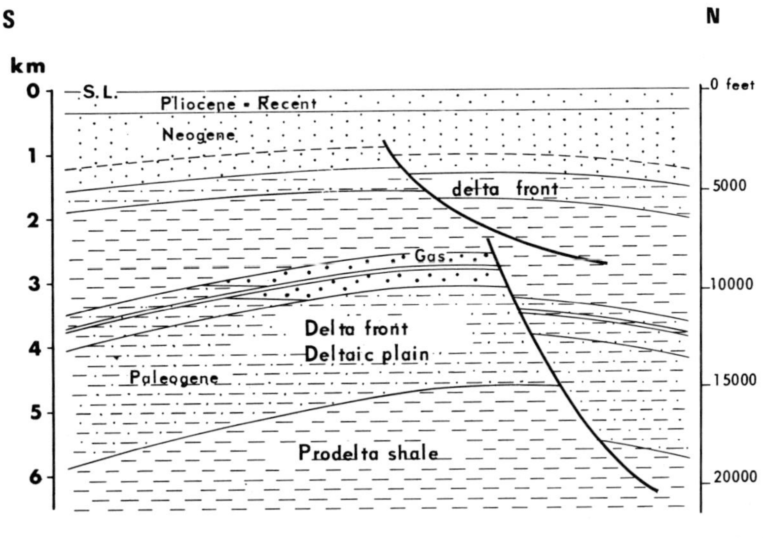

Fig. 90. — Section across Taglu Field, Mackenzie Delta, Canada (Hawkings *et al.*, 1976).

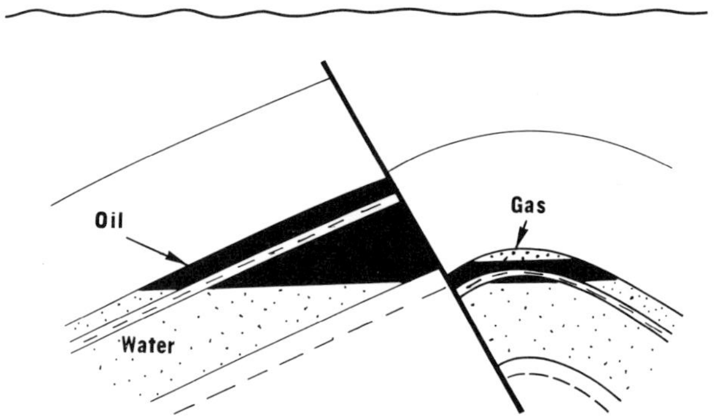

Fig. 91. — Section of Hastings Field, Texas, showing a pool in a fault top compartment and an anticline closed by fault on the lower side (Bleakley, 1968).

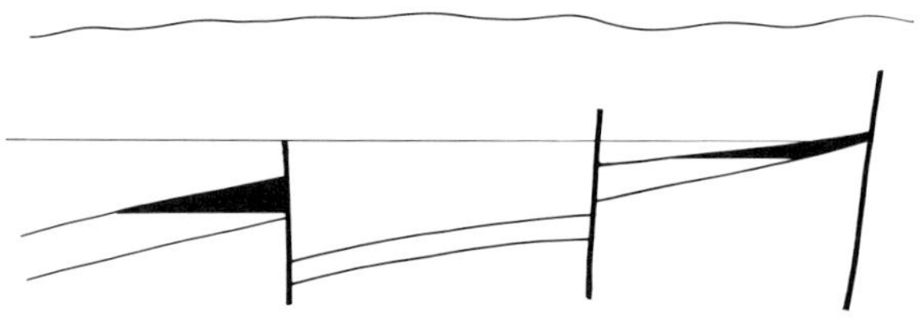

Fig. 92. — Pools associated with high blocks or with a fault/pinchout combination in the Oficina District, eastern Venezuela (Young, 1971).

4.1.2.2. ***Post-sedimentary fault traps***

As a rule in this category, it appears that upthrown blocks have greater possibilities of being productive than downthrown blocks. In the Oficina District, for example, on the south monocline limb of the Orinoco Basin (Venezuela), the pools generally correspond to high blocks on the basin side (Figures 92 and 201). The downthrown blocks on the basin side only produce if a sand pinchout or branched fault is also present : if not, the oil appears to migrate around the end of the fault (Young, 1971).

In the Aquitaine Basin, the Meillon/Saint Faust Gas Field (Figure 93) was discovered in 1965 in the suburbs of Pau, on the southern, unstable margin of the Arzacq Basin. This zone, which is complex from structural and stratigraphic standpoints, corresponds to the northern Pyrenees overthrusting frontal feature that delimits the Senonian flysch trough at the south.

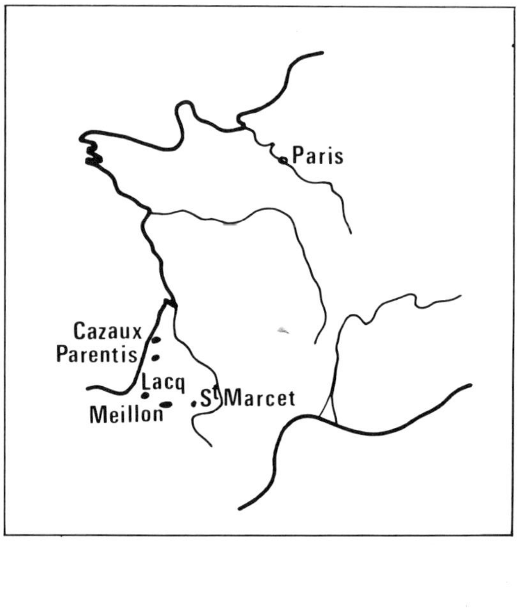

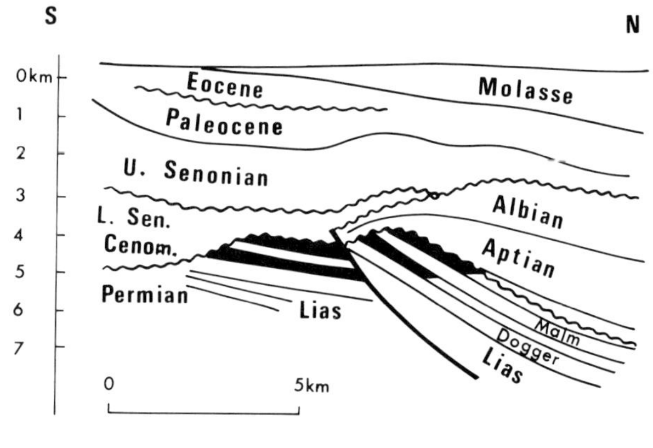

Fig. 93. — Section across Meillon Field (Aquitaine).

In this complex framework, the trap consists of a faulted anticline 22 km long by 2 to 3 km wide, of which practically only the north limb subsists, closed under the Senonian unconformity. The reservoir is formed by the Meillon dolomite (Malm), with porosity of about 4 %, and a smaller part is formed by the Mano dolomite (uppermost Jurassic) whose mean porosity is about 1 %. At a depth of about 4500 m, it originally displayed a pressure of 500 bar and a temperature of 160 °C. Sealing is provided by the Barremian and

Aptian marls. Burial under the Upper Senonian flysch, to some 5000 m, lies at the source of the genesis of gas from a Jurassic/Barremian humic type source rock. This wet, sour gas exhibits various characteristics from one block to another :

C_1	75 to 81 %
C_2	3 to 4 %
C_3	1 %
H_2S	5 to 7 %
CO_2	8 to 20 %
N_2	0.5 %
natural gasoline	15 to 58 cm^3 m^{-3}.

Cumulative production was 33 Gm^3 at the end of 1981. The well productivities are about 1 Mm^3 day^{-1} in the Meillon dolomite (de Chevilly *et al.*, 1969).

The Fahud Field in Oman is a remarkable example of a fault trap under a double unconformity. The fault exhibits a dip of 45 degrees and a throw of 1500 m. The closed zone covers 27 × 4 km, with dips of 15 to 20 degrees and a closure of 960 m. The oil column is as thick as 500 m (Figures 94 and 166). The oil, surmounted by a small gas cap, is located in the limestones of the Cenomanian/Albian Wasia Formation, 420 to 450 m thick, often dolomitized, with fossil or chalky debris, of platform facies, exhibiting porosities

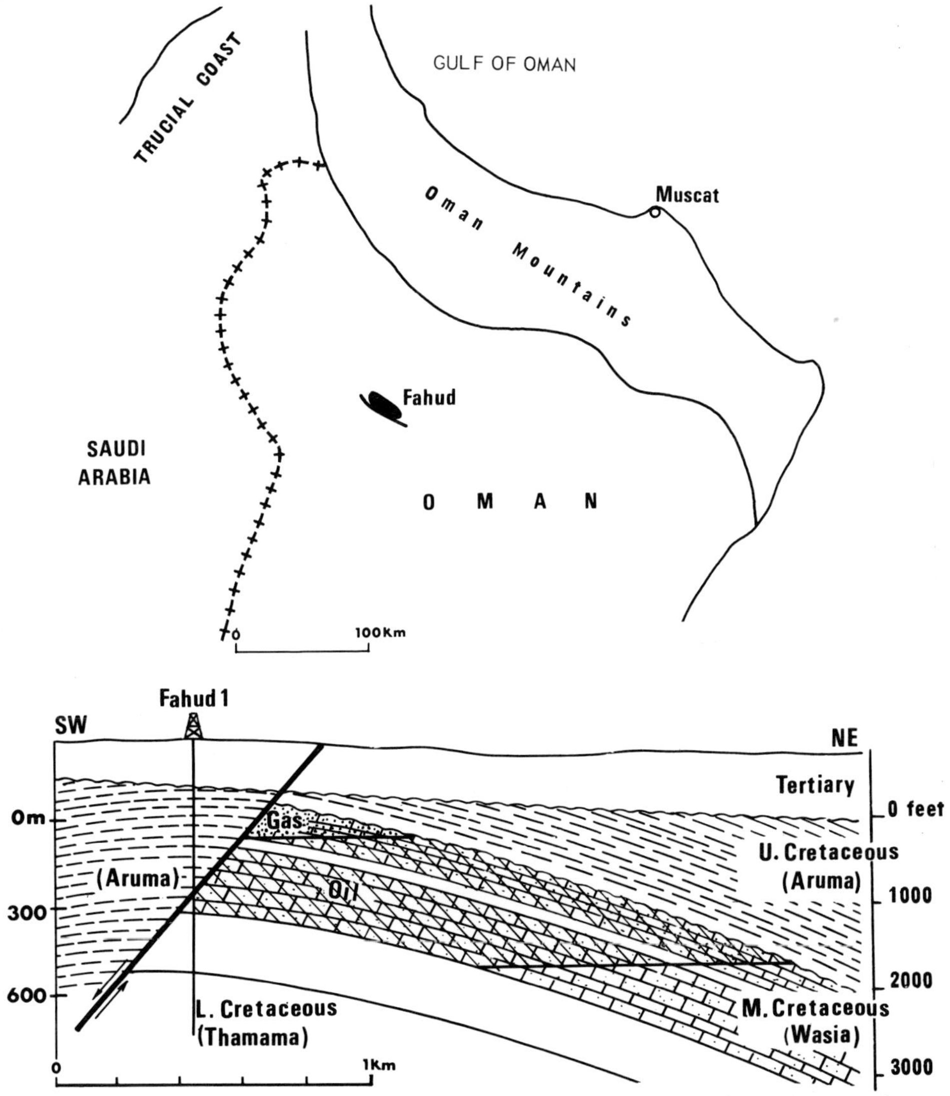

Fig. 94. — Section across Fahud Field (Oman) (Tschopp, 1967).

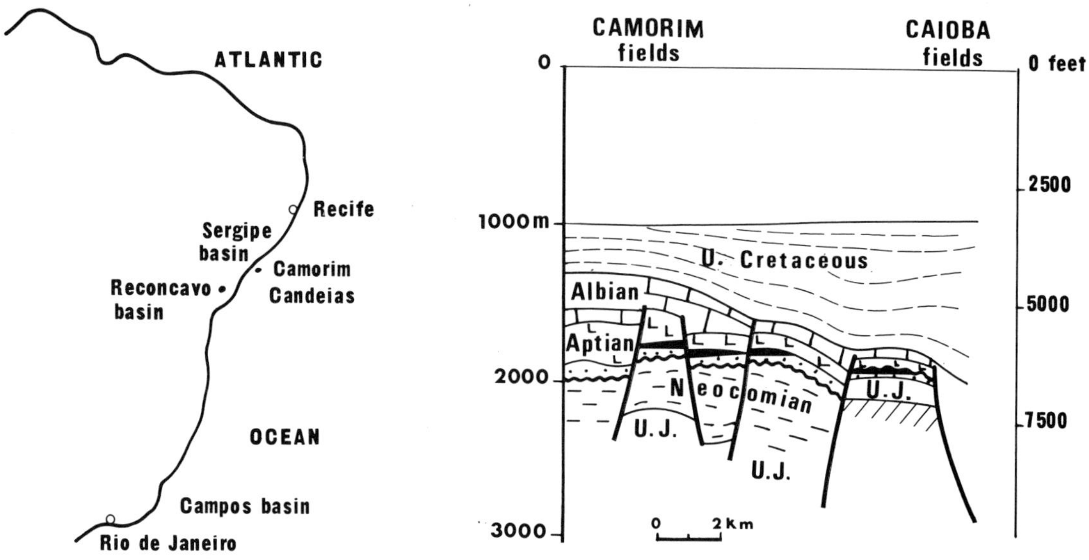

Fig. 95. — Section of Camorim and Caioba Fields, Sergipe/Alagoas Basin, Brazil (Ceiso Ponte, 1977).

around 30 % and permeabilities up to several darcys in the least porous horizons. The oil has a density of 33 °API, with 1.15 % sulfur. The seal is provided by the Campanian/Santonian Aruma shales. The oil probably originates in a bituminous shale and limestone horizon intercalated in the reservoir, or, if not, from the shales of the seal located in favorable maturation conditions in the deep synclines that frame the structure. Migrations probably occurred during the deposition of the Upper Cretaceous. The reserves are estimated at about 125 Mt.

The field was discovered by the second well, because the first, Fahud 1 (1956), crossed the upper block in the lowest part of the reservoir. Only after correlations and comparisons with analogous series of the neighboring structures was the hypothesis considered of a fault eliminating the largest part of the reservoir, and a second productive well was drilled in 1963 1.5 km from the first (Tschopp, 1967).

In rift basins, and especially in passive margin basins, many pools are found in horsts and fault blocks. For example, this is the case of the following :

— the Rankin Field in the Dampier Basin, on the continental shelf of northwestern Australia, where the gas is found in the Trias and Jurassic sandstones, sealed by Upper Jurassic shales (see Section 2.2. of Part 3, and Figure 155).

— several fields of the Sergipe/Alagoas Basin in Brazil, and especially the Caioba and Camorim pools, which are positioned on the major trends in the Uppermost Jurassic sandstones, supplied laterally by lacustrine shales of the Neocomian pro-deltaic facies, and sealed by Aptian salt (Celso Ponte *et al.,* 1977) (Figure 95),

— the Jurassic fields of the Viking Graben, such as Brent in the North Sea, aligned on fault blocks sealed by Kimmeridgian shales (see Chapter 3 of Part 3, and Figure 96).

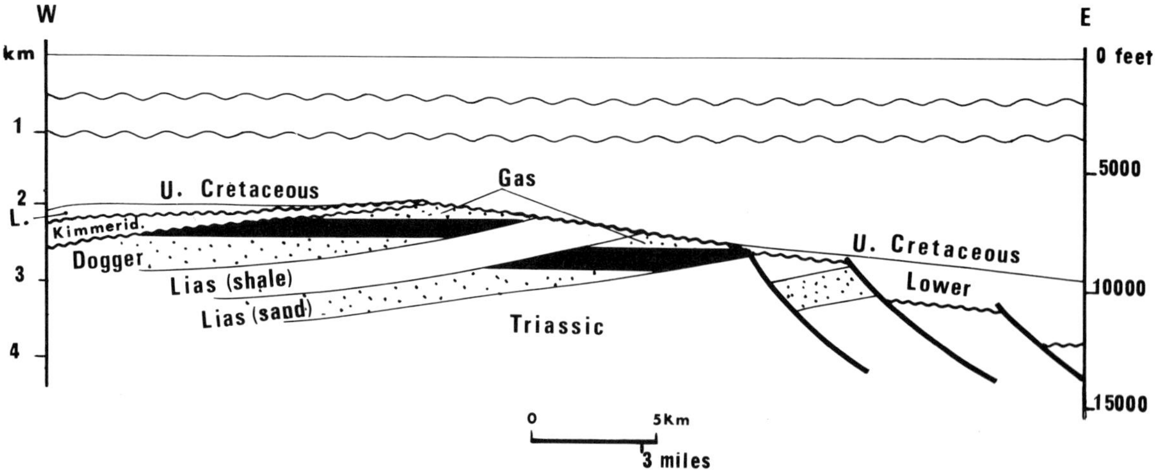

Fig. 96. — Section across Brent Field (Bowen, 1975).

This position of horsts and faulted anticlines exhibits a number of very favorable factors, especially when they are sealed by shale formations, which may play the role of a source rock. Hydrocarbons may often be supplied laterally from synclines or collapsed blocks. An especially favorable case is illustrated by the collapse of an ancient arch later subjected to high subsidence.

4.1.3. **Traps associated with salt and shale domes**

The exceptional plasticity of rock salt (which has a constant specific gravity of 2.17) and undercompacted shales gives rise to the particular tectonics termed salt and shale diapirism. This is relfected by simple intumescences of the salt or shale bed, or by diapirism in zones of distension. Hence salt domes are often aligned with the nodes of a network in the maximum tension, flexure and basement fault areas in particular. The intumescences start with upwarps or ridges that lift the beds, and subsequently pass through them, reaching the surface in many cases. Unlike shale intumescences, salt domes provide excellent conductors of heat, which very substantially improve diagenesis. The thermal effects may thus raise maturation zones locally, by 500 to more than 1000 m.

A large number of traps are associated with these plastic deformations, from the simple anticline to the slender columns which are extinguished at the surface. Finally, the so-called 'turtleback' anticlinal deformations may be produced by the rising of two adjacent intumescences. Pools associated directly with salt domes are very rarely giants, except in the case of deep zones that raise broad arches, as is observed, or assumed, for a number of the biggest fields of the Arabian platform.

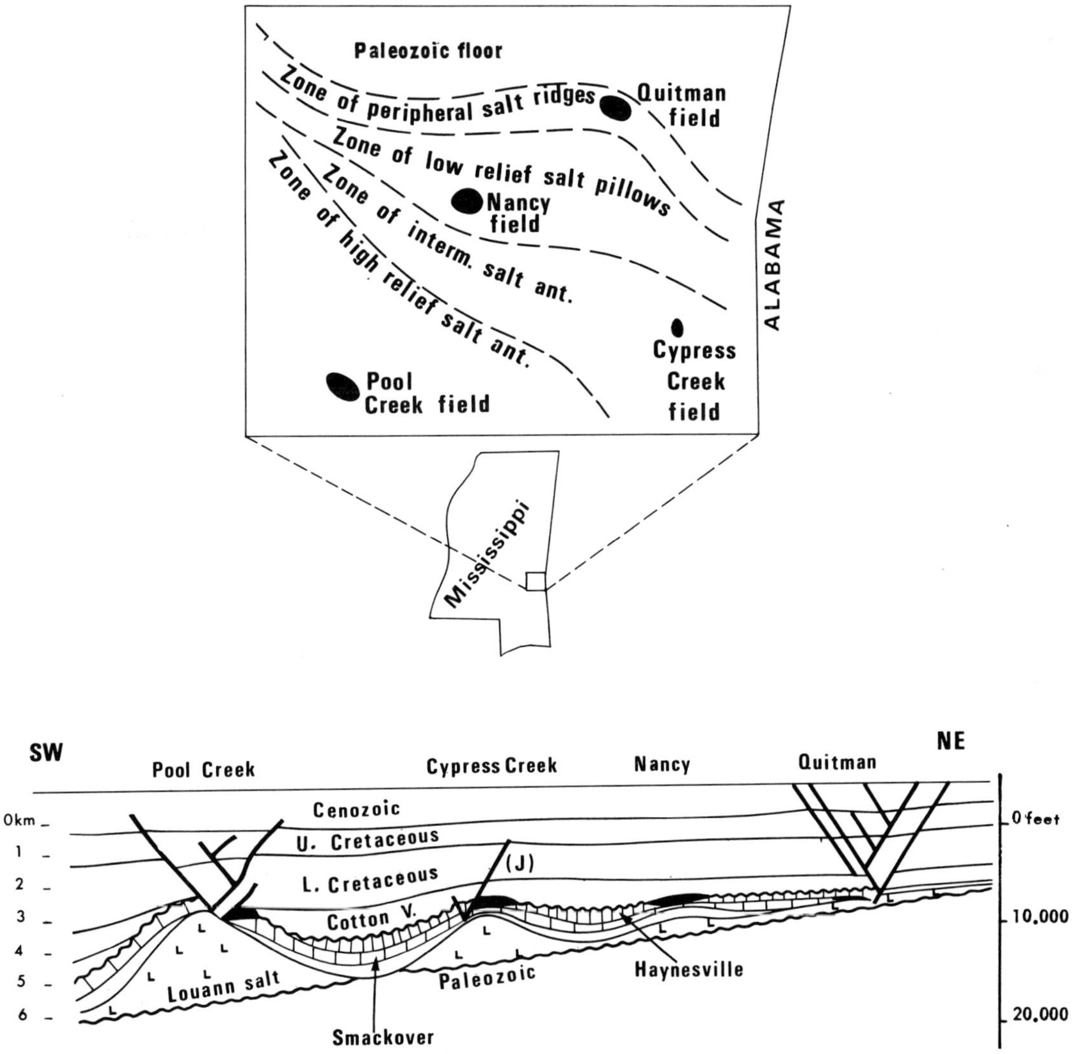

Fig. 97. — Map section of salt trap fields in Mississippi (Hughes, 1968).

4.1.3.1. ***Simple anticlines with saliferous cores***

The Smackover zone, in the eastern part of the Gulf Coast, offers good examples of this type of trap, in which the reservoir is only folded into a relatively simple, sometimes faulted dome. The salt of the Louann series (Jurassic) is developed in certain basins, where its thickness may reach up to 600 m. It is surmounted by overlapping carbonate formations of the Upper Jurassic (Smackover), generally compact, except along a narrow strip, where they display a certain porosity, related to facies containing algae, rudists, and calcarenites. This reservoir is sealed by overlapping compact limestones and dolomites and by anhydrites (Haynesville Formation).

In this area of Mississippi and Alabama, three-quarters of the traps consist of low relief anticlines with salt cores. The latter appear to have become active as soon as the Smackover series was deposited, irrespective of its thickness, but the scale of the deformations depends on the available salt volume. The growth rate was first rapid and then declined with the depletion of the mobilizable salt. Variations in thickness of the initial salt bed, which diminishes towards the boundary of the basin, thus condition the form of the

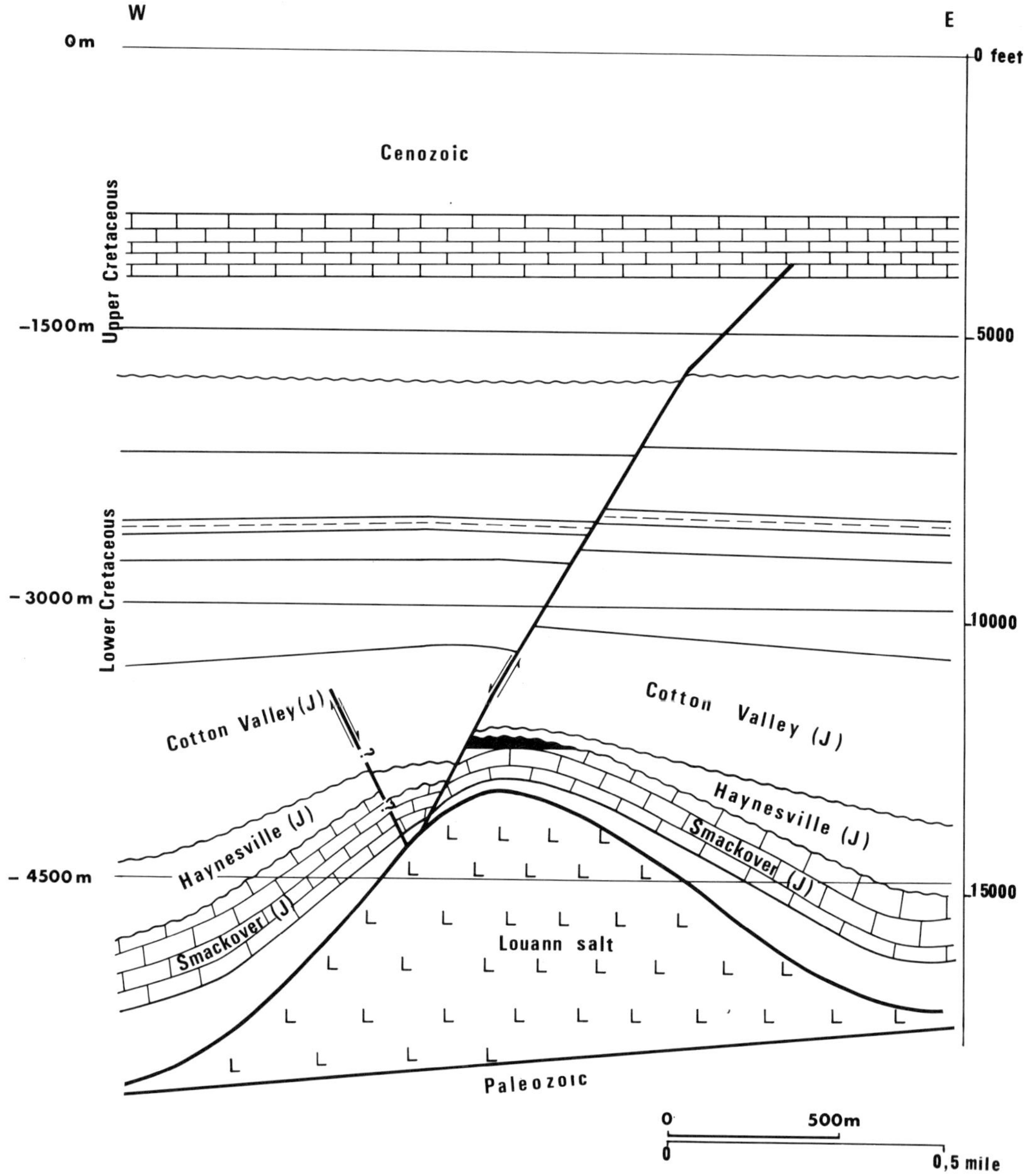

Fig. 98. — Section of Cypress Creek Field (Mississippi) (Hughes, 1968).

intumescences, and going from the edge of the basin to the center, the following are observed in succession (Figure 97) :

- — peripheral ridges, like boundary steps of the salt bed, faulted in the upper part, and illustrated by the Quitman Field,
- — flattened salt pillows lying parallel to the basin edge and not faulted, such as the Nancy Field,
- — intermediate anticlines with a height of 1000 to 2000 m at the salt horizon, such as the Cypress Creek Field (Figure 98),
- — high relief anticlines, reaching 2000 to 4000 m, surmounted by secondary grabens, like the Pool Creek Field (Hughes, 1968).

This scheme recalls the salt arrangement noted below the continental shelf of the Gulf of Mexico up to the Sigsbee Scarp. The salt flows laterally towards the slope, where the thickness of the underlying sediments decreases sharply (Humphris, 1979).

These different types of trap are found in a number of basins, and apart from the Gulf Coast and Mexico :

- — in the Persian Gulf and the Arabian province (deep domes),
- — in the North Sea, Ekofisk for example,
- — in Gabon and Congo (Grondin),
- — in the Gulf of Gabès.

4.1.3.2. ***Traps associated with piercement domes and diapirs***

Piercement domes, which are spectacular in the coastal part of the Gulf Coast but also frequently found in the Emba Basin, and in Russia, North Germany and Gabon, often create closed pinchouts on their flanks, that provide many small pools, as observed in particular in Louisiana and Texas. The rate of rise of the dome, and especially the area of the drainage zone surrounding it, which often depends on its history, largely condition its potential. A remarkable example is offered by the Barbers Hill Field (Texas) (Figure 99). In a productive region, like the Gulf Coast, it has often been observed that the reserves of a dome are directly related to the scale of the uplift and the corresponding drainage area. Hence in 1954 the Barbers Hill dome produced 15 Mt, while the two neighboring domes of Lost Lake and Moss Bluff yielded less than 800,000 t (Halbouty & Hardin, 1959).

With the exception of deep domes, the fields associated with salt domes rarely have large reserves, and very few appear in the list of giants of Halbouty *et al.* (1970), apart from the fields of the Marchand Bay complex offshore Louisiana, and the Fyzabad group in Trinidad, associated with a shale dome. The fields of the Marchand Bay/Timbalier Bay/Caillou Island complex, discovered since 1933, extend around a saliferous ridge over an area 45 km long and 20 km wide. The salt beds rise from a depth of 12 to 15 km to a top level of 600 to 900 m. The reservoirs consist of sands spreading from the uppermost Miocene to the Pleistocene, at depths between 400 and 6000 m. Porosities range from 33 to 15 % and permeabilities from 2400 to 0.5 md. The traps are numerous and formed by anticlines above the dome, closures against faults and against the salt or shale wall, and stratigraphic pinchouts (Figure 100). Reserves are estimated at around 475 Mt. Production, about 14 Mt in 1967, was delivered by over 1900 wells (Frey & Grimes, 1970; St. John, 1980).

Piercement salt domes may also shelter pools in the beds above them, as in the previous examples. This is the case in particular of the Saint-Marcet Gas Field, discovered in 1939, following the establishment of a surface geology unit on an anticline in the Petites Pyrénées. The deep structure corresponds to a diapiric anticline with a Triassic salt core. The main reservoir is represented by a fractured marl/limestone breccia, exhibiting porosity of about 6 % (Figure 101). Initial reserves, nearly completely exhausted today, were 7 Gm^3 of a wet gas, rich in natural gasoline.

Traps associated with shale domes are rarer and/or less well known. Some of these are familiar on the Gulf Coast, offshore Cameroon where they provide the source of many pools including Kolé, Ekoundou, Bétika and Bakassi, and in Trinidad, with the Guayaguayare Field (Figure 102).

Domes, and especially piercement domes, may yield a reservoir and a trap in their upper portion, which is made porous by dissolution under the effect of the aquifers traversed. These are called cap rock pools. Some of them have been described, chiefly in Louisiana and Texas, and many of them also contain sulfur accumulations. Some examples are the following :

- — the Vacherie Field (Louisiana), productive from a conglomeratic series about 40 m thick, dating from the Cretaceous, lying directly on the cap rock and apparently ‘picked up’ by the rise of the dome,
- — the Spindletop Field (Texas), the first discovery in 1901 in the Gulf Coast province, which will soon have produced 25 Mt, some of the wells having yielded very high outputs, up to 10,000 t day^{-1},
- — the Heide Field in North Germany (Boigk, 1981).

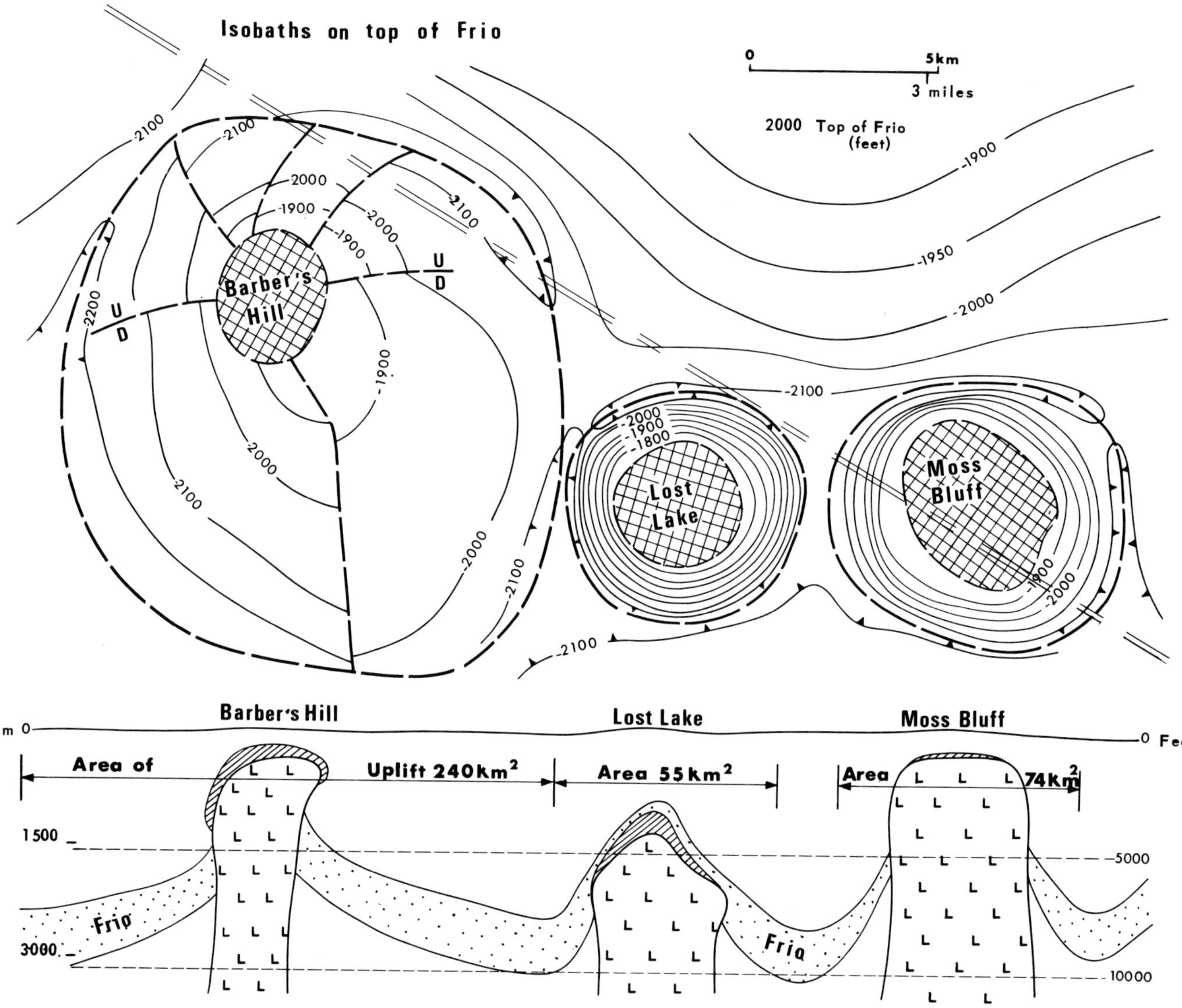

Fig. 99. — Structural map and section of Barber's Hill, Lost Lake and Moss Bluff Domes showing extent of drained area (Halbouty & Hardin, 1959).

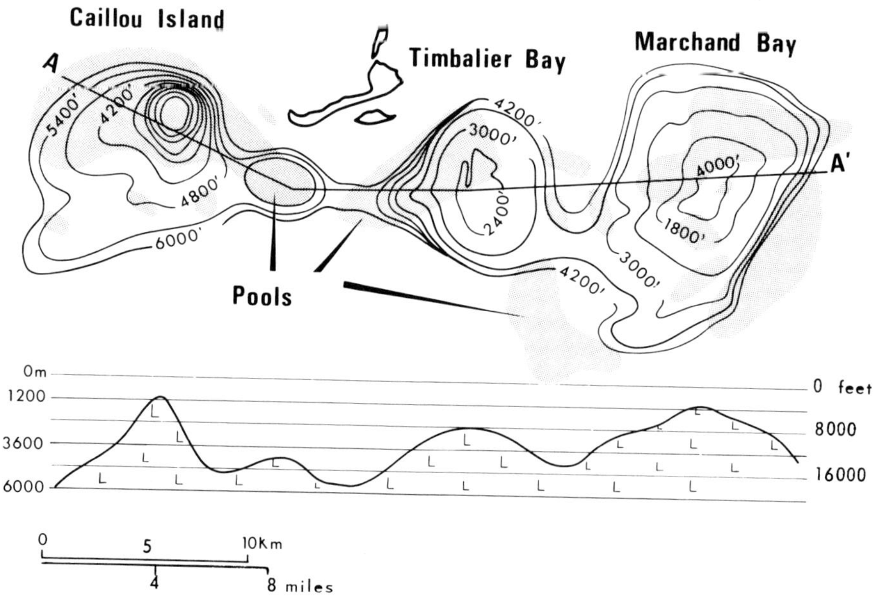

Fig. 100. — Structural map at top of salt and section of Caillou Island and Timbalier and Marchand Bay pools (Frey & Grimes, 1970).

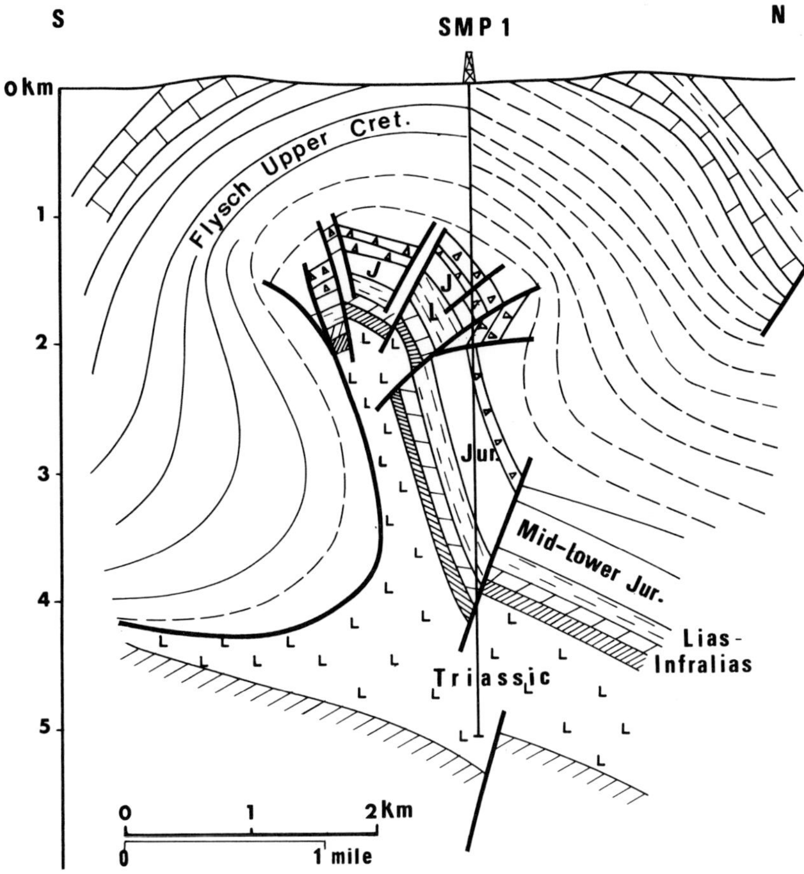

Fig. 101. — Section of Saint-Marcet Field (Aquitaine).

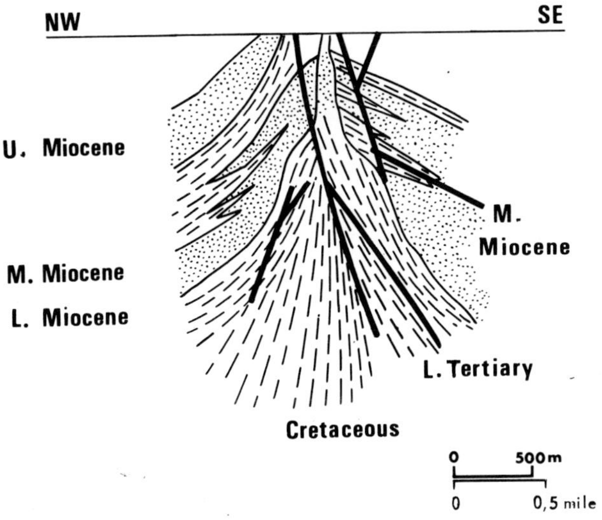

Fig. 102. — Section of Guayaguayare Field (Trinidad), anticline with shale diapir core (Michelson, 1976).

4.1.3.3. ***Intermediate or "turtleback" structures***

A salt dome is rarely isolated in a province, and the causes of their formation generally give rise to a family of comparable domes. 'Intermediate' structures may be formed between two adjacent domes in these conditions.

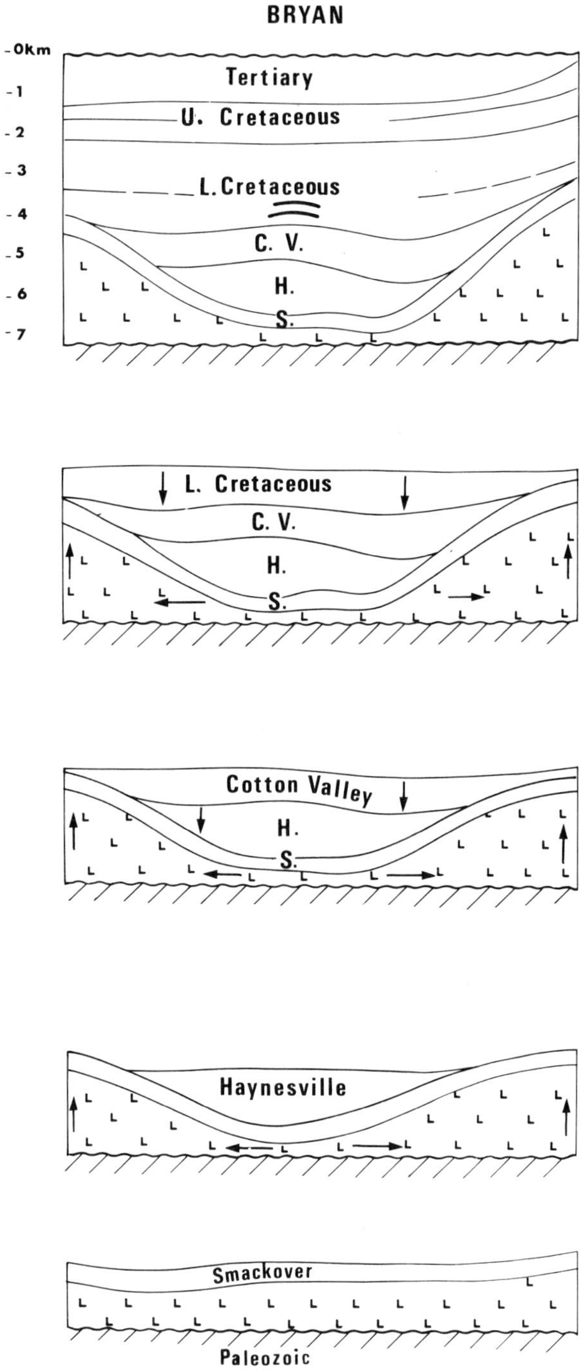

Fig. 103. — Stratigraphic sections showing the geological history of the Bryan 'intermediate' structure, Mississippi (Oxley & Herlihy, 1972)

The analysis of the geological history of the Bryan Field (Mississippi), discovered in 1958, illustrates the method of formation of these traps (Figure 103). Salt migration towards the domes creates a depression between them. This depression is the reason for Jurassic sedimentation (Haynesville Formation) being thicker here than in the neighboring zones. At the end of the Jurassic ('Cotton Valley'), and in the Lower Cretaceous, the growth of the domes causes the formation of rim synclines around the domes, giving rise to a slight upwarp in the central part, previously in a syncline position. Thus a sort of flat bottomed or collapsed mound is formed, an 'intermediate structure', in an anticline situation at certain levels, but remaining in the syncline position for the deepest beds ('turtleback'). The Bryan Field thus contains some 3.5 Mt of oil in the Lower Cretaceous sands (Oxley & Herlihy, 1972).

This type of structure often has relatively large dimensions, allowing the concentration of greater reserves than in the other traps associated with penetrating domes. Among the fields of this family are the following :

— Anguille and Tchengue Océan in Gabon,
— Düste in West Germany (Boigk, 1981).

4.2. STRATIGRAPHIC AND COMBINATION POOLS

Changes in facies and thickness of deposits are a widespread occurrence, very often superimposed on structural deformations, folds and faults. If the scale of their action is equivalent to the latter, the traps formed in this way may be qualified as combination traps. If the structural deformations only play a secondary role, the term 'stratigraphic' (or lithologic) trap is employed. These traps thus offer the common characteristic of a closure provided chiefly by stratigraphic, sedimentological or lithologic mechanisms, such as variations in facies or pinchouts. This new term of stratigraphic trap was created in 1936 by Levorsen for the East Texas Field, discovered 'accidentally' in 1930.

Facies variations, and especially of a reservoir facies passing updip into an impermeable formation, may be the result of depositional or diagenetic processes. They are thus observed both in sand series and in carbonates. Among the latter, reef structures play a major role. These facies variations, and pinchouts in particular, are often related to transgressions and regressions, in other words, with vertical movement of continents and seas. In all these cases, it is obvious that the stratigraphic trapping mechanisms are largely conditioned by structural deformations on the basin scale.

These closures by facies variations, in other words, laterally to the reservoir formation, are often liable to be less effective than the structural closures of anticlines provided by a superposition of permeable and impermeable deposits. Within the shales which prolong it laterally, a detrital horizon tends to leave a few fine detrital horizons which may suffice to allow fluids to pass through. Hence the effectiveness of these traps is often closely dependent on gaps, sedimentary discontinuities and transgressions of impermeable beds, and also on the centripetal hydrodynamic gradients which tend to prevent dysmigration and hydrocarbon seepage.

From the physical standpoint, this type of closure is determined by the notion of interfacial tension or capillary pressure, and should be compared to the concept of an impermeable rock and migration, which have been examined in Sections 2 and 3 respectively of Part 2. Stratigraphic trapping may be considered to begin where migration stops. The lateral variation in facies of the reservoir bed updip, which forms the barrier, corresponds to a rise in the injection pressure. A trap occurs because this displays a higher value than the pressure exerted downdip by the oil or gas column, and opposes its passage. It comes immediately to mind that a closure of this type is relative for a sediment with a given injection pressure. For example, the oil column trapped is higher for a given fluid if the injection pressure is greater, in other words, for a detrital, non-fissured rock, with a fine particle size distribution.

An aquifer bed, which is likely to yield water when tested, exhibiting a certain porosity and permeability, may thus constitute an effective barrier to oil migration and trap an oil accumulation downdip in a more porous and more permeable facies (Berg, 1975). In these conditions, it is easy to understand the importance of hydrodynamic factors which reinforce or diminish the injection pressure depending on the direction of the gradient (Hill *et al.*, 1961). This interpretation is confirmed by the measurement of this injection pressure on various samples (Berg, 1975). The same argument applies to compact fissured rocks for which the interval between the fissure walls determines the value of the injection pressure. Hence the importance of the degree of geomechanical confinement of these facies.

The progressive thinning of a reservoir in a pinchout is often reflected by low productive thicknesses or by mediocre reservoir characteristics. Certain stratigraphic pools, such as Pembina, thus spread wells of mediocre productivity over a vast area. Reefs, on the other hand, concentrate high impregnated thicknesses, often with good characteristics, in small areas, offering especially high productivity indexes and recovery rates. Hence stratigraphic traps represent an aspect that merits further examination. They account for 10 % of the number and of the reserves of the world's giant fields (over 70 Mt), some of them among the biggest. Among these non-reef fields are the following :

Bolivar Coast	4200 Mt oil
Prudhoe Bay	1400 Mt oil
East Texas	840 Mt oil
Pembina	250 Mt oil
Quiriquire	140 Mt oil
Boscan	140 Mt oil
Hugoton	2000 Gm^3 gas
Blanco (North Mexico)	300 Gm^3 gas
Jalmat (North Mexico)	225 Gm^3 gas.

Among the giants of the reef type are the following :

Poza Rica	380 Mt oil
Kelly Snyder	240 Mt oil
Faja de Oro	200 Mt oil
Intissar A	200 Mt oil
Swan Hills	140 Mt oil

The discovery of the first stratigraphic pool, East Texas, which ranked fourth in size at the time, for a certain time turned the search for stratigraphic traps into a sort of elephant hunt. In actual fact, the proportion of stratigraphic fields appears to be substantially higher in countries in which exploration is more intense, as in North America, where it accounts for 31 % of all the 'major' fields (> 14 Mt), with 42 % of hydrocarbon reserves. In certain very stable basins, like those of Alberta and Oklahoma, the proportion of stratigraphic traps is even predominant. At the end of 1967, 2101 stratigraphic fields were counted in this state, compared with 1182 structural fields (Lyons, 1968). Often, especially in the Rocky Mountains, hydrocarbons first gather in primary stratigraphic traps. They are then picked up in structural traps if a fold or fault occurs. This is also the case of the Wilcox Fields in Texas (King, 1973).

A number of criteria appear to be especially favorable to the formation of stratigraphic pools :

— slight instability reflected by unconformities and transgressions, and alternating marine, paralic and fluviatile facies,

— the absence of structural traps : in the Tertiary series of the Gulf Coast, for example, which are thick and unstable, stratigraphic pools account for less than 1 % of the total number of fields, and barely 1 % of the corresponding hydrocarbon reserves.

Major regional events, like the Albian/Cenomanian overlap on vast platforms, often offer an important criterion in the formation of stratigraphic traps. In all cases, the presence and scale of stratigraphic traps depend closely on the characteristics of the sedimentary basin.

4.2.1. **Unconformity traps**

Unconformities play a major role in hydrocarbon migrations and trapping. They lie at the source of some of the world's biggest fields, at least when relatively important subsidence succeeds transgression. The chief factors are the presence of an effective seal in a position overlapping an older and more or less eroded high zone, and a sufficiently thick post-transgression sedimentary series. Depending on the extension of the reservoir/seal unit allowing the formation of the pinchout, on the high zone and from the geometric standpoint, stratigraphic traps or combination traps will exist, without the possibility of determining the exact boundary between these two categories.

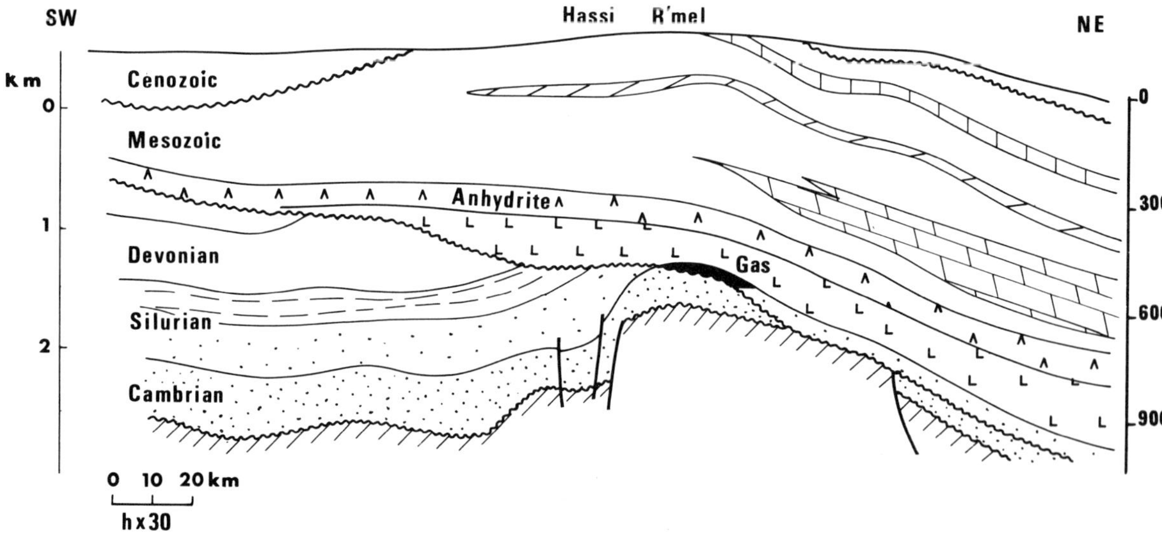

Fig. 104. — Section across Hassi R'Mel pool (Algeria) (Magloire, 1970).

The Hassi Messaoud Field in the Algerian Sahara, for example, perfectly sealed by the saliferous formations of the base of the Mesozoic, is still an anticline, bald-headed at the top (Figures 79 and 204). The Hassi R'mel gas condensate field, located in the plunging Mzab dorsal, and productive in the basal sands of the Triassic transgression, corresponds to a dome whose closure is provided at the southwest, updip, by the pinchout of these sands. This is a combination trap, a 'nose', and most of its area corresponds to an anticlinal movement along strike slip faults (Figure 104).

In the case of the vast East Texas Field, discovered in 1930, the trap is formed by a regional pinchout (elongated north/south) on the west side of the Sabine Arch. The reservoir is represented by the Upper Cretaceous Woodbine Sands that overlap the last components of an eroded Lower Cretaceous. They exhibit a mean thickness of about 10 m, with 25 % porosity and 1500 md permeability, over an area of 520 km^2. These sands are sealed by marine shales of the Eagle Ford Formation, themselves surmounted and closed updip by the impermeable and transgressive Austin Chalk (Figure 105).

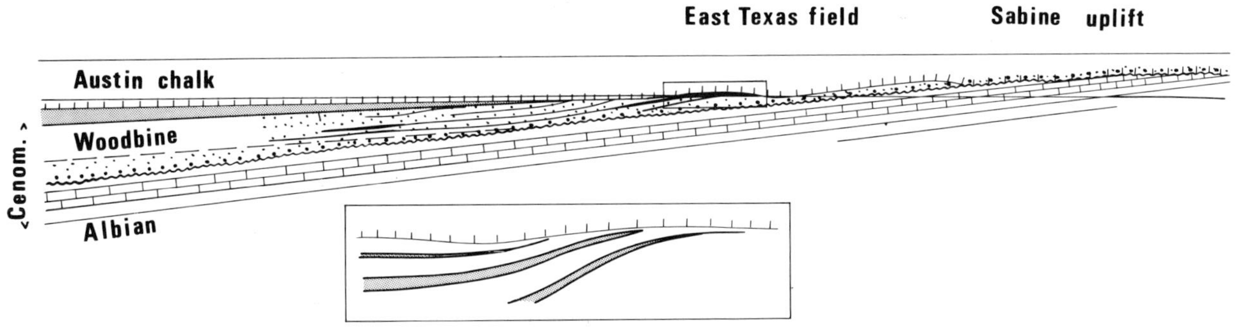

Fig. 105. — Section of Middle Cretaceous passing through East Texas pool (Gussow, 1972).

It is believed that the oil of this field has been in place since the Paleocene. The initial reserves in place are estimated at 1.1 Gt, and the recovery factor is 76 % (Gussov, 1972).

In the same way, but on a far more modest scale, the Cut Bank Field (Montana) lies along a pinchout of Cretaceous sands on the west flank of the Sweetgrass Arch, which plunges by less than 1 degree. The initial reserves of the field are estimated at about 30 million tons (Figure 106).

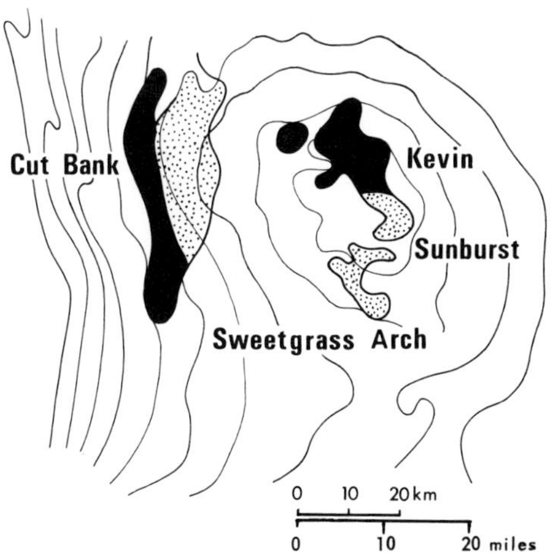

Fig. 106. — Section of Cut Bank and Kevin/Sunburst Field (Montana) (Dobbin, 1956).

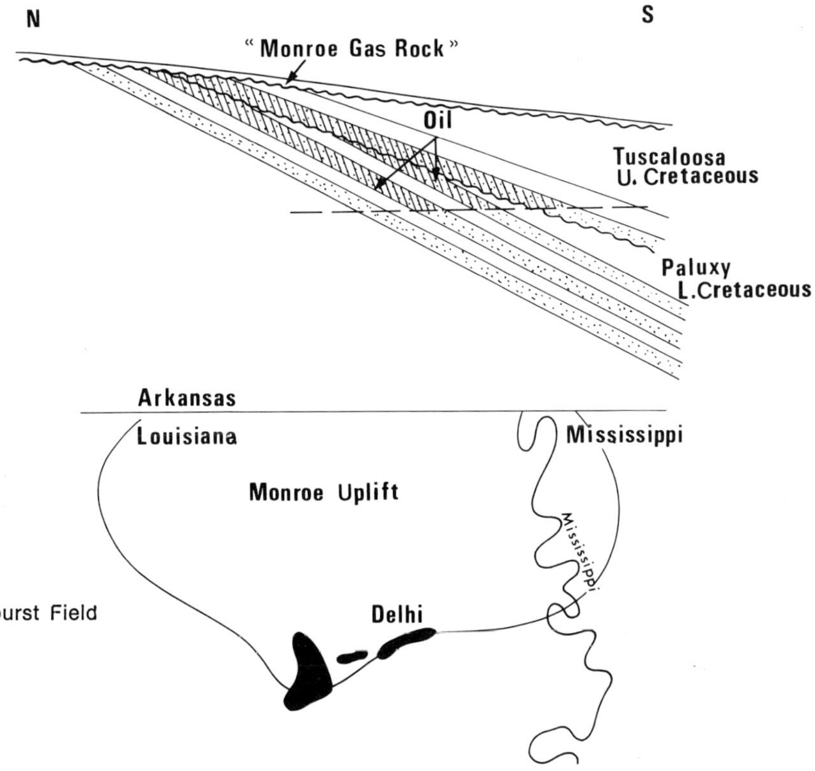

Fig. 107. — Section and map of Delhi Field (Louisiana) (Powel, 1972).

The marginal areas of regional highs are frequently the location of many low amplitude unconformities, corresponding to the recording of various uplift phases. One example of these combinations giving rise to stratigraphic traps is offered by the Delhi Field (Louisiana), discovered in 1944 on the south side of the Monroe Dome (Figure 107). The reservoir consists of Tuscaloosa sandstones (Upper Cretaceous) and Paluxy sandstones of the Lower Cretaceous, separated from the former by a slight unconformity, and cut in an erosion pinchout by the Monroe gas rock that forms the seal. The productive area covers 54 km^2. The reservoir has an average thickness of 6 m, porosity of 32 %, and permeability of 1400 md at a depth of 1000 m. Reserves are estimated at 35 Mt of light oil (41 °API) and nearly 4 Gm^3 of gas (Powell, 1972).

The margins of the basin are also the site of transgressive overlaps often complicated by synsedimentary faults. This is observed, for instance, on the eastern margin of the Eocene basin of Cambay (India) in the Meshana area. The oil is found in Oligocene sands folded and closed by a fault and pinchout under the transgressive Miocene, and is believed to originate in the Cambay shales of the Middle to Lower Eocene (Gambhir, 1976) (Figure 108).

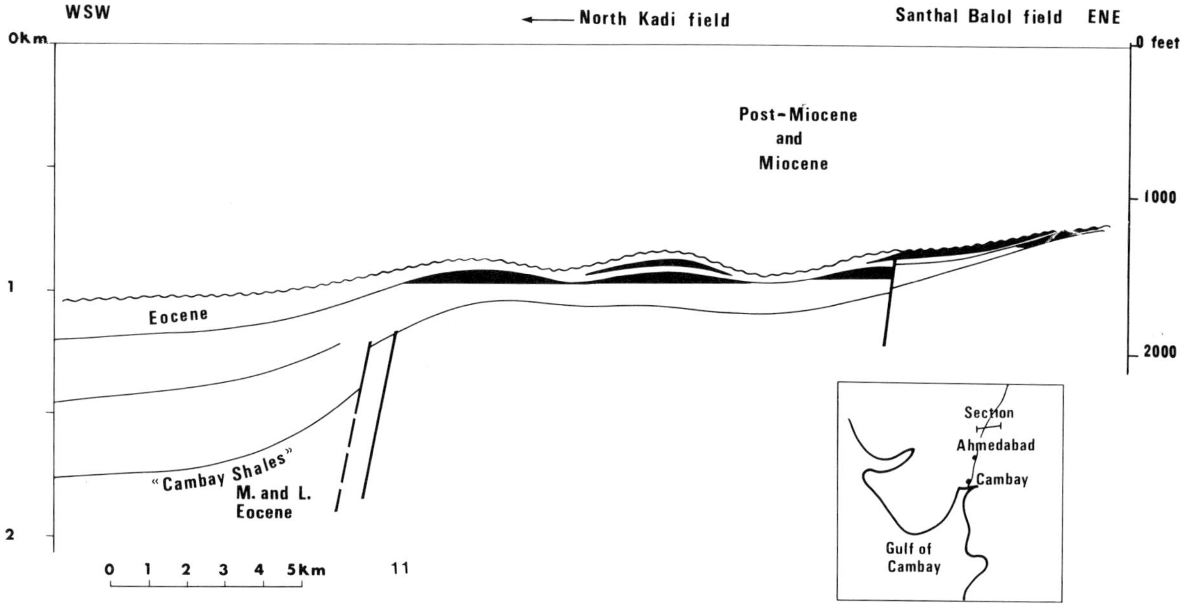

Fig. 108. — Section of North Kadi and Santhal Balol Fields on west flank of Cambay Basin, India (Gambhir, 1976).

The margins of the San Joaquin Basin in California are also characterized by a progressive transgression of the Miocene and the Pliocene, giving rise to the closure of different traps, especially the Coalinga and Midway Sunset Fields (see Section 3.3 of Part 3, and Figure 109).

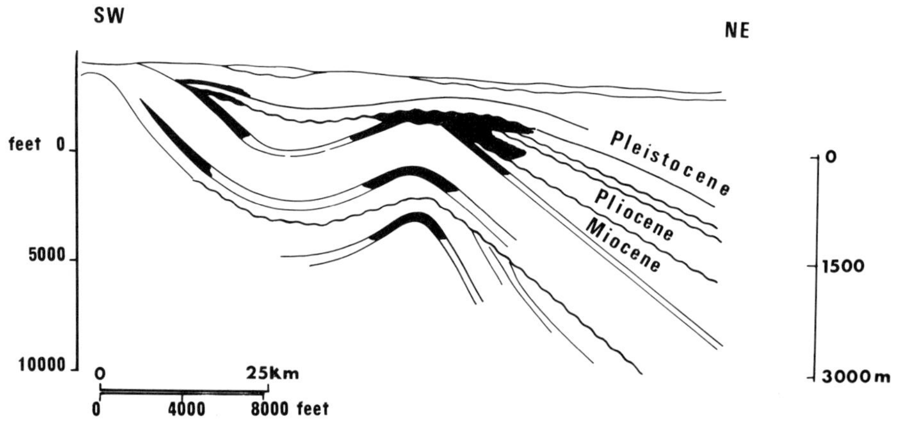

Fig. 109. — Section across Midway/Sunset Field, California (Hill, 1956).

A combination of productive horizons above and below the unconformity plane sometimes gives the appearance of a false anticline. This is the image given by the giant complex of the Bolivar Coast Field, one of the world's largest accumulations with 4.2 Gt. The oil is found in the basal sands of the Miocene onlap, with a southwest dip, and below the unconformity, in the Eocene sandstones subdivided into faulted blocks, with a general dip towards the northeast (Miller *et al.,* 1958).

A highly favorable pattern is provided by the presence of an onlapping series functioning as a source rock. This may supply all the reservoir rocks which it covers, both the basal horizons of the transgression and the series itself, often widely varied and porous due to the alteration of the pre-unconformity formations. This is the model observed in many Libyan fields, and especially in the Sarir Field, as well as the giant Prudhoe Bay Field. The latter, discovered in 1968 at the front of a basin covering 180,000 km^2, is located on the north coast of Alaska, on a regional zone parallel to the shore and separating two syncline areas 8000 to 9000 m deep. The trap is formed by an anticline that is eroded progressively from south to north, exhibiting series ranging from the Mississippian to the Jurassic, and covered unconformably by Barremian black shales. The dips are low to the south, high and cut by faults to the north. The reservoirs consist mainly of the Lisburn limestones (Mississippian/Pennsylvanian) and by the Sadlerochit sandstones (Permo-Trias) with a deltaic facies (Figures 110 and 9).

The seal and the main source rock are provided by the overlapping Barremian black shales, and the Trias and Jurassic also contain source rocks. The structure originated in the Early Cretaceous, and then collapsed towards the southeast and east. The hydrocarbons originate mainly in the Barremian shales, which were buried progressively in the neighboring subsident zones, especially to the south and east, during the Cretaceous. Migrations occurred simultaneously along the unconformity planes. This field, the largest in the United States, has reserves of about 1.4 Gt and nearly 800 Gm^3 of gas (Jones & Speers, 1976; Jamison *et al.,* 1980).

These examples reveal the value of the regional highs, even when they are not productive at the top. In fact, in addition to their ability to shelter giant fields of the dome and anticline type, they often provide the opportunity, on their margins, for pinchouts of all types, Levorsen's permeability fringes, which may constitute numerous pools. These marginal zones always provide a topic of great interest for exploration, whether the top is productive or not.

4.2.2. **Pools by pinchout and facies variation**

The lateral passage from a porous and permeable facies to an impermeable formation updip, or the pinchout of a reservoir horizon within a compact rock, can create trapping conditions. As a rule, a distinction is often drawn between detrital formations and carbonate series.

4.2.2.1. ***Traps and pools related to sand bodies***

The term sand body must be taken here as it is most broadly understood, as a porous and permeable detrital formation. Facies variations which gave rise to the contours of the anomaly exhibiting reservoir qualities may, very schematically, arise from variations in conditions of deposition, or by subsequent changes, while both processes are often closely associated. These detrital deposits often correspond to sedimentation anomalies, sedimentary mounds or bodies of limited size, deposited under certain local and specific conditions of deposition and currents. A statistical analysis dealing with 7241 sandstone formations forming the reservoirs of various stratigraphic traps in the United States gives the following distribution for their origin (Curtis *et al.,* 1961) :

23 % correspond to offshore bars,
19 % correspond to nearshore marine facies,
19 % correspond to beach facies,
7 % corrsepond to channels,
19 % correspond to changes in regional facies,
4 % correspond to unconformities,
9 % correspond to miscellaneous facies

The most frequent reservoirs of the American stratigraphic traps thus belong to sand bar deposits lying on the shelf, and to coastal or beach formations. The latter predominate in the largest stratigraphic accumulations. As a rule, large-scale facies variations only play a secondary role in comparison with sand bodies of local extension (12 % against 88 %) (Curtis *et al.,* 1961).

Many examples are available of old deltaic units, especially in the detrital formations of the North American Cretaceous platforms, where they have been analyzed in detail. An interesting case of a pool

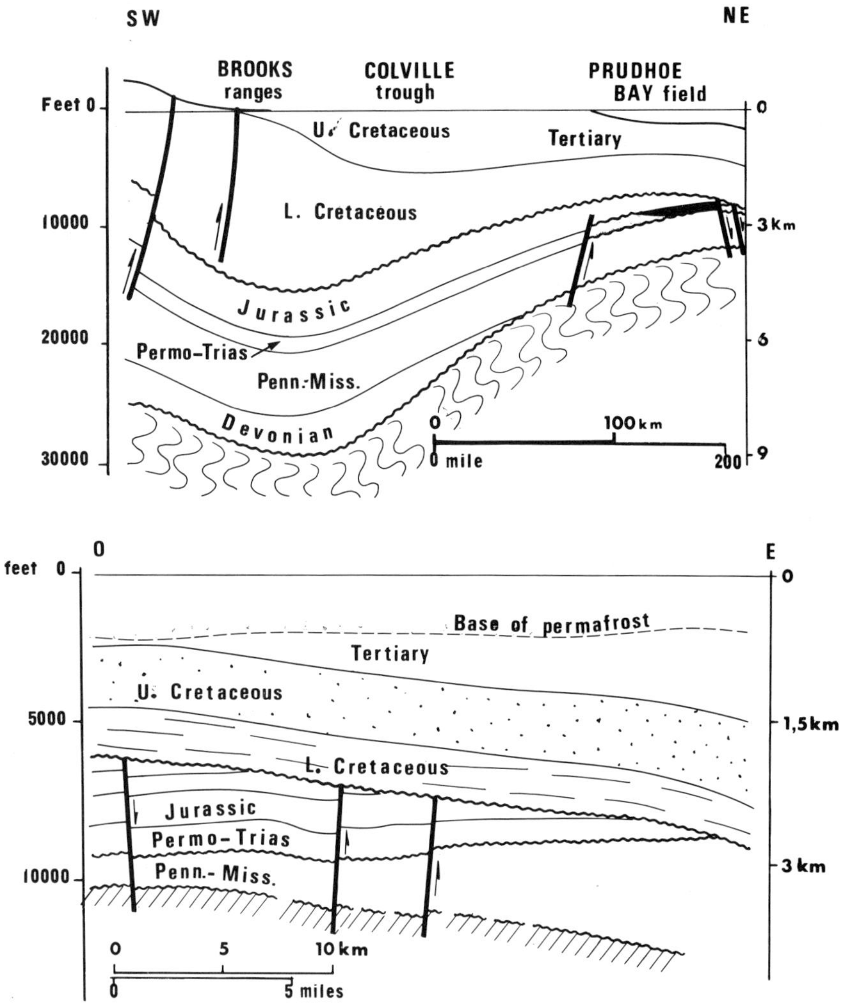

Fig. 110. — Section of North Alaska Basin and Prudhoe Bay Field showing transgression of Cretaceous shale source rocks on underlying reservoirs (Jones & Speers, 1970).

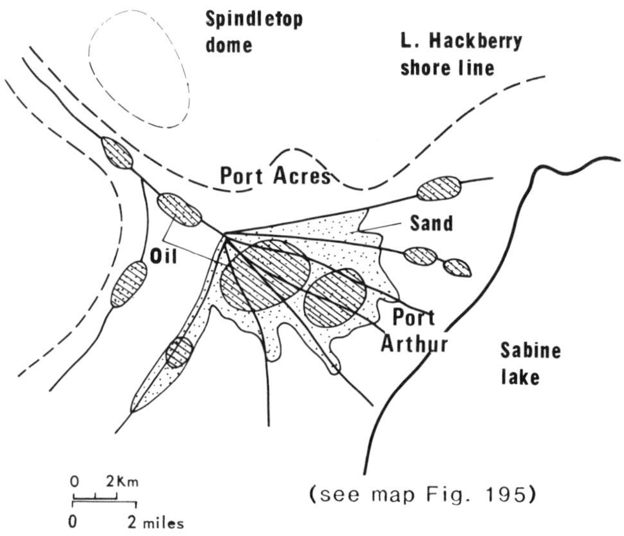

(see map Fig. 195)

Fig. 111. — Paleogeographic sketch showing the situation of Port Acres and Port Arthur Fields in the deltaic unit of the Oligocene Hackberry sands (Halbouty & Barber, 1972).

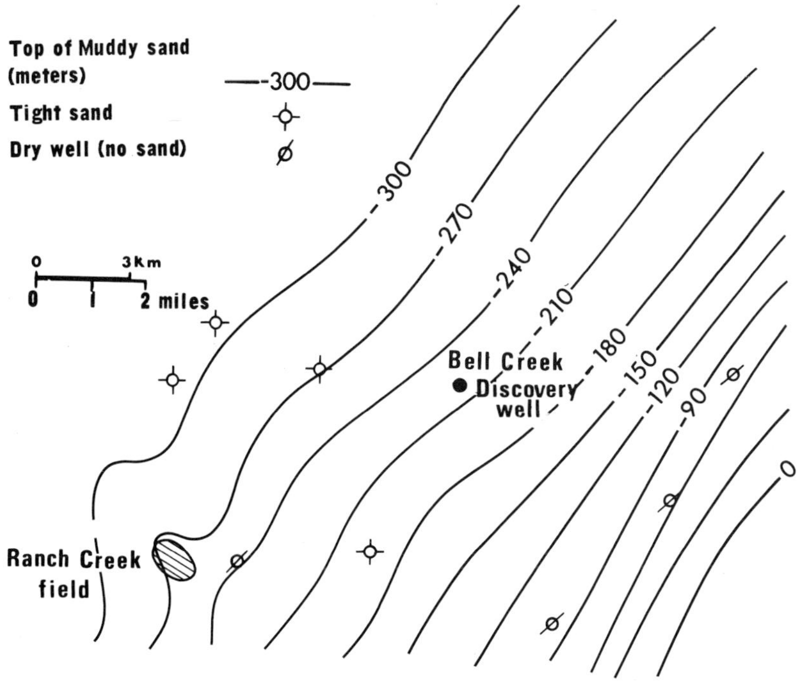

Fig. 112. — Bell Creek Field (Montana) : Structural map showing distribution of wells drilled at time of first discovery (Gregor & Biggs, 1968).

formed by sand bars of the deltaic and littoral type is offered by the Bell Creek Field (Montana), which retained the overall morphology of its depositional conditions. Other examples are found in the Gulf Coast, including the Port Acres and Port Arthur Fields (Texas), partly discussed under fault traps (pages 144 and 164 of this chapter, Figures 81 and 111).

The Bell Creek Field, discovered in 1967, is located on a remarkably uniform monocline sloping northwest 120 m/km of the Powder River Basin (Figures 112 and 113). The reservoir and trap are formed by a sequence of four sand bodies with an average thickness of about 10 m at 1350 m depth. The Muddy sands correspond to an offlap episode between two onlap phases, represented by shales rich in organic matter,

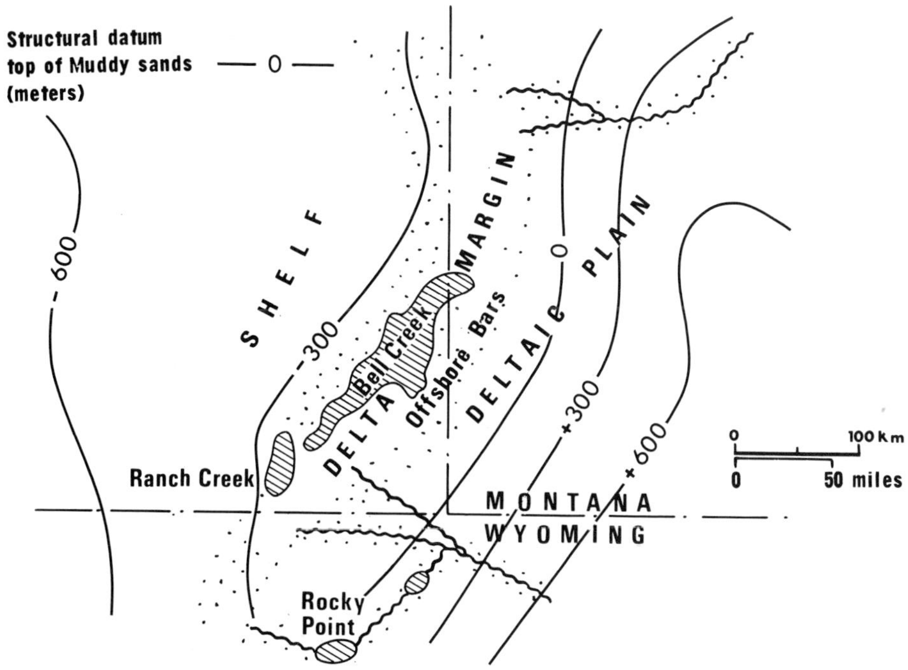

Fig. 113. — Paleogeographic reconstruction of Bell Creek region (McGregor & Biggs, 1968).

which play the role of source rock. The sands which correspond, in accordance with their stratigraphic distribution, to channel, delta front and offshore bar deposits, display excellent reservoir qualities : porosity about 30 % and permeability up to 10 darcys. The wells yielded an average of 70 t day^{-1} originally. Initial reserves amounted to some 30 million tons, with — 200 10^6 bbl — production of 3 Mt per year (60,000 bopd). This stratigraphic field was discovered after the negative drilling of all the structural anomalies of the region, but the geological data provided by these boreholes served to reconstruct the paleogeographic framework and to determine a sand distribution scheme, that led to the drilling of the discovery well (McGregor & Biggs, 1972).

The Candelas Field in the Reconcavo Basin (Brazil) is an original example of a stratigraphic trap in a synclinal position. The reservoir is formed by lenticular sandstones, fractured marls, and limestones of the Lower Cretaceous, at depths between 600 and 2400 m. These sand bodies correspond to fluviatile deposits in synsedimentary fault depressions. The four main reservoirs are regularly offset in relation to each other in accordance with the filling of the basin (Figures 114 and 151, Section 2.2. of Part 3).

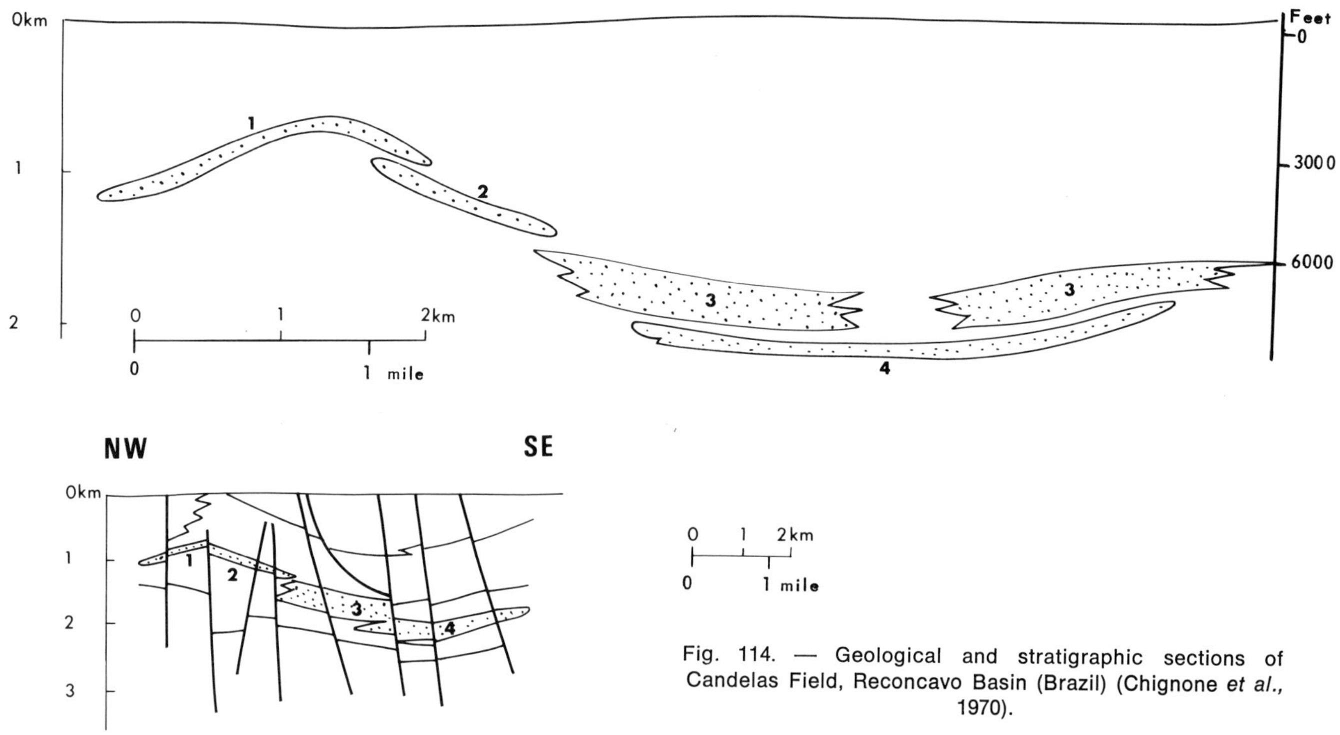

Fig. 114. — Geological and stratigraphic sections of Candelas Field, Reconcavo Basin (Brazil) (Chignone *et al.*, 1970).

Other types of sand body may originate in different habitats, particularly in deeper environments. This is the case of the Frigg Gas Field in the North Sea, which belongs to a sedimentary unit deposited at the base of a slope (see Figure 48 and Section 3.3. of Part 3). Sedimentary mechanisms giving rise to a pinchout trap are often varied and complex, and their combination contributes to the closure of the traps.

The Quiriquire Field (eastern Venezuela) discovered in 1928 offers a very specific example of a sub-outcropping stratigraphic pinchout trap. The oil is found in lens-shaped Miopliocene sands with shale beds, corresponding to old alluvial fans, which pinch out against a Cretaceous unconformity. The characteristics of these reservoirs are extremely variable, but with very good horizons exhibiting porosities of 17 to 20 %

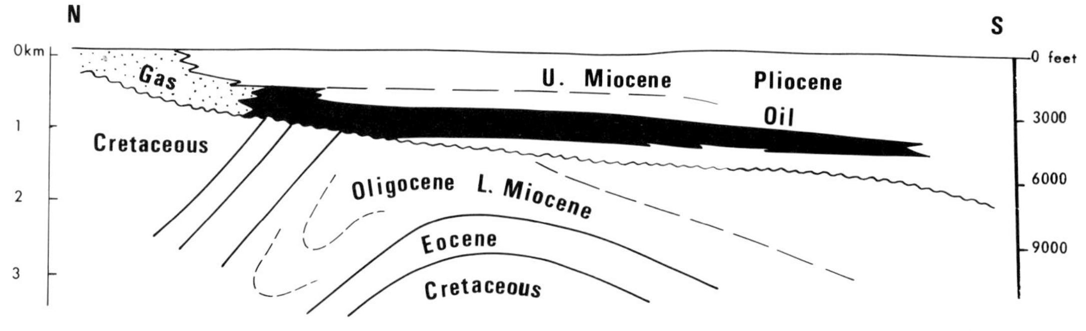

Fig. 115. — Section of Quiriquire Field (eastern Venezuela) (Borger, 1952).

and permeabilities up to 11 darcys (Figures 115 and 201, and Section 3 of Part 3). These sands are sealed and closed updip by shale seams, by sandstones impregnated with bitumen, asphalt plugs, and possibly by a fault. This closure is effective because gas-bearing sands (about 30 Gm^3) occupy the top portion of the traps. The crude is relatively heavy, 18 °API, with 13 % S and an initial GOR of 20 $m^3 t^{-1}$. The impregnated area covers 80 km^2, and the depth of the sands produced extends from 370 to 1320 m. Initial reserves in place amounted to some 350 Mt, of which a little more than 100 Mt have been produced.

4.2.2.2. ***Fields related to carbonate bodies***

Facies variations may be as interesting and valid with similarly effective closures in a carbonate environment as in a detrital area. In this family, the following will be examined in succession :

— traps closed by a lateral updip facies variation,

— the particular anomalies that consist of reef structures.

Pools by facies variations

The closure in this group is provided by the passage updip, for example on a nose, to compact and impermeable carbonate formations. These facies variations may be due in particular to differences in depositional environments, as well as subsequent transformations, either by the improvement of porosity and permeability downdip, or by the cementation and loss of these characteristic updips. In actual fact, these synsedimentary and postsedimentary actions are intimately linked, and the initial characteristics of the rock largely determine the circulation of the brines and the resulting transformations. While it is sometimes possible to identify a major feature, such as dissolution and dolomitization, for example in the Lima Indiana and Poza Rica Fields, most often only the final result can be recorded. The Jay Field is a good example.

In an initial view, one can group together the reservoirs formed by the dissolution of a previously compact or relatively non-porous rock downdip. These porosities, sometimes consisting of genuine cavities, may be caused by the leaching of meteoric waters on emerged carbonate surfaces. They are especially encountered under old unconformity planes. This is the case of the Lima Indiana Fields (70 Mt), in the Ordovician limestones along the Finley Dome (Perrodon 1966, p. 192). Porous and permeable zones are often related to dolomitization mechanisms, which can transform a compact limestone into a reservoir. This may be observed in the Poza Rica Field in Mexico, a dolomitic reservoir of the Tamabra facies (Upper Cretaceous), closed updip by the compact formation of the Tamaulipas limestones (Barnetche & Illing, 1956 and Section 3.3.5. of Part 3).

In a second view, one may group together the traps created by updip cementation. The Jay Field, discovered in 1970 by a seismic survey yielding data up to 7600 m at the eastern tip of Florida, offers a brilliant illustration of this. Intense drilling campaigns in this basin generally failed due to the poor qualities of the reservoir. Thirty-four fields, making a total of only 35 Mt, were discovered between 1937 and 1950 in these carbonate reservoirs at depths from 1800 to 3600 m. Another 38 small pools were found between 1960 and 1968 at depths from 3600 to 5500 m. The Jay trap is formed by a variation of the Smackover facies on a nose, at a depth of 4700 m. The reservoir consists of dolomitic limestones containing pellets and algae passing laterally into micrites. The reservoir characteristics are improved by leaching processes. Porosity ranges from 14 to 18 % and mean permeabilities from 20 to 50 md. The trap is sealed by evaporites (Figures 116 and 195).

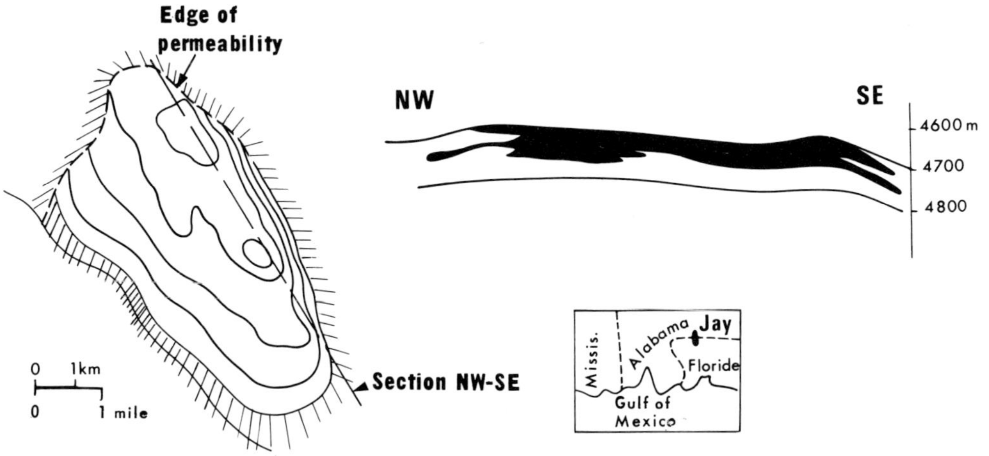

Fig. 116. — Structural map and section of Jay Field (Florida) (Ottmann, 1976).

The field covers an area of 58 km^2 — 14,400 acres — average "pay" thickness is 30 m (100 ft). Recoverable reserves are estimated at about 50 Mt and 10 Gm^3 (350 Bcf) of gas. The crude oil contains 9 % H_2S, 3 % CO_2 and N_2, and 88 % hydrocarbons. Free sulfur has been found in the impermeable zones (Ottmann *et al.*, 1976). The Jay Field is a remarkable example of the possibilities of stratigraphic trapping in a zone that appeared disappointing before its discovery.

Reef pools

— *Characteristics*

Reef traps and pools represent an important family of stratigraphic traps. Like some sand bodies with limited extension, but more precisely, they illustrate a category of traps which are identified with reservoirs, and which may be termed trap reservoirs. They generally consist of structures built by animal and plant organisms, marine and lacustrine, of the bioherm type, and by their destruction products. Among the most frequently encountered constructing organisms, and depending on the geological eras, are corals, algae (lithothamnium), rudists, echinoderms and stromatopores. These structures are often strongly dolomitized, in which case they lose their original character.

They are organized in structures of varying size, ranging from simple pinnacles to large continuous barriers extending over several hundred kilometers, and including atolls, fringing reefs, etc. The characteristics of the basin and the depositional environmental have a strong influence on the morphology and volume of these traps. The reefs constructed in the open sea, like those of the Silurian in Illinois, develop largely by surrounding themselves with detrital slopes that withstand the action of the waves, while a lagoon with fine deposits is established in the central part (Ingels, 1963; Felber, 1964). Reef formations built in a transgression period, like those of the Leduc Devonian, generally exhibit good seal conditions (De Mille, 1970). In lower energy environments, the structures often assume the conical form of a pinnacle, as in the Rainbow district. The presence of an environment which is frequently richer in salts facilitates dolomitization.

The analysis of the reef structures of the Devonian of Alberta shows the following general scheme :

— at the base, a transgression on an often eroded platform,
— slight subsidence favoring the development of a broad carbonate platform, rich in organisms,
— rapid increase in the subsidence rate causing the construction of bioherms on certain areas in low relief on the sea floor,
— gradual decrease in subsidence allowing the establishment of a reef border protecting the back-reef environment,
— halted reef growth and filling of the basin (Klovan, 1974).

The petrophysical characteristics of reef reservoirs are often excellent. While porosities display mean values of around 10 to 15 %, the structure of the porous environment is generally characterized by a predominance of large pores and slight water saturation, giving high outputs and recovery rates (often 40 to 50 % in primary recovery).

These reef structures often reach a height of 100 m, more rarely 300 m, and exceptionally 1000 to 2000 m, as in some of the great structures of Mexico or West Texas. The impregnated heights vary substantially from one trap to another, with a tendency to grow with decreased trap volume, as if each reef was supplied by a roughly equivalent quantity of hydrocarbons (Figure 67). The heights of the oil column average some 40 m in Ontario, and about 60 m (with a maximum of 200 m) in the Rainbow zone, and 300 m at Intisar in Libya.

Like the structures themselves, the dimensions of these traps exhibit a wide variety of types and sizes. In the Leduc region, for example, the Devonian reef traps display an average area of about 5 km^2, ranging from a few acres for the smallest, to 150 km^2 for the Redwater Field, where production extends to the lithothamnium sole. In the Keg River Formation of the Rainbow/Zama district, the average area is only 2.5 km^2 for the pinnacles, 8 km^2 for the atolls (Hriskevich, 1971), and falls to 0.8 km^2 for the 25 fields of the pinnacle type of the Silurian in Ontario.

In Alberta, the 35 Devonian D3 pools have average recoverable reserves of about 10 million tons, with a maximum of 80 Mt and reserves per km^2 ranging from 0.3 to 9 Mt. In the Rainbow/Zama region, the volume of oil in place ranges from 1.5 Mt for the small pinnacles to 40 Mt for the atolls (Hriskevich, 1971).

— *Examples of reef pools*

The Devonian pools of western Canada, which have been thoroughly analyzed, offer outstanding illustrations of this type of trap, and especially of the porosity distribution.

— The Redwater Pool, discovered in 1949 in Alberta, corresponds to a vast limestone bioherm lying on a carbonate platform encased in shales. Its area is 520 km^2 and it is 245 m thick. It resembles an atoll, surrounded by organism debris. The entire lower part corresponds to very finely stratified and compact

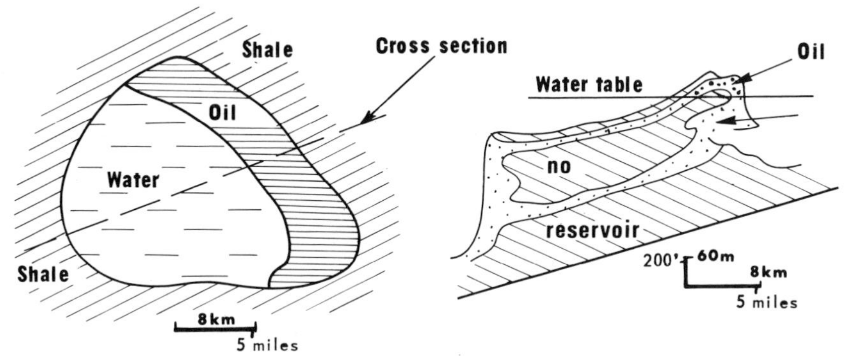

Fig. 117. — Map and section of Redwater oil Field (Canada) showing distribution of reservoir facies and fluids (Jardiné *et al.*, 1977).

lagoon facies (Figure 117). The oil is concentrated on the upper edge, and only occupies 5 % of the total volume of the reef. Thanks to an active aquifer, recoverable reserves amount to 120 Mt (65 % of the oil in place, of which 80 Mt were extracted by the end of 1975) (Jardiné *et al.*, 1977).

— The Judy Creek Field, discovered in 1959, also in Alberta, covers an area of 120 km^2 with a better developed height, largely enclosing the lagoon core facies, and the oil occupies three-quarters of the reef volume. Reserves in place are estimated at 112 Mt, but the recovery factor is less than 50 % in the absence of a large aquifer (Figure 118) (Jardiné *et al.*, 1977).

— The Horseshoe Atoll Pools (West Texas) on the northwest border of the West Texas Basin, mark the high points of an old atoll, extending in an arc over 280 km and covering an area of some 15,000 km^2. This vast reef structure was built in the Pennsylvanian on a carbonate platform devoid of detrital deposits. It then grew irregularly in the Lower Permian, until its growth was halted by a deepening of the environment by some 300 m, and the arrival of terrigenous deposits advancing from the northeast to the southwest. A westward tilt then affected it in the Pennsylvanian (Figures 119 and 190). The oil is distributed in about 15 fields sealed by marls of the Wolfcamp Formation, whose absence in certain locations appears to mark the sterility of certain high zones. The overall reef complex contains 355 Mt of recoverable oil. Only three fields, all located on the east flank, have reserves greater than 14 Mt. A total of more than 5500 wells have been drilled.

The largest accumulation of this complex is the Scurry Field, which covers 295 km^2 (73,000 acres), and has an oil column of 230 m (765 ft) with an anticipated recovery factor of 50 %. The field is produced by 1825 productive wells, established in a 16 hectare grid. The reservoir exhibits mean porosity and permeability of 10 % and 30 md respectively. The oil density is 42 °API, and it contains 0.18 % sulfur. The first field was discovered in 1948, nine years after the first well drilled in a sterile trap (Vest, 1970).

The Intisar D Field in Libya, discovered in 1967, is formed by a circular Paleocene algae and coral reef, 5 km in diameter and 385 m high. The mean porosity of the reservoir, which is very homogeneous, is 22 %, and its average permeability is 85 md with horizons at 500 md. The trap is filled with an oil column of 290 m. The crude is of the paraffinic type, with a density of 40 °API. The initial GOR was 90 m^3 per m^3. Original reserves in place were estimated at 250 Mt, of which 100 Mt had been produced by the end of 1978, and the final recovery factor is estimated at 75 % (Brady *et al.*, 1979).

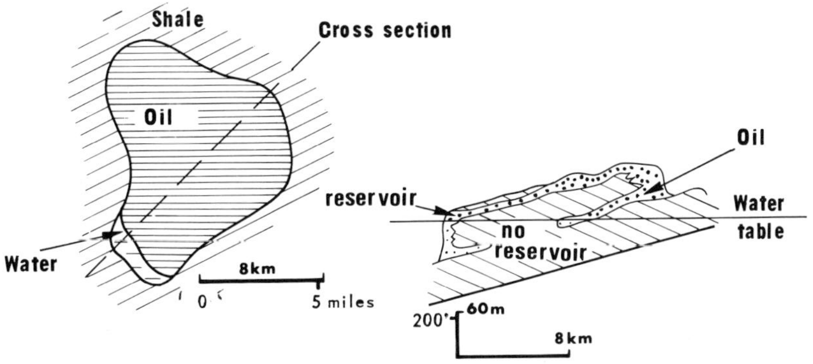

Fig. 118. — Map and section of Judy Creek oil Field (Canada) showing distribution of reservoir facies and fluids (Jardiné *et al.*, 1977).

4.3. MISCELLANEOUS POOLS AND UNCONVENTIONAL RESOURCES

This category includes accumulations unrelated to structural traps or stratigraphic systems, but resulting from various geological occurrences. Two main categories can be distinguished, more or less continuous. The first category includes miscellaneous traps, a term used to define accumulations whose trapping factors are specific and are generally related to diagenetic mechanisms or differential pressure, often associated with fractures and characterized by pressure anomalies. One particular case is that of basin floor gas accumulations, which may exhibit a regional extension and considerable reserves. The second category includes different oil and gas accumulations considered to be unconventional, and often extending on a regional scale.

The tremendous value of these different types of trap does not need to be emphasized, and they can offer completely new paths to future exploration. It is very well known that mineral concentrations, and

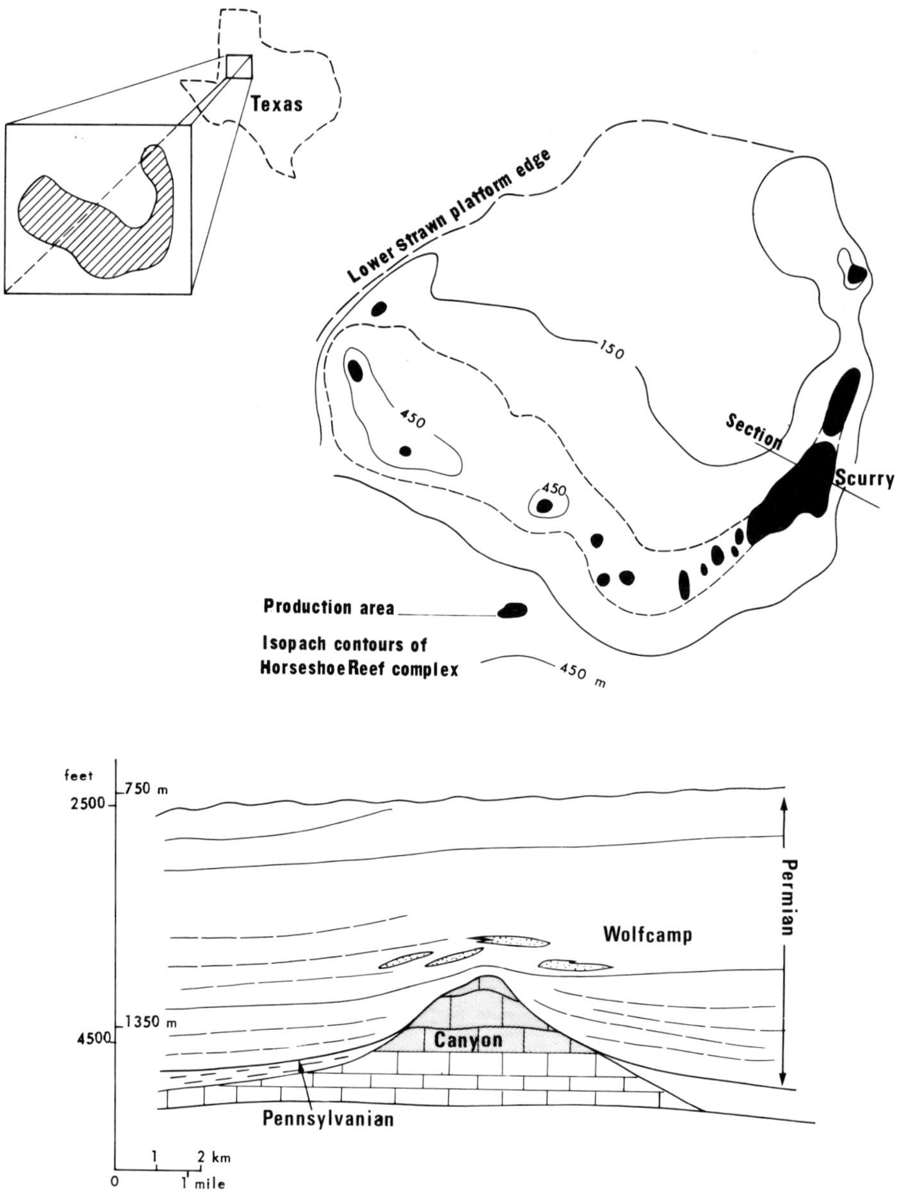

Fig. 119. — Map and section of Scurry/Horseshoe Atoll Field (West Texas) (Vest, 1970).

hydrocarbons are no exception to the rule, are more extensive and larger if the concentrations are lower. A large share of the resources of the twenty-first century could belong to these new accumulations.

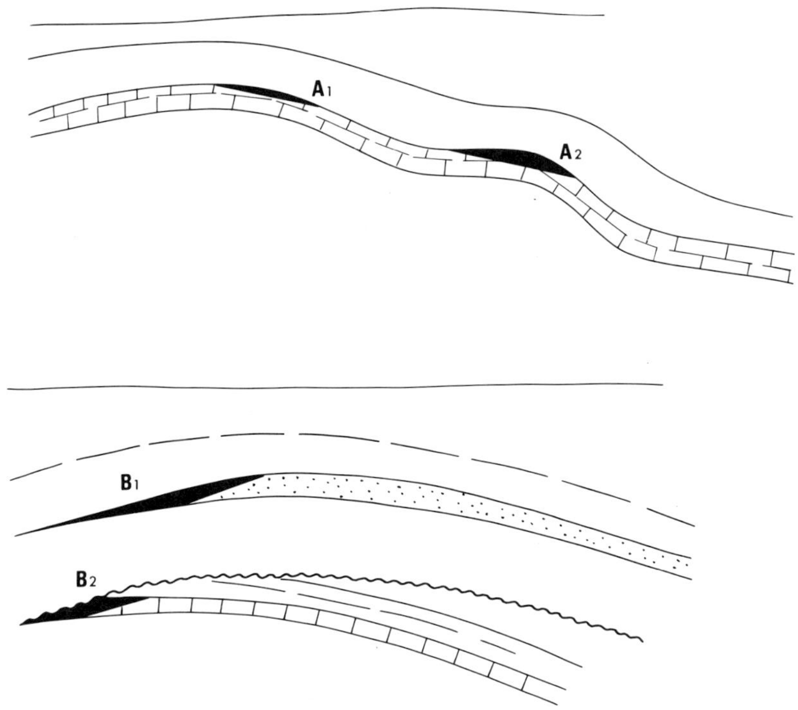

Fig. 120. — 'Diagenetic' traps.

4.3.1. **Miscellaneous pools**

The following will be discussed in succession :
— 'diagenetic' traps,
— fracture traps,
— basin floor gas pools.

4.3.1.1. ***'Diagenetic' traps***

Sudden variations in the cementation of certain reservoirs between the aquifer parts and zones occupied by hydrocarbons may create closures, likely to keep the hydrocarbons imprisoned in various structural situations, following movements or bed deformations. The resulting pools constitute a special family of stratigraphic or combination traps, qualified by certain authors as 'diagenetic' traps (Wilson, 1977). These traps are generally built in three stages :
— initial trapping, stratigraphic or structural,
— subsequent cementation of the aquifer part of the reservoir surrounding the pool,
— subsequent structural deformation, moving the sealed pool and placing it in an anomalous position.

Different schemes may thus be imagined. The most frequent appear to be caused by a tilt of an anticline trap, yielding a 'tilted' or slightly flanking accumulation, independent of a movement of the aquifer (Figure 120 A_1 and A_2). Other cases may be related to stratigraphic or unconformity pinchouts, which are subsequently tilted (Figure 120 B_1 and B_2).

A particular case of diagenetic traps is represented by pools closed by an asphalt plug. These accumulations, found quite frequently in provinces rich in seeps, such as California, Venezuela and Trinidad, have become extremely rare with the exhaustion of this type of province. In the oil-rich basins, permeable monoclines may be 'closed' by oxidized products creating an asphalt plug, and forming what is called an open pool. Examples of this type are the Coalinga and Midway Sunset Fields in the San Joaquin Basin in California.

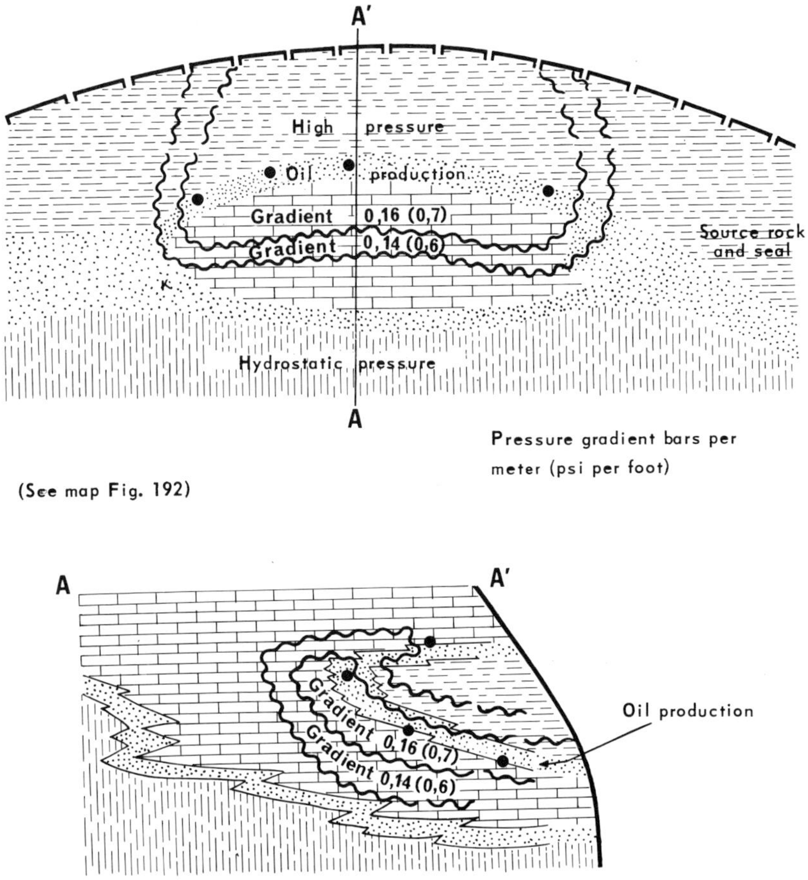

Fig. 121. — Map and section of Altamont/Bluebell pool in the Uinta Basin (Utah) (Lucas, 1976).

4.3.1.2. ***Fracture traps***

Many fields produce from fractured reservoirs in different types of trap. In some cases, the fracturation that gave rise to the reservoir also forms the trap, in the absence of any structural deformation, just as a stratigraphic trap consists essentially of the discontinuous development of a porous and permeable stratigraphic horizon. Hence in the small Florence Pool (Colorado), the oil is trapped in a fractured zone of the Pierre shales of the Cretaceous, in a region where the beds are substantially flat and appear to be unaffected by tectonic deformations or stratigraphic variations. The Spraberry Field in West Texas, whose initial reserves were about 40 million tons of oil, includes a complete series of pools spread over an area measuring about 240 km by 120 km (2500 km^2). The reservoir is formed by a sandy black argillite, siltstones, and a few fine sand intercalations of the Lower Permian. The porosity is generally less than 10 %. Average permeability is about 0.5 md, but this is improved by fractures, chiefly in the vertical direction.

Reservoirs formed by the intrusion of volcanic rocks may be assigned together with fracture pools. This is the case of the Lytton Springs and Chapman Fields in Texas, whose reservoirs are formed by an altered serpentine. In California, the Edison Field is formed by the alteration of the metamorphic basement at the core of an anticline. Similarly, in Morocco, the reservoir of the Oued Beth Pools is formed by a Paleozoic metamorphic schist fractured by faults.

Another of these miscellaneous and partly fractured traps is the Altamont/Bluebell Field in the Uinta Basin of Utah (Figure 121). The trap of this unconventional pool is related to a permeability pinchout of fractured compact calcareous sandstones, at the base of the Tertiary, in the midst of a compact overpressured unit forming the source rock. The many productive zones are distributed between 2400 and 5200 m depth within an area of 70 by 25 km. The reservoir qualities are guaranteed by a series of open vertical fractures in the most compact limestone horizons, and by a matrix porosity of the sandstones of around 5 %. The reservoir is sealed by high pressure compact sandstones. The oil originates in the surrounding lacustrine shales. The accumulation is the result of a simple primary migration and is distributed in accordance with the diagenetic conditions : heavy products (30 °API), naphthenic and immature in the upper parts, lighter paraffinic oils (46 °API) in the deeper parts. The geothermal gradient is low, 0.87 °C per 30 m. The field, which represents an example of a dispersed accumulation, with an area of 900 km^2, produced 360,000 tons in 1974 with 210 wells, and is expected to produce 35 Mt (Lucas & Drexler, 1976).

4.3.1.3. *Deep basin gas trap*

In certain not quite clear conditions, large accumulations of gas may be found in the deep, synclinal parts of certain basins. The gas is concentrated in sandstone formations with low porosity and permeability, displaying relatively high water saturation and a pressure lower than the hydrostatic pressure. The gas beds pass laterally updip without a structural feature or permeability barrier to more porous and permeable deposits, entirely of the aquifer type. A transition zone several km wide separates these two areas. Everything occurs as if the gas were trapped in the finest pores, at low pressure in the 'gas kitchen', bordered by aquifer series with a higher potential (Figures 122 and 123).

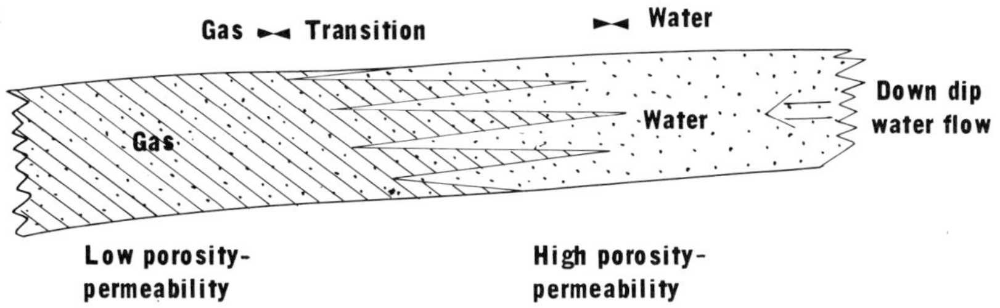

Fig. 122. — Diagram of gas entrapment in open pinchout of Alberta syncline (Masters, 1979).

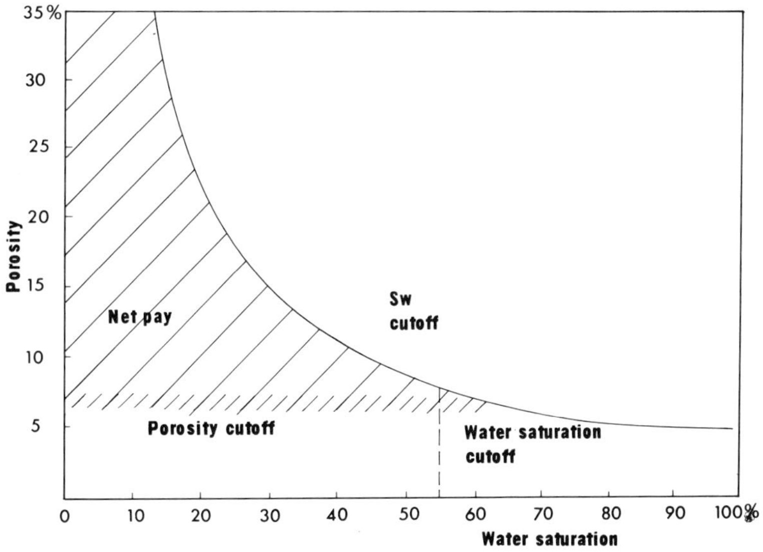

Fig. 123. — Limits to the variation in gas production in the Alberta syncline as a function of porosity and water saturation (Masters, 1979).

The pools, identified today only in the Cretaceous basins of North America (San Juan, Denver, Alberta), are characterized by considerable gas volumes and low production rates.

San Juan Basin

The central part of this basin, in a synclinal position, contains about 700 Gm^3 of gas in an area of about 10,000 km^2, in a Cretaceous sandstone unit up to 1500 m thick. The reservoir sandstones display porosity of 10 % and permeability of 1 to 2 md in the productive zone, passing updip, through a transition zone 8 to 16 km wide, to porosities of 20 % and permeability of 15 md in the marginal aquifer zone. Some authors claim that a favorable hydrodynamic gradient on the south limb could contribute to this trapping (Figure 124).

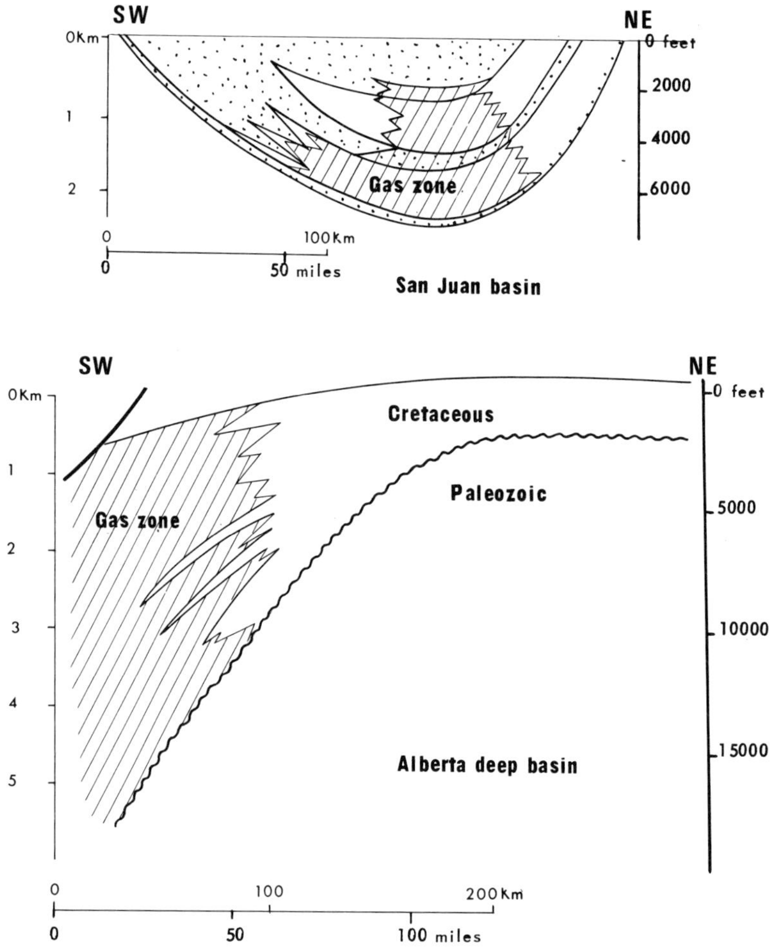

Fig. 124. — Sections showing the situation of gas zones in the San Juan and Alberta Basins (Masters, 1979).

Alberta syncline

The Mesozoic series passes rapidly from east to west from thin platform deposits (300 m), relatively porous (20 to 25 %) and permeable (more than 50 md) to a thick shale series (over 4000 m) with low porosity (10 to 15 %) and permeability (1 to 10 md). Simultaneously, the resistivity of the beds rises from an average of 20 ohms in the aquifer horizons to values up to 200 ohms in the deep gas-saturated series, varying inversely with porosity. With rare exceptions, the overall Mesozoic series of basin facies appears to be saturated with gas below a depth of about 1000 m. This impregnated volume represents over 125,000 km^3, over an area of nearly 70,000 km^2. Variations in the reservoir characteristics give rise to more favorable concentration and production zones which can be qualified as a field. The gas volumes thus trapped amount to some 12,000 Gm^3, including 4000 Gm^3 producible at a price of \$2 per 1000 ft^3. Well productivity averages 4 million cfd (120,000 m^3 per day) and may be multiplied by four (450,000 m^3 day^{-1}) in the best zones.

The Milk River Field, in the southeast part of Alberta, which corresponds to a trap of the same type, has reserves of 250 Gm^3 in a reservoir with average values of 15 % porosity, 1 md permeability, and 45 % water saturation (Masters, 1979).

4.3.2. **Unconventional hydrocarbon concentrations**

Oil and gas are generally found in pools, the only natural concentrations allowing their production today. It is customary to qualify the hydrocarbons of these accumulations, which are of limited dimensions, as conventional. But in exceptional cases, oil and gas are also found in concentrations that are considerably spread out and liable to occupy huge volumes.

4.3.2.1. *Tar sands*

The largest and relatively most accessible accumulations are the heavy crude oils, and especially those of the tar sands, characterized respectively by a density of 10-20 °API and viscosity < 10,000 cP for the heavy crudes and < 10 °API and > 10,000 cP for tar sands. These oils are often immature products, but also hydrocarbons degraded by their nearness to the surface. These accumulations appear to be trapped, like conventional hydrocarbons, by the combination of stratigraphic and structural factors, and especially in vast deltaic units.

Most of the reservoirs correspond to cases of river sands with good characteristics. In the main accumulations, and especially in Alberta and Venezuela, these sands are sealed by an onlap of marine shales. The extension of the deltaic system allows vast accumulations to form by secondary migrations that may extend over a hundred kilometers.

Main accumulations

By the latest count, some sixteen accumulations contain about 320 Gt of oil in place, in other words substantially the same order of magnitude as proved reserves in place of conventional crude oil. About nine-tenths of these resources are concentrated in Northern Alberta and in Venezuela in the Orinoco asphalt belt. The West Canadian Peace River, Cold Lake, Lloydminster, Wabasca and Athabasca accumulations, trapped in a Cretaceous sand monocline, represent some 100 Gt in place (see Section 3 of Part 3).

The Orinoco asphalt belt in Eastern Venezuela appears to be very similar to that of Canada on the edge of the Guyana shield, in a Cretaceous deltaic complex. The oil volumes in place are estimated at 100 to 260 Gt.

The third accumulation is located in the Melekess depression, in the Volga-Urals province of the USSR. Quantities in place amount to about 18 Gt, distributed over some 11,000 km^2 between Kazan and Kuybyshev. Three-quarters of this oil are trapped in an Upper Permian deltaic complex on the south flank of the Tatar Arch. The USSR has two other large concentrations in Siberia :

— at Olenek, also in Permian sands (1.2 Gt),

— at Sihgir, in Cambrian carbonates (2 Gt).

In the United States, and especially in Utah, tar sands are found in :

— Permian sands of the 'asphalt triangle' (2.4 Gt) probably originating in the 'Phosphoria' Formations,

— Eocene and Cretaceous deltaic sands (1.7 Gt) : they are also found in California, in the Mio-Pliocene sands of the Santa Barbara Channel.

Other accumulations have been identified in the Canadian Arctic, Melville Island, in the Triassic sands at Bemolanga (Madagascar) (within Liassic sands), and in Miocene lenticular sands in Selenizza (Albania) (Demaison, 1977; and Chilingarian & Yen, 1978).

4.3.2.2. *Oil shales*

This term applies to rocks which are generally shaly but sometimes calcareous, especially rich in organic matter which is more or less evolved. This kerogen can be transformed into petroleum products by pyrolysis. This operation, which requires about 250 calories per gram of rock, sets the lower limit of their organic matter content, approximately 2.5 %, or the equivalent of the quantity of fuel necessary for pyrolysis, since the peak capacity of kerogen is about 10,000 calories per gram. Due to their volume and their wealth of organic matter, these rocks offer very large potential hydrocarbon resources. The quantities of hydrocarbons in place that they may contain has been estimated at some 530 billion m^3 (Burger, 1973).

The main deposits correspond to Paleozoic or Mesozoic epicontinental sea or lake deposits, chiefly in the Tertiary. The largest are those of the Eocene lacustrine deposits of Green River in the Piceance (Colorado), Green River and Washakie (Colorado), and Uinta (Utah) Basins, which cover an area of 65,000 km^2, with thicknesses ranging from 1 to 600 m. These are dolomitic shales containing 8 to 50 % organic matter, mainly of algal origin. Potential reserves in place are estimated at about 350 Gt of oil.

The Diatom shales of Monterey and Sisquoc, dating from the Miocene and Pliocene in California, that extend over nearly 8000 km^2, appear to contain a potential of about 7 Gt of oil, including 140 Mt outcropping and of high quality. The Devonian and Mississippian black shales that cover 750,000 km^2 in the eastern and central United States, between the surface and a depth of 3000 m, contain 5 to 25 % organic matter over thicknesses up to 250 m. These low potential oil shales could contain some 30 Gt of potential oil in the richest zones, and about 100 tons in the poorest formations.

The world's second largest accumulation is represented by the Permian formations of Irati in Brazil, amounting to about 120 Gt of potential oil. The Lower Paleozoic on the periphery of the Scandinavian shield contains rich horizons of oil shales in Sweden and especially in the Baltic countries, where they are produced, including the kukersite which contains up to 40 % organic matter. Large Cretaceous/Jurassic deposits are worked in Manchuria, together with coal. Other smaller deposits have been found in :

— Saharan Silurian,
— the Devonian of the eastern basins of the United States,
— the Permo-Carboniferous in Australia (kerosene shales and tasmanite) and South Africa,
— the Trias in Zaire,
— the Lower and Upper Jurassic of Western Europe,
— the Jurassic and Cretaceous in Alaska and Canada.

4.3.2.3. *Geopressure gas*

Other unconventional resources are beginning to be identified for gas also. The main discovery is the gas contained in deep high-pressure aquifers, also called geopressure gas. It is well known that the water of deep aquifers contains quantities of methane that increase rapidly with pressure. When brought to the surface, this water releases its gas, giving the classic 'lemonade'. It has been calculated that the volumes of gas in solution could amount at saturation to 17 m^3 per m^3 (100 cf per barrel) at 6000 m, 170 m^3 (1000 cf) at 10,000 m. Hence basins such as that of the Gulf Coast could contain resources amounting to about 1400 Tm^3 and perhaps more over an area of 400,000 km^2, if it is confirmed that the free gas, in bubbles in the fissures, is individualized within these aquifers. Already wells at 4000 m are yielding about 9 m^3 per m^3 of water, or outputs of around 2500 m^3 per day of gas for a daily discharge of 280 m^3 per day of water.

4.3.2.4. *Methane hydrates*

Other large gas resources are stored in certain cold zones in a solid state in the form of hydrates. They are found in sedimentary beds on the floor of certain oceans. They are mainly known in the Arctic zones, where the permafrost provides an effective and continuous seal. Initial estimates point to considerable volumes of around $50 \cdot 10^{15}$ m^3! (Hodgson, 1978).

***

These unconventional accumulations can probably only be worked in conditions that are different from those prevailing in conventional fields. The often considerable size of these resources must be considered from the standpoint of very low production rates.

Summary and conclusion

The trap, a key component of petroleum exploration, is a closed zone, generally exhibiting relatively low pressure in comparison with the neighboring zones. All pools correspond to a trap, but the contrary is obviously untrue. It is customary to distinguish between structural traps and other traps, among which the stratigraphic traps are the most frequent. Structural traps, including domes and anticlines, are both the most numerous, the largest, and perhaps for this reason the easiest to find. But it is increasingly emerging that entrapment is far more a question of dynamics than geometry, and the sedimentological, hydrodynamic and geodynamic factors, the place in the basin as well as the time factor, often play a more important role than the structure itself.

— Structural traps obviously appear in folded and/or faulted basins, whatever the origin of these deformations. In basins which are relatively stable during sedimentation, and which consequently have homogeneous reservoirs and aquifers, hydrocarbons normally occur in a concentrated habitat, within structural traps. In a basin which is unstable from the sedimentary period, the hydrocarbons are normally in a dispersed habitat, and structural traps are often predominant.

— Stratigraphic traps, on the other hand, predominate in relatively stable basins with a low structural relief, and which exhibit, both in sand/shale areas and in carbonate series, significant variations in facies or a geomechanical framework that favors a closed or pressurized system. In both cases, the dimensions of the traps condition the type of habitat. The absence or rarity of stratigraphic traps in a basin must not be taken to mean that stratigraphic factors are unimportant in trapping. On the contrary, it is very rare that sedimentary mechanisms fail to influence significantly the location of the oil or gas, and it is obvious that in both cases sedimentary analysis is a prerequisite to structural analysis.

— Miscellaneous and unconventional traps are by definition more difficult to place within a geodynamic context. The mere fact that they form pools, which are sometimes large, should not only remind us that hydrocarbons can gather outside the conventional traps that currently form the main target of our explorations, but also that we must always display imagination and boldness in prospection.

Hydrocarbons are trapped in the first closed zones that they encounter, and this is one of the advantages of stratigraphic traps. If the basin is very subsident, synsedimentary structural traps are numerous and the oil and gas are trapped in the neighborhood of the source rocks. If the basin is relatively stable and exhibits pinchouts (regional pinchouts accompanied by unconformities are especially favorable), the latter may concentrate primary accumulations, which are sometimes very large, and which may preserve this potential in the absence of renewed fluid movements caused by subsequent tectonic deformations.

If a basin of this type is subsequently folded, the oil and gas are again set in motion and trapped in the structures. In this case, stratigraphic factors have certainly played an important, but not exclusive role, and have served to direct the hydrocarbons towards structures that are better situated from the paleogeographic standpoint. In fact, pools correspond to the coexistence of a trap and a flow of hydrocarbons. Hence it is important for the trap to be located on a migration path, and to offer a closed zone at the time that these fluids are transferred. If traps that have been closed early are more likely to be better placed at the time of the first migrations, it may also be said that younger traps have better chances of preserving hydrocarbons generated later, as the risks of leaching, leakage and seepage with the passage of time can be blamed for the sterility of many traps.

In other words, general concepts must be adapted to the specific case of each zone. The value of a trap can only be assessed by situating it in space in relation to the source rocks and in time in relation to the periods of oil and gas genesis, and by attempting to make the most accurate possible analysis of the relative chronology of the different stages in the formation of a pool.

REFERENCES

ABRAHAM, H. (1960). – Asphalts and allied substances, 1. – Van Nostrand, Princeton N. J., 6th ed., 370 pp.

ALESKESEYEV, F. A. (1974). – Zonality in oil and gas formation in the Earth's crust based on isotope studies. – *Geol. Nefti i Gaza,* **4,** 62-67.

ALLEN, G. P., LAURIER, D. & THOUVENIN, J. (1979). – Etude sédimentologique du delta de la Mahakam. – *Notes Mém. Comp. franç. Pétroles,* **15,** 156 pp.

ARPS, J. J. (1964). – Engineering concepts useful in oil finding. – *Bull. amer. Assoc. Petroleum Geol.,* **48,** 2, 157-165, 7 fig.

ATWATER, G. I. & MILLER, E. E. (1965). – The effect of decrease in porosity with depth on future development of oil and gas reserves in South Louisiana. – *Bull. amer. Assoc. Petroleum Geol.,* **49,** 334 (abstract).

BAILEY, N.J.L., JOBSON, A.M. & ROGERS, M.A. (1973). – Bacterial degradation of crude oil : comparison of field and experimental data Chem. – *Geology,* **2,** 203-221.

BAKER, E.G. (1962). – Distribution of hydrocarbons in petroleum. – *Bull. amer. Assoc. Petroleum Geol.,* **46,** 1, 76-84.

BALDUCCHI, A. & POMMIER, G. (1970). – Cambrian oil field of Hassi Messaoud, Algeria. – *Mem. amer. Assoc. Petroleum Geol.,* **14,** 477-488, 19 fig.

BARK E., van den & THOMAS, O. (1980). – Ekofisk : First of the giant oil fields in Western Europe. – *Mem. amer. Assoc. Petroleum Geol.,* **30,** 195-224, 26 fig.

BARKER, C. & KEMP, M.K. (1979). – Natural gas stability in deep subsurface. – *Bull. amer. Assoc. Petroleum Geol.,* **63,** 3, 414 (abstract).

BARNETCHE, A. & ILLING, L.V. (1956). – The Tamabra limestone of the Poza Rica oilfield. – *Int. geol. Congr.,* 20th Sess. (Mexico).

BARTENSTEIN, H. & TEICHMÜLLER, R. (1974). – Inkohlung und Erdöl. – Fortschr. Geol. Rheinld. Westf., **24.** Trad. franç. P. ROBERT : Houillification et pétrole. – *Bur. Rech. géol. min.,* n° 5477, 220 pp.

BEBOUT, D.G. (1979). – Secondary carbonate porosity. Zelten field, Libya. – *Bull. amer. Assoc. Petroleum Geol.,* **63,** 3, 416 (abstract).

BERG, R.R. (1975). – Capillary pressures in stratigraphic traps. – *Bull. amer. Assoc. Petroleum Geol.,* **59,** 6, 939-956.

BERG, R.R. (1979). – Reservoir morphologies for Gulf Coast turbidite Sandstones, Texas. – *Bull. amer. Assoc. Petroleum Geol.,* **63,** 3, 418-419 (abstract).

BESSAGUET, J. & MARTIN, B. (1977). – Évolution de l'interprétation du champ albien de Cazaux au cours de son exploitation. – *Pétrole et Techniques,* févr., **241,** 21-35.

BILY, C. & DICK, J.W.L. (1974). – Naturally occurring gas hydrates in the Mackenzie Delta NWT. – *Bull. Canad. Petroleum Geol.,* **22,** 3, 340-352, 12 fig.

BLEAKLEY, W.B. (1968). – How an old giant is faring. – *Oil and Gas J.,* 1st July, 101-109, 8 fig.

BOIGK, H. (1981). – Erdöl und Erdölgas in der Bundesrepublik Deutschland. – Enke, Stuttgart, 313 pp.

BORGER, H.D. (1952). – Case history of Quiriquire field, Venezuela. – *Bull. amer. Assoc. Petroleum Geol.,* **36,** 12, 2291-2330.

BOWEN, J.M. (1975). – The Brent field. *In* : WOODLAND, A.W. (ed.) : Petroleum and the continental shelf of northwestern Europe, **1,** *Geology,* 353-362. Applied Science Publ.

BRADY, T.J., CAMPBELL, N.D.J. & MAHER, C.E. (1979). – Intisar « D » oil field. Libya. – *Bull. amer. Assoc. Petroleum Geol.,* **63,** 3, 423 (abstract).

BRUCE, C.J. & PARKER, E.R. (1975). – Structural features and hydrocarbon deposits in the Mackenzie Delta. – Proc. 9th World Petroleum Congr. (Tokyo), **2,** 251-261, 8 fig.

BUNGENER, J.A. (1969). Le champ de gaz de Groningen. – *Rev. Assoc. franç. Techn. Pétrole,* **196,** 19-32, 14 fig.

BUSCH, D.A. & GOVELA, S.A. (1978). – Stratigraphy and structure of Chicontepec turbidites, southeastern Tampico-Misantla Basin, Mexico. – *Bull. amer. Assoc. Petroleum Geol.,* **62,** 2, 235-246, 9 fig.

CASSAN, J.P. (1978). – Sédimentogenèse et diagenèse des réservoirs anciens. – Rapp. interne SNEA(P).

CELSO-PONTE, F., DOS REIS FONSECA, J. & CAMARRA MORALES, R. (1977). – Petroleum Geology of Eastern Brazilian continental margins. – *Bull. amer. Assoc. Petroleum Geol.,* **61,** 9, 1470-1482, 19 fig.

CHAPMAN, R.E. (1972). – Primary migration of petroleum from clay source rocks. – *Bull. amer. Assoc. Petroleum Geol.,* **56,** 11, 2185-2191, 5 fig.

CHEVILLY, F. de, COLO, G., HENRY, G., NICOLAÏ, A., QUOIX, J. & WINNOCK, E. (1969). – Le champ de gaz de Meillon-Rousse et la structure profonde de l'anticlinal de Pau. – *Rev. Inst. franç. Pétrole,* **24,** 5, 564-578, 14 fig.

CHIARELLI, A. (1970). – L'hydrodynamisme, ses applications dans la recherche pétrolière. – *Rev. Inst. franç. Pétrole,* **25,** 9, 1015-1036, 10 fig.

CHIARELLI, A. (1973). – Étude des nappes aquifères profondes. – Contribution de l'hydrogéologie à la connaissance d'un bassin sédimentaire et à l'exploration pétrolière. – Thèse Univ. Bordeaux I.

CHIARELLI, A. (1979). – Évaluation des possibilités de conservation des gisements d'hydrocarbures par utilisation conjointe de données hydrodynamiques et hydrochimiques. – *Bull. Centres Rech. Explor.-Prod. Elf-Aquitaine,* **3,** 1, 213-221, 1 fig.

CHIARELLI, (1980). – Evaluating basin potential : don't forget fluid geology. – *World Oil,* **191,** 5, 145-153.

CHIARELLI, A. & ROUCHET, J. du (1977). – Importance des phénomènes de migration verticale des hydrocarbures. – *Rev. Inst. franç. Pétrole,* **32,** 2, 189-203, 14 fig.

CHILINGARIAN, G.H. & YEN, T.F. (ed.) (1978). – Bitumens, asphalts and tar sands. – Elsevier, Amsterdam, 331 pp.

CLARET, J., TCHIKAYA, J.B., TISSOT, B., DEROO, G. & VAN DORSSELAER, G. (1977). – Un exemple d'huile biodégradée à basse teneur en soufre : le gisement d'Emeraude, Congo. – *In :* CAMPOS, R. & GONI, J. (ed.) : Advances in organic geochemistry 1975, 509-522. Madrid.

CLARET, J., JARDINÉ, S. & ROBERT, P. (1981). – La diversité des roches mères pétrolières. – *Bull. Centres Rech. Explor.-Prod. Elf-Aquitaine,* **5**, 2, 383-417, 13 fig., 1 tabl., 2 pl.

CLAYPOOL, G.E. & KAPLAN, I.R. (1974). – The origin and distribution of methane in marine sediments, 99-140. – Plenum Press, London.

CLAYTON, J.L. (1977). – Preliminary report : petroleum geochemistry of the Denver Basin. – Rocky Mt. Assoc. Geol. Fall Field Conf. Guidebook, 223-233.

CLAYTON, J.L. & SWETLAND P.J. (1980). – Petroleum generation and migration in Denver basin. – *Bull. amer. Assoc. Petroleum Geol.,* **64**, 10, 1613-1633, 18 fig.

COLEMAN, M.L., CURTIS , C.D. & IRWIN, H. (1979). – Burial rate, a key to source and reservoir potential. – *World Oil,* March, 83-92, 4 fig.

COLMAN-SADD, S.P. (1978). – Fold development in Zagros simply folded belt, SW Iran. – *Bull. amer. Assoc. Petroleum Geol.,* **62**, 6, 984-1003, 16 fig.

COMBAZ, A. (1973). – La matière algaire et l'origine du pétrole. – Actes 6e Congr. int. Géoch. organ., 423-438, 5 fig.

COMBAZ, A. & MATHAREL, M. de (1978). – Organic sedimentation and genesis of petroleum in Mahakam Delta, Borneo. – *Bull. amer. Assoc. Petroleum Geol.,* **62**, 9, 1684-1695, 11 fig.

COMBAZ, A. (1975). – Essai de classification des roches carbonées et des constituants organiques des roches sédimentaires. – *In :* Pétrogr. mat. org. (Colloque Paris, 1973), 93-101. Centre nat. Rech. sci., Paris.

COMBAZ, A. (1980). – Les Kérogènes vus au microscope. – *In :* DURAND, B. (éd.) : Kerogen, 55-111, 8 fig., 10 pl. *Technip,* Paris.

CONNAN, J. (1974). – Time-temperature relation in oil genesis. – *Bull. amer. Assoc. Petroleum Geol.,* **58**, 12, 2516-2521, 3 fig.

CONNAN, J. (1981). – Un exemple de biodégradation préférentielle des hydrocarbures aromatiques dans les asphaltes du bassin Sud-Aquitaine. – *Bull. Centres Rech. Explor.-Prod. Elf-Aquitaine,* **5**, 1, 151-171, 9 fig.

CONNAN, J. & VAN DER WEIDE, B. (1969). – Intérêt géochimique de l'étude des paraffines dans les pétroles et extraits de roche. – Rapp. inédit SNPA.

CONNAN, J., CASSOU, A.M. & VAN DER WEIDE, B. (1978). – Kerogen maturity and gas/oil properties in shale – sandstone sequences. – Indonesian Petrol. Assoc. ann. Conv. (preprint).

CONNAN, J. & VAN DER WEIDE, B.M. (1978). – Thermal evolution of natural asphalts. – *In :* CHILINGARIAN & YEN (ed.) : Bitumens, asphalts and tar sands, 27-55, 10 fig. Elsevier, Amsterdam.

CONNER, D. C., PETERSON, K. E. & WILLIAMS, J. J. (1975). – The Piper oil fiel, UK North Sea : a fault block structure with Upper Jurassic beach/bar reservoir sands. – *In :* WOODLAND, A. W. (ed.) : Petroleum and the Continental shelf of NW Europe, **1**, 363-78. Applied Science Publ., London, Barking.

CONYBEARE CEB (1976). – Geomorphology of oil and gas fields in sandstone bodies. – *In :* Development in Petroleum Science, **4**. Elsevier, Amsterdam.

COOK, T.D. (1979). – Exploration history of South Texas Lower Cretaceous carbonate platform. – *Bull. amer. Assoc. Petroleum Geol.,* **63**, 1, 32-49, 22 fig.

COPELIN, E.C. (1979). – Crude oil-like hydrocarbons in recent sediments of Santa Barbara Basin. – *Bull. amer. Assoc. Petroleum Geol.,* **63**, 3, 436 (abstract).

CORDELL, R.J. (1972). – Depths of oil origin and primary migration : a review and critique. – *Bull. amer. Assoc. Petroleum Geol.,* **56**, 10, 2029-2067, 5 fig., 6 tableaux.

CORDELL, R.J. (1973). – Colloidal soap as proposed primary migration medium for hydrocarbons. – *Bull. amer. Assoc. Petroleum Geol.,* **57**, 9, 1618-1643.

CORDELL, R.J. (1976). – How oil migrates in clastic sediments. – *World Oil;* Part 1 : Nov., 207-226, 3 fig.; Part 2 : Dec., 75-80, 5 fig.

CORDELL, R.J. (1977). – How oil migrates in clastic sediments. – *World Oil;* Part 3 : Jan., 97-200, 3 fig.; Part 4 : Feb., 36-38, 3 fig.

CURTIS, B.F. and Sandstone Res. Comm. (1961). – Characteristics of sandstone reservoirs in United States. – *In :* PETERSON, J.A. & OSMOND, J.L. (ed.) : Geometry of sandstone bodies, 208-219, 8 fig., *Mem. amer. Assoc. Petroleum Geol.*

CURTIS, D.M. (1980). – Paleotectonic settings for petroleum source rocks. – Int. geol. Congr., 26th Sess., Paris, Coll. **C 2**, 215-222.

DAILLY, G.C. (1976). – Pendulum effect and Niger Delta prolific belt. – *Bull. amer. Assoc. Petroleum Geol.,* **60**, 9, 1543-1549.

DEBYSER, J. & DEROO, G. (1969). – Faits d'observation sur la genèse du pétrole. – *Rev. Inst. franç. Pétrole,* **24**, 1, 21-48, 10 fig., 3 tabl. et **24**, 2, 151-174, 20 fig.

DEMAISON, G.H. (1977). – Tar sands and supergiant oil fields. – *Bull. amer. Assoc. Petroleum Geol.,* **61**, 11, 1950-1961, 9 fig.

DICKEY, P.A. (1975). – Possible primary migration of oil from source rock in oil phase. – *Bull. amer. Assoc. Petroleum Geol.,* **59**, 2, 337-344.

DOW, W.G. (1978). – Petroleum source beds on continental slopes and rises. – *Bull. amer. Assoc. Petroleum Geol.,* **62**, 9, 1584-1606, 22 fig.

DUNNINGTON, H.V. (1958). – Generation, migration, accumulation and dissipation of oil in northern Iraq. – *In :* WEEKS, L.G. (ed.) : Habitat of Oil, 1194-1251, 22 fig., Amer. Assoc. Petroleum Geol.

DURAND, B. & OUDIN, J.L. (1979). – Exemple de migration des hydrocarbures dans une série deltaïque : le delta de la Mahakam, Kalimantan, Indonésie. – Proc. 10th World Petroleum Congr. (Bucharest), **2**, 3-11, 8 fig. (publ. 1981).

DURAND, B. (1980). – Kerogen. *Technip, Paris,* 519 pp.

ELF AQUITAINE (1977). – Essai de caractérisation sédimentologique des dépôts carbonatés. 2. Éléments d'interprétation. 231 p., 76 fig. – Elf-Aquitaine, Boussens et Pau.

ELZARKA, M.H. (1975). – Geochemical relations of fluids in oil fields of Gulf of Suez. – *Bull. amer. Assoc. Petroleum Geol.,* **59**, 9, 1667-1675, 4 fig., 6 tableaux.

ERDMAN, J. G., MORRIS, D. A. & SMITH, J. E. (1971). – Migration, accumulation and retention of petroleum in the earth. – Proc. 8th World Petroleum Congr. (Moscow), **2**, 13-26.

EVAMY, D.O., HAREMBOURE, J., KAMERLING, P., KNAAP, W.A., MOLLOY, F.A. & ROWLANDS, P.H. (1978). – Hydrocarbon habitat of Tertiary Niger Delta. – *Bull. amer. Assoc. Petroleum Geol.,* **62**, 1, 1-39, 28 fig.

EVANS, C.R., McIVOR, D.K. & MAGARA, K. (1975). – Organic matter, compaction history and hydrocarbon occurrence, Mackenzie Delta. – Proc. 9th World Petroleum Congr. (Tokyo), **2**, 149-157, 8 fig.

EXUM, F.A. & HARMS, J.C. (1968). – Comparison of marine-bar with valley-fill stratigraphic traps, Western Nebraska. – *Bull. amer. Assoc. Petroleum Geol.,* **52**, 10, 1851-1868, 20 fig.

FREY, M.G. & GRIMES, W.H. (1970). – Bay Marchand-Timbalier Bay, Caillou Island, salt complex, Louisiana. – *Mem. amer. Assoc. Petroleum Geol.,* **14**, 277-291, 11 fig.

GAMBHIR, S.C. (1976). – Pattern of accumulation and composition of oils in Mehsana area of Cambay Basin, India. – *Bull. amer. Assoc. Petroleum Geol.,* **60**, 9, 1550-1561, 6 fig.

GATEWOOD, I.E. (1970). – Oklahoma city field – anatomy of a giant. – *Mem. amer. Assoc. Petroleum Geol.,* **14**, 223-254, 27 fig.

GHIGNONE, J.I. & ANDRADE, G. de (1970). – General geology and major oil fields of Reconcavo Basin, Brazil. – *Mem. amer. Assoc. Petroleum Geol.,* **14**, 337-358, 12 fig.

GILLESPIE, J. (1967). – The geology of the Sarir oil field, Sirte basin, Libya. – Proc. 7th World Petroleum Congr. (Mexico), **2**, 181-193, 7 fig.

GUSSOW, W.G. (1954). – Differential entrapment of oil and gas : a fundamental principle. – *Bull. amer. Assoc. Petroleum Geol.,* **38**, 5, 816-853, 17 fig.

GUSSOW, W.C. (1972). – The East Texas field – a new interpretation. – *Int. geol. Congr., 24th Sess., Canada (Montreal),* **5**, 82-91, 1 fig.

HALBOUTY, M.T. & HARDIN, G.C. (1959). – A geological appraisal of present and future exploration techniques on salt domes of the Gulf region. – Proc. 5th World Petroleum Congr. (New York), **2**, 95-108, 7 fig.

HALBOUTY, M.T., DOTT, R.H., SR., HALBOUTY, M.T., KING, R.G., KLEMME, H.D. & MEYERHOFF, A.A. (1970). – World's giant oil and gas fields, geologic factors affecting their formation and basin classification. – *Mem. amer. Assoc. Petroleum Geol.,* **14**, 502-525, 9 fig., 15 tabl.

HACQUEBARD, P.A. (1977). – Rank of coal as index of organic metamorphism for oil and gas in Alberta. – *Bull. amer. Assoc. Petroleum Geol.,* **61**, 5, 791 (abstract).

HAWKINGS, T.J., HATLEDID, W.G., BOWERMAN, J.N. & COFFMAN, R.C. (1976). – Taglu gas field, Beaufort basin, NW Territories. – *Mem. amer. Assoc. Petroleum Geol.,* **24**, 51-71, 30 fig.

HEDBERG, H.D. (1964). – Geologic aspects of origin of petroleum. – *Bull. amer. Assoc. Petroleum Geol.,* **48**, 11, 1755-1803.

HEDBERG, H.D. (1967). – Geologic controls on petroleum genesis. – *Proc. 7th World Petrol. Congr. (Mexico),* **2**, 3-11.

HEDBERG, H.D. (1974). – Relation of methane generation to undercompacted shales, shale diapirs and mud volcanoes. – *Bull. amer. Assoc. Petroleum Geol.,* **58**, 4, 661-673.

HEDBERG, H.D. (1979). – Methane generation and petroleum migration. – *Oil and Gas J.,* **77**, 19 (May 7), 186-192, 5 fig.

HERITIER, F.E., LOSSEL & WATHNE, E. (1980). – Frigg field – Large submarine – fan trap in Lower Eocene rocks of the Viking graben. – *Mem. amer. Assoc. Petroleum Geol.,* **30**, 59-79, 18 fig.

HILL, M.L. (1956). – Oil in California. – *In :* Symposium yacimientos petroleo. – Int. geol. Congr., 20th Sess., Mexico, **3**, 157-188, 15 fig.

HOBSON, G.D. & TIRATSOO E.N. (1975). – Introduction to petroleum geology. – Scientific Press Ltd., Beaconsfield, 300 pp.

HOOD, A., GUTJAHR, C.C.M. & HEACOCK, R.L. (1975). – Organic metamorphism and the generation of petroleum. – *Bull. amer. Assoc. Petroleum Geol.,* **59**, 6, 986-996, 7 fig.

HORTENBACH, R. von (1977). – Drucklösungserscheinungen in Karbonaten und ihre Bedeutung. – *Z. geol. Wiss.,* **5**, 5, 617-621.

HRISKEVICH, M.E. (1967). – Middle Devonian reefs of the Rainbow region of northwestern Canada, exploration and exploitation. – Proc. 7th World Petroleum Congr. (Mexico), **3**, 733-763.

HUBBERT, K. (1953). – Entrapment of petroleum under hydrodynamic conditions. – *Bull. amer. Assoc. Petroleum Geol.,* **37**, 6, 1954-2026, 44 fig.

HUGHES, D.J. (1968). – Salt tectonics as related to several Smackover fields along the northeast rim of the Gulf of Mexico Basin. – Gulf Coast Assoc. geol. Soc., 18th ann. Meet., **18**, 320-330, 8 fig.

HUMPHRIS, C.C. Jr. (1979). – Salt movement on continental slope, Northern Gulf of Mexico. – *Bull. amer. Assoc. Petroleum Geol.,* **63**, 5, 782-798, 15 fig.

HUNT, J.M. (1977). – Distribution of carbon as hydrocarbon and asphaltic compounds in sedimentary rocks. – *Bull. amer. Assoc. Petroleum Geol.*, **61**, 1, 100-104.

HUNT, J.M. (1979). – Petroleum geochemistry and geology. – Freeman, San Francisco, 615 pp.

JAMISON, H.C., BROCKETT, L.D. & McINTOSH, R.A. (1980). – Prudhoe Bay – a 10 year perspective. – *Mem. amer. Assoc. Petroleum Geol.*, **30**, 289-314, 22 fig.

JARDINE, D., ANDRELOS, D.P., WISHART, J.W. & YOUNG, J.W. (1977). – Distribution and Continuity of Carbonate Reservoirs. – *J. Petroleum Technol.*, July, 873-885.

JONES, H.P. & SPEERS, R.G. (1976). – Permo-Triassic reservoirs of Prudhoe Bay field, North Slope, Alaska. – *Mem. amer. Assoc. Petroleum Geol.*, **24**, 23-50, 26 fig.

JONES, R.W. (1981). – Some mass balance and geological constraints on migration mechanisms. – *Bull. amer. Assoc. Petroleum Geol.*, **65**, 1, 103-122, 11 fig.

KANTSLER, A.J., SMITH G.C. & COOK A.C. (1978). – Lateral and vertical rank variation : implication for hydrocarbon exploration. – *Austral. Petroleum Explor. Assoc. J.*, **18**, 143-156.

KATZ, D.L., CORNELL, D., KOBAYASHI, R., POETTMANN, F.H., VARY, J.A., ELENBAAS, J.R. & WEINAUG, CH.F. (1959). – Handbook of natural gas engineering. – MacGraw Hill, New York.

KENT, R.E. & WALMSLEY, P.J. (1970). – North Sea progress. – *Bull. amer. Assoc. Petroleum Geol.*, **54**, 1, 168-181, 11 fig.

KENT, P. & WARMAN H.R. (1972). – An environmental review of the world's richest oil bearing region : The Middle East. – Int. geol. Congr., 24th Sess., Canada (Montreal), **5**, 142-152, 1 fig.

KING, R.E. (1971). – Prospects for large increases in world hydrocarbon reserves from stratigraphic traps. – Proc. 8th World Petrol. Congr. (Moscow), **2**, 221-226.

KIRLAND, D.W. & EVANS, R. (1981). – Source rock potential of evaporitic environment. – *Bull. amer. Assoc. Petroleum Geol.*, **65**, 2, 181-190, 4 fig.

KLEMME, H.D. (1972). – Heat influences sizes of oil giants. – *Oil and Gas J.*, July 1 : 136-144 and July 24 : 76-78.

KLOVAN, J.E. (1974). – Development of western Canada Devonian reefs and comparison with Holocene analogues. – *Bull. amer. Assoc. Petroleum Geol.*, **58**, 5, 787-799, 17 fig.

KOZERUBA, V.V. (1969). – The formation of pools due to the lateral migration of fluids. – *In :* Selected lectures on petroleum exploration, **1**, 62-77, 6 fig. Dehra Dun. (India).

KUBLER, B., BETRIX, M.A. & MONNIER, F. (1979). – Les premiers stades de la diagenèse organique et de la diagenèse minérale. – *Bull. Ver. Schweiz. Petroleum-Geol. u. Ing.*, **45**, 108, 1-22.

KÜBLER, B. (1980). – Quelques aspects de la géochimie organique dans la prospection des hydrocarbures. – Int. geol. Congr., 26th Sess., Paris, Colloque **C 2**, 301-315.

LANGTON, J.R. & CHIN, G.E. (1968). – Rainbow member facies and related reservoir properties. Rainbow Lake, Alberta. – *Bull. amer. Assoc. Petroleum Geol.*, **52**, 10, 1925-1955, 24 fig.

LE BLANC, R.J. (1977). – Distribution and continuity of sandstone reservoirs. – *J. Petroleum Technol.*, July, 776-804.

LEYTHAEUSER, D., SCHAEFER, R.G. & YUKLER, A. (1982). – Role of diffusion in primary migration of hydrocarbons. – *Bull. amer. Assoc. Petroleum Geol.*, **66**, 4, 408-429, 15 fig.

LINK, W.L. (1952). – Significance of oil and gas seeps in world oil exploration. – *Bull. amer. Assoc. Petroleum Geol.*, **36**, 8, 1505-1540.

LOPATIN, N.V. (1971). – La température et le temps géologique comme facteur de carbonification. – Izvestiya Akademia Nauk SSSR, Serv. Geol., **3**, 95-106 (traduction).

LUCAS, P.T. & DREXLER, J.M. (1976). – A major naturally fractured stratigraphic trap, Uinta Basin, Utah. – *Mem. amer. Assoc. Petroleum Geol.*, **24**, 121-135, 16 fig.

LYONS, P.L. (1968). – The potential and the challenge - stratigraphic traps. – Unpubl. paper presented at Soc. Explor. Geophysicists, 38th ann. Meeting, Denver.

McAULIFFE, C.D. (1979). – Oil and gas migration – Chemical and physical constraints. – *Bull. amer. Assoc. Petroleum Geol.*, **63**, 5, 761-781, 4 fig.

McGREGOR, A.A. & BIGGS, C.A. (1968). – Bell Creek field Montana : a rich stratigraphic trap. – *Bull. amer. Assoc. Petroleum Geol.*, **52**, 10, 1869-1887, 23 fig.

McGREGOR, A.A. & BIGGS, C.A. (1972). – Bell Creek oil field Montana. – *Mem. amer. Assoc. Petroleum Geol.*, **16**, 367-375, 7 fig.

McKENZIE, D.B. (1972). – Primary stratigraphic traps in sandstone stratigraphic oil and gas fields. – *Mem. amer. Assoc. Petrol. Geol.*, **16**, 47-63.

MAGARA, K. (1968). – Compaction and migration of fluids in Miocene mudstone, Nagaoka Plain, Japan. – *Bull. amer. Assoc. Petroleum Geol.*, **52**, 12, 2466-2501.

MAGARA, K. (1974). – Compaction, ion filtration, and osmosis in shale and their significance in primary migration. – *Bull. amer. Assoc. Petroleum Geol.*, **58**, 2, 283-290.

MAGARA, K. (1978*a*). – The significance of the expulsion of water in oil primary migration. – *Bull. canad. Petroleum Geol.*, **26**, 1, 123-131, 4 fig.

MAGARA, K. (1978*b*). – Time of oil generation and migration and oil window. – *Bull. canad. Petroleum Geol.*, **26**, 1, 152-155, 3 fig.

MAGARA, K. (1978*c*). – Compaction and fluid migration. – Elsevier, Amsterdam : 319 pp.

MAGARA, K. (1978*d*). – Petroleum migration and accumulation. – *In :* HOBSON G.D. (ed.) : Developments in petroleum geology, 83-126, 24 fig. Applied Science Publ., London.

MAGARA, K. (1980). – Primary migration of oil and gas. – *Mem. canad. Soc. Petroleum Geol.,* **6,** 173-192.

MAGARA, K. (1981). – Mechanisms of natural fracturing in a sedimentary basin. – *Bull. amer. Assoc. Petroleum Geol.,* **65,** 1, 123-132, 19 fig.

MANNHARD, G.W. & BUSCH, D.A. (1974). – Stratigraphic trap accumulation in southwestern Kansas and northwestern Oklahoma. – *Bull. amer. Assoc. Petroleum Geol.,* **58,** 3, 447-463, 13 fig.

MASSON, P. (1977). – La Géologie du champ d'Ekofisk. – *Pétrole et Technique,* **244,** mai, 7-19, 13 fig.

MASTERS, J.A. (1979). – Deep basin gas trap, Western Canada. – *Bull. amer. Assoc. Petroleum Geol.,* **63,** 2, 152-181, 21 fig.

MAYUGA, M.N. (1970). – Geology and development of California's giant Wilmington oil field. – *Mem. amer. Assoc. Petroleum Geol.,* **14,** 158-184, 23 fig.

MEISSNER, F.F. (1978). – Petroleum geology of the Bakken formation Williston Basin - N Dakota and Montana. – Montana geol. Soc. Williston Basin Symp., 207-227, 24 fig.

MICHELSON, J.E. (1976). – Miocene deltaic oil habitat, Trinidad – *Bull. amer. Assoc. Petroleum Geol.,* **60,** 9, 1502-1519, 14 fig.

MOORE, H.C. & DRUCKMAN, Y. (1981). – Burial diagenesis and porosity evolution, Upper Jurassic Smackover, Arkansas and Louisiana. – *Bull. amer. Assoc. Petroleum Geol.,* **65,** 4, 597-628, 13 fig.

NEGLIA, S. (1979). – Migration of fluids in sedimentary basins. – *Bull. amer. Assoc. Petroleum Geol.,* **63,** 4, 573-597, 19 fig.

OTTMANN, R.P., KEYES, P.L. & ZIEGLER, M.A. (1976). – Jay field, Florida - a Jurassic stratigraphic trap. – *Mem. amer. Assoc. Petroleum Geol.,* **24,** 276-286, 15 fig.

OXLEY, M.L. & HERLIHY, D.E. (1972). – The Bryan field - a sedimentary anticline. – *Geophysics,* **37,** 1, 59-67, 12 fig.

PARKER, J.R. (1978). – Deep sea sands. – *In :* HOBSON G.D. (ed.) : Developments in petroleum geology, **1,** 225-242, 6 fig. Applied Science Publ., London.

PELET, R. (1977). – Géochimie organique des sédiments marins profonds de la mer de Norvège. Vue d'ensemble. – Orgon 1 – Mer de Norvège; 281-296, 12 fig. Centre nat. Rech. Scient., Paris.

PERRODON, A. (1966). – Géologie du Pétrole. – Presses univ. France, Paris, 440 pp.

PERRODON, A. (1973). – Les gisements stratigraphiques, mythes ou objectifs majeurs ? – *Rev. Assoc. franç. Tech. Pétrole,* **222,** 39-40.

PETERSON, J.A. & OSMOND J.C. (1961). – Geometry of sandstone bodies. – *Mem. amer. Assoc. Petroleum Geol.,* 240 p.

PHILIPPI, G.T. (1965). – On the depth, time and mechanism of petroleum generation. – *Geochim. cosmochim. Acta,* **29,** 9, 1021-1049.

PHILIPPI, G.T. (1977). – On the depth, time, mechanism of origin of the heavy to medium-gravity naphtene crude oils. – *Geochim. cosmochim. Acta,* **41,** 33-52.

POULET, M. (1968). – Problèmes posés par la migration verticale des pétroles et sa mise en place dans les gisements. – *Rev. Inst. franç. Pétrole,* **22,** 2, 159-173, 4 fig.

POWELL, J.T. (1972). – Exploration history of Delhi field, northeastern Louisiana. – *Mem. amer. Assoc. Petroleum Geol.,* **16,** 13, 548-558, 10 fig.

POWELL, T.G. & McKIRDY, D.M. (1975). – Geologic factors controlling crude oil composition in Australia and Papua, New Guinea. – *Bull. amer. Assoc. Petroleum Geol.,* **59,** 7, 1176-1197, 26 fig., 2 tables.

POWERS, M.C. (1967). – Fluid release mechanisms in compacting marine mudrocks and their importance in oil exploration. – *Bull. amer. Assoc. Petroleum Geol.,* **51,** 7, 1240-1254.

PRICE, L.C. (1976). – Aqueous solubility of petroleum as applied to its origin and primary migration. – *Bull. amer. Assoc. Petroleum Geol.,* **60,** 2, 213-244, 19 fig., 9 tables.

PRICE, L. C. (1980). – Shelf and shallow basin oil as related to hot-deep origin of petroleum. – *J. Petroleum Geol.,* **3,** 1, 91-116.

PURSER, B. H. (1975). – Sédimentation et diagenèse précoce des séries carbonatées du Jurassique moyen de Bourgogne. – *Bull. Bur. Rech. géol. min.* (4), **2,** 19-47.

PURSER, B. H. (1978). – Early diagenesis and the preservation of porosity in Jurassic limestones. – *J. Petroleum Geol.,* **1,** 2, 83-94.

PUSEY, W. C. (1973). – Paleotemperatures in the Gulf Coast using the ESR-Kerogen method. – *Trans. Gulf Coast Assoc. geol. Soc.,* **23,** 195-202.

RAYNAUD, J. F. & ROBERT, P. (1976). – Les méthodes d'étude optique de la matière organique. – *Bull. Cent. Rech. Pau-SNPA,* **10,** 1, 109-127, 1 fig., 4 pl.

REYRE, D. (1975). – Pays et opérations autour de Kangan; première découverte du groupe Egoko en Iran. – *Rev. Assoc. franç. Techn. Pétrole,* **231,** 15-23, 5 fig.

RICE, D. D. & CLAYPOOL, G. E. (1981). – Generation, accumulation and resource potential of biogenic gas. – *Bull. amer. Assoc. Petroleum Geol.,* **65,** 1, 5-25, 12 fig.

RICHE, P. & PRESTAT, B. (1981). – Paléogéographie du Crétacé moyen du Proche et du Moyen-Orient et sa signification pétrolière. – Proc. 10th World Petroleum Congr. (Bucharest), **2,** 57-75, 11 fig.

RITTENHOUSE, G. (1961). – Problems and principles of sandstone body classification. – *In :* PETERSON, J. A. & OSMOND, J. L. (ed.) : Geometry of sandstone bodies, 3-12, 8 fig. *Mem. amer. Assoc. Petroleum Geol.*

RITTENHOUSE, G. (1972). – Stratigraphic trap classification. – *Mem. amer. Assoc. Petroleum Geol.,* **16,** 14-28, 14 fig.

ROBERT, P. (1973). – Analyse microscopique des charbons et des bitumes dispersés dans les roches et mesure de leur pouvoir réflecteur. – Actes 6e Congr. int. Géoch. org., 549-569, 5 fig.

ROBERT, P. (1979 a). – Classification des matières organiques en fluorescence, application aux roches mères pétrolières. – *Bull. Centres Rech. Explor.-Prod. Elf-Aquitaine,* **3,** 1, 223-263, 5 fig., 2 tabl., 6 pl.

ROBERT, P. (1979 b). – Kerogen optical evolution applied to oil and gas exploration and geothermal history. – *In :* DURAND, B. (ed.) : Kerogen, chapitre 12. Technip, Paris.

ROBERTS, W. H. & CORPEL, R. J. (ed.) (1980). – Problems of Petroleum migration. – *Bull. amer. Assoc. Petroleum Geol.,* Studies in Geology, **10,** 278 pp.

ROGERS, M. A. (1979). – Application of organic facies concepts to hydrocarbon source rock evaluation. – Proc. 10th World Petroleum Congr. (Bucharest), **2,** 23-30, 7 fig.

ROUCHET, J. du (1974). – Introduction à une sédimentologie de la matière organique. – Rapp. interne ELF.

ROUCHET, J. du (1978 a). – Indices chimiques pour l'évaluation de l'état diagénétique des huiles et des roches sapropéliques. – *Rev. Inst. franç. Pétrole,* **33,** 1, 33-45, fig. 11.

ROUCHET, J. du (1978 b). – Explication paléoclimatique des grands gisements superficiels de gaz sec du nord de la Sibérie occidentale. – Rapp. inédit SNEA(P).

ROUCHET, J. du (1978 c). – Eléments d'une théorie géomécanique de la migration de l'huile en phase constituée. – *Bull. Cent. Rech. Explor.-Prod.* Elf-Aquitaine, **2,** 2, 337-373, 11 fig.

ROUCHET, J. du (1981). – Stress fields, a key to oil migration. – *Bull. amer. Assoc. Petroleum Geol.,* **65,** 1, 74-85, 6 fig.

RUMEAU, J. L. & SOURISSE, C. (1972). – Compaction, diagenèse et migration dans les sédiments argileux. – *Bull. Centre Rech. Pau*-SNPA, **6,** 2, 313-345, 13 fig.

RUMEAU, J. L. & SOURISSE, C. (1973). – Un exemple de migration primaire en phase gazeuse : la formation des gisements de Lacq et Meillon - Pont d'As. – *Bull. Centre Rech. Pau*-SNPA, **7,** 1, 53-67, 3 fig.

RUMEAU, J. L. & SOURISSE, C. (1975). – Quelques mécanismes possibles pour la formation de gisements d'hydrocarbures dans les stades précoces de la diagenèse. – *Bull. Centre Rech. Pau*-SNPA, **9,** 1, 129-146, 10 fig.

ST-JOHN, B. (1980). – Sedimentary basins of the world and giant hydrocarbon accumulations. – *Amer. Assoc. Petroleum Geol.* 1 map.

SCHOLLE, P. A. (1977). – Chalk diagenesis and its relation to petroleum exploration. Oil from chalks, a modern miracle ? – *Bull. amer. Assoc. Petroleum Geol.,* **61,** 7, 982-1009, 13 fig.

SCHOLLE, P. A. (1978). – Porosity prediction in shallow versus deep water limestone. – 53rd ann. SPE of AIME Fall. Techn. Conf. Preprint n° SPE 7554, 6 p.

SCHOWALTER, T. T. (1979). – Mechanics of secondary hydrocarbon migration and entrapment. – *Bull. amer. Assoc. Petroleum Geol.,* **63,** 5, 723-760, 35 fig.

SECOR, D. T. Jr. (1965). – Role of fluid pressure in jointing. – *Amer. J. Sci.,* **263,** 633-646.

SEIFERT W. K., MOLDOWAN, J. M. & JONES, R. W. (1981). – Application of biological marker chemistry to petroleum exploration. – Proc. 10th World Petroleum Congr. (Bucharest, 1979), **2,** 425-438, 8 fig.

SELLEY, R. C. (1978 a). – Porosity gradients in North Sea oil bearing sands. – *J. geol. Soc.* (London), **135,** 119-132.

SELLEY, R. C. (1978 b). – Deltaic facies and petroleum geology. – *In :* HOBSON (ed.) : Developments in Petroleum Geology, **1,** 197-224. Applied Science Publ., London.

SERRA, O. (1979). – Diagraphies différées - bases de l'interprétation. Tome 1 : Acquisition des données diagraphiques. – *Bull. Centres Rech. Explor.-Prod. Elf-Aquitaine,* Mém. 1.

SHIBAOKA, M., SAXBY, J. D. & TAYLOR, G. H. (1978). – Hydrocarbon generation in Gippsland Basin, Australia. Comparison with Cooper Basin, Australia. – *Bull. amer. Assoc. Petroleum Geol.,* **62,** 7, 1151-1158, 9 fig.

SILVERMAN, S. R. (1971). – Influence of petroleum origin and transformation on its distribution and redistribution in sedimentary rocks. – Proc. 8th World Petroleum Congr. (Moscow), **2,** 47-54.

SILVERMAN, S. R. (1978). – Geochemistry and origin of natural heavy oil deposits. – *In :* CHILINGARIAN & YEN (ed.) : Bitumens, asphalts and tar sands, 17-25. Elsevier, Amsterdam.

SNARSKY, A. N. (1964). – Relationship between primary migration and compaction of rocks. – Petroleum Geology (trad. *Geologiya Nefti i Gaza),* **5,** 7 (1961), 362-365.

SOKOLOV, V. A., ZHUSE, T. P., VASSOYEVICH, N. B., ANTONOV, P. L., GRIGORIYEV, G. G. & KOZLOV, U. P. (1964). – Migration processes of gas and oil, their intensity and directionality. – Proc. 6th World Petroleum Congr. (Frankfurt), **1,** 493-505.

STAHL, W. J. (1977). – Carbon and Nitrogen isotopes in hydrocarbon research and exploration. – *Chem. Geol.,* **20,** 121-149, 28 fig.

STANLEY, T. B. (1970). – Vicksburg fault zone, Texas. – *Mem. amer. Assoc. Petroleum geol.,* **14,** 301-308.

STAUBE, A. J. & MILIUS, G. (1970). – Geology of Groningen gas field, Netherlands. – *Mem. amer. Assoc. Petroleum Geol.,* **14,** 359-369, 8 fig.

STOUDT, E. L. (1979). – Smackover formation of Gulf Coast region. – *Bull. amer. Assoc. Petroleum Geol.,* **63,** 3, 534-535 (abstract).

SULLWOLD, H. H. (1961). – Turbidities in oil exploration. – *In :* PETERSON, J. A. & OSMOND, J. L. (ed.) : Geometry of sandstone bodies, 63-81, 9 fig. *Mem. amer. Assoc. Petroleum Geol.*

TAYLOR, J. L. M. (1978). – Sandstones as reservoir rocks. – *In :* HOBSON, G. D. (ed.) : Developments in Petroleum Geology, **1,** 147-196, 2 fig. Applied Science Publ., London.

TISSOT, B. (1966). – Problèmes géochimiques de la genèse et de la migration du pétrole. – *Rev. Inst. franç. Pétrole,* **23**, 1621-1671, 20 fig.

TISSOT, B. & PELET, R. (1971). – Nouvelles données sur les mécanismes de genèse et de migration du Pétrole - Simulation mathématique et application à la prospection. – Proc. 8th World Petroleum Congr. (Moscow), **2**, 35-46, 8 fig.

TISSOT, B. (1973). – Vers l'évaluation quantitative du pétrole formé dans les bassins sédimentaires. – *Rev. Assoc. franç. Techn. Pétrole,* **22**, 27-31, 11 fig.

TISSOT, B. (1978). – The application of the results of organic geochemical studies in oil and gas exploration. – *In :* HOBSON, G. D. (ed.) : Developments in Petroleum Geology, **1**, 53-82, 12 fig. Applied Sciences Publ., London.

TISSOT, B., DURAND, B., ESPITALIE, J. & COMBAZ, A. (1974). – Influence of nature and diagenesis of organic matter in formation of petroleum. – *Bull. amer. Assoc. Petroleum Geol.,* **58**, 3, 499-506, 12 fig.

TISSOT, B., DEROO, G. & ESPITALIE, J. (1975). – Etude comparée de l'époque de formation et d'expulsion du pétrole dans diverses provinces géologiques. – Proc. 9th World Petroleum Congr. (Tokyo), **2**, 159-169, 8 fig.

TISSOT, B. & WELTE, D. H. (1978). – Petroleum formation and occurence. – Springer, Berlin, 538 p.

TISSOT, B. & BESSEREAU, G. (1982). – Géochimie des gaz naturels et origine des gisements de gaz en Europe occidentale. – *Rev. Inst. franç. Pétrole,* **37**, 1, 63-77, 16 fig.

TREMOLIERES, P. & REULET, J. (1978). – Influence des déformations tectoniques sur les caractéristiques pétrophysiques matricielles des réservoirs calcaires. – *Rev. Inst. franç. Pétrole,* **22**, 3, 331-348, 13 fig.

TROFIMUK, A. A., CHERSKII, H. V., MAKAGON, Y. F. & TSAREV, V. P. (1972). – Sur un mécanisme possible de formation des gisements de gaz naturel (trad.). – Akademia Nauk SSSR, Geologia i geofisica, sect. sibir., **8**, 3-9.

TROFIMUK, A. A., CHERSKII, N. V. & TSAREV, V. P. (1976). The role of continental glaciation and hydrate formation on petroleum occurence - Future supply of nature made petroleum and gas. – Pergamon, London.

TSCHOPP, R. H. (1967). – Development ot the Fahud field. – Proc. 7th World Petroleum Congr. (Mexico), **2**, 243-250, 2 fig.

TYLER, A.N. & ERWIN, W.L. (1976). – Sunoco-Felda field, Hendry ans Collier Countries, Florida. – *Mem. amer. Assoc. Petroleum Geol.,* **24**, 287-299, 12 fig.

VAN ELSBERG, J. N. (1978). – A new approach to sediment diagenesis. – *Bull. canad. Petroleum Geol.,* **26**, 1, 57-86, 16 fig.

VAN KREVELEN, D. W. (1961). – Coal. – Elsevier, Amsterdam.

VEST, E. L. (1970). – Oil fields of Pennsylvanian. – Permian Horseshoe atoll, W Texas. – *Mem. amer. Assoc. Petroleum Geol.* **14**, 185-203.

VIDAL, J. (1980). – Geology of Grondin field, Gabon. – *Mem. amer. Assoc. Petroleum Geol.* **30**, 577-590, 11 fig.

WAPLES, D. W. (1979). – Simple method for oil source bed evaluation. – *Bull. amer. Assoc. Petroleum Geol.,* **63**, 2, 239-244, 11 fig.

WAPLES, D. W. (1980). – Time and temperature in petroleum formation. – *Bull. amer. Assoc. Petroleum Geol.,* **64**, 6, 916-926.

WARDLAW, N. C. (1976). – Pore geometry of carbonate rocks as revealed by pore casts and capillary pressure. – *Bull. amer. Assoc. Petroleum Geol.,* **60**, 2, 245-257, 14 fig.

WARDLAW, N. C. & CASSAN J.-P. (1978). – Estimation of recovery efficiency by visual observation of pore systems in reservoir rocks. – *Bull. canad. Petroleum Geol.,* **26**, 4, 572-585, 12 fig.

WARDLAW, N. C. (1980). – The influence of pore structures in rocks on the entrapment of oil. – *Mem. canad. Soc. Petroleum Geol.,* **6**, 193-208.

WEBER, V. V. & MAXIMOV, S. P. (1976). – Early diagenetic generation of hydrocarbon gases and their variations dependent on initial organic composition. – *Bull. amer. Assoc. Petroleum Geol.,* **60**, 2, 287-292.

WEN, C. S., CHILINGARIAN, G. V. & YEN, T. F. (1978). – Properties and structure of bitumens. – *In :* CHILINGARIAN & YEN (ed.) : Bitumens, asphalts and tar sands, 155-190, 17 fig. Elsevier, Amsterdam.

WIJHE, D. H. van, LUTZ, M. & KAASSCHIETER, J. P. H. (1980). – The Rotliegend in the Netherlands and its gas accumulation. – Geol. Mijnbouw, **59**, 1, 3-24, 18 fig.

WILSON, H. H. (1977). – « Frozen-in » hydrocarbon accumulations or diagenetic traps - Exploration targets. – *Bull. amer. Assoc. Petroleum Geol.,* **61**, 4, 483-491, 1 fig.

YOUNG, A., MONAGHAN, P. H. & SCHWEISBERGER, R. T. (1977). – Calculation of ages of hydrocarbons in oils - Physical chemistry applied to petroleum geochemistry I. – *Bull. amer. Assoc. Petroleum Geol.,* **61**, 4, 573-600, 11 fig.

ZIEGLAR, D. L. & SPOTTS, J. H. (1978). – Reservoir and source bed history of Great Valley, California. – *Bull. amer. Assoc. Petroleum Geol.,* **62**, 5, 813-826, 16 fig.

PART III

PETROLEUM PROVINCES

'The past is a prologue.'
Shakespeare.

Part 3 is devoted to petroleum provinces, the remarkable associations of pools that characterize a basin or part of a sedimentary basin. The following four chapters deal with :

— the concept of petroleum systems and provinces, and especially, the main geological criteria favorable to the formation of pools,

— the main types of petroleum provinces,

— the world's chief petroleum provinces,

— some viewpoints on petroleum exploration and ultimate resources.

1. – THE CONCEPT OF PETROLEUM SYSTEMS AND PROVINCES

> 'Oil and gas are like a flock of sheep. They meet in herds or clusters. But rather than instinct, these clusters obey a profound logic of sedimentary basin geodynamics, even if this logic is often long and difficult to decipher, so difficult that one is often tempted to qualify it as "hazard".'
>
> E.W. Owen, 1975.

Every mineral concentration is a geological anomaly, but this anomaly is the result of a more or less complex logic. Hence the concentrations are not distributed at random, but constitute familes which are ordered in accordance with a number of stratigraphic, sedimentological, hydrogeological and structural parameters. The geological criteria governing the distribution of pools, and in particular, the combined presence of source rocks, reservoirs and seals, generally exhibit a certain geographic extension which is reflected by the formation of a family of pools, or even better, a petroleum system, a structured set of natural elements of the same species or having the same function. From the geographic standpoint, and according to their dimensions and complexity, these sets are reflected by the existence of a petroleum zone or province.

The latter may be defined by :

— geographic criteria, especially by their proved or 'useful' sedimentary volume or area, in other words, of petroleum interest,
— geological criteria, primarily related to the type of basin and sedimentation, serving as a basis for their classification,
— petroleum criteria, including the number and types of pools, volumes of oil and gas discovered, namely, the results of exploration : among these are two criteria that specifically characterize these provinces :
  — potential,
  — type of habitat.

We shall first examine the latter, and then go on to a longer analysis of the different geological criteria that form the basis for petroleum provinces. A combination of these criteria, of which one or two often play a predominant role, generally stamps the distributions of pools in a province with their logic, which we might be tempted to call their geo-logic.

1.1. MAIN PETROLEUM CRITERIA

A petroleum province or area is mainly defined in terms of potential and type of habitat, which reflect two independent and different realities.

1.1.1. The concept of potential

Potential, or relative wealth, may be defined as the quotient of cumulative discoveries divided by the area of the province, that is to say, in terms of density. Discoveries can also be related to the 'useful' sedimentary volume, but the latter seems more difficult to evaluate. Cumulative discoveries or initial reserves are at any time the sum of the cumulative production and remaining proven recoverable reserves.

The definition of the area to be considered is often a more delicate matter. At the start of exploration, the temptation is to think big. As the borehole results come in, the contours become clearer and the area considered has a tendency to shrink. From a pragmatic standpoint, and by analogy with a new province to be explored, it is nevertheless necessary to consider, not the strict envelope of the accumulations, but rather the overall sedimentary area within which the productive areas are individualized. The overburdens often play an active role in the formation of a petroleum province. Hence a distinction can be drawn between the following :

— very rich provinces, with an average of over 10,000 tons of oil per km^2,
— rich provinces, with 1000 to 10,000 tons,
— poor provinces, with less than 1000 tons.

1.1.2. **The concept of habitat**

The habitat of oil and gas is the spatial distribution of the pools in a basin or province. It may be characterized by :

— the number of pools distributed over a given area, 10,000 km^2 for example,
— the percentage of reserves in the largest field(s).

$$\frac{G_n}{G_{n-1}} = \frac{n}{n-1} \cdot K$$

where G represents the size of a pool of rank n and K a constant generally not equal to 1. For $K = 1$, the distribution observes a log-normal pattern which is also called Zipf's Law (Ivanhoe, 1976 ; Folinsbee, 1977).

This is reflected empirically by the observation that the number of pools varies inversely with their size. In the case of a log-normal distribution, this means that if the first has a value of 100 :

— the second is equivalent to 50,
— the third is equivalent to 33.3,
— the fourth is equivalent to 25,
— the fifth is equivalent to 20, etc.

A simple representation is given by a log-log graph in which the rank of the accumulation is plotted on the x axis, with their size or reserves on the y axis (Figure 125).

Different types of habitat

In actual fact, the geological criteria mark this distribution, which is reflected by values of K that are different from 1. A dispersed habitat corresponds to a distribution of reserves in a large number of pools. This is shown on the log-log diagram by a line with a gentle slope or a value of $K < 1$, generally $1/2 < K < 1$ (Figure 125). This type of habitat corresponds with minor variations to provinces in which the same types of sedimentary or structural anomaly are repeated, forming traps. As may be observed in deltaic basins and in certain stable provinces, reef and others, it is not rare for the biggest field to contain only 5 to 10 % of total reserves. Some examples of this are (Figures 126 and 127) :

— the coastal Gulf Coast,
— the Niger Delta, where the largest field, Jones Creek, only accounts for 6 % of cumulative discoveries, and the five biggest fields only 16 %,
— the Michigan Basin, where slightly more than 100 Mt are distributed in over 600 fields, of which 500 have less than 140,000 t (1 M barrels),

and also :

— the Paris Basin, which has a factor $K = 0.76$ (11 Mt in 18 fields),
— The Po Valley Basin, with $K = 0.24$ (260 Gm^3 of gas in 60 fields),
— the Emsland Basin (Germany/Netherlands), with $K = 0.22$ for gas (50 Gm^3 in 26 fields) (from Coustau, 1979, unpublished, 1980).

A concentrated habitat is the inherent characteristic of provinces in which the bulk of the reserves is concentrated in a very small number of pools. Hence these pools exhibit wide differences in size. This type of habitat is represented by a line with a high slope and a value of $K > 1$, generally $1 < K < 2$. It corresponds from the geological standpoint to highly contrasting structural and/or stratigraphic conditions, such as one or a few vast high regional zones, offering few but very large trapping conditions. In these basins, the biggest, or the two or three largest fields, may concentrate more than half of the total reserves. Some illustrations :

— the Vienna Basin in Austria, with a coefficient $K = 1.65$ (150 Mt of oil or gas equivalent in 35 fields),
— the Parentis Basin, with $K = 2.1$ (45 Mt in 10 fields),
— the Algerian Sahara, where the largest field, Hassi Messaoud, contains nearly 80 % of all the oil, and Hassi R'Mel more than one-third of all gas reserves,
— the Volga-Urals Basin with some 40 % of the oil discoveries concentrated in Romashkino Field alone, nearly two-thirds in the five largest fields, hence in less than 1 % of the total number of fields,

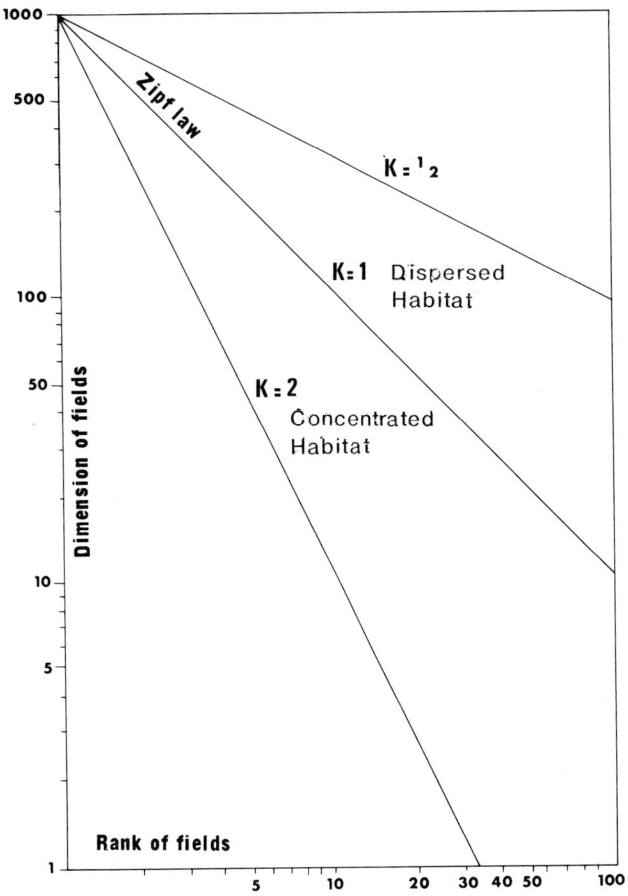

Fig. 125. — Theoretical rank of fields by dimension and different types of habitat (Coustau, 1979).

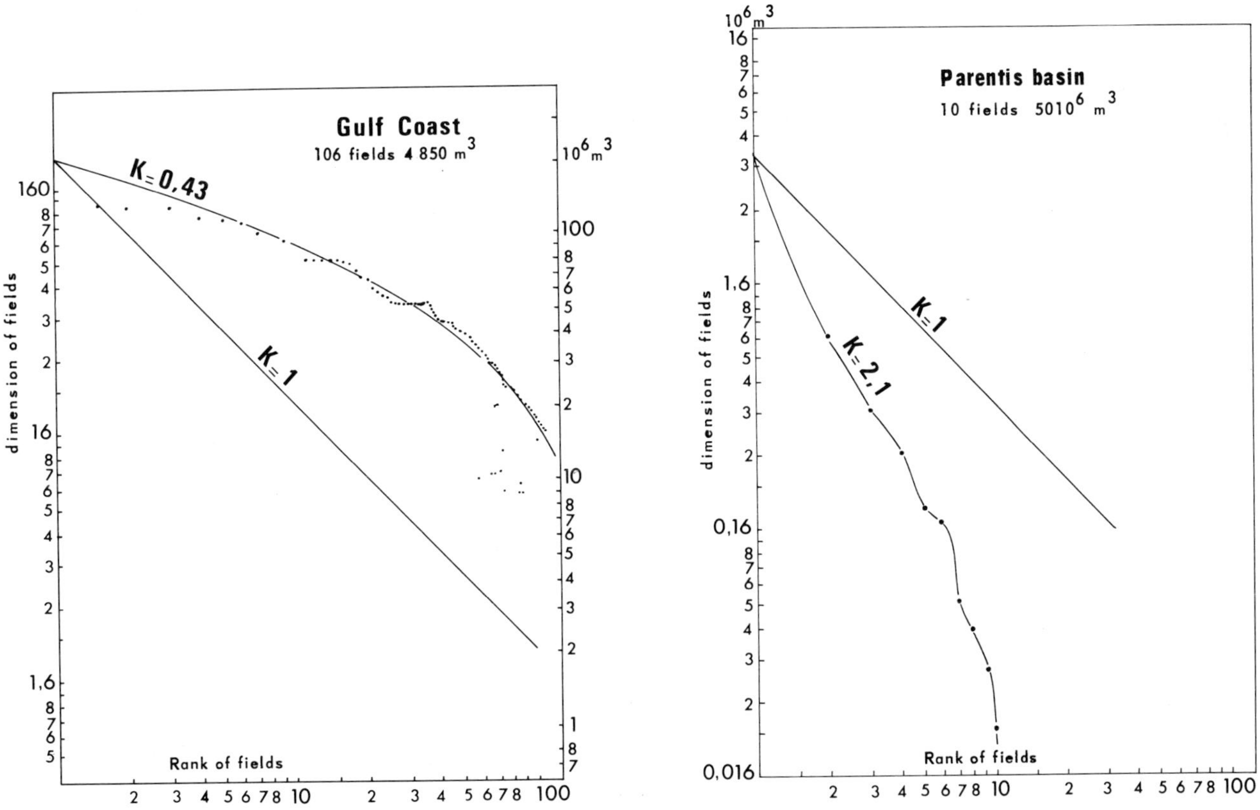

Fig. 126. — Rank of Gulf Coast and Parentis Basin fields according to dimension (Coustau, 1979).

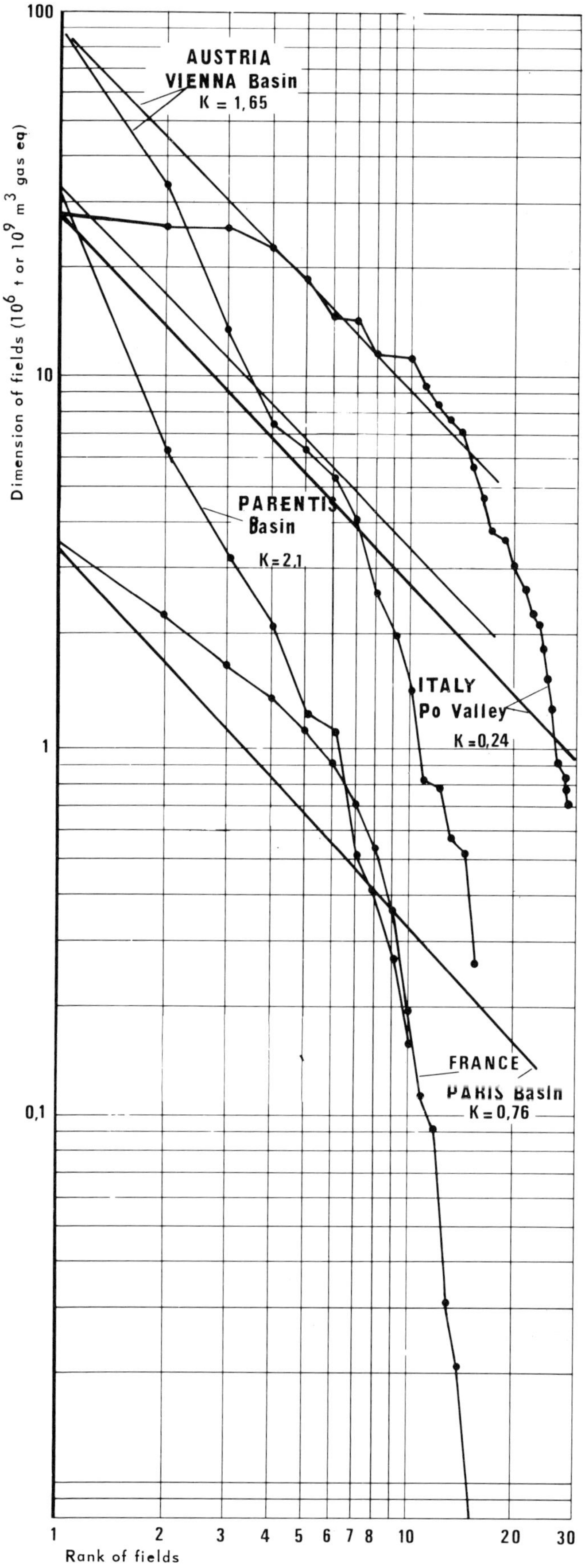

Fig. 127. — Rank of fields according to dimension in the Paris, Parentis, Vienna and Po Valley Basins (Coustau, 1979).

This case in particular reveals the influence of the degree of knowledge, namely of exploration, on the image of the habitat and on that of potential. Since the biggest fields are generally discovered first, the earliest images tend to suggest :

— a concentrated pattern, if the size of these fields rapidly decreases, i.e. if they fall within the log-log diagram above Zipf's line : note that in this case, the largest fields must be individualized more easily and more rapidly on the geophysical maps, so that consequently the ten biggest fields are generally discovered in a relatively early stage of exploration ; the value of K, corresponding to the mean line passing through the ten leading fields, is stabilized at a value close to its final position,

— a dispersed pattern, if the size of these fields decreases gradually, i.e. if they fall on the log-log diagram below Zipf's line : note that in this case the biggest fields are individualized less clearly and less quickly on the geophysical maps; consequently, during a relatively long exploration period, the ten largest fields may be cast in doubt at any time, and the value of K corresponding to the mean line of these ten biggest fields will constantly decline towards lower values ($K < 1$) until the ten largest fields have finally been discovered; K will then be stabilized at its final value characterizing the habitat.

All other things remaining equal, provinces with a dispersed habitat are attractive for a longer time than provinces with a concentrated habitat. This method offers a better means of characterizing the habitat and classifying provinces according to the types of habitat, namely, according to the value of K, rather than those involving the percentage of reserves mobilized by the first or first ten fields. The latter case often occurs in 'Lorenz' type representations (cumulative percentage of reserves corresponding to percentage of fields on a linear scale) or the 'log-normal' type (percentage of fields larger than or equal to a given size). The concentrated/dispersed appearance varies according to the size of the population analyzed. Hence according to Zipf's Law, the first field concentrates :

— 33 % of reserves for a population of 10,
— 20 % of reserves for a population of 100,
— 13 % of reserves for a population of 1000,
— 10 % of reserves for a population of 10,000.

This means that the larger the number of fields, the closer the habitat approaches the concentrated type, owing to the increasing weight of the small accumulations in the total number accounted for, compared to their very slight importance in comparison with cumulative reserves.

To avoid, or at least to reduce significantly the influence of the number of fields in this type of representation, in other words of the size of the unit taken into account and/or its degree of exploration (and hence knowledge), two alternatives are available :

— to eliminate fields cumulating less than 1 % of reserves, thus introducing a distorsion for poor provinces,

— more satisfactorily, to consider only fields in a size range within the ratio of 100,000 to 1 (Coustau, 1979).

It is also obvious that the plot of these curves varies according to the degree of knowledge and exploration of a basin. Different analyses have demonstrated that the final tendency emerges when half of the discoveries have been achieved (Coustau, 1979).

The number of small fields increases in accordance with the exploration effort (Figures 128 and 129). However, the distribution curve of the number of fields as a function of their dimensions, in logarithmic coordinates, generally exhibits a bell shape, whereas the theoretical distribution should yield a descending concave curve (Figure 130). Vinkovetsky (1979) showed that this divergence could be ascribed to economic factors, and more specifically to the fact that the economic limits to exploitation hinder exploration and falsify the knowledge of the population of the smaller accumulations, the number of which increases in inverse proportion to their size.

1.2. **MAIN GEOLOGICAL CRITERIA**

The characteristics of a petroleum province are closely conditioned by the geology of the sedimentary basin and by its general framework. The fundamental prerequisite is the existence of a sedimentary basin with a fairly large volume, covering a minimum area of several thousand km^2, with thicknesses greater than 2000 m, and a duration of several tens of millions of years as a rule. Geological criteria have a different effect on the main petroleum characteristics of a province, and in particular its potential and its type of habitat.

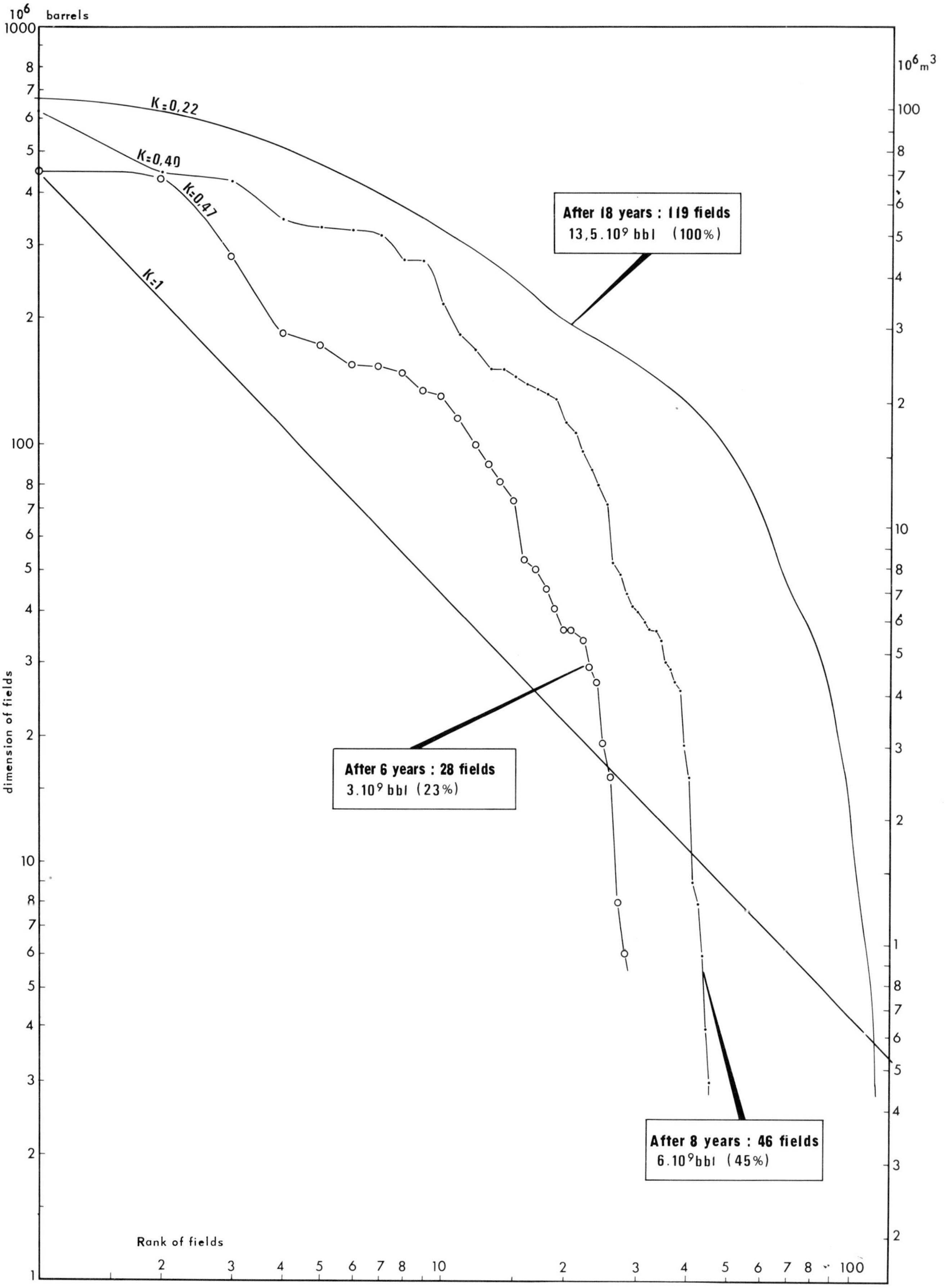

Fig. 128. — Curve showing rank of fields in Nigeria with degree of exploration (Coustau, 1979).

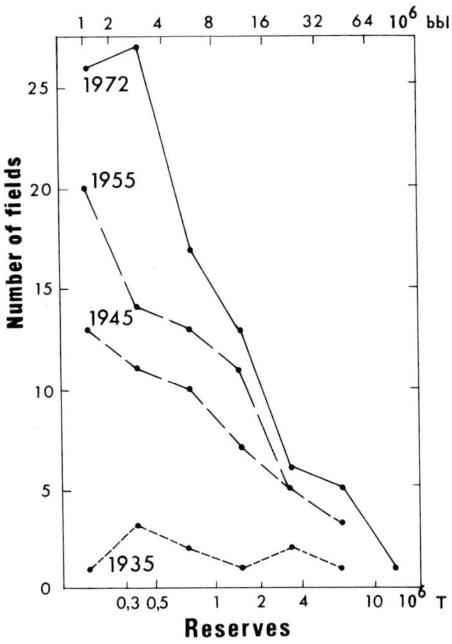

Fig. 129. — Variation in average size of fields discovered in the Michigan province (Vinkovetsky, 1979).

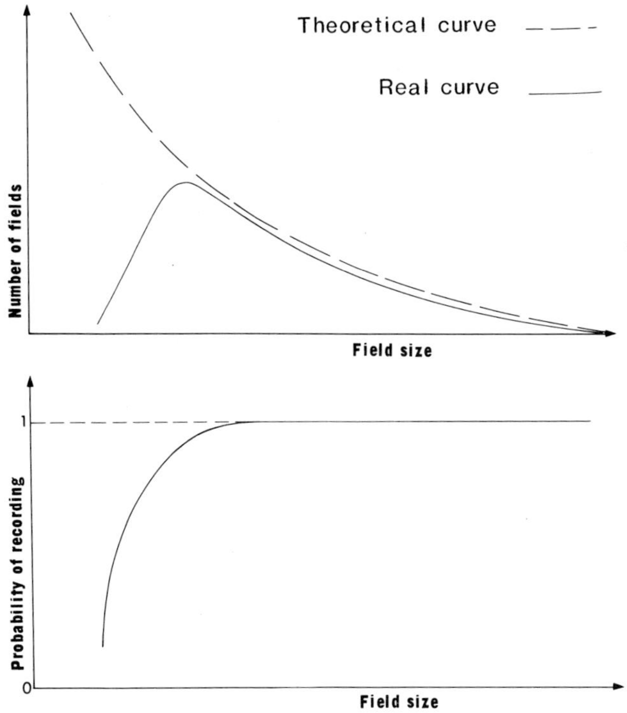

Fig. 130. — Distribution of fields in a basin : theoretical and observation data. Influence of probability of recording of fields (Vinkovetsky, 1979).

The fact that a petroleum province is rich or poor stems from the geological history of the sedimentary basin, namely the distribution and succession of its depositional environments, its deformations, its aquifers, its geothermal characteristics, and the relations in time between these different mechanisms. These parameters narrowly condition the composition of the hydrocarbons, and especially the gas or oil-bearing nature of the different provinces.

The type of habitat is more specifically dependent on the characteristics of the depositional environment and the structure of the basin. These overall geological criteria may be divided more or less arbitrarily and very schematically into two chief families :

— paleoenvironmental criteria in the broad sense of the term, namely sedimentation, hydrogeology and geothermal mechanisms (Section 1.2.1.),
— criteria related to the historical approach and which give rise to the structural framework (Section 1.2.2.).

1.2.1. **Paleoenvironmental criteria** (s.l.)

1.2.1.1. ***Conditions of deposition***

A petroleum zone is produced by the combination of different depositional environments, including fine, confined sediments necessary for the formation of source rocks, and clastic or carbonate rocks having a reservoir function. As we have seen above, the conditions of sedimentation primarily condition the quality of the hydrocarbons formed. Environments with a brackish or littoral tendency, containing organic matter which is mainly of the humic type, and often related to humid climates, for example, are especially rich in gas and paraffinic oils. Environments with an evaporitic tendency, generally in a desert climate, and rich in sapropelic organic matter, preferably yield oils of the naphtheno-aromatic type, often with a high sulfur content. Many other interim types obviously exist between these two examples, which in a way mark the extremes of a long series of different provinces.

One of the prerequisites for the formation of pools is the bringing into contact of the opposing environments required for the deposition of source rocks and reservoirs. This contact can be produced in two ways and especially by :

— lateral facies variations,
— superpositions of different facies.

There are good reasons for believing that the effectiveness of source rocks depends on their thickness, as excessively thick beds exhibit a poor yield from the standpoint of the expulsion of the hydrocarbons formed. Some authors feel that a thickness of about 30 m is an optimum. This means that alternating beds of sands and source rock shales offer an effective hydrocarbon expulsion/concentration solution. The best yields appear to correspond to sand percentages of 30 to 40 %, more rarely 20 to 60 %, explaining why hydrocarbon reserves are often proportional to the square of the mean thickness of the sandstone reservoirs (Magara, 1979).

Lateral facies variations

The transition from source rock facies to reservoir horizons may occur on marginal zones of basins, in deltaic and lagoonal environments, and also in deeper zones, especially in marginal flexure zones.

A sand/shale area often displays the presence of pools in these facies transition and repetition zones, including the following examples:

— Tertiary series of Nigeria and Maracaibo,
— Cretaceous and Cenozoic series of the Gulf Coast,
— Cretaceous formations of the Rocky Mountains, West Canada, Gabon, Mangychiak in Central Asia,
— the Pennsylvanian (Atoka Formation) in Oklahoma.

In a carbonate area and especially in reef provinces, a similar pattern may be found, as in the following:

— the Miocene of the Saliwati Basin (Indonesia),
— the Eocene/Paleocene of the Sirte Basin,
— the Cretaceous of the Persian Gulf,
— the Lower Cretaceous (Viking formation) of Alberta,
— the Permian of West Texas,
— the Pennsylvanian of the Paradox Basin,
— the Devonian in Canada.

Facies superpositions and unconformities

Contact between different facies is facilitated by the presence of hiatuses and onlaps. The overlapping of reservoirs by impervious formations forming a seal, and sometimes source rocks, is encountered in particular in certain unconformity and transgression zones (Figure 131). The case of transgressive source rocks playing the role of a seal is illustrated by:

— the sand/shale units of the neogene basins in Indonesia,
— the shale series of the Senonian in the Sirte and Labrador Basins, and of the Lower Cretaceous in Prudhoe Bay and Rankin (on the Northwest Shelf of Australia),
— the bituminous black shales, often phosphatic, of the Pennsylvanian and Permian in the Mid-Continent and in Wyoming,
— marls and black limestones of the Ordovician, the Upper Devonian, the Pennsylvanian and the Permian in West Texas, especially rich in unconformities and in hydrocarbons.

Unconformities and transgressions thus play a major role in the potential and habitat of a province, chiefly by facilitating contact between source rocks and reservoirs. Their importance has been stressed many times, especially in the provinces of the Mid-Continent, West Texas, Maracaibo, Sirte, the Orinoco, West Canada, Labrador, North Alaska and the North Sea. Their action may be exerted in different ways:

— by producing contact between the source rock and the reservoir, as in the above examples,
— by facilitating the development of porous and permeable surfaces serving as a preferential migration path for hydrocarbons, such as the Hercynian unconformity in the Algerian Sahara, that drains the products formed in the Silurian and Devonian shales, the Cimmerian unconformity in the North Sea, the Upper Devonian in Michigan,
— possibly by giving rise to pinchout traps, as in the Maracaibo and Mid-Continent Basins.

This drainage function, which often enhances the concentration of the habitat, is reinforced by the presence of continuous sand horizons at the base of the transgressive series. But for a pool to exist, this pattern presupposes the presence of structural traps exhibiting a sufficient closure, because these continuous horizons are also the seat of pronounced hydrodynamic mechanisms and seepage. Unconformities may be considered as negative elements if they correspond to a long lacuna, and if the overlying series is not sufficiently thick to offer, either directly or indirectly from the underlying series, a new possibility of hydrocarbon genesis.

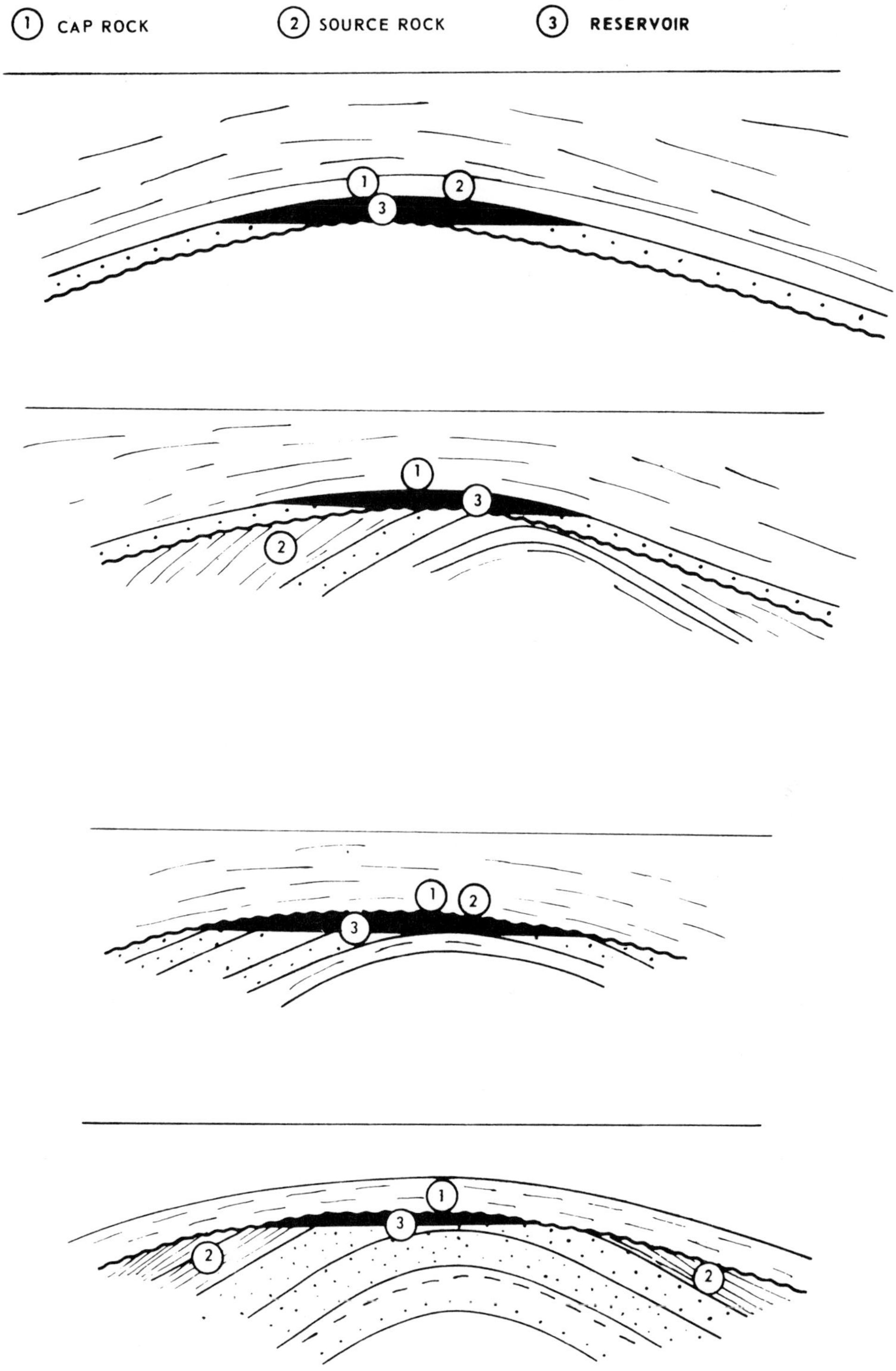

Prudhoe Bay
Athabasca Sands

NE Sahara
(Hassi R'Mel/Gassi Touil)
Bolivar C
North Sea

Mid-Continent
Sirte (Sarir)
Dampier (Rankin)
Prudhoe Bay
Gippsland

NE Sahara
(Messaoud)
Gippsland
Bolivar C

Fig. 131. — Sections of different types of pool related to unconformities.

Sedimentary logic

It is tempting to assume that the most favorable stratigraphic schemes consist in particular of :

— positive megasequences covered by fine, impermeable end of sequence deposits, such as evaporites,

— frequent transgressions and regressions of the sea terminating in the regional advance of an impermeable formation.

These are undeniably some of the best conditions for the regional closure of a basin. These patterns may be illustrated by :

— Jurassic cycles and especially the Hith anhydrite, on the Arabian platform, the saliferous Trias in the Algerian Sahara, the Middle Devonian in West Canada, the Permian of the Permian Basin for the former type,

— sand/shale series with transgressive marls of the Cretaceous/Tertiary in the Gulf Coast and the Neogene of California and Indonesia, for the latter.

Also noteworthy are cases in which impermeable facies, especially evaporites, provide a lateral closure on the margins of the basin, like the Mississippian in the Williston Basin. It is obvious that these stratigraphic and sedimentological arrangements reflect the history and evolution of certain types of basin. These stratigraphic criteria impose their logic in the distribution of pools, by the distribution of reservoirs, source rocks and/or seals. This denotes the predominating mark of the reservoir facies in many detrital trends, as in the Gulf Coast (Wilcox, Yegua, Frio) and reef trends, especially in Alberta, West Texas and Mexico.

A cap rock logic is observed, notably in the Persian Gulf, with prospects sealed respectively by the Fars, Hith and Sudair evaporites. The Fars cap rock in particular appears as one of the prerequisites for the presence of oil in the foothills structures (Figures 132 and 166). On the other hand, irregularities in the stratigraphic distribution of accumulations in the Congo are often related to discontinuities in the saliferous cap rock, and a deep source rock logic is more probable in this case. Similarly, in the Paris Basin, the distribution of Toarcian source rocks conditions a certain source rock logic in this province. A comparable logic appears to prevail in the Michigan and Illinois Basins.

Since these three criteria, source rocks, reservoirs and seals, constitute three regional prerequisites for any accumulation, we generally have a three-component logic. This is the classic case today of the Permian gas fields in the southern part of the North Sea, all situated in an area inscribed within an envelope common to the extension of the cal measures, the Rotliegendes sandstones, and the Zechstein salt (Figure 133).

1.2.1.2. ***Hydrogeological conditions***

The hydrogeological, and especially hydrodynamic, conditions that largely govern depositional environments and paleogeographies, profoundly affect the characteristics of a petroleum province, starting with its potential. This action occurs mainly through the physical, pressure gradient, and chemical and biochemical criteria. Hydrocarbon migrations follow the direction of decreasing pressure gradients. Hence it is logical to find their accumulations in lower potential zones. It is worth noting the value of zones protected by an overlying higher pressure bed, forming an effective seal. These sheltered zones are also distinguished, from the hydrochemical standpoint, by salinities which are often higher than in waters in the more or less invaded marginal areas, or in contact with meteoric waters. In the case of very slightly subsident and stabilized platform basins, hydrogeologically favorable zones are encountered preferably in the central areas, better protected from the meteoric invasions. Good examples of this are offered by the Paris, Parentis, Williston and Michigan Basins, and the Great Artesian Basin of Australia (Chiarelli, 1970 ; Coustau *et al.,* 1975).

On the other hand, in the case of young basins with high subsidence, both recent and present, high potential zones occupy the central and deep areas. The valuable lower pressure zones are situated at medium or shallow depth, in relatively marginal regions, where the pressure is sub-hydrostatic. In young basins with deep undercompacted beds, it has been observed that the bulk of the accumulations is generally located in the transition zone above the high pressure series. In the Mackenzie Delta Basin (North Canada) for example, hydrocarbon accumulations, principally gas, are found at the top of the undercompacted zone, characterized by a slight increase in the sonic speed, which may be read in the seismic sections (Van Elsberg, 1978) (Figure 34). In the Gulf Coast province, Timko and Fertl (1971) indicate that 90 % of the commercial oil and gas fields are situated just above the high pressure beds also identified on the resistivity curves (Figure 134).

Different relationships have been observed between crude oil and water characteristics. As a rule, it has generally been observed that the density of the crude increases with lower salinity, as generally occurs on basin margins. The eastern margin of the subandean or Putumayo Basin offers a good example of this (Figure 135). Brackish waters are often related to homogeneous and continuous reservoir horizons, while waters which are more salty and/or display rapid variations in salinity are associated with more irregular and mediocre reservoirs. In the former case, the pressures are generally hydrostatic or hydro-

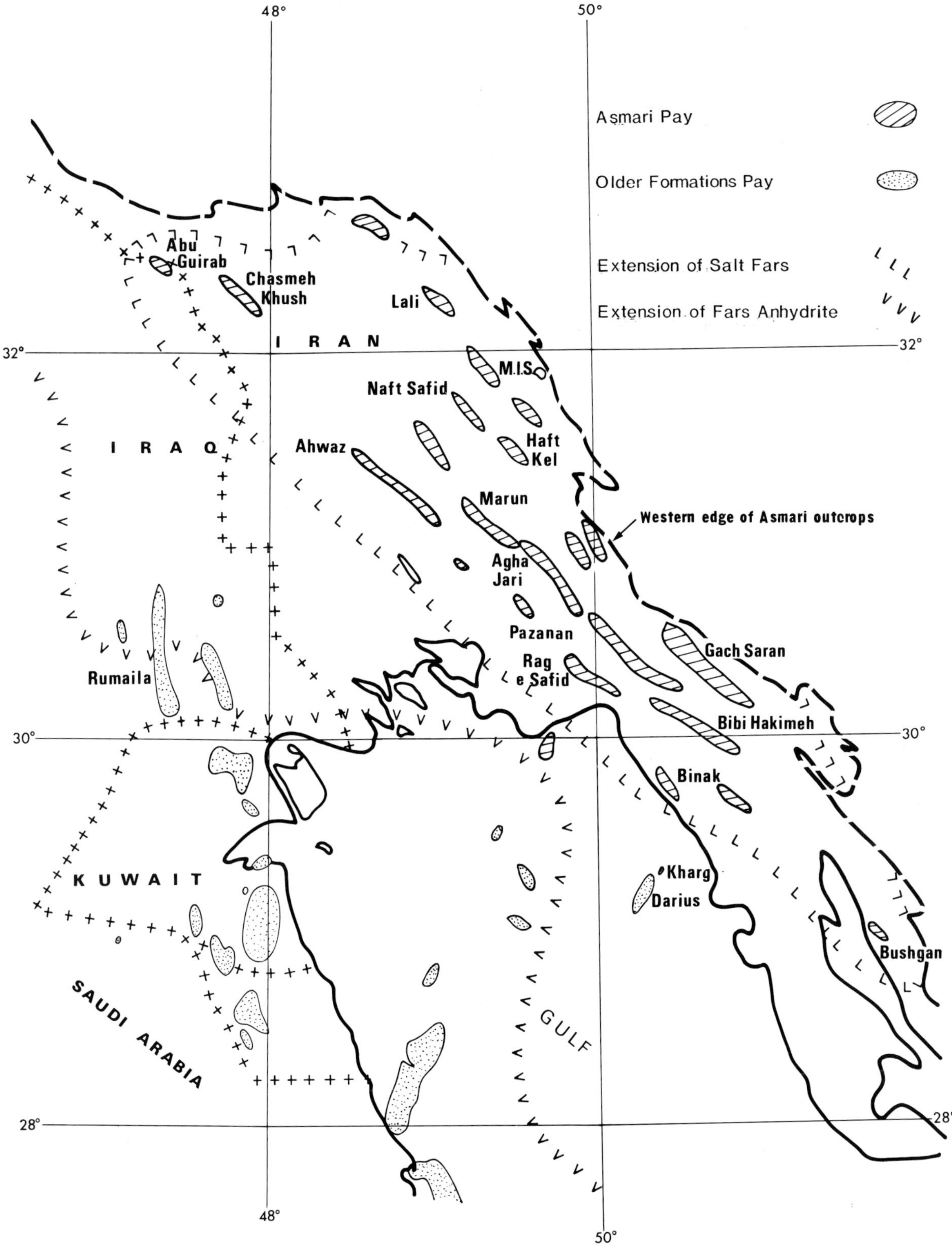

Fig. 132. — Map of foothills fields of Iran and Iraq showing the extent of the Fars seal.

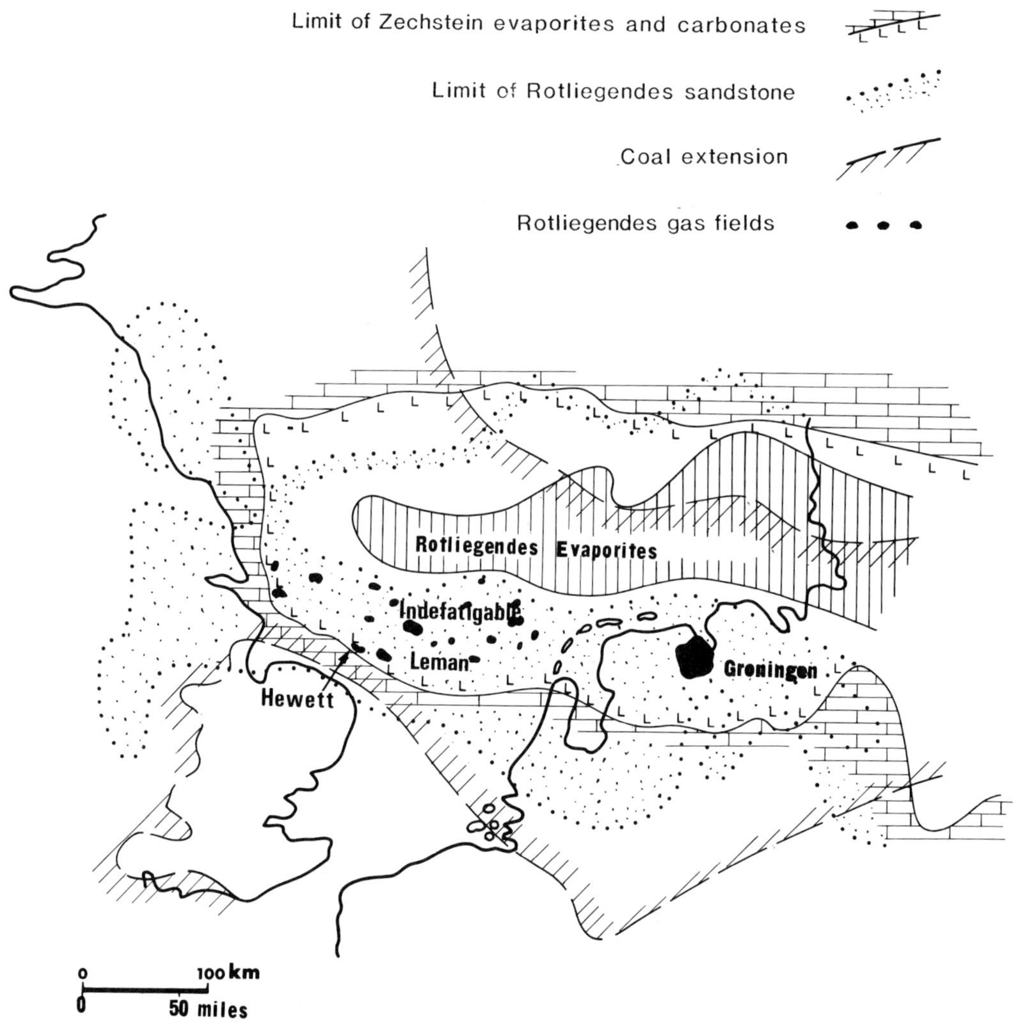

Fig. 133. — Situation of gas fields in the southern North Sea in the triple limit of coal extensions, Rotliegendes sandstones and Zechstein salt.

dynamic, but always uniform. In the latter, they are very likely to be higher and heterogeneous. Structural traps normally predominate in the former, and stratigraphic traps are more likely to be encountered in the latter.

Waters containing sodium chloride are often associated with light paraffinic crudes, and waters containing calcium chloride are associated with heavier naphthenic or aromatic crudes, as the latter appear to favor the existence of bacteria that eliminate the paraffins. Brackish, high-pressure aquifers often indicate the nearness of undercompaction mechanisms. In this case, as in the Gulf Coast, the pools are very often located at the top of the abnormal pressure horizons. High pressures and high salinities mark the vicinity of salt beds. Abnormally low pressures are frequently the sign of regional uplifts, as the expansion of the pores is greater than that of the water. High pressures often indicate the existence of good seals (Figure 136).

In a large number of petroleum provinces, one can thus identify a hydrogeological, hydrodynamic, and hydrochemical logic in the distribution of the accumulations. These zones, protected from external invasions, at relatively low pressure, and hydrochemically quite different from the neighboring regions, constitute what Coustau *et al.* (1975) call the 'petroleum redoubt', an area common to the 'hydrodynamic redoubt' and to the 'geochemical redoubt' This 'redoubt' is especially well circumscribed in basins subjected to hydrodynamism by gravity, in which the meteoric waters more or less deeply invade the marginal and sometimes internal areas. Its boundaries are routinely marked by pools in the process of degradation, and more rarely by seeps, including the well-known asphalt belts of certain basins. The concept of petroleum redoubt casts doubt on certain set ideas and exploration policies. In fact, if structural considerations traditionally tend to attract us to the highest zones, geochemistry and fluid geology, on the other hand, point to deeper zones, better protected from external destructive action. They suggest that the hydrocarbons, which are delicate and unstable entities, show an equal need for a confined environment for their formation and preservation (Coustau *et al.*, 1975).

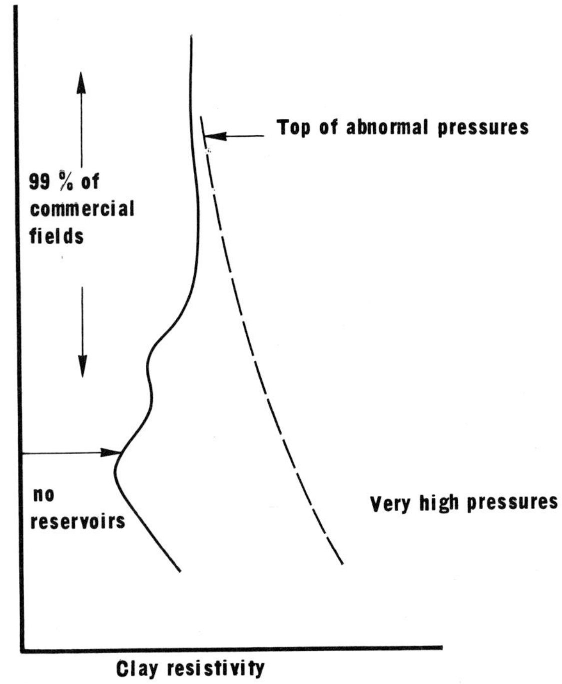

Fig. 134. — Statistical position of Gulf Coast Fields at the top of undercompacted zones (Timko & Fertl, 1971).

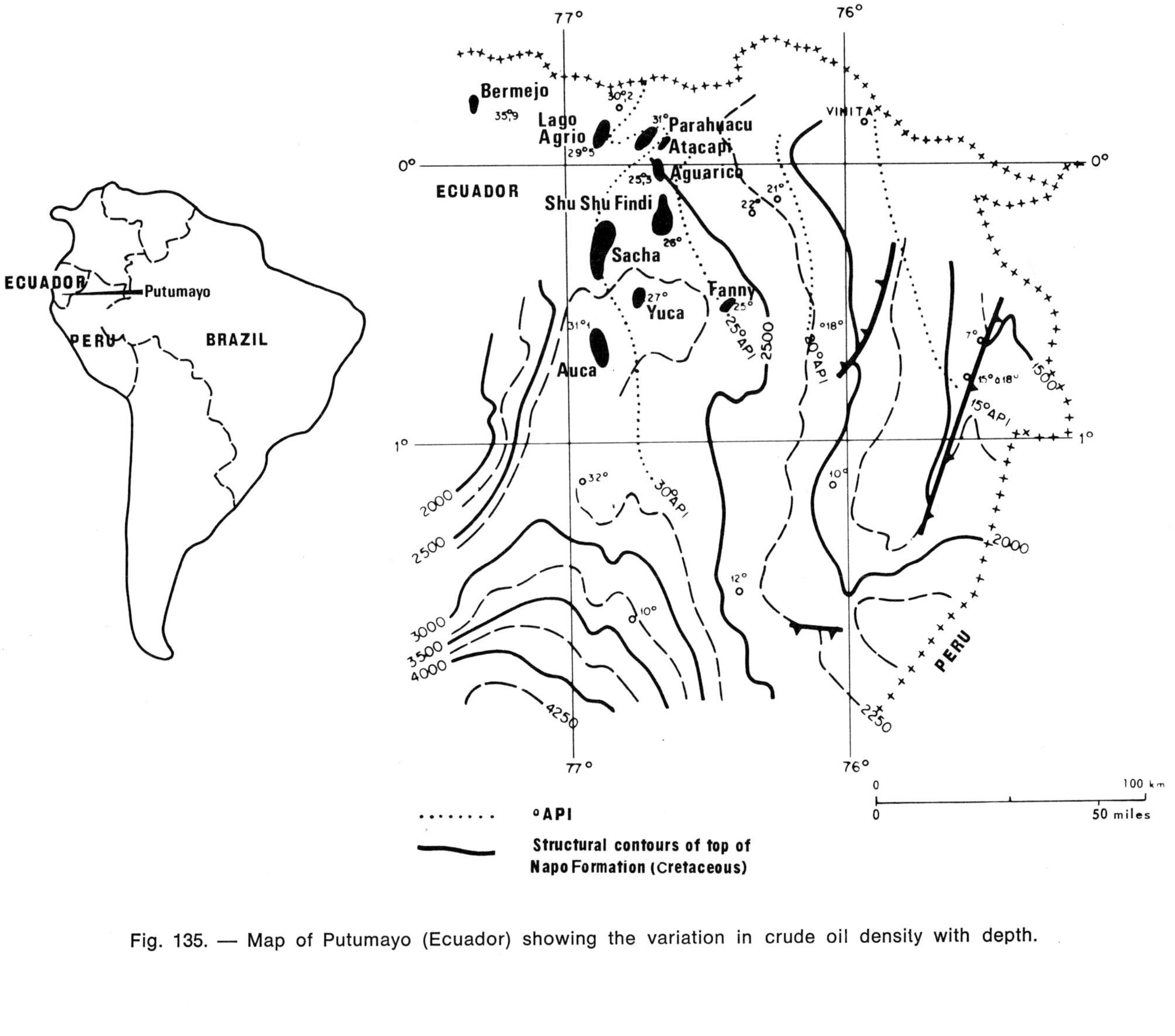

Fig. 135. — Map of Putumayo (Ecuador) showing the variation in crude oil density with depth.

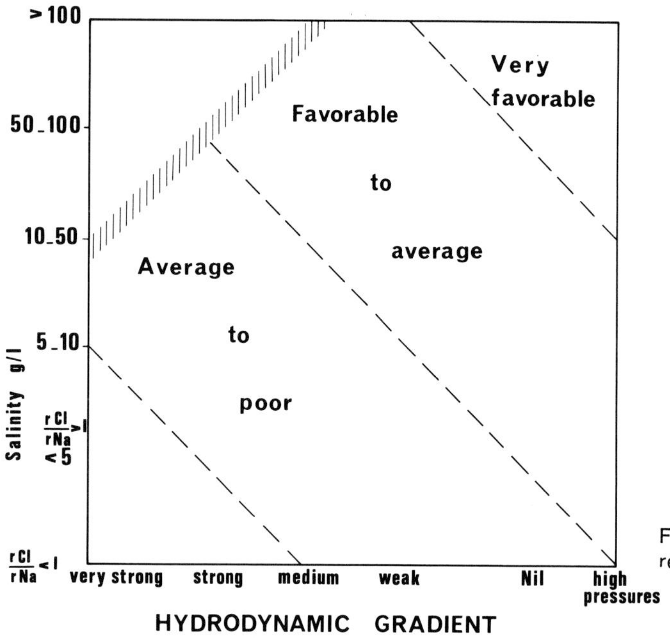

Fig. 136. — Diagram showing the petroleum value of a reservoir in accordance with its hydrodynamic and hydrochemical characteristics (Chiarelli, 1980).

These considerations cannot condemn brackish reservoirs, since a number of pools exist in these conditions, especially in the Paris, Illizi and Cooper Basins. But these provinces are relatively poor. If these pools are located on the aureole undergoing invasion, between the 'redoubt' and the invaded zone, they are frequently affected by a tilt of their water table, as rapid variations in salinity can cause fresh water and salt water to co-exist. This aureole initially appears as opposed to most of the rich provinces, or the rich parts of a province. In Alberta, for example, a statistic dealing with half of the reserves shows that two-thirds are associated with waters displaying salinity over 180 g l^{-1}. In practice, these three zones (redoubt, intermediate aureole and totally invaded margins), which are characterized by means of pressure measurements and fine hydrochemical analyses, are encountered in all basins, to the extent that the prolongation of these zones is highly variable, and sometimes zero from one basin to another.

It has been shown (Chapter 2 of Part 1) that the waters are of the sodium chloride type as a rule. In very rare cases, 'abnormal' aquifers have been mentioned, such as sodium carbonate aquifers in the Taglu Field (Chapter 4 of Part 2), and also in the Cooper Basin. In general, the waters associated with a lacustrine or continental depositional environment display salinities which are lower and of different types. More often, the waters associated with marine depositional environments are much more saline. The degree of salinity and the composition of the water finally depend on the alteration processes of the original sea-water, following the water/rock interactions, the filtration mechanism through the shales, and the solubility products (Coustau, 1976). Excellent illustrations are found, especially in Chiarelli's dissertation (1973), of the role of the hydrogeological framework in the basins of the Algerian Sahara, the Persian Gulf, Gabon/Congo and Cameroon.

1.2.1.3. ***Geothermal conditions***

As we have shown (Chapter 1 of Part 2), geothermal conditions govern the formation and preservation of hydrocarbons. Migrations may sometimes substantially modify this geothermal logic, but never misrepresent it completely. While migrated and somewhat allochtonous crudes are sometimes found above the top of the oil kitchen, in most cases a certain harmony exists between the respective positions of the oil kitchen and the pools (Figure 142). In the province of Alberta, for example, 88 % of the oil discoveries are found in the sedimentary series lying between RV values of 0.5 to 0.9 %, approximately corresponding to paleotemperatures of 68 to 116 °C, and no oil has been found in the areas exhibiting RV values greater than 1.3 %, or temperatures around 140 °C (Haquebard, 1977). In the coastal province of the Gulf Coast, most of the pools are found at the top of the undercompacted shales, in the neighborhood of the 120 °C isotherm.

Vertical zoning

As observed with respect to the conditions of the genesis of oil and gas, a vertical zoning of the different hydrocarbon categories is identified as a function of temperature, and especially of depth. In the

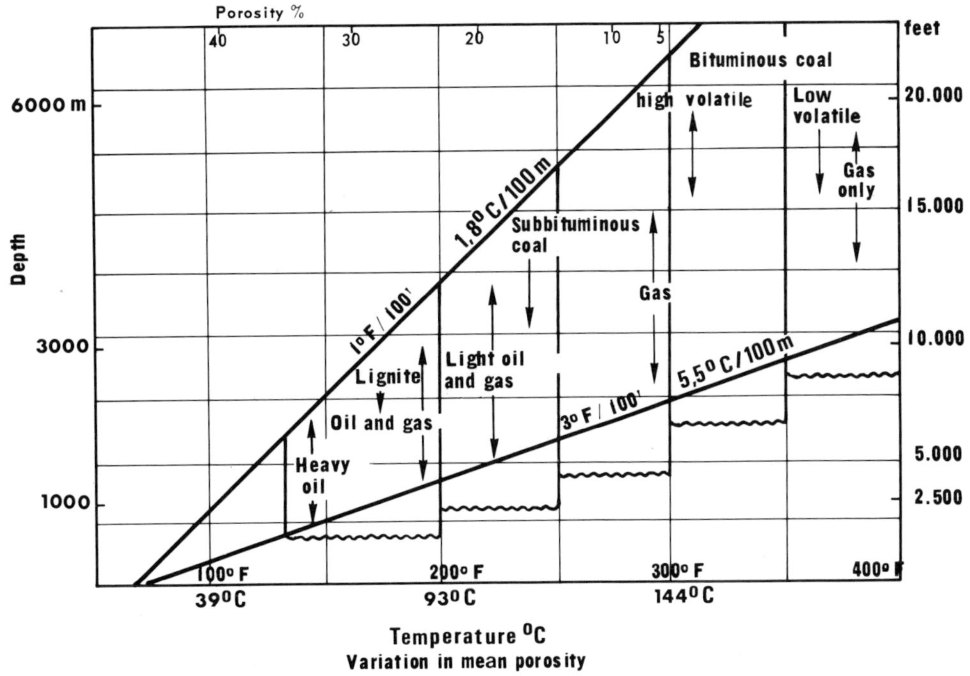

Fig. 137. — Diagram showing the distribution of hydrocarbons and coals versus depth and temperature (Klemme, 1972).

subsident basin of the Niger Delta, for example, where present temperatures are very probably at their maximum value due to the fact that subsidence is continuing, the following are found (Figure 210) :

— mainly oil up to the 100 °C isotherm,
— oil and gas between 100 and 130 °C,
— a predominance of gas above 130 °C.

This pattern is often supplemented by the presence of heavy oils and early gases in the upper horizons, with methane in the deeper zones. It can be seen that the action of temperature progressively tends to nullify the imprint of the other geological criteria, and especially that of the depositional conditions.

For over half a century, precise statistical relationships have been established between the degree of diagenesis or catagenesis, initially evaluated by the carbon ratio, and the oil and gas compositions (especially Russel, 1925). Today, the major hydrocarbon zones, like the different species of coal, can be defined by several methods based either on the state of the organic substances (carbon ratio, vitrinite reflectance, level of organic metamorphism or LOM), or on the state of the mineral species, especially by the crystallinity of illite. It should also be noted that fluorescence, which is very important to the geologist, disappears in the neighborhood of a vitrinite reflectance of 1.

This distribution as a function of temperature obviously varies in terms of depth according to the geothermal gradient. The depth/temperature and age/depth diagrams (Figure 137) illustrate the mean extension limits of coal and hydrocarbons. The latter demonstrate the development of heavy oils in recent series, light products and wet gases in older series or in the deeper parts of young basins, methane being more omnipresent. This pattern may be reinforced by the action of alteration and degradation mechanisms, especially by infiltration waters. The latter, chiefly by the action of bacteria, cause a loss of light hydrocarbons and principally of paraffins, with a general increase in heaviness that is often characteristic of marginal and poorly protected zones of the basins. This evolution thus coincides with the action of diagenetic mechanisms.

Regional gradients

As a rule, generally higher than average geothermal gradients are observed in most rich provinces, while in the poor regions, except for the great paralic basins, low gradients frequently prevail. The following gradients may be mentioned :

— very high (5 to 7 °C per 100 m) in the basins or in certain zones of the basins of Sumatra, Sirte, California, Central Asia and Arabia,
— high (3 to 5 °C per 100 m) in the Prudhoe Bay, Ghawar, Gippsland, Orinoco and Oklahoma areas, in the Romanian pre-Carpathian depression of Edjeleh, in certain sectors of Western Siberia, and in the Gulf of Gabes.

This confirms that geothermal gradients vary considerably according to the type of basin, and even within a single basin. In the Volga-Urals province, for example, high zones have been observed exhibiting higher geothermal gradients and generally containing the largest accumulations of oil and gas, while sectors with lower gradients are poorer or contain only heavy oil (Valitov, 1972). Furthermore, there is every indication that the excellent thermal conductivity of salt domes favors the maturation and genesis of hydrocarbons in their immediate vicinity.

1.2.2. **Historical criteria** (s.l.)

It is rather arbitrary to try to individualize historical criteria, while stratigraphy is already a succession of deposits and paleogeographies. However, it is important to stress the major role of the time factor in the geological conditions contributing to the formation of petroleum provinces. This factor plays a fundamental, although often poorly understood, role in the distribution of hydrocarbon flows and in the formation of accumulations. It is certainly easier to include in our reasoning and undoubtedly appears to be more reliable, since the availability of chronological scales associating the quantitative measurement of time units by isotopic or absolute datings, and the broad and classic correlations of biostratigraphy.

This assessment of chronological conditions is normally carried out from the analysis of geological sections, especially as they are seen in high resolution seismic sections, referred to today as the analysis of seismic facies or stratigraphic seismics. This stratigraphic and structural approach naturally leads to the analysis of the structure and framework of the basin, which is our first subject.

This analysis is supplemented by the construction of time/depth diagrams (or burial, or another criterion), allowing a quantitative assessment, as a function of age, of the variations in subsidence, sedimentation and erosion; in other words, the possibility of characterizing the type of basin, or the functioning of a trap or a petroleum system. This will be our second subject.

1.2.2.1. ***Structural conditions***

The framework of a basin, the culmination of its geological history, may be characterized very schematically by the juxtaposition and succession in time of negative and positive areas. We shall analyze the latter for greater clarity, first from the regional standpoint, and then from the local standpoint.

Regional level

On this level, which corresponds to that of the petroleum zone, subsident areas must be considered as sources of hydrocarbon production :

— by depositional environments favorable to source rock and seal facies,
— by the burial of these deposits, favorable to their compaction and maturation, and to the expulsion of the fluid,
— more exceptionally, by the deposition of facies favorable to reservoirs, especially turbidites and chalk.

On the other hand, positive areas exhibit interesting characteristics, such as the deposition of facies favorable to reservoirs, and the formation of relatively low pressure beds, allowing the convergence of hydrocarbons, but also involving negative aspects, in particular by the probability of condensation of series and the absence of sediments by non-deposition and/or erosion, and by the risk of invasion and degradation.

Many uplifts (too high) are liable today to be bald or unsealed. The most interesting uplifts are finally those found sufficiently within the basin, and/or which have been the subject of pulsations or slight advances and withdrawals of the sea, allowing the development of alternating facies, and in particular the deposition of seals. Transition zones between positive and negative areas are very often the most interesting, the former more specifically favorable to storage, and the latter to the genesis of hydrocarbons. It has been confirmed that the permeability fringes that frequently border positive areas constitute an especially favorable area for accumulation.

Regional highs, not necessarily the earliest or the most permanent, have often provided the most interesting sites of the first hydrocarbon accumulations. In the case of stable areas, these primary traps may preserve their hydrocarbons until the present, yielding giant fields. This appears to be the case of the Hassi R'Mel and Hassi Messaoud pools in the Algerian Sahara, the Panhandle/Hugoton accumulations in the United States, and the Ghawar Fields on the Arabian platform. If the positive area is subsequently fragmented by tectonics, folded or intersected by faults, it could supply a series of fields. This suggests the fields of the central platform or the northern margin of the Permian Basin, and of the El Biod-Ghassi Touil Ridge in the Sahara. This is also undoubtedly the case of certain rift provinces, as grabens are often the descendants of early uplifts.

Often noted is the outstanding value of uplifts of long duration which have played the role of permanent hinges between two or more subsident areas. This is the case for example of :

— the Prudhoe Bay Field on a high, located first on the northern margin of a Paleozoic-Jurassic basin, and then on the southern edge of a second Cretaceous/Cenozoic basin,
— the Hassi R'Mel Field at the boundary between the Paleozoic basins of the Western Sahara and the Secondary and Tertiary subsident area of the Lower Sahara, to the east (Figure 144),
— the Bolivar Coast Fields, straddling the hinge between the Paleogene and Neogene basins.

Structural factors, especially the alignment of pools along certain regional highs, condition a structural logic, well illustrated by the Nesson and Cedar Creeks Trends in the Williston Basin, the La Salle and du Quoin Trends in the Illinois Basin, the Nemaha Ridge in the Mid-Continent, the Zelten-Defa and Dahra-Raguba Trends in Libya, and Meillon/Saint Faust in Aquitaine.

Local level

On the local, or trap level, the structural conditions act in many ways. By considerably simplifying the matter, two clearly divided cases may be distinguished, folds and faults. Many observations have been made on folds. The first consideration is purely geometric : the cause is irrelevant, because any form may be interesting as soon as it offers a closed and sealed surface turned downward. It may be noted, however, that the largest structures or those exhibiting the largest closures are not necessarily the most valuable. The intensity of folding often gives rise to fracture systems detrimental to sealing qualities, at least above the neutral point. Very frequent and vast accumulations are sometimes found within domes and anticlines with thickness variations along the same vertical.

The historical aspect is always very important. The older structures are not necessarily the most favorable, but as we have seen, the essential condition is a close geographical and chronological relationship between the existence of the trap and the hydrocarbon flow paths and periods. Very often, and rightfully, the frequency of local unconformities on anticline pools has been observed and the repetition of supply phases identified. This applies in particular to the Oklahoma City Field and other neighboring accumulations of the Mid-Continent, a number of Pliocene pools in the Po Plain, starting with the largest, Cortemaggiore, and several horst fields of the Viking Graben. Certain advantages of recent tectonics have perhaps been less well analyzed, at least in areas or provinces that are still very recently subsident. This applies in particular to Iran, to the vast anticlines of the Zagros foothills, formed since the Miocene. This youth has not allowed erosion to attack them deeply, and they exhibit a remarkable succession of competent beds separated by pinched synclines. On the contrary, deeper erosion would have resulted in vast synclines and pinched anticlines (Colman Saad, 1978).

The situation is probably more complex, and hence more controversial, concerning faults, simultaneously and often alternately barriers and seepage paths. With considerable simplification, and hence by accepting inexactitudes and even errors, one is tempted to consider that faults are better sealed if they are older and if they have not been active recently. In fact, they are primarily discontinuity planes and hence water circulation surfaces, where the water is progressively mineralized and cements the fractures. Hence synsedimentary faults, particularly the great basin boundary faults, like the Steinberg Fault in the Vienna Basin, marked by a string of pools, prove their effectiveness. The Cook Inlet Basins in Alaska, the Reconcavo and Suez Basins, also very probably owe their potential to large boundary faults.

Conversely, recent or active faults are undeniable seepage paths. In the gas-bearing structures of the Kangan region in Iran, intersected by large recent transverse faults, some of which are active during earth quakes, the closure is often limited by this type of slip.

***

Entrapment conditions make a significant imprint on the final distribution of oil and gas accumulations. As shown previously, differential trapping in a folded area with tight seals results in a higher proportion of gas pools in the internal areas of the basin, and oil in the marginal and shallower parts. This pattern thus coincides with the diagenetic distribution of oil and gas.

On the contrary, differential trapping by vertical migration, through a fault or seal, results in a majority of gas pools in the shallow beds, while the oil remains in the underlying and more internal series. This pattern corresponds to the oil/early gas distribution (Figure 139). In reality, geological conditions are rarely simple or clear enough to impose a single pattern at the expense of all others, especially throughout a whole basin. Hence it is generally difficult to identify a single distribution logic through the combined or opposed action of the different possible patterns.

1.2.2.2. ***Geodynamic conditions***

The historical study of a petroleum zone, or the evolution in time of its geodynamic framework, serves to represent the chronology of the different mechanisms contributing to the formation and preservation of

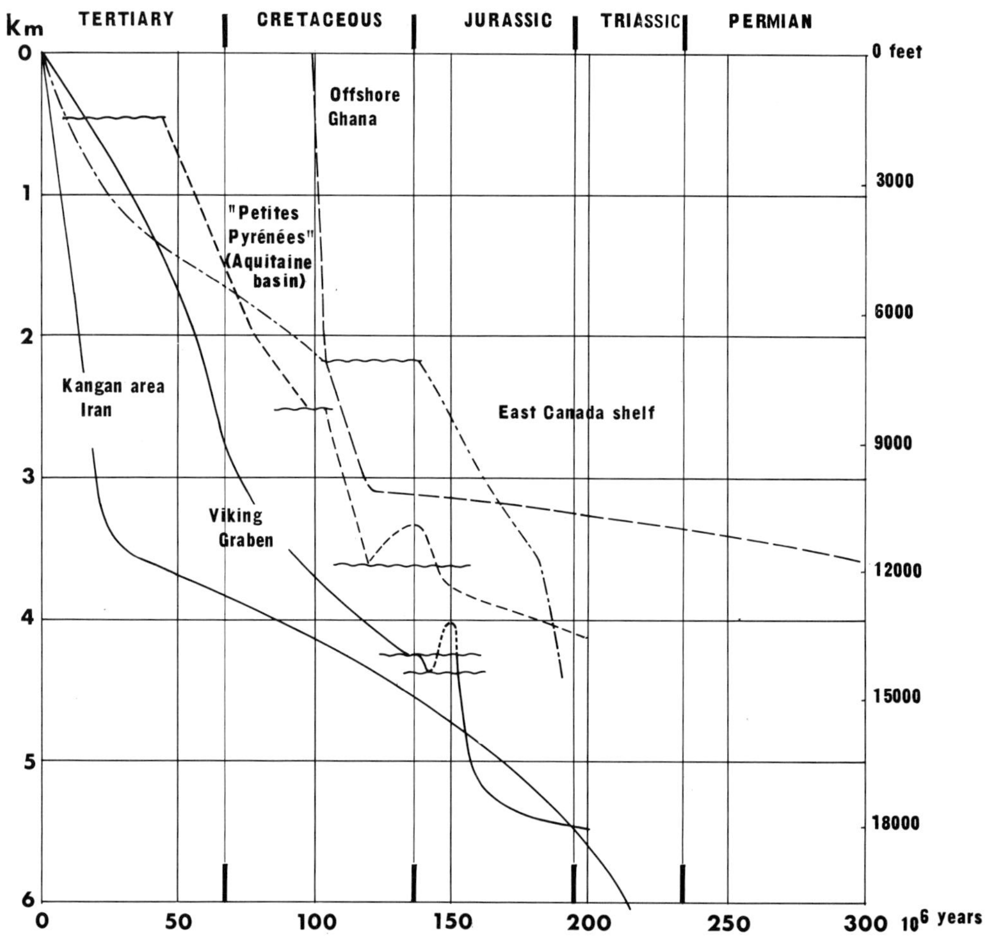

Fig. 138. — Age/depth graph of some wells in Iran, the North Sea, Aquitaine, East Coast of Canada and the Ghana Coast.

accumulations. This representation is mainly offered by diagrams that illustrate the evolution of a parameter as a function of time, especially burial depth or sedimentary thickness. These 'age/depth' or 'geohistorical' diagrams (van Hinte, 1978) help to reconstruct the variations in a parameter, an area, and a mechanism as a function of time, and offer a quantitative assessment of the variations in subsidence, sedimentation, erosion, and the duration of a hiatus, and to identify the type of basin and, from the petroleum standpoint, the performance of a trap or a system (Figure 138).

Another representation is provided by the age/angle of sedimentation diagrams that allow a comparison between two sections, and a clear representation of the variations in sedimentation and especially onlaps, offlaps and subsidence on the scale of the basin. In this case, the thickness is replaced by the sediment column of a given time interval in comparison with an apex corresponding to the non-deposition line at the margin of the basin (Rona, 1973).

The principle of these different diagrams is not a new one. The earliest representations are undoubtedly due to Lemoine and his Geology of the Paris Basin, published in 1911. The revival and increased value of these methods of expression result from advances in dating techniques, which help to quantify the mechanisms and add a new dimension to the time factor, both with respect to relationships between the mechanisms and the absolute aspect of the measurement. In these analyses, the time factor must be perceived in three complementary ways :

— the relative positions of the different geopetroleum mechanisms in relation to each other,
— the continuity or non-continuity of the major geological events,
— their chronological situation in relation to the present.

Relativity

It has been shown in different basins that maturation processes of organic matter and the expulsion of hydrocarbons were very often recent events in geological history, events related to periods (and zones) of the pronounced subsidence of the basins (Tissot *et al.,* 1975). In these conditions, as has been shown throughout this work, it is not the 'absolute' age of the structures that counts, but rather their relative age or their situation, in time and in space, in relation to the hydrocarbon flows. Hence the earliest traps are not necessarily the best, and care must be taken to avoid condemning the most recent, even if they very often show a greater likelihood of containing gas. The rich Cook Inlet Basin in Alaska, for example, like those of California, displays very young Plio-Pleistocene structures, but a high Cenozoic subsidence of about 7500 m.

Duration and continuity

Another important and less emphasized aspect of the time factor is the notion of duration and continuity. It is very well known for the mechanisms of genesis and maturation of organic matter that the temperature must be higher if the duration of 'cooking' is shorter, and *vice versa* (Chapter 1 of Part 2). This means that the young or Neogene basins must be subjected to a higher heat flow than the older basins, or, in other words, the productivity of a basin at constant temperature is proportional to its duration. This is illustrated by the work of Lopatin (1971) on coals, and Zieglar & Spotts (1978) on the San Joaquin Basin in California, as we have shown (Section 1.2. of Part 2) (Figure 41).

The coastal province of Louisiana displays a fairly traditional pattern, in which the accumulations are situated slightly above the oil window, due to vertical migrations (Figure 139).

It is then necessary to analyze the continuity or non-continuity of the events that occupy this time, and break it down into phases. It is obvious that, for a given period, the number of phases is inversely proportional to their own duration. The larger the number, the shorter the phases, the more discontinuous the history, and the lower the 'degree of symmetry'. Provinces with a high 'degree of symmetry' and a relatively simple history often turn out to be the richest. One example is the province of the Zagros foothills : a sedimentary series 12,000 m thick, the principal part consisting of a competent Permian/Miocene unit 6000 to 7000 m thick with a duration of 250 million years, yielding a series of folds with a wavelength of 10 to 20 km and amplitude of 2 to 3 km, formed late since the Miocene, thus very slightly eroded, and consequently offering simple anticlines separated by synclines pinched at the surface. Simple, big and rich.

The Arabian province exhibits a larger network of folds (50 km wavelength) and a slightly different geological history, but also of long duration and relatively undiversified. Western Siberia also illustrates a continuous pattern from the Trias to the present. These provinces are the world's three richest provinces containing giant fields.

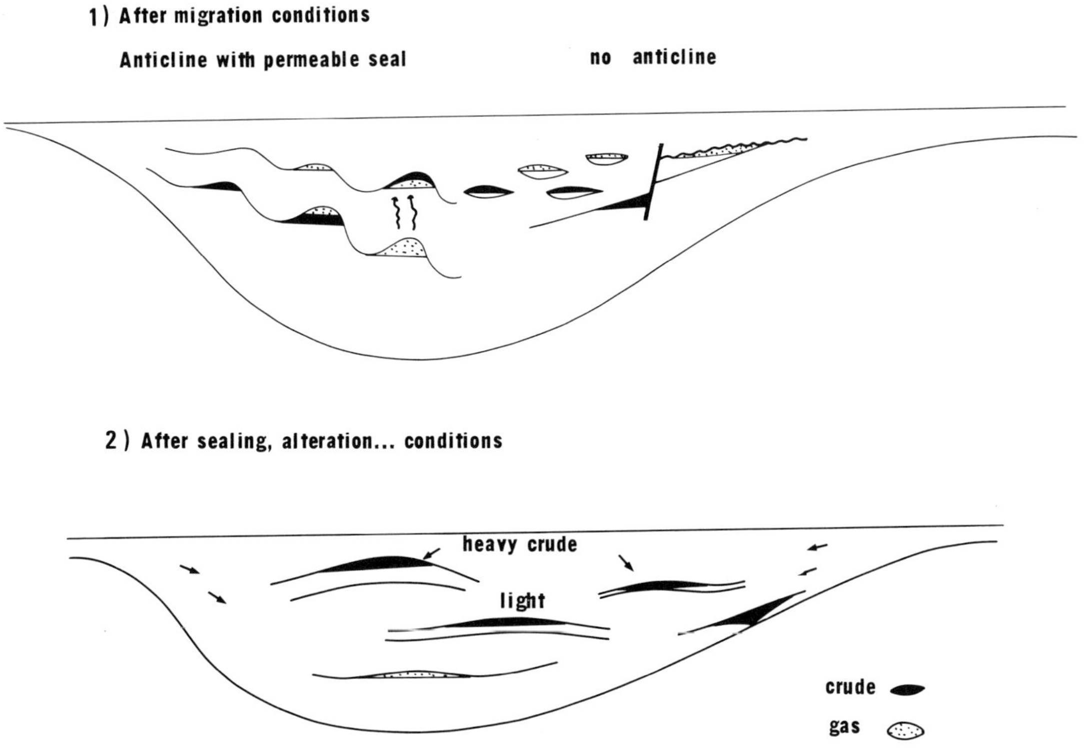

Fig. 139. — Distribution of hydrocarbons in a basin according to type of entrapment.

On the contrary, many phases generally yield a dispersed habitat. All other things remaining equal, a long history with repetitive episodes, but with local 'events', testifies to a rich province with a dispersed habitat, like the deltaic provinces of the Gulf Coast and Nigeria, and reef provinces. A complex and eventful history, a land mass emerged several millions or tens of millions of years ago, means a poor province with a dispersed habitat. Typical examples are the Aquitaine Basin, the Rocky Mountains, Morocco and the Appalachians.

In this history, phases of non-deposition and erosion are, as may be guessed, particularly unfavorable. Uplifting first results in decreases in temperature and pressure, fractures, fissures, invasions of meteoric water and alteration processes, seepage and finally the elimination of deposits. Provinces whose history is cut by a long lacuna, and whose transgressive series are not sufficiently thick, are rarely favorable, like a large part of the Canadian Arctic except the Sverdrup basin, and the northern margin of the Aquitaine Basin. In other words, positive movements that often reflect gaps are destructive episodes that may exert negative action, as opposed to subsidence episodes, which represent periods of the 'manufacture' of pools. These aspects of geological history, which may appear to contradict traditional structural concepts, offer new elements to supplement our argument.

Situation with respect to the Present

It is important to situate the history of a field and its time-table in relation to the present. In petroleum exploration, it is necessary to rehabilitate the viewpoint of the 'spectator', so rightly criticized, in the analysis of geological processes. Wegman (1950), in his comparison between the history of the sedimentary basins and racing, emphasized the periods of filling of the sedimentary basins, namely, the entire progress of the race, and not only the unconformities which in some way represent the finishing line. For the oilman, the whole story that culminates in the formation of pools, the entire race, is interesting, but even more the viewpoint of the spectator at the finishing line, who records the presence or absence of hydrocarbons in the trap.

As we have seen above (Section 1.2 of Part 2), the genesis of oil may occur over a relatively short period (on the geological scale), less than 10 million years in the case of periods of active subsidence, and over a long period, perhaps around 100 million years, in the platform basins. In this case, however, it may be felt that the genesis and expulsion of hydrocarbons occur intermittently over one or more periods of intense subsidence. Thus Tissot & Welte (1978) note that nine out of ten oil and gas fields existing today were formed in the Cretaceous and/or Tertiary, in other words in the last 100 million years, and often much shorter periods, irrespective of the age of their source rocks. The remaining fields are primarily associated with the deeps of the Hercynian orogeny.

What counts to us as prospectors is the relationship between geological history and the presence today of oil and gas. Apart from exceptional cases of perfectly tight seals, it must be recognized that the richest zones are those which have retained their hydrocarbons, in other words, those which have succeeded in compensating for their losses by inflows. Subsidence causes an increase in pressure and temperature, which results in the production of new quantities of hydrocarbons, and these zones are characterized by a subsidence rate (rapid sedimentation during the Tertiary). In other words, sedimentation rates of around 40 to 50 m per million years offer a highly favorable factor. This remark is backed up by the number of rich provinces subjected to recent if not present subsidence, as illustrated by the age/depth graphs.

This is at least partially the case of the biggest and richest provinces : the Persian Gulf, West Siberia, Maracaibo, the Gulf Coast, Tampico/Chiapas, Sirte, Southern California, Nigeria, the North Sea, North Slope, Baku, the continental shelf of Labrador (Figure 138). In Certain provinces, like those of the Po and Niger Deltas, and the continental shelf of Brunei/Sabah, some accumulations have been dated as very recent : the Tembugo Field, for example, in the latter province, discovered in 1971, productive in the turbidite type sands of the Upper Miocene, at the top of a faulted anticline, folded less than a million years ago (Whittle & Short, 1978).

It must not be inferred that all basins displaying this pattern are prolific : this is only one prerequisite among others. Nor is it true that all basins displaying the other type of diagram can be discarded : one can only conclude that the chances of finding productive pools are rarer and the risk of failure greater.

1.3. HYDROCARBON DISTRIBUTION

1.3.1. Distribution in the basin framework

The petroleum distribution in the basin obeys a series of geochemical, hydrogeological, sedimentary and structural mechanisms, of which the complex result often masks the logical pathways.

1.3.1.1. *Geochemical and hydrochemical criteria*

Genetic criteria

We have already shown that, as a rule, with an increase in burial depth and temperature, in other words, under the action of subsidence, the following are encountered :

- heavy crude oils, immature or slightly matured, at depths of a few hundred meters, corresponding to low degrees of catagenesis, and to average vitrinite reflectance values less than 0.5 to 0.6,
- below, and down to depths corresponding to RV of 1 to 1.2, increasingly light crude oils,
- above an RV of 1.2 to 1.3, gas condensates followed by dry gases.

Early gases, mainly represented by methane, may also be found at shallow depths in immature zones. Methane thus appears both as a product of formations subjected to strong catagenesis, and immature zones that contain, or that previously contained clathrates during the last glacial periods. In the latter case, as observed in Siberia and Canada, methane is a predominant product in the Arctic and probably Antarctic zones, and possibly of certain deep offshore basins.

The oil gathered in a reservoir is always sensitive to variations in temperature caused in particular by a subsidence movement. An increase in temperature is reflected by the cracking of the molecules, yielding a growing percentage of light fractions and gas, and finally methane and pyrobitumens. This is a process of aggradation which may be qualified as transformation or thermal alteration. The West Canada Basin offers a good example of the evolution of crude oil characteristics with depth.

An increase in pressure, especially under the effect of subsidence, also leads to desulfurization and 'deasphalting', which may proceed as far as the deposition of bitumen. The dissolution of gas or light products in crudes associated with increased pressure gives rise to the precipitation of asphaltenes. This action, which occurs in the same manner as the cracking of hydrocarbons under the action of temperature, is often difficult to distinguish from the latter. This mechanism has been investigated closely in the reef pools of the Upper Devonian in West Canada (Evans *et al.*, 1971; Rogers *et al.*, 1974; Bailey *et al.*, 1975).

Alteration criteria

Hydrocarbons are extremely fragile products which are only preserved in anaerobic environments, and at moderate temperatures. Hence they are particularly sensitive to physical, chemical and biological action, which is liable to affect them throughout the geological history of the basin. These different actions have the result of gradually obliterating the imprint of the depositional conditions. One can distinguish, apart from the transformations caused by subsidence or the increase in temperature, and which may be compared to aggradation mechanisms, a whole series of processes more or less associated with uplifts and more directly related to alterations and degradations.

These positive movements are reflected by erosions, alteration of seals, fracturing, drops in pressure, and invasions of meteoric waters, causing the escape of the lightest fractions, so that the preserved fractions become progressively heavier. The deterioration of the seals also diminishes the protection of the pools, causing alteration and oxidation of the crudes. This action, especially in contact with meteoric waters, results in polymerizations and yields dark colored, highly viscous products with a pasty or solid appearance. This is the facies of seeps. It is also the process that culminates in the tar mats, layers of asphalt formed in certain pools in contact with the water table, which may considerably restrict the action of the aquifer during exploitation. The attack of external waters is marked simultaneously and strongly by the action of microorganisms, and especially the bacteria which they convey (Evans *et al.*, 1971; Bailey *et al.*, 1973; Connan *et al.*, 1975). This biodegradation results in the disappearance of light fractions, the breakage of the weakest bonds, and consequently heavier remaining products and an increase in $\delta^{13}C$. It may also result in the liberation of methane.

It must be kept in mind that hydrocarbons, which are fragile and unstable chemical substances, unceasingly adapt to the variations in the physicochemical conditions of the environment, both in the reservoirs and in the source rock. These different transformations, and especially alteration mechanisms, exert an influence on the distribution of hydrocarbons in the basin. They reinforce the distribution determined by genetic mechanisms, significantly increasing the number of heavy products in shallow beds, and especially on the basin margins.

The age of an oil

Several answers may be given to the question : how old is a crude oil ? It may be the age of the reservoir in which it is found. This is a simple, easy answer, often given by statistics. It may also be characterized by the age of the source rock in which the oil originated. This answer is more hypothetical, and correlations are problematic. Above all, it is inaccurate, because the oil is practically never contemporaneous with its source rock, and the time gap may be considerable.

One can also define the age of a crude oil by the era of its genesis, namely of the 'cooking' of the organic matter in the source rock and its expulsion. This is certainly the most accurate answer. As demonstrated by Tissot *et al.*, (1975), the interval between the deposition of the source rocks and the principal oil formation phase may vary from a few million to 300 million years. 'In particular, the oil in the bulk of present fields was formed in the Cretaceous and the Tertiary, whatever the age of the source rock.' (Tissot *et al.*, 1972). This re-emphasizes the importance of the mechanisms of subsidence and the downwarping of the sedimentary basins.

1.3.1.2. ***Sedimentary and climatic criteria***

The facies condition the characteristics of the hydrocarbons formed and preserved therein. For example, we have seen that facies rich in organic matter of the humic type, often corresponding to clastic shale environments such as deltaic zones, exhibit a large share of gas and paraffinic oils rich in waxes. On the other hand, evaporitic and carbonate facies are more favorable to naphthenic oils and sulfur-bearing crudes. Temperature appears to have a greater effect on hydrocarbons in shaly environments than in carbonate facies. On another level, it has been observed that gas is especially abundant in cold climates, due to the action of clathrates in reducing the solubility of methane in water.

1.3.1.3. ***Structural and hydrodynamic criteria***

Through the mechanism of migration, structural criteria very closely determine the trapping conditions. The oil and gas migrate, often separately, from the high pressure zones to lower potential zones, generally more shallow zones. These migrations comply with two main patterns, depending on the rigorously impermeable or permeable character of the seals and closures (Figure 139). If migration occurs laterally, under an unfaulted seal, a differential trapping pattern occurs, giving rise to the retention of the gase in the internal zones, with oil in the shallower marginal structures. This scheme, well-known in the case of a succession of anticlines, may also apply to reef traps, as illustrated by the Silurian reef zone in northwestern Michigan (Figure 53) (Gill, 1979). This zone displays a close convergence with the previous hydrocarbon distribution models. However, the presence of active hydrodynamism may expel the oil and only allow gas accumulations to remain at relatively shallow depths.

If migration passes through a succession of stratigraphic traps, in a sort of series arrangement, or if it follows a vertical path through the beds, and in particular the seals, through fractures and fissures, light products, and chiefly gas, reach the upper stages. It remains trapped if it encounters an impermeable seal. If not, it is lost by seepage through the overlying beds, and from there into the atmosphere. In this scheme, the oil may thus remain in the deep zones, while the gas may be found higher up in the shallow beds. In actual fact, this vertical migration (or seepage) often occurs preferentially from accumulations, and especially from anticlinal pools. In this case, it may be superimposed on Gussow's lateral flow path.

In case of the joint supply of gaseous and liquid hydrocarbons, we know that the former occupy the upper part of the closed arches. If the closures are of low amplitude, the gas occupies most or all of the trap, yielding predominantly gas provinces or zones in this case.

1.3.2. **Hydrocarbon distribution in the stratigraphic framework**

The geological characteristics of the major steps in the Earth's history stamp their particular features on the great petroleum provinces and the stratigraphic distribution of the hydrocarbons. The Silurian black shales, rich in organic matter, which have undergone considerable extension, appear to constitute the earliest major hydrocarbon source rocks. The fluids formed often collect in the detrital beds of the Lower Devonian.

An initial major concentration of gas appears in the Permian detrital formations, often of eolian facies, especially developed in this period with a desert tendency, succeeding an era of high rainfall. The reservoirs are largely supplied with gas from the coal deposits of the Carboniferous, which concentrate tens of billions of tons of organic carbon. They are often protected by evaporites, well developed in the Permian and the Trias. Deposits of black shales, often bituminous, reappear in strength in the Upper Triassic/Lower Jurassic in Mediterranean Europe, and throughout the Jurassic in more northerly areas, offering a second outstanding store of organic matter.

The Lower Cretaceous, a period of regression and basins with detrital and reducing facies, often rich in coal deposits, appears to be another major period of hydrocarbon genesis. The Upper Cretaceous, a period of vast transgressions, shallow deposits, often carbonates containing rudists, in a relatively warm climate, is a remarkable period of the formation of reservoirs and traps that concentrate the largest reserves on the entire stratigraphic scale. The presence of vast deposits of black shales and the extension of anaerobic events and of the major reducing environments of the Middle Albian to the basal Turonian in the Atlantic areas, seem to provide the source of the many giant fields originating in this era.

The largest quantities of hydrocarbons were generated around the end of the Mesozoic and the Early Cenozoic, and the bulk of these reserves derives from the Middle Cretaceous. It should be noted that this period does not appear to be characterized by continental dislocations, but by an acceleration of the rate of oceanic drift and the absence of magnetic inversions (Larson, 1979). The Tertiary is characterized by a disintegration of the Earth's structure, and especially by an increase in the number of basins, at the expense of size. These basins often provide excellent conditions for the genesis and trapping of hydrocarbons. The Eocene and Miocene appear to be especially favorable periods.

Taking the analysis of the world's oil and gas fields as a basis (Halbouty, 1970), this distribution can be illustrated by various presentations. The distribution of oil discoveries in the giant fields, amounting to around 100 Gt, shows the predominance of the Cretaceous, the reservoirs of which contain more than one-third of the total, despite a loss of pools in Iran to the Tertiary, a second peak in the Neogene with 16 Gt, followed by the Paleogene (8 Gt) and the Devonian (5.6 Gt) (Figure 140).

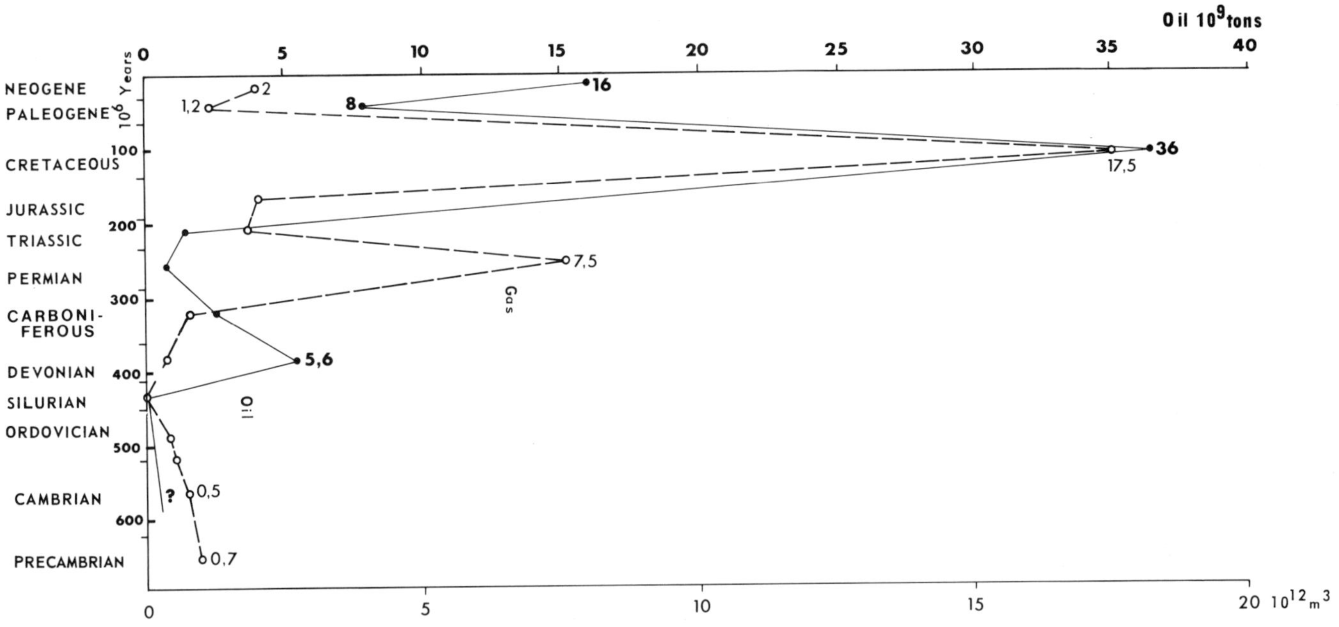

Fig. 140. — Distribution by age of proven Giant Field oil and gas reserves as of 1971 (Meyerhoff, 1979).

The landscape is substantially different for gas. The Cretaceous always occupies a predominant position with slightly over 50 % of cumulative discoveries, owing to the potential of Western Siberia. This is followed by the Permian, with about 7.5 Tm^3, or nearly one-quarter (North Sea, United States, Middle East, USSR). Note that in these two stratigraphic units, which account for three-quarters of the world's gas discoveries, are generally located in the neighborhood of extensive measures, in the Carboniferous/Permian and the Cretaceous. From the geodynamic standpoint, these two periods correspond to a certainly stability (Meyerhoff, 1979).

The distribution of giant fields as a function of age and depth (Figure 141) for an average geothermal gradient of 27 °C km^{-1}, also demonstrates the importance of Jurassic, Cretaceous and Cenozoic formations, and, to a certain degree, a shallower depth for Paleozoic accumulations. These graphs, which tend to emphasize the weight of the Cenozoic and Mesozoic periods, illustrate the recent character of hydrocarbon formation and the losses of older series. A more thorough analysis reveals the importance of zones in periods of active subsidence which were not too intensely folded and fractured subsequently.

Oil and gas pools are nevertheless found in the Lower Cambrian and Precambrian, in areas between 500 and 800 million years old. Pools have been discovered, the first in Siberia in the Irkutsk Basin. The Markovo Field, discovered in 1962, contains a million tons of oil in the Lower Cambrian, and about 17 Gm^3 of dry gas in the Upper Proterozoic. Sredne-Butuobin, some 1000 km further north, has reserves of 166 Gm^3 of gas containing 88 % C_1 and 4 % C_2 and 22 Mt of light oil, also in the Lower Cambrian and the Upper Proterozoic. Verkhenevilyuy, discovered in 1975, contains about 300 Gm^3 and 36 Mt of condensate (Meyerhoff, 1981). Furthermore, hydrocarbons and kerogen have been found in northern Michigan, southern Africa and Australia, in rocks dating from 2.3 to 2.8 billion years (Murray *et al.*, 1980).

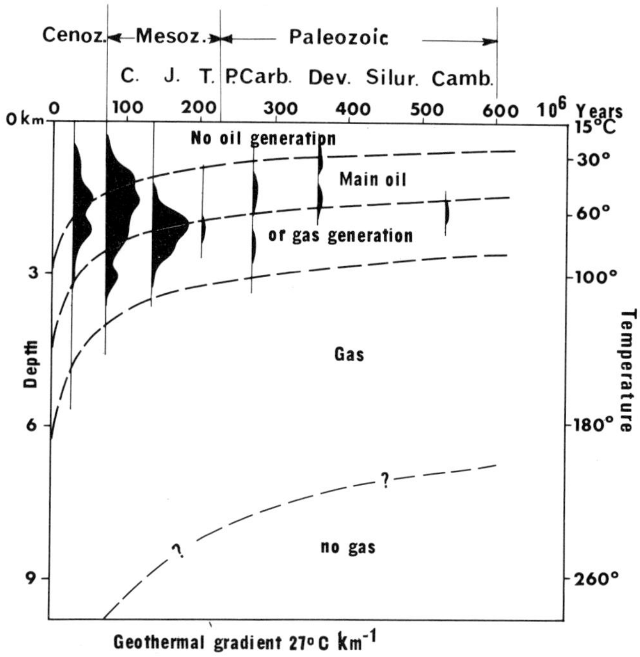

Fig. 141. — Diagram showing distribution as a function of age and depth/temperature of giant oil fields and oil and gas genesis zones (Magara, 1978).

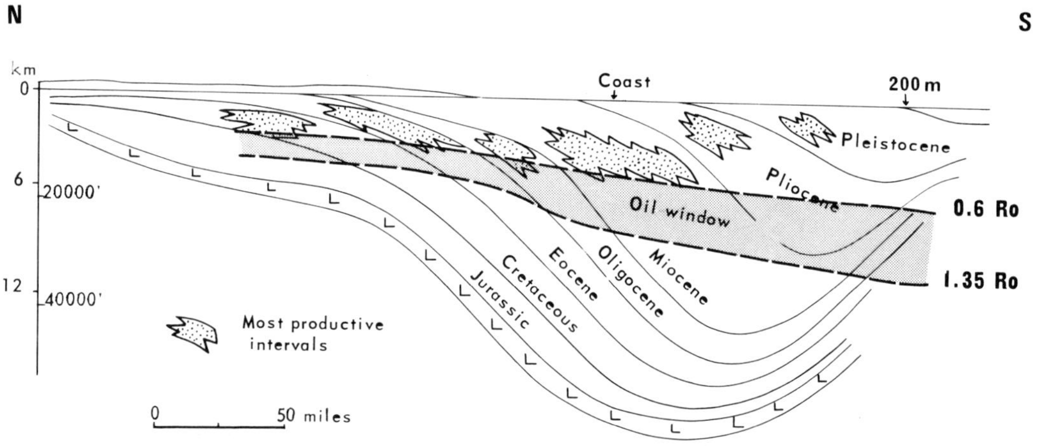

Fig. 142. — Distribution of productive oil and gas zones in the Louisiana province (Dow 1978).

Summary and conclusions

The concept of petroleum province does not mesh very closely with a specific type of basin or sedimentation. The richest provinces display a broad diversity of geodynamic, paleogeographic and sedimentological features, and convergences are many. The really important factor is fairly pronounced subsidence at the beginning. Without subsidence there is no basin, and without a sedimentary basin, no possible petroleum province. Subsequent structural deformation must remain moderate to avoid losing all the hydrocarbons under the action of erosion and seepage.

However, not every basin necessarily contains oil. It is essential for the distribution of sediments, both in space and in time, to allow certain combinations and contacts necessary for hydrocarbon genesis, concentration and protection. In certain periods, it is especially necessary for diversified paleogeographies to occur, associating confined environments and porous facies, with the unit largely sealed by impermeable formations. It is then important for these favorable facies to be conveyed into favorable maturation zones, as observed in the Gulf Coast (Figure 142). The final determining factor is the need for all these mechanisms to proceed according to a tightly programmed chronology.

2. – THE MAIN TYPES OF PETROLEUM PROVINCES

« Les vestiges du Passé que nous découvre la science ne sont qu'une ombre de ce qui a réellement existé. » .. ..

'The vestiges of the Past that Science reveals to us are no more than ghosts of what really existed.'

Teilhard de Chardin.

Not every sedimentary basin is a petroleum province, but every petroleum province is superimposed on an area, over one or more sedimentary basins. For a set of fields to exist, it is important to have a subsident area in which the various sediments necessary for the genesis and preservation of oil and gas pools are trapped, accumulated and transformed. Among the geological requirements for the creation of such an accumulation, three generally exhibit a certain geographic extension. These are the prerequisites for the formation of source rocks, seals, and in most cases, reservoirs. The coexistence of these three factors, and even more, certain chronological relationships between their deposition, constitute one of the keys to the formation of a petroleum province. The distribution of these remarkable facies and their development are obviously closely conditioned by the features of the sedimentary basin. Any classification of the different families of petroleum provinces thus reflects a classification of basins, to which sedimentological considerations are often added.

The petroleum province very closely approaches the concept of a basin, but the scale of the depositional and diagenetic processes, which occur on a different scale, make it imperative to associate the latter. As observed by Meyerhoff (1974), these classifications must remain fairly open, since the features of a basin may be altered during its history, and a number of provinces often straddle two distinct types. As an example, this author notes that three of the largest petroleum provinces, the Persian Gulf, the Volga-Urals region and Western Siberia, which contain over 100 Gt of oil and 70,000 Gm^3 of gas in cumulative, proved and probable reserves, do not fall within an obvious classification, especially the first two.

The convergence of features between provinces that allow groupings must never obscure the fact that each province has an individual personality, which always more or less differentiates it from any province that appears to be its twin. The descriptions of provinces presented below must therefore not be considered as standard types, but only as illustrations of a multi-faceted reality. Going from the simplest to the most complex, from the most stable to the most mobile, the following may be distinguished :

- platform provinces, with special stress on reef provinces,
- rift provinces : rift, graben, passive margin or pull-apart basins to which the large family of deltaic provinces must be added,
- provinces in an orogenic belt, intermontane, strike slip and foredeep provinces, related to convergent movements, on the margin of cratons and also associated with cordilleras and island arcs.

As a rule, provinces located in an orogenic belt are usually Tertiary, while the older provinces belong to more stable zones, and are more susceptible to destruction by erosion.

2.1. PLATFORM BASINS

These provinces, generally located in a cratonic regime, on lithospheric plates, although they are often superimposed on narrow precursory rift valleys, are characterized by great overall stability which was reflected by the following :

- on the structural level, by a slight subsidence and rarity of deformations, represented by large radius folds and faulted blocks, often with important strike slip movement,
- on the sedimentary level, by relatively thin and generally homogeneous series from the twofold standpoint of facies and thickness; this applies in particular to the area of the great inflows and outflows of the sea, and especially the major transgressions with the possibility of regional stratigraphic pinchouts.

From the petroleum point of view, the continuity of the reservoir facies implies the existence of effective structural closures to trap the oil and gas that could migrate over long distances. These conditions favor the birth of giant fields in a concentrated habitat. However, the presence of certain sedimentary conditions and especially transgressions, or certain temporary accelerations of subsidence, may give rise to sand bars or reef structures that often provide the source of rich pools, but in a frequently dispersed habitat (Chapter 4 of Part 2). Periods of transgression are also favorable to the deposition of source rocks.

These provinces taken as a whole account for :
— nine-tenths of Paleozoic hydrocarbons,
— three-quarters of gas reserves,
— more than one-quarter of all light crude oil discoveries (Klemme, 1977).

Depending on the dimensions and complexity of these basins, a distinction can be drawn between :
— stable platform provinces,
— more unstable platform provinces, with special emphasis on reef provinces.

2.1.1. **Stable platform provinces**

These provinces are characterized by generally thin sedimentary series, only slightly or not folded, often corresponding to a single megacycle, and relatively rich in reservoir rocks. The shallow depth favors neither hydrocarbon genesis nor reservoir protection. Pressures are generally poor and characterized by a dispersed habitat, usually limited to the better protected central areas, at lower pressure, as observed in the Paris Basin (Figure 143).

The pools are generally small (only one giant field has been found in this class of basin, in Illinois), and are principally determined by :
— domes and anticlines with a small closure, such as the Williston, Michigan, Paris and West Libyan Basins,
— fault blocks, as on the south flank of the Orinoco Basin and the northern margin of the extra-Alpine Molassic Basin,
— stratigraphic pinchouts, such as the Denver, Powder River and Oklahoma Basins.

In the Michigan Basin, for example, at the end of 1967, out of 413 oil pools, 352 had reserves less than 10^6 b (140,000 t). In the Denver Basin, the average pool contains only 130,000 t and the largest (Adena) 10 Mt. But a few years ago, an exploration well at 2000 m only cost $ 30,000 if it was dry, twice as much if completed, and a discovery of 5000 tons of oil was profitable in these conditions (McGinnis, 1958).

The provinces of this type developed especially in the Paleozoic (ante-Mississippian) and the Mesozoic. The crude oils are generally light. The presence of reef formations, that correspond, at least temporarily, to a certain acceleration of subsidence, may offer remarkable exceptions of richer provinces with a dispersed habitat, as noted in the Middle and Upper Devonian in Alberta, and in the Silurian in Michigan, for example.

2.1.2. **Unstable or complex platform provinces**

These provinces are not only characterized, like the foregoing, by the presence of relatively thick sedimentary series, mainly Paleozoic, very extensive and with continuous facies, often intersected by gaps and regional unconformities, but in addition by the existence of vast large undulations (West Siberia, Arabian platform, West Texas, Great Artesian Basin, Volga-Urals, Central Asia), horsts and grabens, or even intra-cratonic ranges (North Sea, Northeast Sahara, Sirte). These characteristics allow the action of hydrocarbon migrations over medium or long distances, which facilitate the formation of giant fields, as observed on the Arabian platform with Ghawar, Burgan and Safaniya, in the Volga-Urals province with Romashkino and Arlan, in Western Siberia with Samotlor and Urengoï, in the Algerian Sahara with Hassi Messaoud and Hassi R'Mel, and in the Sirte Basin with Sarir. The giant fields of these types of province often exhibit areas ranging from 1000 to 3000 Km^2, at depths of around 1500 to 2500 or 3000 m. The reservoirs are usually sandstone with facies variations. Recoverable reserves amount to 1 to 2 Gt of oil or gas equivalent.

As in the more stable platform basins, the richer zones on the regional level are generally located in the better protected central aras, with waters richer in salts, organic matter and CH_4, as shown by the studies of the Western Siberian Basin.

An often interesting variation is provided by what may be termed 'nesting' basins. The often marginal stable zone of an initial basin becomes subsident in the following period, the basin tilts about a hinge, near which the principal fields are usually concentrated. This type is illustrated by the Maracaibo Basins before and after the Miocene, North Alaska before and after the Upper Cretaceous, and the Algerian

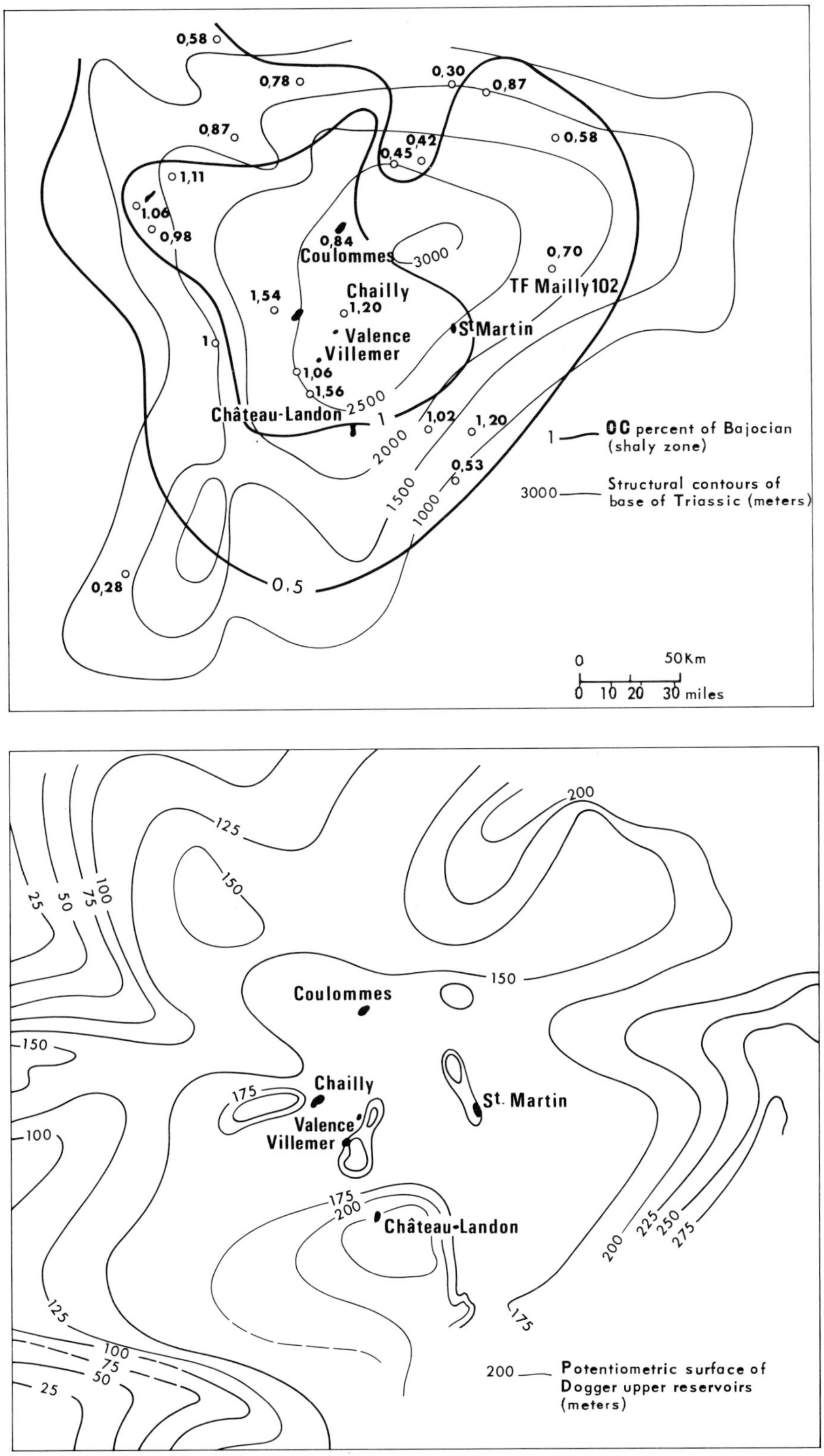

Fig. 143. — Maps of the Paris Basin showing distribution of pools in the deepest and most sheltered part.

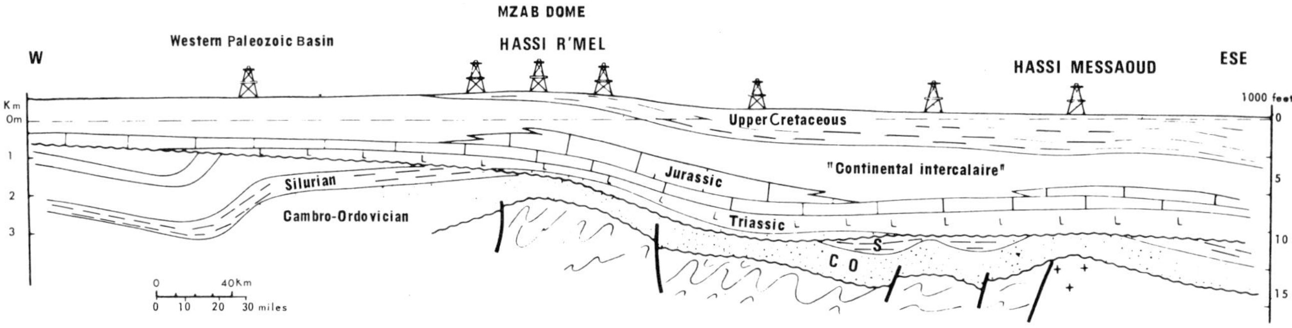

Fig. 144. — Section across northern Sahara basins showing nesting of Primary and Secondary Basins and hinge position of the Mzab Dome.

Sahara in the Paleozoic and Mesozoic (Figures 9 and 144). In these platform basins, three-quarters of the reservoirs are of the sandstone type, one-quarter carbonate, often associated with evaporites. The crude oil is generally light, and gas is frequent (this type of province today contains two-thirds of the world's gas reserves).

This class includes some twenty basins with more than eighty giant fields, containing nearly a quarter of world reserves. Two basins of this type out of three contain giant fields, and there is one chance out of two of making a commercial discovery. The average potential is generally over 2000 t km^{-2} and very often greater than 7000 (from Klemme, 1971, 1975 and 1977). These basins appear to be more interesting if they are vaster, more subsident and deformed along a few large radius folds. The structural logic, which often governs the distribution of pools, is generally rapidly defined by seismics, especially if an overall view of the general structure can be obtained from the early surveys. It is in this type of basin that the yields are most often the highest, while the success rates may be relatively modest. It is here also that discoveries are made at the lowest cost (Klemme, 1971). A number of examples are mentioned in the next section among the giant provinces.

2.1.3. **Carbonate and reef provinces**

Carbonate areas and especially reef provinces deserve special emphasis within the group of platform provinces. Although carbonates account for only 20 % of sedimentary units, they contain some 40 % of world oil reserves, which are found in only 5 % of these facies and more specifically in dolomitic facies. In North America, for example, the statistics show that 80 % of the hydrocarbons of the carbonate provinces are found in dolomitic reservoirs. Gas is found less frequently in these provinces than in sand/shale series. In the United States, they contain only 28 % of non-associated gas discoveries, against 72 % in sands (Beebe & Curtiss, 1968). Considering the world's giant fields, only 13 % produce gas in exclusively carbonate reservoirs, while oil occupies over 50 % of this type of reservoir. The gas in carbonate formations is also richer in H_2S and CO_2 than the gas of clastic/shale series.

One of the characteristic features of carbonate provinces is often the development of platform landscapes bordered by depressions and islets or shoals of the reef type. These diversified environments appear to be especially favorable to the juxtaposition of source rock and reservoir facies. Apart from their inherent value, particularly from the seal standpoint, the presence of evaporites often leads to the migration of salts, especially Mg, which contributes to dolomitization and the development of reservoir facies. Carbonate source rocks play a major role in a number of provinces with exclusively carbonate sequences, as in the Mississippian of the Williston Basin, the Devonian of the Swanhills and Rainbow regions in Alberta, and the Michigan Basin, in the Cretaceous (Luna limestone) of the Maracaibo Basin. In many cases, however, the oil appears to originate in fine marl or shale intercalations, often bituminous and phosphatic. This appears to be the case in the Permian basin and in the Mississippian of the Paradox Basin in the Rocky Mountains.

Reefs, which offer rich pools, are more specifically developed on carbonate platforms of certain basins of the cratonic type, and also on the shallows of more mobile troughs. They are often aligned or split up in front of a reef barrier or platform, more continuous and generally invaded by water. The frequent combination of carbonates and evaporites allows for excellent seals and closures, especially if the latter overlap the margins of the basin. This is the case and one of the reasons for the potential of the Middle Devonian in northwestern Alberta and Michigan, the Permian in West Texas and the fore-deep of the Urals, the Upper Jurassic of the Arabian platform, and the Miocene (Fars) of the Zagros province. Lateral closures may also be provided on the regional level by lateral transitions from limestone facies to anhydrites, as observed in the

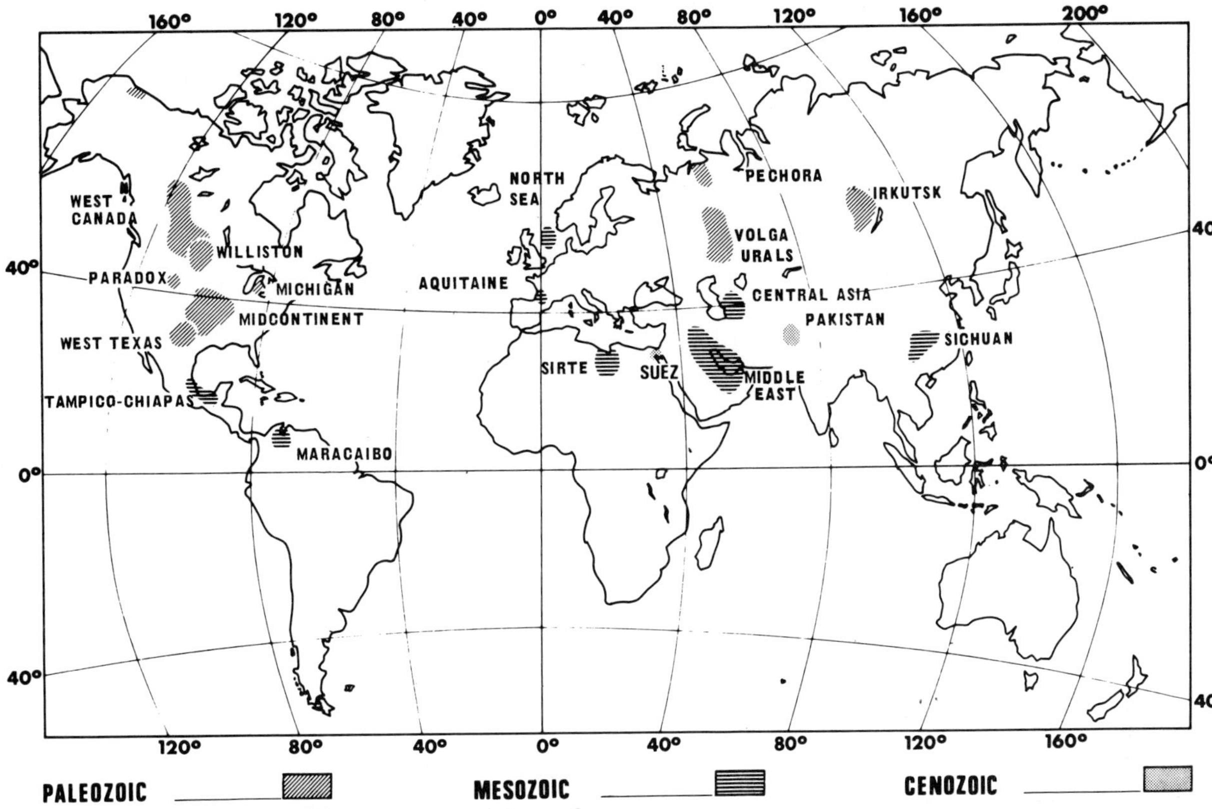

Fig. 145. — General map of main carbonate petroleum provinces.

Upper Devonian in Alberta, the Mississippian in the Williston Basin, and the Pennsylvanian in the Northern Rockies.

Reef provinces generally exhibit a dispersed type of habitat. In rare cases, the oil is concentrated in vast structures copiously supplied with hydrocarbons. Among the richest reef and more generally carbonate provinces are the following (Figure 145) :

— in the Silurian, the Michigan Basin and part of the Mid-Continent,
— in the Devonian, West Canada,
— in the Mississippian, the Paradox Basin (Aneth) and Prudhoe Bay,
— in the Pennsylvanian/Permian, West Texas and part of the Mid-Continent,
— in the Permian, the pre-Urals regions of Pechora and Belaiya, and the Persian Gulf (gas),
— in the Triassic, the Sichuan Basin in China,
— in the Jurassic, Central Asia and the Arabian province,
— in the Cretaceous, the Faja de Oro and Reforma District in Mexico, the Emirates zone in the Persian Gulf, the Maracaibo Basin, *pro parte,*
— in the Paleocene/Eocene, the Sirte Basin, and Pakistan (gas),
— in the Miocene, the Salawati Basin in West Irian, and the marginal province of the Zagros.

Some of these provinces clearly correspond to stable cratonic basins, like those of Michigan, Alberta and West Texas. Others, however, represent the relatively stable margin of more complex basins, which are more subsident on the whole. This is the case of the basins of Mexico, Sirte, Suez and Salawati. The Silurian province of Michigan and the Miocene province of Salawati are discussed below as examples of these two groups. The following section discusses West Canada, West Texas, the Sirte Basin and Mexico.

An example of a reef province : **The Michigan Basin**

The cratonic basin of Michigan represents a stable Paleozoic province of average potential, where reef developments constitute small but prolific pools. The basin was formed during the Ordovician, on a zone of older grabens. It was slightly folded in the Late Mississippian/Early Pennsylvanian. The main productive horizons are located in :

— faulted and dolomitized zones of the Ordovician,

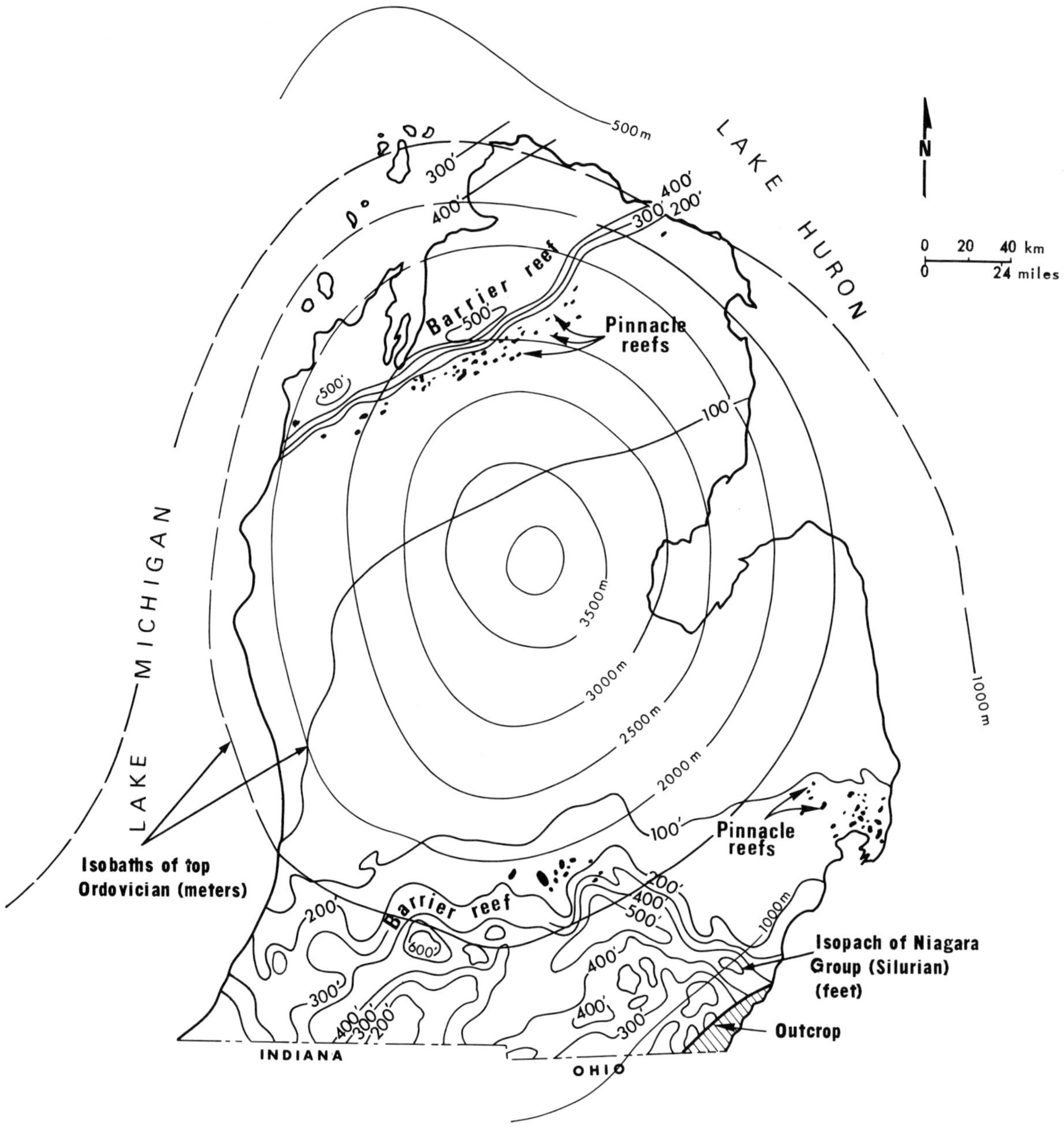

Fig. 146. — Map of Michigan Basin showing distribution of Silurian reefs and reef barriers (Mesolella *et al.*, 1974).

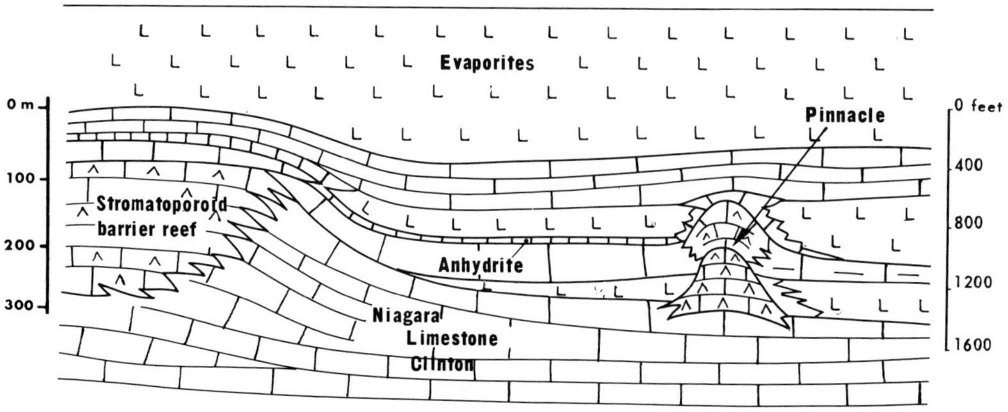

Fig. 147. — Section of Silurian of the northern margin in the Michigan Basin, showing the situation of reefs in front of the reef barrier (Mesolella *et al.*, 1974).

— anticlinal traps with porosity variations of the Devonian,
— Silurian reefs.

The reef zones correspond to two belts 15 to 30 km wide by 270 and 160 km long respectively, extending on the northwest and southeast margins of the basin. These zones, which plunge from 6 to 26 m km^{-1}, represent the outer edge of an ancient continental shelf at the basin boundary. The reefs are developed in the form of pinnacles in front of a reef barrier containing stromatoporoids (Figure 146). The reefs consist of algae, corals, stromatoporoids and debris of echinoderms wrapped in a compact carbonate and evaporite series. Their development corresponds to oscillations of the sea level and to accelerated subsidence. They exhibit reduced areas ranging from 12 to 344 hectares (30 to 850 acres), with a mean of 32 hectares (80 acres) for the northwest zone and a height of 180 m with slopes of 30 to 45 degrees. They are located at a depth of 900 to slightly more than 2000 m. Their porosity lies between 3 and 37 % (Figure 147).

The pools are uniformly distributed on the northwest margin. The outermost reefs are aquifers. They are bordered on the basin side by a belt of oil pools, passing in the low position to a trend of gas pinnacles, according to a logic complying with Gussow's differential trapping scheme. The innermost reefs of the basin are often cemented by the salt of the Salina Formation, which provides the seal. Simultaneously, the oil density declines from the edge towards the internal zone, varying on the average from 0.81 to 0.72 (Figure 53). From 1969, when the first discovery was made, to the end of 1976, 360 productive reefs were discovered on the northwest margin, 221 containing oil and 139 containing gas, with 72 sterile. These successes correspond to reserves estimated at some 50 million tons (700 10^6 bbl) of oil and 80 to 140 Gm^3 (3 to 5 Tcf) of gas, representing average pools of 0.20 Mt (1.4 10^6 bbl) and 0.5 to 1 Gm^3 (18 to 36 Bcf). Thanks to advances in seismics, and often backed by gravimetry (the reefs are marked by an anomaly of 0.2 milligal), the success rate amounted to 85 % during this period. These pools are generally exploited with one or two wells only, and up to ten in rare cases (from Mesolella *et al.*, 1974; Gill, 1979).

The Salawati Basin

The Neogene Salawati Basin, which spreads over the western part of Irian Jaya (Indonesia), features a Neogene series more than 4500 m thick (Figure 148). This sequence consists chiefly of Miocene carbonate formations, dark colored and containing planktonic foraminifera, intercalated with black shales with a basin facies, rich in organic matter. The deposits pass from south to east into platform carbonates, in front of which Upper Miocene reef structures are developed. These reefs are progressively covered by the onlapping shales on the basin margins. Shales with sand and silt horizons, up to 1500 m thick, fill the basin in a regression context throughout the Pliocene. Movements are mainly vertical.

The traps of this province are formed by Miocene reefs which are sometimes faulted reefs. They are formed of foraminiferal limestones with debris of algae, corals, echinoderms and molluscs. Their porosity is as much as 30 to 40 %, with a mean of 20 to 30 % in the impregnated zones, dropping to less than 20 % in the aquifer. Permeabilities are excellent, allowing unit outputs of 2800 to 4500 tons per day (20,000 to 32,000 bopd). The dimensions of these reefs, which are generally rounded, appear to vary widely, with diameters ranging from 2.4 to 19 km and heights from 200 to 500 m. The seals are provided by Pliocene marine marls. The oil appears to originate in intercalated black marls, with relatively deep facies. The crude oil is of the paraffinic type, with an average density of 27 to 44° API, and poor in sulfur.

The main fields are those of Kasim and Jaya, located on the same reef unit, and Walio, probably the largest (Figure 149). The discoveries of this province amount to about 75 Mt. The first discovery, in 1936, revealed a small heavy oil accumulation in the Miocene limestones of a surface anticline, at a depth of 130 m. This 'success' led to the drilling of 25 dry wildcats. The data provided by these drillings nevertheless helped to conduct a survey of the basin, which led in 1972 to the first discovery in a reef. Advances in seismics and a detailed photogeological survey subsequently led to accurate identification of the reef traps (Vincelette & Soeparjadi, 1976).

2.2. RIFT BASINS

The rift basins, situated in certain weak cratonic zones or on their margins, are characterized by predominantly vertical movements acting along flexures or large faults. They are generally narrow and elongated, sometimes triangular or rounded, with sediments up to 5 to 6 km thick and sometimes more. The basins of this type are generally located in a cratonic or pericratonic area. Three main groups are distinguished :

— provinces of the rift or graben type, corresponding to rift valleys in a cratonic area, often marking the

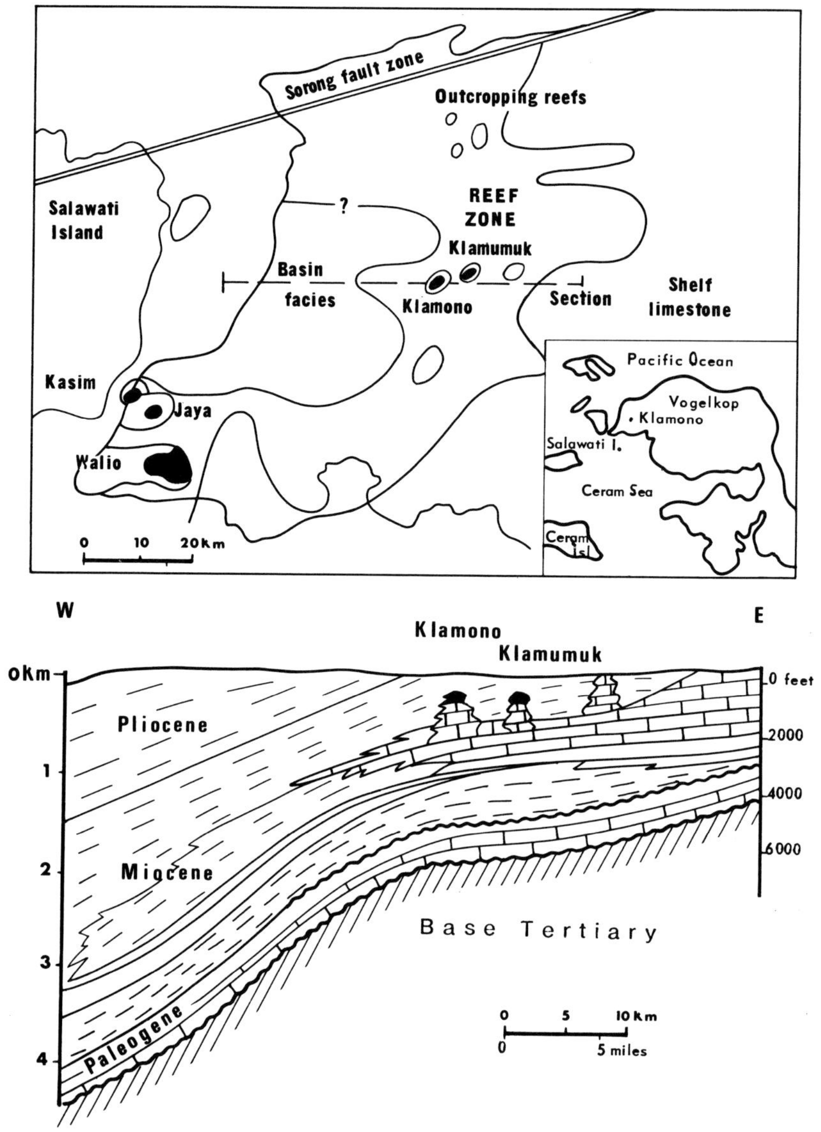

Fig. 148. — Map and section of Salawati Basin (Irian Jaya) showing distribution of reefs (Vincelette & Soepardaji, 1976).

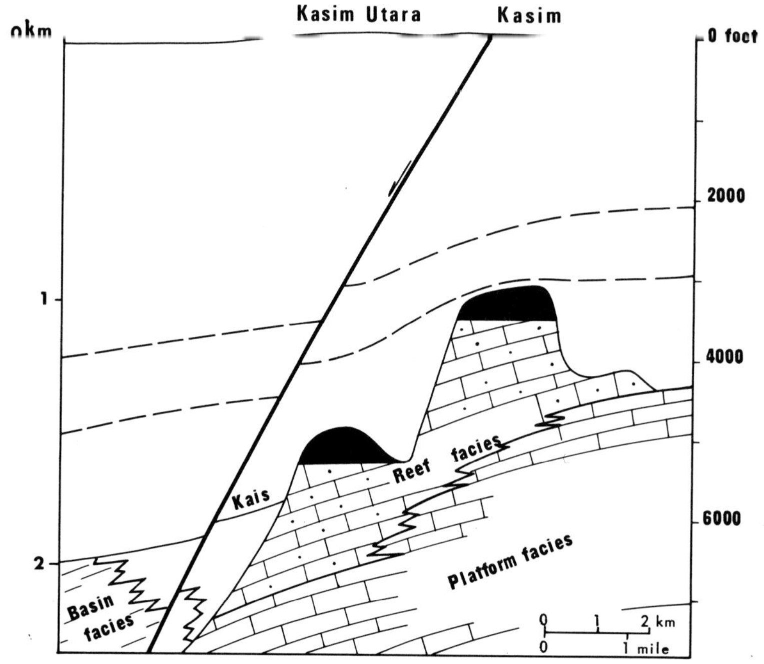

Fig. 149. — Section of Kasim field, Salawati Basin (Irian Jaya) (Vincelette & Soepardaji, 1976).

initial stage of the downwarping of a basin : examples are the Neogene province of the Gulf of Suez and the Cretaceous/Cenozoic Reconcavo Basin in Brazil,

— passive margin or pull-apart basins are often superimposed on older rifts split by an oceanic drift : they are filled by two major ante- and post-rift sedimentary units, with the latter possibly lying on the oceanic crust; these basins display a fundamentally asymmetrical transverse profile, but they may be symmetrical in the basin of the opposite oceanic shore. We shall pay special attention below to the Atlantic basins of Equatorial Africa (Gabon, Congo, Angola), of Campos on the Brazilian Atlantic coast, and Dampier and Gippsland on the northwest and southeast shelves of Australia,

— Cenozoic deltaic provinces represent a specific case of a passive margin basin filled with very thick deltaic series : we shall examine here the Mahakam Delta Basin in Kalimantan, leaving the Niger Delta and Gulf Coast Basins for the section on giant provinces.

2.2.1. **Basins of the rift type**

Basins of this type generally exhibit a relatively complex structure of horsts and grabens. The traps thus consist of fault blocks, but also of developments of reservoir facies on high zones and as a result of unconformities. The reservoirs are both clastic and carbonate, but often of limited extension. The source rocks and seals are more variable, however, they are often related to confined facies and especially to evaporites. The geothermal gradient is generally high and tends to increase the share of gas. The crude oil is most often light.

This family contains some ten provinces including nearly forty giant fields. The latter are generally rich to very rich, with an average of 8000 t km^{-2}. The four biggest fields of a basin contain an average of 10 to 20 % of the reserves (Klemme, 1971). To date, some 38 Gt of oil and gas equivalent have been found in this type of province, especially in the giant fields ranging from 300 to 400 Mt (Huff, 1980). Among the rich provinces are the Cenozoic graben of Suez, the Meso-Cenozoic Viking Graben in the North Sea, the Boukhara-Tadjik Graben in Central Asia, and the Dnepr-Donets and Pripet Basins in the Devonian/ Carboniferous of the USSR.

The Suez and Reconcavo Basins are analyzed briefly below, to provide illustrations of a marine model with evaporites and a non-marine model respectively. Short monographs on the Viking Graben and the Sirte Basin are provided further on in the text (Chapter 3 of Part 3).

The Gulf of Suez

The Gulf of Suez is a Neogene rift valley bordered by large normal faults, with an area of 12,000 km^2. It was downwarped on a relatively stable area occupied by Carboniferous and Mesozoic epicontinental formations (Figure 150) (El Ayouti, 1980). At the top, the Miocene productive series contains thick formations of Globigerina marls with sand and reef limestone horizons. It is surmounted by an evaporite series which is particularly thick in the central part, covered by lithothamnium limestones. The total thickness of the Miocene is 3300 m, and the Plio-Pleistocene series, which are chiefly clastic, may be up to 1500 m thick.

The Gulf of Suez is a rich petroleum province in which over 700 Mt of oil have been discovered, distributed over some twenty fields : it had produced 28 Mt of oil by 1981. The traps consist mainly of fault blocks, draping anticlines, sand lenses, and more rarely reef limestones, all of these associated with horsts. The reservoirs are normally represented by fairly thick sand bodies up to 100 m thick in the Miocene shale series, and secondly by Eocene limestones, and more rarely by older formations. The porosity of the Miocene sandstones (Kareen and Belayim Formations) is generally high, 20 to 25 %, while that of the fractured Eocene limestones ranges from 10 to 15 %. The seals are provided by shales often intersected by anhydrite horizons. The source rocks appear to be formed mainly of Globigerina marls, in which some levels contain up to 3 % bitumen. These rocks were subsquently subjected to intense subsidence (several thousand meters) until the present time, with a fairly high geothermal gradient. The oil is generally paraffinic, undersaturated, density 35 to 20 °API, often rich in sulfur (1 to 2 %), and sometimes with a naphthenic base. The density varies inversely with the salinity of the waters and the thickness of the shale seals. It decreases from the northwest to the southeast, corresponding to an increase in thickness of the Miocene evaporites. It also decreases with stratigraphic depth, as the upper horizons are degraded by waters from the surface (Elzarka, 1975).

The largest field is that of El Morgan, discovered in 1965. The structure is formed of an elongated anticline with 46 km^2 of productive area. Production is obtained chiefly from the Miocene sands of the Kareen Formation, 60 m thick at 1670 m. Recoverable reserves are estimated between 140 and 210 Mt

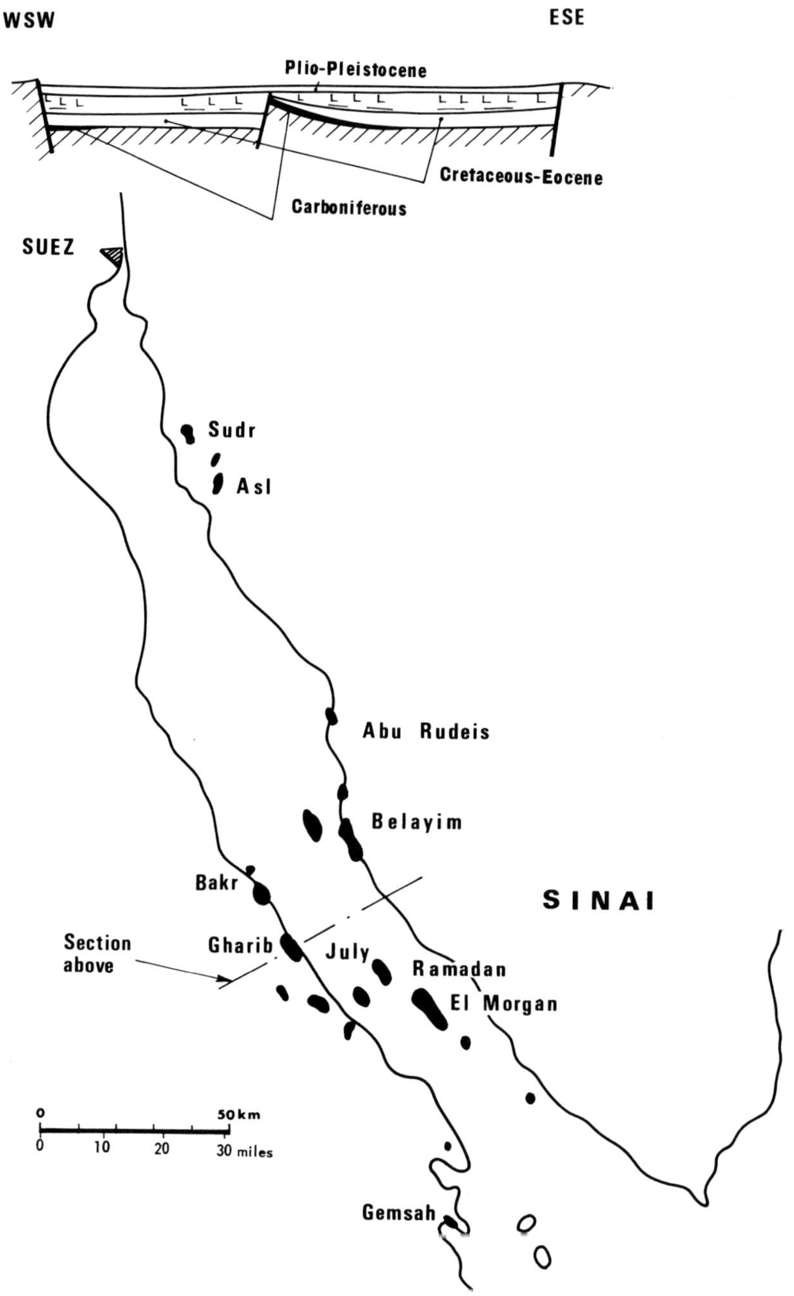

Fig. 150. — Map and section of Gulf of Suez (Egypt) (Said, 1962).

(St. John, 1980; Brown, 1980). The second largest well is Ramadan (1974), with 135 Mt, productive in the Lower Cretaceous at 3400 m, and the third is the July Field (1973) with 100 Mt, productive in the Miocene, the Upper Cretaceous and the Paleozoic between 1550 and 1950 m. The two Belayim fields (1955 to 1961) rank fourth, with 70 and 100 Mt in the Miocene. With Rudeis, which prolongs them northward, the fields produced 4.5 Mt in 1978.

To summarize, this province of the Gulf of Suez appears to owe its potential in particular to :

— a wide variety of structural and stratigraphic traps,

— depositional conditions, often confined, pre-evaporitic, favoring the formation of source rocks,

— a wide variety of reservoirs, often thick and with good characteristics,

— an excellent regional evaporitic seal,

— large boundary faults guaranteeing good lateral closures,

— intense subsidence in the Neogene.

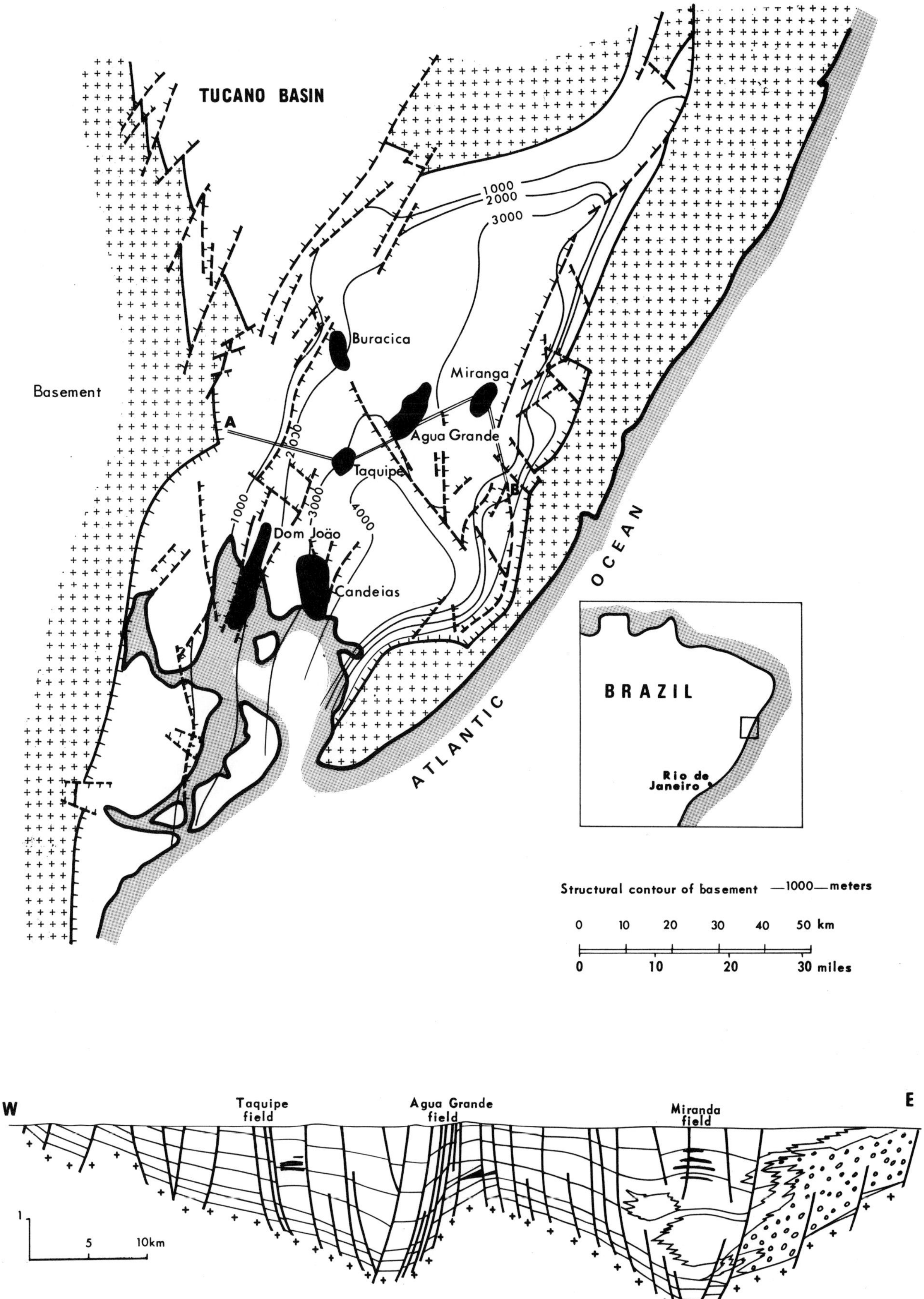

Fig. 151. — Map and section of Reconcavo Basin (Brazil) (Chignone *et al.* 1970).

The Reconcavo Basin in Brazil

The Reconcavo Basin, stretching slightly obliquely to the Atlantic coast, along a meridian direction, is a good example of a rift province of the complex semi-graben type. It extends over about 200 by 40 km and covers an area of 10,000 km^2 (Figure 151). The stratigraphic series is mainly formed by deposits of the uppermost Jurassic and Lower Cretaceous, mainly non-marine, and its thickness is up to 6000 m on the southeast boundary. The reservoirs are essentially clastic, represented by fine poorly sorted sandstones, exhibiting porosities of 15 to 20 % on the average and permeabilities up to 1 darcy, and yields amount to 100 t day^{-1}. The largest are the sandstones of the Uppermost Jurassic/Lower Cretaceous (Sergi Formation). The most continuous and best formation is the Itaparica, which contains three-quarters of the reserves. The remainder is more or less equally shared by the Candeias and Ilhas (Wealdian) sandstones. The shale seals appear to be effective, considering the rarity of seeps. The source rocks are probably located in the bituminous and carbonaceous shales containing ostracods and fish, with a non-marine Wealdian facies. The crude oil is of average density, 25 to 40 °API, low in sulfur, but rich in paraffin (15 to 25 %) and with high viscosity. It contains very little gas.

The traps are mainly structural and faulted :
— simple horst, as at Dom Joao,
— complex horsts, such as Agua Grande,
— fault block with pool at the base of the faults, such as Araças.

But the following are also encountered :
— traps formed by paleoreliefs, such as Fazenda Imbe,
— stratigraphic traps consisting of sand bodies, as in Candeias.

Main discoveries

The Reconcavo Basin contains about 130 Mt of oil and about 30 billion m^3 of gas, distributed among some forty fields. 95 % of the reserves are concentrated in six accumulations.

— Agua Grande, discovered in 1951, is the largest field with 47 % of the production of the basin in 1976, and reserves of about 40 Mt. It is formed by a complex horst situated in the central part of the basin. It is productive in the Sergi sands at 1000 to 1300 m, giving initial yields of about 100 t day^{-1} per well. 1981 production amounted to 480,000 tons with 105 wells.

— Dom Joao (1947), also productive in the Sergi sands between 160 and 370 m, on the west flank of the basin, could contain some 35 Mt of reserves. The average yield of the wells fell to 4 t day^{-1} in 1972.

— Miranga (1965) corresponds to a faulted elongated dome in the deep part of the basin, productive in different Wealdian sands. It produced 0.7 Mt in 1981; cumulative production 21 Mt.

— Candeias (1941) is a stratigraphic trap located in a partly synclinal position in the central part of the basin, with reserves of 13 Mt (see Chapter 4 of Part 2 and Figure 114). The average productivity of the wells was 4 t day^{-1} in 1981.

— Buracica (1959) corresponds to a fault block on a transverse high zone, and had produced 16 Mt, mainly in the Sergi, at the end of 1981.

— Taquipe (1958) is a flattened anticline with reserves estimated at about 12 Mt. Production in 1981 was 700 t day^{-1} with 50 wells.

The Reconcavo Basin is an outstanding illustration of a rich rift province in a principally non-marine environment. Its potential is about 13,000 t km^{-2}, and the habitat is fairly concentrated. At the end of 1963, these discoveries were achieved with the drilling of 255 wildcats, representing an average yield of 0.5 Mt per wildcat, with a success rate of 17 %. This potential is associated with :
— a rift basin, well protected by boundary faults and cut into horsts and grabens,
— rapid sedimentation, displaying frequent alternations of sands and shales.

In this well-sheltered basin sealed by thick shale series, the bulk of the reserves is found in a continuous sand horizon. Finally, in this confined basin, some shales, although of non-marine origin, clearly appear to have played the role of prolific source rocks (Ghignone & de Andrade, 1970; Celso Ponte *et al.*, 1980).

2.2.2. **Passive margin basins**

The provinces of this type are of considerable petroleum value, and have already given rise to the discovery of some 3.6 Gt of oil and gas equivalent (Huff, 1980). The size of the fields is generally modest, less than about 40 million tons. The productive series are found both in the pre-drift formations and in

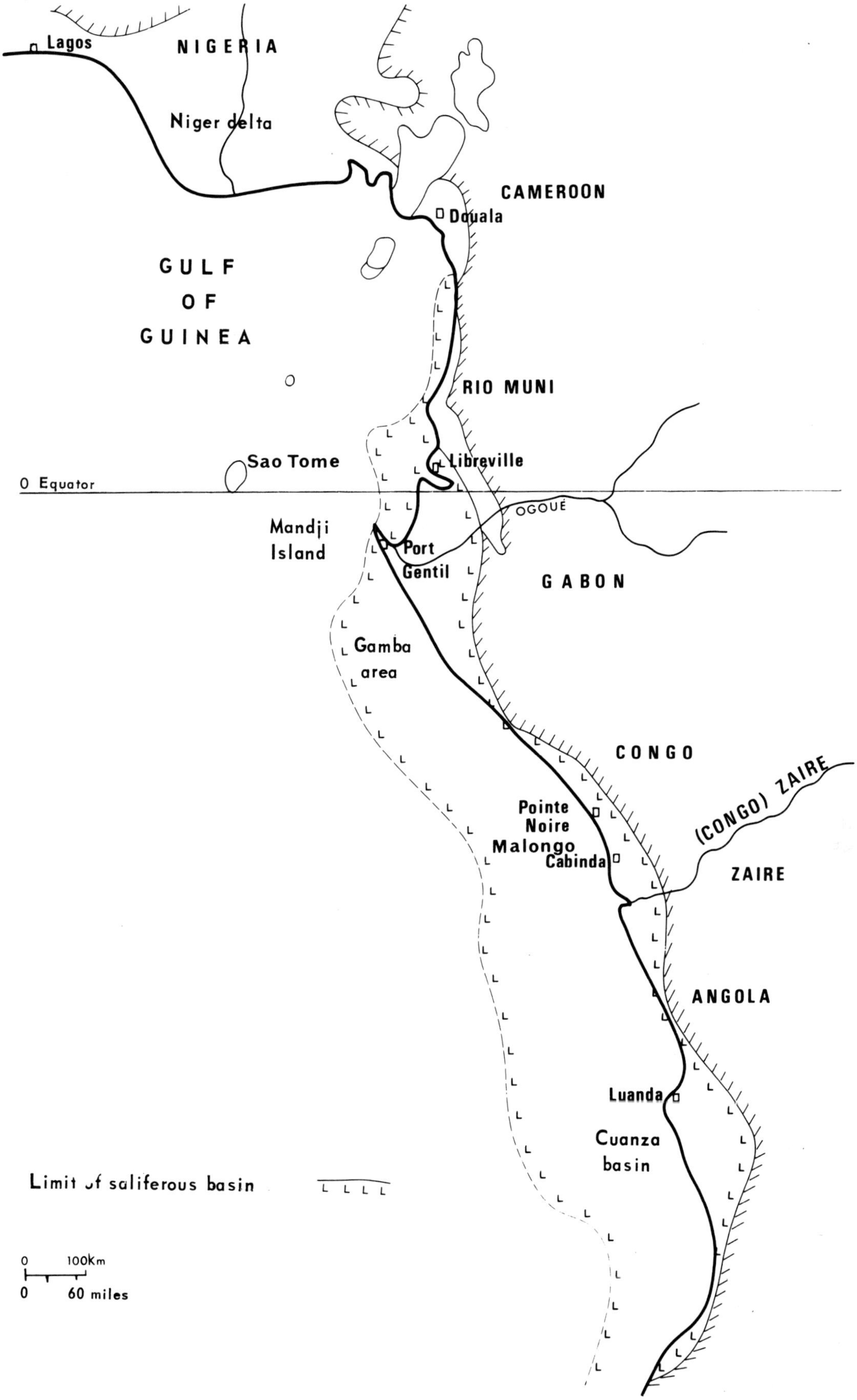

Fig. 152. — Atlantic sedimentary basins of Equatorial Africa.

subsequent deposits. Source rocks are generally abundant in most of these series. Sand and carbonate reservoirs are found in both units, and the bulk of production generally originates in boundary phases reservoirs associated with the early phases of oceanic drift, or with turbidite deposits. The structural traps are usually related to basement horsts or to salt or shale tectonics. Folds are relatively rare and of low amplitude.

Among the main petroleum provinces associated with this type of basin are the following :

— on the Atlantic shore of South America, the Brazilian basins and particularly those of Campos, Santos and Sergipe-Alagoas,
— on the symmetrical African shore, the basins of Gabon, Congo and Angola,
— on the Canadian Atlantic margin, the offshore Labrador Basin,
— on the Australian coast, the Carnavon (Dampier) and Gippsland Basins.

The examples given below offer data on the provinces of Campos, Gabon/Congo, Dampier and Gippsland.

The Campos Basin (Brazil)

The Campos Basin, situated on the Brazilian continental shelf, offshore the State of Rio de Janeiro, is a typical passive margin province with an area of 30,000 km^2. Its history began with the formation of an intracratonic rift in the Early Cretaceous, filled with detrital sediments of poor reservoir characteristics, shales with source rock facies, and lavas. A transition stage appeared in the Aptian with the deposition of evaporites and carbonates, followed in the Albian by the appearance of thick open marine carbonates of shallow water facies.

The Upper Cretaceous corresponds to the deposition of deeper marine series, including turbidites in the Paleocene. An overlapping shale horizon offers a potential source rock. Salt diapirism mechanisms are active from the Upper Cretaceous, forming different traps. The main fields are :

— Garoupa, discovered in 1974, by the ninth wildcat drilled in the basin,
— Namorado (1975), productive at 3000 m, in 166 m water depth, in Albian turbidite sandstones exhibiting very good characteristics : porosity 30 %, permeability 1 darcy, output 6000 bopd, reserves around 40 Mt,
— Enchova,
— Cherne.

In all, a total of eighteen discoveries producing 4 Mt in 1981 and containing about 90 Mt of oil and 12 Gm^3 of gas (Bacoccoli *et al.*, 1979; Celso Ponte *et al.*, 1980; Bacoccoli *et al.*, 1980).

Gabon-Congo-Cabinda Province

The Gabon/Congo province covers most of the coastal sedimentary basins of these countries. They are classic ocean spreading basins of the passive margin type. Their history is divided into two major periods corresponding to the following :

— prior to ocean spreading, rift basins with lacustrine or brackish sedimentation,
— after the oceanic spreading, passive margin basins, with marine series affected by saliferous tectonics.

These two units are separated by major Aptian salt deposits corresponding to a vast enclosed basin, extending over nearly 2000 km, from the Walvis Ridge in the south to the shores of Cameroon in the north (Figure 152). This saline episode, which plays a major role in the petroleum geology of these basins, especially as a seal for the rift series, and as a driving force for the structural deformations in the overlying formations, creates two main petroleum systems, especially distinct in Gabon (Figure 153).

The Pre-Saliferous system

The rift basins existing prior to the oceanic spreading are characterized from the structural standpoint by a series of horsts and grabens. They are filled with Lower Cretaceous lacustrine to brackish sediments, locally intersected by unconformities and periods of non-sedimentation :

— black shales, rich in algal organic matter, forming excellent source rocks,
— lenticular or continuous sandstones exhibiting porosities up to 30 % and permeabilities up to 500 to 1500 md,
— very locally, lacustrine coquina, forming good reservoirs.

The system is capped regionally by the Aptian salt that forms an effective barrier to migration, except in some zones in the southern portion in particular, where the intensity of salt diapirism causes discontinuities in these beds and hence dysmigration paths. The geothermal gradients were generally high but showed

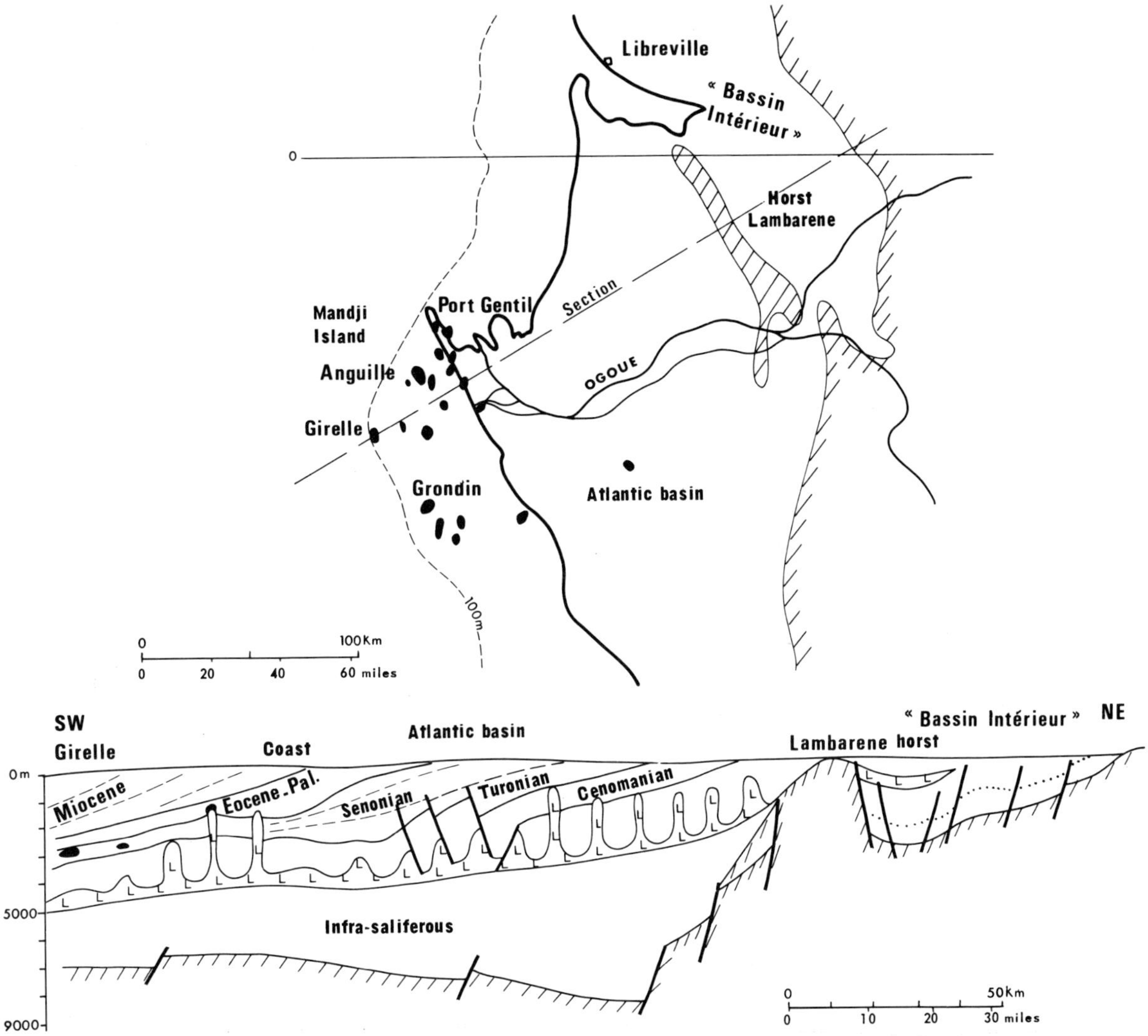

Fig. 153. — Map and section of internal basin and Mandji Island zone (Gabon) (Vidal *et al.*, 1975).

rapid local variations : they are now normal. Catagenetic effects are detrimental to reservoir quality. The traps are mainly structural, consisting chiefly of fault blocks and anticlines. Trangressive overlaps are developed on the flanks of certain high zones (e.g. Lucina). The oil of these series is paraffinic (10 to 15 % paraffin) with fairly high specific gravity (0.87 = 31 °API) and very high viscosity.

The Post-Saliferous system

The Post-Saliferous pools are distributed among different horizons of the stratigraphic series. The reservoirs are chiefly represented by sand or silt units with a deltaic, littoral or deep deposit facies stretching from the Cenomanian and Turonian (Emeraude) to the Senonian (Anguille and Batanga sandstones), the Eocene and the Miocene. But oil is also found in Albian detrital and carbonate reservoirs and in Eocene silicified shales, which display good porosity but very low permeability. The source rocks in the Mandji Island area consist of Turonian/Santonian black shales, rich in sapropelic organic matter. In Congo, the hydrocarbons appear to originate in Infra-Saliferous black shales in all zones where the saliferous cap rock is discontinuous. In this case, vertical and horizontal migrations may occur to a significant extent. Many seeps have been observed on the margin of the basin.

The cap rocks are shales belonging to different stratigraphic horizons. By preventing vertical migrations, they directly control the distribution logic of the accumulations. The crude oils are of the intermediate paraffinic type, poor in sulfur, and generally heavy (specific gravity 0.86 — 33 °API (Anguille) to 0.98 — 12 °API) with high viscosity, up to several hundred cP, especially in the Eocene oils. The GORs are low. The traps are structural and conditioned by salt diapirism. They may be :

— anticlines, sometimes affected by a collapsed central zone, formed by a deep salt dome, like Grondin (see Section 4.1 of Part 2),

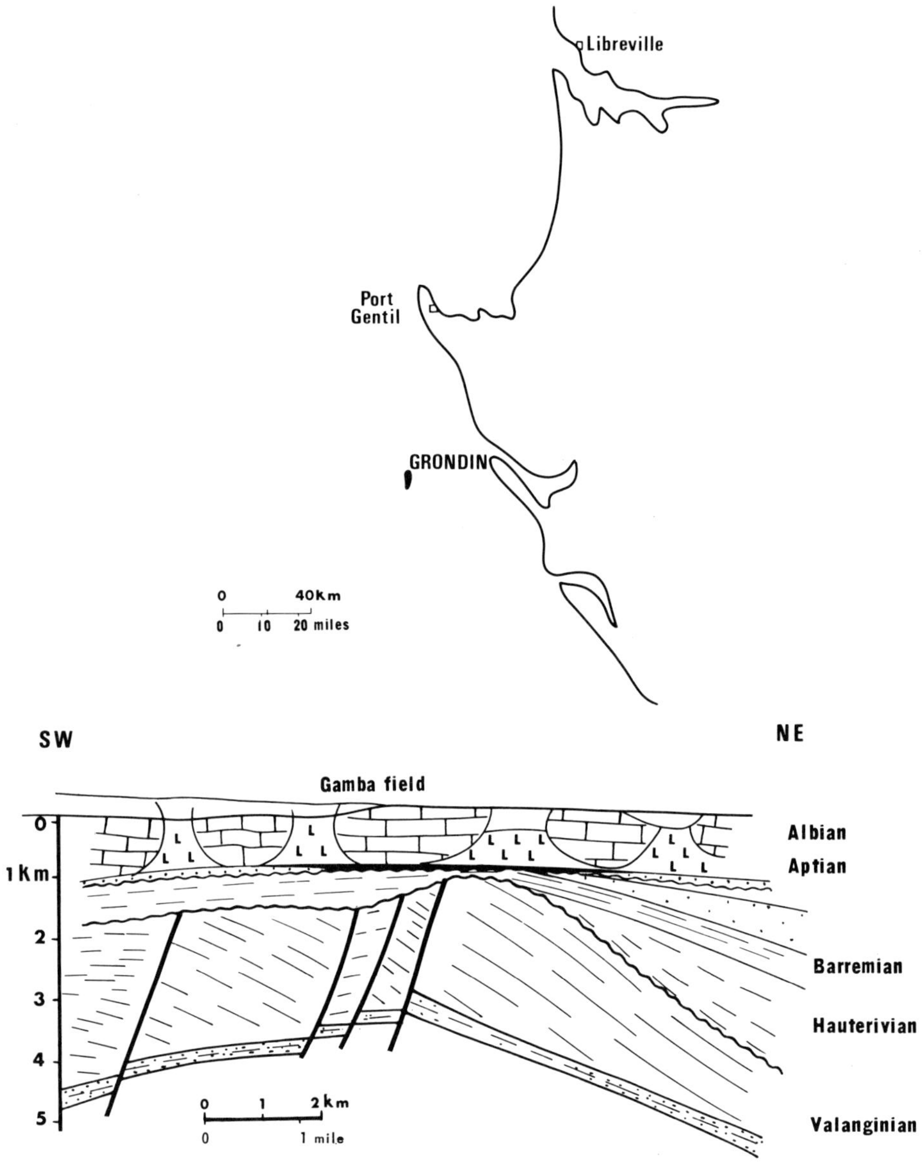

Fig. 154 a. — Section of Gamba Field (Gabon).

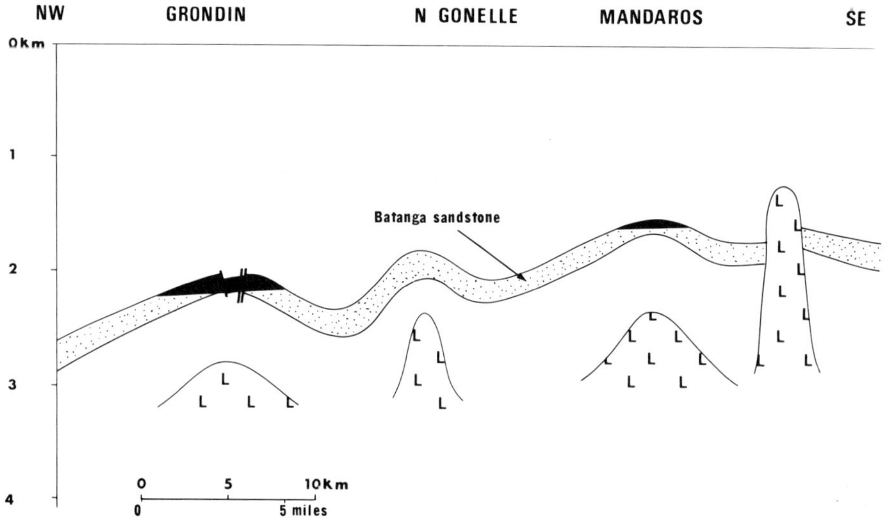

Fig. 154 b. — Section of Grondin and Mandaros Field (Gabon) (Vidal, 1979).

— intermediate structures between domes such as Anguille,
— partly dome flanks such as Barbier, a complex structure wihch is productive in the Senonian and Miocene,
— stratigraphic pinchouts of Miocene sands, as at N'Tchengue.

Pressures are generally hydrostatic. The pools of these Post-Saliferous series in Gabon exhibit a dispersed habitat, with 36 accumulations containing some 110 Mt of recoverable reserves (Reyre, 1966; Brink, 1974; Vidal *et al.,* 1975; Wenger, 1975).

Main discoveries

Oil and, to a lesser extent, gas have been discovered in the Pre- and Post-Saliferous series in Gabon,
— the Mandji Island area in the Upper Cretaceous,
— the Gamba Infra-Saliferous area, in the southern part and extending into Congo.
Congo, Zaire and Angola. In Gabon, two petroleum areas or plays have been identified :

The Mandji Island play is known today to contain some thirty pools, mainly offshore, representing about 120 Mt, of which two-thirds are located in the Batanga sands trend.

The main fields are :
— Grondin (1971) (see Section 4 of Part 2) (Figure 154 b),
— Anguille (1962) and its satellites, formed by Saliferous and intermediate type structures, productive in the Coniacian/Santonian sands and silts at a depth of 2100 to 2500 m : the reserves amount to about 20 million tons.

The Gamba play, in the southern section, contains half a dozen productive accumulations in different Lower Cretaceous sands in Gabonese territory, both onshore and offshore. The main field is Gamba (1963), a simple anticline under diapiric salty formations, productive at 900 m in continuous Aptian sandstones, yielding a paraffinic oil with high viscosity (30 cP at 40 °C). Initial reserves are about 25 Mt (Figure 154 a). Among the remaining fields is the offshore Lucina Field (1971), where the trap is a pinchout below an unconformity, productive in discontinuous sandstones, earlier than those of Gamba, with reserves of about 6 Mt.

In Congo, since the discovery of the Pointe Indienne Field (1957) in Pre-Saliferous formations, a dozen other modest discoveries have been made on the continental shelf, in the Post-Saliferous series. Among these are the Emeraude Field (1969), productive around a depth of 350 m in fine Turonian/Cenomanian sandstones and limestones, yielding a heavy (22 °API), viscous (200 to 400 cP at 15 °C), low sulfur oil (less than 0.8 %), exhibiting a pour point below — 30 °C. The associated gas contains 20 % CO_2 and H_2S. The recovery rate is low and production has not exceeded 2 Mt per year (Gainette, 1972).

The Angolan enclave of Cabinda has turned out to be the richest zone of these coastal basins, with the Malongo North (1966), South (1966) and West (1969) Fields, with cumulative reserves of about 380 Mt. These fields are productive in two distinct horizons, below and above the salt, with three-quarters of the reserves located in the former series. The deep Lower Cretaceous reservoirs, between 2100 and 2700 m, are represented by coarse sandstones of the fluviatile type (Lucula) and limestones containing lacustrine algae (Toca) 50 to 180 m thick, porous and permeable (average 350 md). These often dolomitized limestones pass laterally into black bituminous shales with a lacustrine facies (Bucomazi) which is most probably the source rock. The oil is paraffinic, with medium density (34 °API). It freezes at + 30 °C and has a gas cap. The well outputs are variable, up to 2100 t day^{-1} (15,000 bopd) for the best wells, with 150 t day^{-1} (1130 bopd) on the average, but only 60 t day^{-1} (420 bopd) for 80 % of the wells. The Upper Senonian horizons above 750 m consist of vuggy calcarenites and contain a heavy oil (26 °API), undersaturated, with a downhole viscosity of 10 to 30 cP and some H_2S. The wells yield between 55 and 125 t day^{-1} (400 to 900 bopd) (Caflisch, 1975).

The Dampier Province (northwest Australia)

The Dampier sub-basin, in the Carnavon Basin, which spreads over the continental shelf of northwestern Australia over an area of some 50,000 km^2, is an outstanding example of a petroleum province corresponding to a marginal basin on a passive margin. This basin was developed in the Late Carboniferous and took shape in the Middle Jurassic, with the splitting of the continent Gondwana and the birth of the Indian Ocean (Figure 155). The stratigraphic series stretches from the Trias to the Present, and is marked by a general overlap of Upper Jurassic and basal Cretaceous on older rift series. During the Callovian, the latter were split up into a succession of horsts and grabens. The marine Upper Cretaceous and Tertiary formations are up to 6000 m thick, especially in the Lewis Deep to the southeast. The basin is thus characterized by an alignment of horsts on a general uplift of older beds : the Rankin Platform, at the edge of the present shelf,

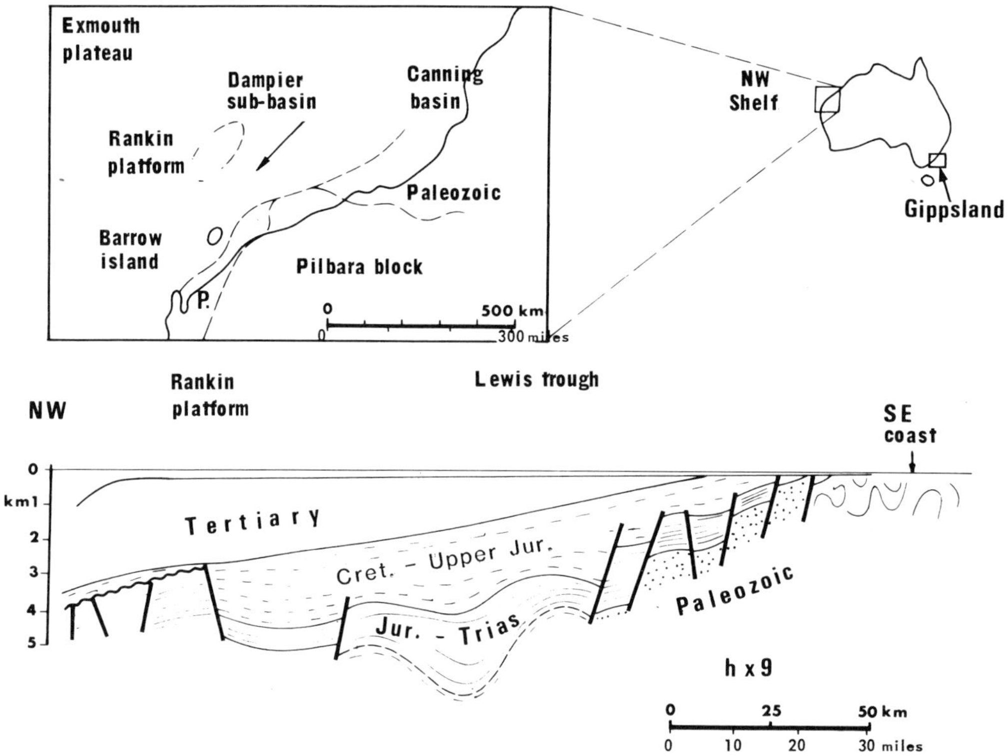

Fig. 155. — Map and section of Dampier Sub-Basin (Carnavon Basin) (Lofting *et al.,* 1975).

limited offshore and near the coast by troughs. The Dampier Basin is a rich gas condensate province. Since 1971, some 380 Gm^3 have been discovered there, together with about 50 Mm^3 of condensates and a slightly larger amount of oil. Eleven fields have been identified, including four major ones. The traps are formed mainly by Callovian fault blocks and anticlines, well located on trough margins, and also by overlying draping anticlines. The faults, which are often tight, complete the lateral closure of the structures. The reservoirs, which display good characteristics, and which are possibly clastic, are represented by shallow fluviatile and marine sandstones, of the Upper Triassic and Middle Jurassic respectively. These formations are capped by shales and siltstones which also constitute gas source rocks. The Upper Jurassic shales and, to a lesser degree, those of the Trias and Permian, appear to be the main source rock of the oil and partly of the condensate rich gas. The oil is light (specific gravity 0.81) and paraffinic, and the gas is very rich in condensates, containing 60 to 340 g/l.

This example of a passive margin province is typical of this kind of basin due to its horst traps, and its abundant source rocks through a broad range of the stratigraphic series. It is more atypical by its large number of good reservoirs, formed by fluviatile and littoral series, well closed by an overlapping shale series (Crostella & Chaney, 1978).

The Gippsland Province (southeast Australia)

This sedimentary basin, which lies mostly offshore, covers 40,000 km^2 on the continental shelf of southeastern Australia (Figure 156). The series, which is unconformable on the Paleozoic, includes a first Lower Cretaceous continental unit corresponding to graben deposits preceding the dislocation of the continent of Gondwana. It continues with a megasequence stretching from the Upper Cretaceous to the Present. This second unit begins with a thick sand/shale series with coal seams, about 4500 m thick, from the Upper Cretaceous/Lower Eocene (Latrobe Valley Group), corresponding to a thick deltaic unit. This series is overlaid by a hiatus of some 25 million years, approximately in the Eocene, characterized by deep erosion channels. The overlying unconformable Oligocene to Recent Seaspray Group, a marine series 2000 m thick, consists of shales deposited rapidly at a rate of about 100 m per million years in the central part, with carbonates on the margins. The series terminate in continental formations. This sedimentary unit thickens steadily offshore, where it becomes up to 7000 m thick.

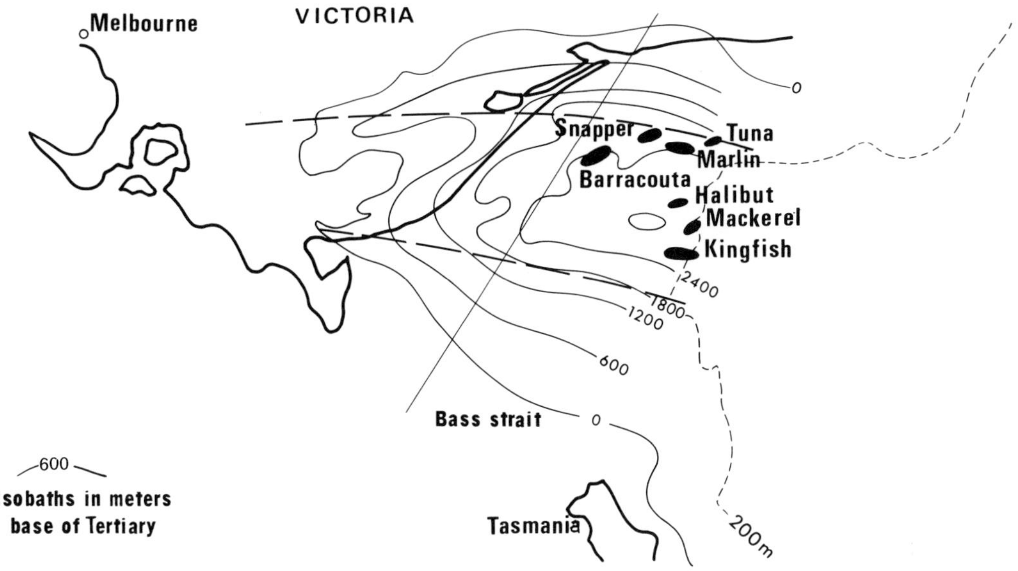

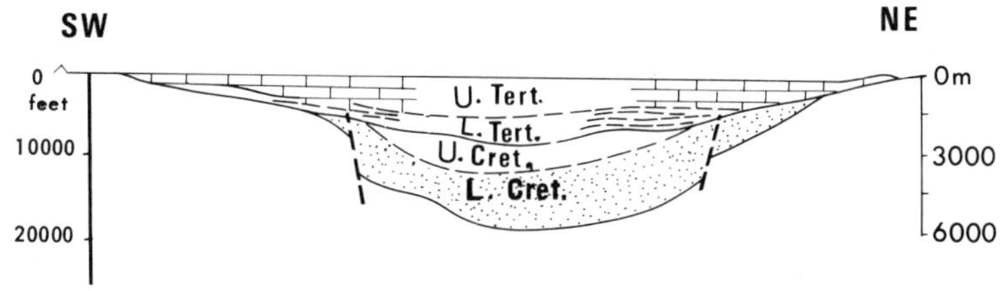

Fig. 156. — Map and section of Gippsland Basin (southeast Australia).

The traps are mainly structural, associated with major shear mechanisms, and occupy the higher parts of the anticlinal axes. They correspond to often faulted fold sequences formed in the Late Eocene and in the Miocene. Their closure covers 100 km^2 and has a height of 300 m. These folds in the Latrobe Valley are often complicated by the unconformity and erosion of the overlying deposits, and certain traps, such as Halibut, correspond to a fault block under an unconformity plane (Figure 157). The sand reservoirs belong to the deltaic system of the Latrobe Group and correspond to fluviatile channels and offshore bars. Their characteristics are generally excellent, with thickness from 100 to 200 m, porosity 25 to 30 %, and permeability of 450 md and up to 5 d. The cap rocks are provided by the Lake Entrance marls and Gippsland limestones. The rocks are represented by the Latrobe Group black shales, often carbonaceous, but fairly rich in lipids. These formations, buried deeper than 4 km, at a temperature of about 130 °C, are currently in the 'cooking' stage.

The geothermal gradients range from 40 to 45 °C km^{-1} near the coast, to some 20 °C at the edge of the shelf : they range from 30 to 40 °C in the productive zone (Figure 14). The oil is paraffinic and light (44 to 49 °API), and gas is frequently found, with some CO_2. The uppermost structures (Barracouta, Snapper) contain gas, and the lowest (Kingfish, Halibut) contain oil. Since 1975, some 280 Mt of oil (450 Mt according to Nehring, 1980) and a nearly equivalent quantity of gas (225 Gm^3) have been discovered in this province, distribute among four oil fields and five gas fields. The main fields are :

— Marlin (1966) 100 Gm^3 of gas,

— Kingfish (1967) with 175 Mt of oil and 110 Gm^3 of gas,

— Halibut (1967) 120 Mt of oil,

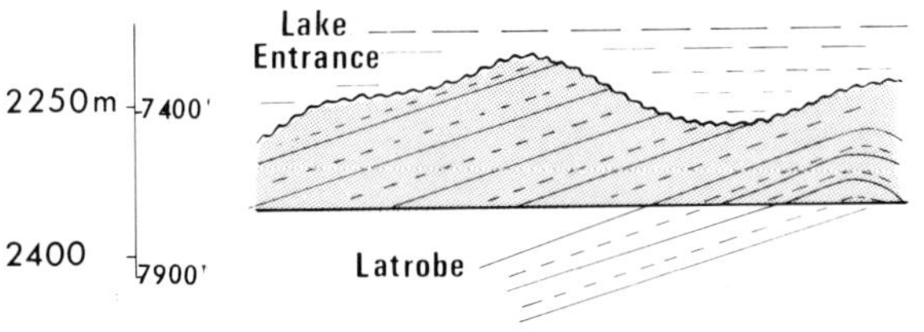

Fig. 157. — Section of Halibut Field (southeast Australia) (Griffith & Hodgson, 1971).

— Snapper (1967) 85 Gm^3 of gas,
— Barracouta (1968) 60 Mt of oil,
— Mackerel (1969) 40 Mt of oil,
(Wallis, 1967; Hocking, 1972; Shibaoka *et al.,* 1978).

2.2.3. **Deltaic provinces**

These provinces, which are mainly Cenozoic, correspond to vast subsident areas filled with the massive deposits of the great permanent rivers.

They are usually active depositional basins characterized by relatively continuous subsidence. Two of the largest are situated, almost symmetrically, on either side of the Central Atlantic, the Gulf Coast province, on the northern margin of the Gulf of Mexico, and the Niger Delta, at the entrance of the Benue Trough. Added to these are the basins of the Mackenzie Delta and the Beaufort Sea, the deltas of the Nile, the Ganges and the Mahakam, as well as the Cretaceous series of the Rocky Mountains, the Jurassic series in the North Sea, and the Permian series of Anadarko, included in other basin types and usually classified in accordance with the criteria characterizing the latter.

Deltaic provinces are of special interest due to a number of their geological characteristics. In particular, they are preferential zones for detrital deposits in a marine environment, which combine all the geological characteristics contributing to the formation of oil and gas accumulations. They primarily correspond to unstable areas allowing the accumulation of a very thick sedimentary series, up to about 10 km, and marked by incessant movements of the sea, in a general regression situation.

Deltaic complexes inherently possess their own tectonics, characterized by :
— synsedimentary listric faults, growth faults and associated roll overs,
— intumescences and domes of salt or undercompacted shale.

The growth faults are normal arcuate faults, concave towards the center of the basin and exhibiting a dip of about 45 degrees, gradually diminishing with depth. This type of fault is called listric due to its geometric shape. The throw is a maximum in the central part and diminishes laterally and vertically. The fault is bordered on the downthrown internal side by a slight roll over, exhibiting a thicker sedimentary series than that on the upthrown side.

These features demonstrate different characteristics in accordance with their position in the basin :
— in the updip zones of a deltaic complex, the shale sediments are generally thin, alternating with many sand horizons : the roll overs exhibit very offset tops and the accumulations are generally small,
— in the median zones, the series are thicker and more shaly, often undercompacted, and the tectonics are highly complex and often diapiric : the pools are broken up by faults.

The faults are probably due in a general tensile context to variations in loading between a thinner outer zone richer in sands, and a deeper internal zone which is more shaly and more subsident (sliding on a titled basement undoubtedly plays a significant role). The faults are perpendicular to the thickness gradients and parallel to the lines of facies change.

Traps

The traps in this type of province are mainly indirectly or directly related to the characteristic growth faults of these basins. As a rule, the trap is formed by a roll over at the edge of the growth fault, without the fault contributing to the closure. In these multibedded fields, the water tables are generally located above or in contact with the main fault (Figures 88, 89 and 90). In the case of a system in which the main fault is accompanied by secondary synthetic or antithetic faults, the closure is often completed, if not totally assured, by the minor faults. In the case of a complex faulted system, as observed in the downdip part of the Nigerian Basin, for example, the oil and gas accumulations are distributed in an apparently anarchical manner, in accordance with the play of the different faults, which sometimes act as barriers and sometimes as migration paths.

The oil and gas pools appear to be located equally in the upper and lower fault blocks, the clastic series of the latter being the most developed. Due to the discontinuity of the reservoirs and the barrier or drain role played by the faults, vertical migrations predominate. This results in a heterogeneity of the hydrocarbon characteristics, often related to the variety of depositional environments. These paralic provinces are also frequently the location of shale intumescences, due to the developments undergone by the undercompacted or evaporitic beds. These domes and diapirs provide a large number of often complex traps with many possibilities.

The rapid accumulation of shale deposits leads to the formation of plastic beds under high pressure, which often take advantage of low pressure zones to penetrate through the overlying formations to form shale diapirs and, if they reach the surface, mud volcanoes. These structures are well known in the Baku area, in Indonesia, in the Mississippi Delta, in Venezuela, and in Trinidad. The Guayaguayare Field on this island illustrates this type of trap (Figure 102) (Michelson, 1976).

Reservoirs

Deltaic basins offer a preferential site for sand deposits of different facies, fluviatile, beach, submarine dunes and offshore bars, and even deep detrital fans in front of the deltaic unit itself (see Section 2 of Part 2) (Allen *et al.,* 1979). The most favorable zones correspond to alternations of sand and shale. In the updip zones, where the proportion of sands exceeds 50 %, the latter are generally not sealed and are filled with fresh water. In the downdip zones, shale predominates and the reservoir possibilities disappear. The most favorable zones are located around a sand/shale ratio of 30 to 50 %. The productive facies in the Upper Miocene in Louisiana mainly correspond to deltaic plane and delta front landscapes.

Seals

Subsident deltaic basins are the site of frequent advances and retreats of the sea. The overlapping marine shale horizons, which may play a major role in the genesis of oil, generally provide excellent seals.

Source rocks

Deltaic basins are distinguished by abundant gas pore organic matter of the humic type, and are therefore often rich in gas. However, marine shales often exhibiting Type II kerogen may play a major role in the genesis of liquid hydrocarbons. The deltaic basins, installed in highly unstable zones, generally receive high heat flows, which are reflected by relatively high geothermal gradients at the basin margins. The presence of thick shale series, often more or less undercompacted with increasing depth, constitute barriers to these flows, resulting in low gradients in the central areas of deposition, often lower than 26 °C/km^{-1}. These thermal and lithologic characteristics allow the presence of oil pools at relatively great depths. The dispersion of source rocks in a relatively thick oil genesis zone suggests different types of migration, and particularly the possibility of migration in solution in water and in methane.

One example of a high grade source rock is offered by the bituminous zone of the Mackenzie Delta. These Upper Cretaceous overlapping marine shales are especially rich in organic matter, which is formed of a mixture of 65 % lipid materials and 35 % humic materials at the base, changing to 90 % lipids at the top. Its thickness varies from 0 to over 500 m in the delta environment. In the richest part, it contains about 3 % of pyrolysable hydrocarbons, probably making it the richest source rock in Canada. It can yield up to 25 tons of oil and 1.4 Gm^3 of gas per km^3 (Bruce & Parker, 1975).

Hydrocarbons

The crude oils from deltaic basins are generally light, paraffinic and low sulfur. Gas is three times more abundant than the world average (Klemme, 1971, 1975 and 1977). About 15 Gt of oil and gas equivalent, nine-tenths in the Gulf Coast and Nigeria, have been discovered in this type of province. The many pools are generally small (Huff, 1980). The two great provinces of the Gulf Coast and the Niger delta have about a dozen giant fields, which nevertheless only concentrate about 4 % of all the reserves. Provinces of this type are typical of the dispersed habitat, and the largest field rarely contains more than 5 % of total reserves. The following section contains a monograph on these two great provinces, and a sketch of the Mahakam Basin is given below (Figure 158). Some data concerning the Mackenzie Delta Basin (Canada) was provided in the discussion of the Taglu field (Section 3.4.1.2. of Part 2).

An example of a deltaic province : **the Mahakam Delta (Kalimantan)**

The Kutei Basin, and especially the Mahakam Delta Basin, situated on the eastern coast of Kalimantan, identified over an area of about 10,000 km^2, exhibit thick Neogene sedimentary series, up to 6 to 8 km thick. A high pressure zone is found below 2900 m at Handil (Figure 158).

About ten discoveries were made in the past ten years, following that of Sanga Sanga (1980), chiefly the oil fields of Attaka in 1968, Bekapai in 1972, and Handil in 1974, located between 1000 and 2500 m, and the Badak (200 Gm^3 and 20 Mm^3 condensate) (1971) and Nilam gas pools (170 Gm^3). The overall unit contains about 350 Mt of oil (Nehring, 1980).

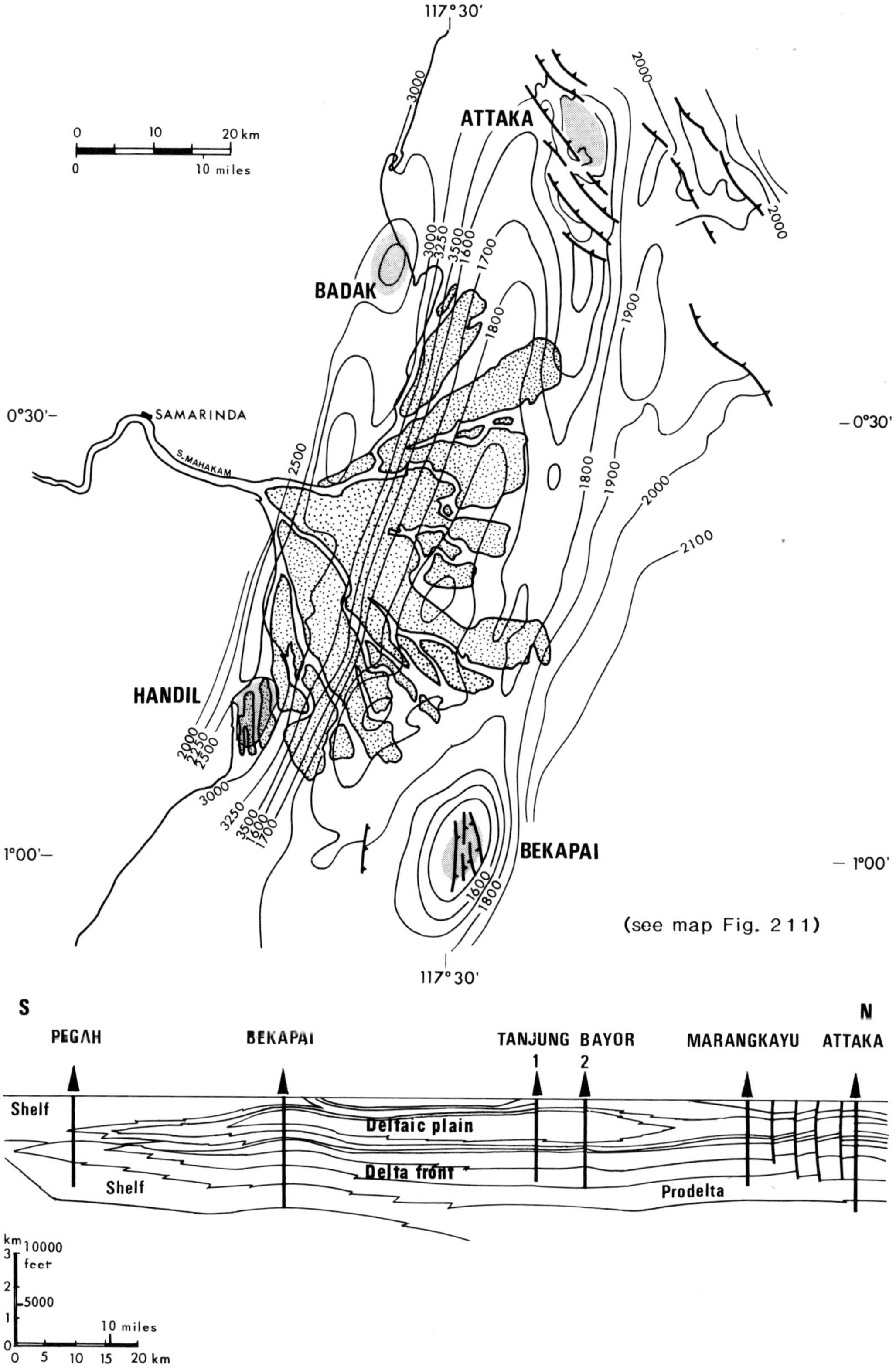

Fig. 158. — Structural map and section of Mahakam Basin (East Kalimantan) (Matharel, 1976).

These accumulations are distributed along two trends substantially parallel to the coast, corresponding to the same depositional environments, delta front and alluvial plains with fluviatile formations. The internal trend, with the Handli and Badak fields, is productive in the multibedded sands of the Middle Miocene, and the external trend with the Attaka and Bekapai fields, in the Upper Miocene/basal Pliocene. The paraffinic oil is very poor in sulfur and cyclanes, as is generally the case in this type of basin, and relatively heavy (35 °API), with a high pour point, (32 °C) at Handil. The vertical distribution of the oil also exhibits the special feature of a slight but real increase in density with depth. This can be explained by an easier and more extensive vertical migration of the lighter products. The crude oil is lighter (42 °C) and only "congeals" at — 5 °C in the outer fields.

The hydrocarbons originate in source rocks mainly formed of type III kerogen, located deeper than 2600 m. Migrations occur along the faults, or through productive horizons from adjacent synclines, over a distance which is longer for the lighter hydrocarbons and vice versa. Traps are provided by a combination of sand bars oblique to the folding. The often large traps (35 km^2 of closed area at Handil) seem to be completely filled on the internal trend, but not on the external trend, where the big Bekapai structure is only partly impregnated (Magnier *et al.,* 1975; Combaz and de Matharel, 1978; Durand and Oudin, 1979; Verdier *et al.*, 1979; Allen *et al.,* 1979; Verdier *et al.,* 1980; Huffington & Helmig, 1980; de Matharel *et al.,* 1980).

2.3. OROGENIC BELT PROVINCES

Provinces situated in an orogenic belt, often on the edge of a tectonic plate, and always in weak zones of the Earth's crust, are subject to convergent displacements or shear/transcurrent fault actions. They are distinguished by considerable tectonic instability, high subsidence and an often complex structuration. However, along with diversity, they demonstrate a number of common characteristics. From the structural standpoint, many large traps are often the result of faulting or folding, but the latter may also result in vertical dismigrations. From the sedimentary standpoint, these provinces, often qualified as mobile, are characterized by thick series, frequently of the progradating type, rich in variations of facies and thickness. Depending on the type and rate of sedimentation, undercompacted beds may develop and give rise to active syn-sedimentary tectonics.

From the petroleum standpoint, the sedimentary volume, the considerable variations in facies, compaction and catagenetic mechanisms, as well as the frequency of structural deformations, are all favorable factors in the context of an often dispersed habitat.

These provinces are principally Mesozoic and Cenozoic, the fields belonging to orogenic belts tending to a relatively early destruction. They contain about half of the world's hydrocarbon reserves (Klemme, 1971, 1975 and 1977). They may be subdivided into three main groups, according to their geodynamic context, and in the order of increasing complexity :

- provinces of the intermontane type or post-orogenic rift basins, often in a transverse or oblique position in relation to the earlier structural directions, characterized by predominating vertical movements, and thus forming a transition with the provinces of the previous family : one example is the Vienna Basin,
- strike slip provinces associated with shear movements : the Californian basins for example, especially the Los Angeles Basin,
- provinces in convergent zones, especially foredeep basins, affected by more or less pronounced tangential movements, the most important from the petroleum standpoint, with the province of Ploesti taken as an illustration : a number of giant provinces are associated with this group.

In fact, many terms of transition exist between these different families, and classifications are often debatable.

2.3.1. Provinces of the intermontane type

These provinces, often complicated by transcurrent faults, exhibit close analogies with provinces of the ritf type :

- elongated subsident troughs, triangular or rounded, limited by large fractures or flexures,
- thick sedimentary series often corresponding to confined environments, at least during certain periods,
- horst and graben structuration,
- generally high thermal gradients.

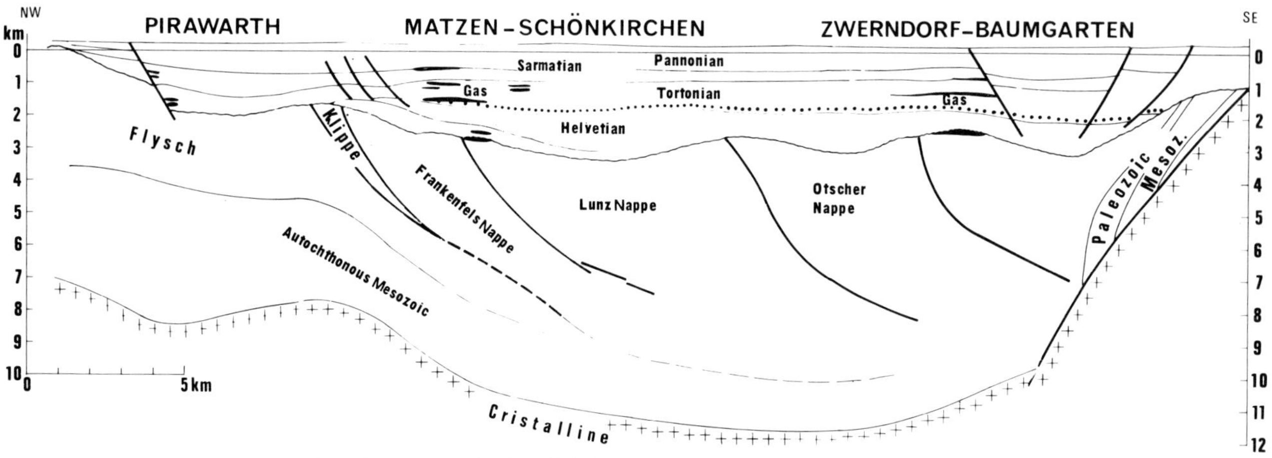

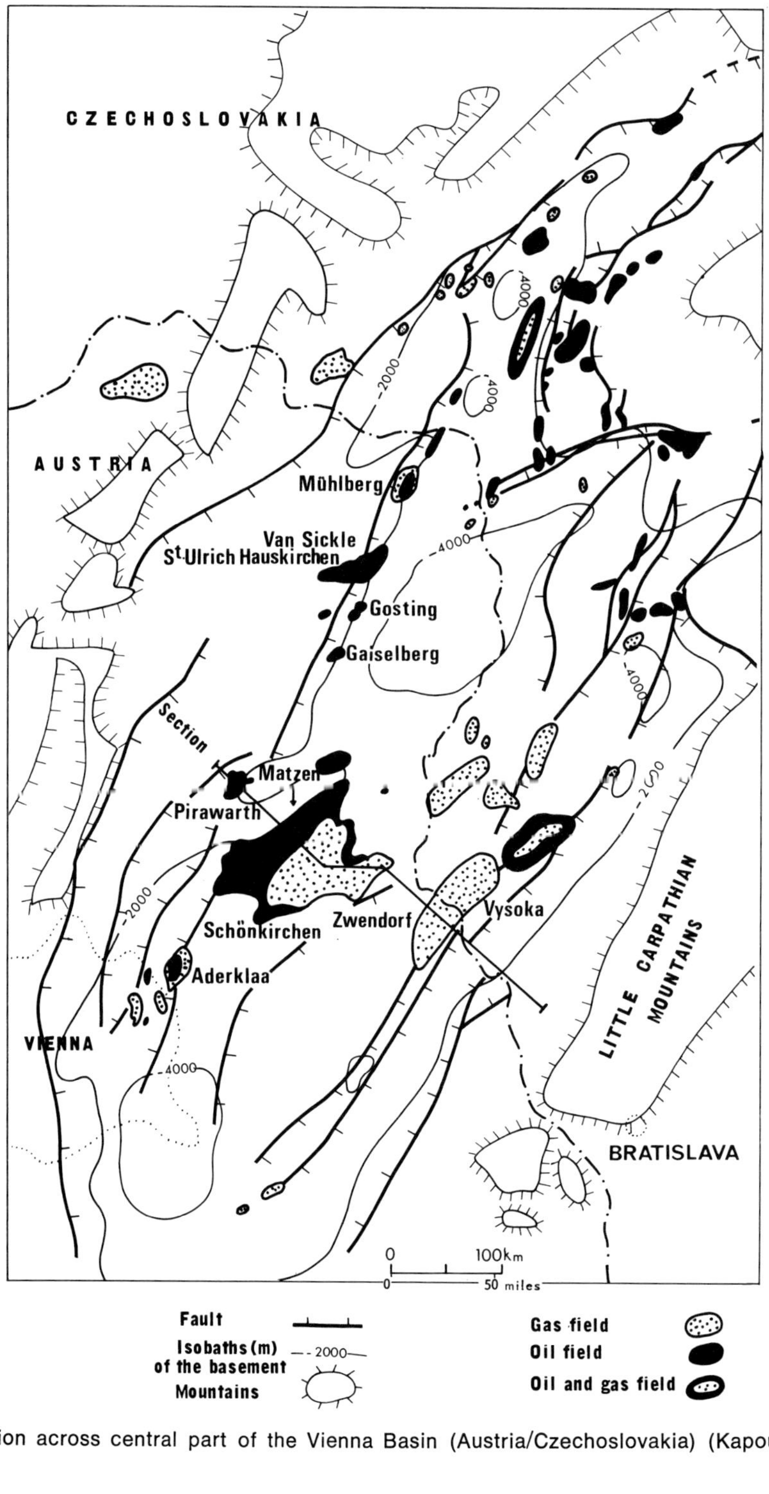

Fig. 159. — Section across central part of the Vienna Basin (Austria/Czechoslovakia) (Kapounek *et al.*, 1963).

The main differentiating feature from troughs and rifts is their geotectonic situation, and in particular their association, or lack of association, with an orogenic belt. These provinces lie on a sedimentary basement, usually tectonized, but nevertheless not devoid of petroleum potential.

These geological characteristics often constitute excellent petroleum criteria. The example of the Vienna province in Austria is discussed below. We shall also examine the Cook Inlet basin in Alaska, whose more complex structure reflects tangential movements related to an older island arc. Also worth mentioning are :

— The Middle Magdalena Basin in Colombia,
— the Transylvania Basin, essentially gas-bearing, in Romania,
— the Pannonic Basin in Hungary.

The Vienna Basin

The Vienna Basin, in Austria but also extending into Czechoslovakia, is a classic example of an intermontane basin. Elongated, dissymmetrical, in the form of a semi-graben, it covers an area of 7000 km^2. It is essentially filled by a Neogene series up to 5000 m thick, and including :

— sand/shale series at the base, merging into carbonates on certain Burdigalian/Lower Helvetian overlapping shoals,
— overlaid unconformably by finely bedded sand/shale series with a marine facies, passing into an Upper Helvetian lacustrine environment to the northeast, overlaid by alternating Tortonian sands, marls and marine lithothamnium limestone,
— above, the Sarmatian formed of marls and sands of brackish facies, and a Pannonian sand/shale horizon which is increasingly lacustrine.

This Neogene series lies on a Mesozoic basement represented by flyschs and carbonates which are found in the different Alpine overthrust nappes (Figure 159). From the tectonic standpoint, this province corresponds to a post-orogenic rift basin in a slightly oblique position to the Alpine trends. It is characterized by large NE/SW striking synsedimentary faults, tilted 50 to 75 degrees towards the center of the basin, and sometimes exhibiting major throws.

This petroleum province may be qualified as rich, and has given rise to the discovery of about 100 Mt of oil and 100 Gm^3 of gas, or 14,000 t km^{-2} of oil and an equivalent amount of gas, distributed among about sixty pools, seven of which contain 84 % of the oil reserves and 62 % of the gas reserves. The main field is Matzen/Schönkirchen, discovered in 1949, with estimated reserves of 65 Mt of oil and 20 Gm^3 of gas, including 50 and 18 respectively in the Miocene, and the remainder in the underlying Alpine Triassic. The oil of this province mainly originates in the Helvetian and Tortonian marls, while the gas originates in the brackish beds of Upper Tortonian/Sarmatian age, and also possibly from the Mesozoic series of the basement. The average geothermal gradient is 3 °C per 100 m.

In the Neogene, and chiefly in the Tortonian, the reservoirs consist of porous (20 to 25 %) and permeable (100 to 1000 md) sandstones, generally lenticular and 5 to 25 m thick, at between 500 and 2000 m depth. The Mesozoic reservoirs consist of fractured carbonate horizons. The cap rocks consist of shale. The traps are mainly structural and varied. The Matzen trap, the largest, is formed by a draping anticline on a paleorelief dotted with pinchouts and sand lenses. A number of others are associated with faults, and especially the large syngenetic Steinberg fault. In the basement overthrust nappes, complex traps are closed by unconformable Secondary impermeable beds (Schönkirchen deep). The crude oils are generally paraffinic in the older formations, naphthenic in the younger series with a lower sulfur content (less than 1 %). The Neogene gas is very rich in C_1, and the Mesozoic gas contains CO_2 (up to 11 %) and a little H_2S.

The Vienna Basin offers an especially concentrated habitat, with a factor K of 1.65 for the Austrian section (see Section 1 of Part 3). Apart from Matzen/Schönkirchen, which contains nearly two-thirds of the oil discoveries, the main fields are relatively small :

— Zwerndorf (1950) straddling the Austro-Czech border, 22 Gm^3 of gas, largely exploited,
— Sankt Ulrich-Hauskirchen-Plattwald (Van Sickle Group) (1938), 7.5 Mt of oil and less than 2 Gm^3 of gas,
— Mühlberg (1941), 5.6 Mt and 2 Gm^3,
— Gaiselberg (1938), 4.5 Mt and 0.5 Gm^3,
— Aderklaa (1939/1950), 2.5 Mt and 6 Gm^3 (Janoschek, 1958; Kapoubek *et al.*, 1963).

The Cook Inlet Basin

The Cook Inlet petroleum province corresponds to an intra arc rift basin with the Mesozoic lying on an oceanic crust. It is limited by large faults and affected by tensile and compressive movements, the latter

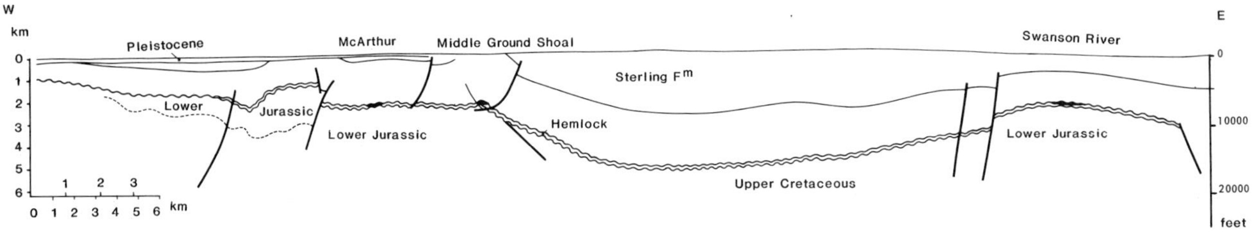

Fig. 160. — Cross-section of Cook Inlet Basin through Middle Ground Shoal and Swanson River Fields (Boss *et al.,* 1976).

reflected by reverse faults. It has an area of 40,000 km² (40 % offshore) spreading over a length of 320 km and a width of 100 km (Figures 160 and 165).

The sedimentary thickness is 18,000 m, including 8000 m in the deepest part of the Cenozoic, which represents the bulk of the productive series.

Overlying the Jurassic, or, in the central area, a deep facies Upper Cretaceous partly made up of turbidites, is an uncomformable Cenozoic series of largely paralic and non-marine facies.

The productive reservoirs are mainly the Hemlock sandstones, probably Oligocene, and fluviatile facies that contain 90 % of the oil discoveries. They exhibit porosity of 10 to 25 % and permeability of 20 to 100 md, giving average productivities per well of 170 t day^{-1}. The Pliocene sandstones of the Sterling Formation, also of continental facies, constitute the main gas reservoir, containing 85 % of the total 110 Gm³ of methane in the basin. They have excellent characteristics, porosity up to 30 % and permeabilities of 2 to 3 d, with thicknesses of 100 m and more. The traps are structural, simple and faulted anticlines with frequent reverse or asymmetric faults, aligned along three main regional trends. They are the result of a single phase of Plio-Pleistocene folding. In the Hemlock sandstones, they contain a paraffinic oil with density 30 to 38 °API, poor in S, generally undersaturated, which could originate in Middle Jurassic marine shales, that reached good maturation conditions in the Late Tertiary.

The Upper Miocene gas (Beluga Formation) is generally wet, with 83 % C_1 and 7 % C_{6+}. The Pliocene gas (Sterling Formation) is dry, often containing 99 % C_1. These gases originate in the coal seams interbedded in the series, and could largely be of biochemical origin (Rice & Claypool, 1981). Cumulative discoveries amount to 150 Mt of oil distributed among four main fields, and 200 Gm³ of gas distributed among some fifteen pools. The main oil fields are :
— McArthur River (1965), a gentle anticline, 70 Mt,
— Swanson River (1957), also a simple dome, 32 Mt,
— Middle Ground Shoal (1982), a narrow and faulted anticline, with reserves estimated at 26 Mt,
— Granite Point, 15 Mt.

The main gas field, Kenai (1959), contains 140 Gm³ (Crick, 1971; Osment *et al.,* 1967; Boss *et al.,* 1976; Magoon & Claypool, 1981).

2.3.2. **Strike slip basins**

Petroleum provinces formed by strike slip basins are associated with the action of large shear fractures applied at the level of the oceanic crust and stretching up to a distance of several hundred kilometers.

Basins of this type often lead to rift basins and all possible variants between these two families, depending on the extent of horizontal throw.

The major characteristic of shear and wrench faults is the generation of ‘en echelon’ folds at their contact, with an axis slightly oblique to the main fault strike. These ‘en echelon’ anticlines constitute excellent, well-closed traps, often in clusters, and formed fairly early during subsidence. However, these folds frequently appear to be cut by tension faults which are liable to compromise the structural closure. The action of distensive and compressive forces throughout the development of the shear faults may be reflected by vertical movements of different blocks. The uplifted compartments can thus be eroded and improve the sedimentation of downthrust zones.

The reservoirs of these unstable basins are usually detrital and irregular, and deep deposits of the turbidite type are not infrequent (Wilcox *et al.,* 1973; Harding, 1973). The formation of depressions, often favorable to confined environments, is a positive factor for the formation of source rocks, whose maturation is facilitated by the presence of geothermal gradients generally higher than normal.

Briefly, basins of this type are characterized in particular by :
— high thermal gradients during their subsidence phase,

— short duration of genesis, about 20 million years,
— elongated, narrow and deep form,
— rapid sedimentation,
— early structuration, intersected by faults.

In this context, the hydrocarbon genesis and entrapment conditions are generally good, but the faults incur the risk of vertical migration. Provinces belonging to this type are fairly frequent and are often rich. A total of nearly 20 million tons of oil and an equivalent of gas are attributed to them (Huff, 1980). This family includes the Californian provinces of Ventura and San Joaquin, Baku (Section 3 of Part 3), and Los Angeles, analyzed below.

The Los Angeles Basin

This Neogene intermontane basin, triangular in shape and covering only 3500 km^2, is one of the world's richest petroleum provinces (Figures 161 and 193). The sedimentary basin is essentially represented by alternating Miocene and Pliocene sands and shales, 4000 to 5000 m thick, corresponding mainly to relatively deep formations often deposited in more than 1000 m water depth, at about 50 km from the shore. In this respect, the area constitues a model of an ancient deep offshore basin.

The main reservoirs are formed by the Lower Pliocene sands of the Repetto Formation, considered to be overlapping deposits of turbidites and deep sea fans with a high sedimentation rate (400 m per million years), in a period of active subsidence and deepening of the basin, which lasted about five million years. These sands occur in the form of successions of often thick beds, accounting for 50 % of the series in the productive zones. Their porosity and permeability characteristics are excellent and the volumes of recoverable oil are especially high, up to an average of 4.3 Mt km^{-2} for the overall basin, 9 Mt for the richest fields. The relatively deep characteristics of the depositional environment favor the spread of these sands in the central part of the basin at the expense of the margins, thus in zones which are well protected and well supplied with hydrocarbons.

The cap rocks are provided by Upper Pliocene shales, which are well developed, especially on the basin margins. The source rocks are represented by Upper Miocene (Mohnian) overlapping black shales, rich in organic matter, phosphatic and diatomaceous, corresponding to more or less closed basin deposits, in rapid subsidence (about 240 m per million years). Maturation of the organic matter was facilitated by a fairly high thermal gradient (2 to 3 °F per 100 ft, or 36 to 55 °C km^{-1}), and migrations occurred vertically through frequent fractures.

The Los Angeles Basin lies on the boundary of the great shear fractures of the American Pacific margin, which caused the formation of a series of 'en échelon' anticlines with a slightly oblique trend in relation to the main faults. Thus up to 90 % of the traps are formed by fairly well aligned and fairly pronounced anticlines, often faulted but always effectively closed by thick Upper Pliocene shale series. The productive areas range from a few square kilometers to 31 km^2 for the Wilmington Field, the largest (Figures 162 and 82).

Main discoveries

About 1.2 Gt of oil and 200 Gm^3 of gas have been discovered in this province, representing an exceptional potential of 350,000 t km^{-2}.

These reserves are distributed among some forty fields aligned on either side of the subsident axis in four successions of 'en échelon' anticlines, intersected by shear faults. The pools cover a productive area of 200 km^2. The four main fields, the giants, are :

— Wilmington (1932), 390 Mt (Section 4 of Part 2),
— Huntington Beach (1920), a faulted anticline productive between 570 and 2000 m, with initial reserves of 150 Mt,
— Long Beach (1921), 125 Mt, a faulted anticline with an area of 7 km^2,
— Santa Fe Springs, discovered in 1919 by a morphological survey, formed by an unfaulted dome, containing 85 Mt distributed among nine thick pay sands between 1000 and 3000 m deep (14 Mt per km^2).

*
**

The exceptional potential of this province is ascribed in particular to :
— the intensity of the sedimentation rate,
— the variety of deposits, placing rich source rocks at the base, broad developments of porous and permeable sands above, the whole system covered by a thick shale series,

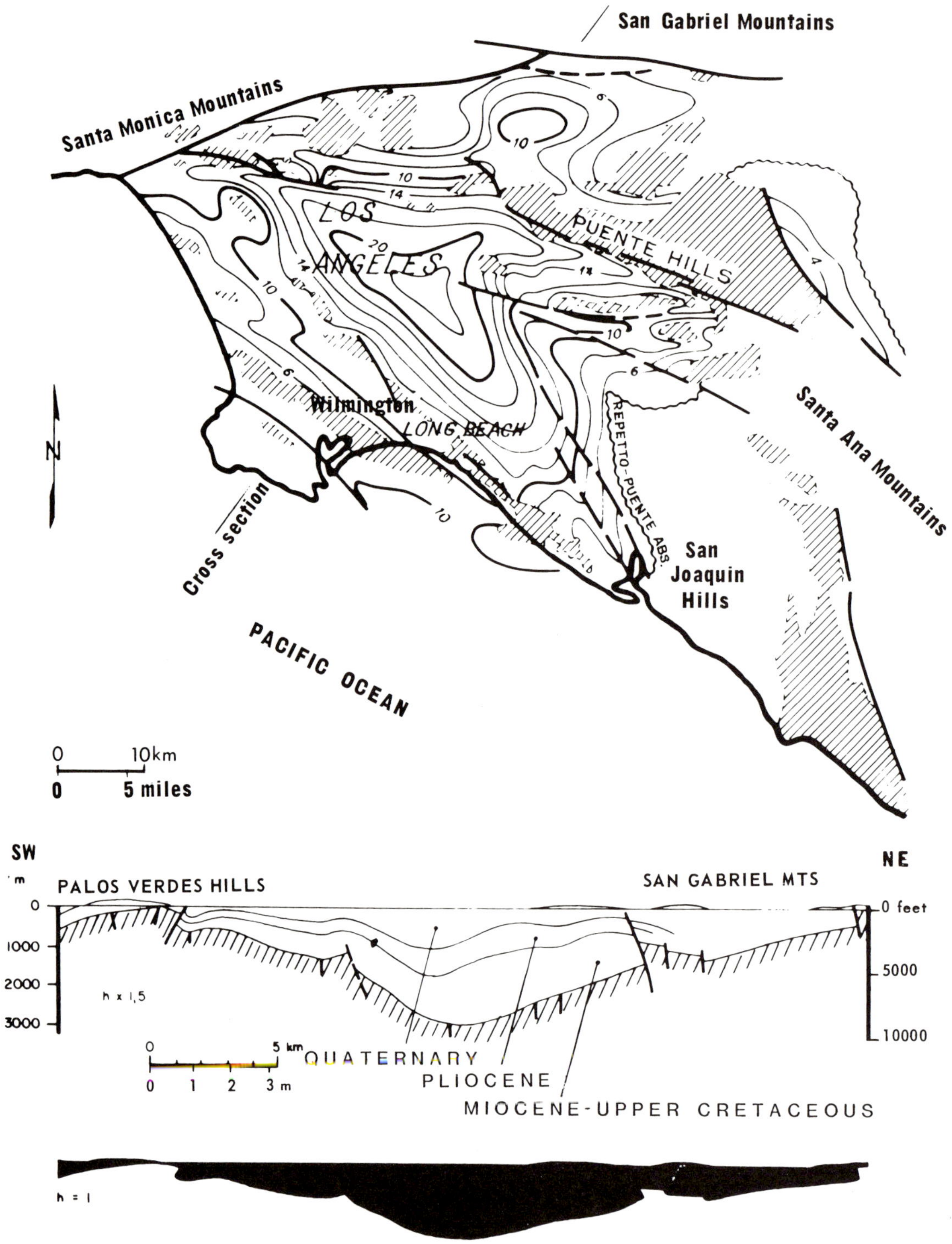

Fig. 161. — Isopach map of Upper Miocene and Pliocene and section of Los Angeles Basin (Gardett, 1971).

— high geothermal gradients,
— faults allowing effective migration,
— folds providing excellent structural traps.

This set of specific conditions appears to be the result of a short but eventful geological history, characterized by intense and recent subsidence, progressive filling of the basin ensuring an effective closure on the margins, and active tectonics (Barbat, 1958; Garnet, 1971; Harding, 1973; Zieglar & Spotts, 1978 especially).

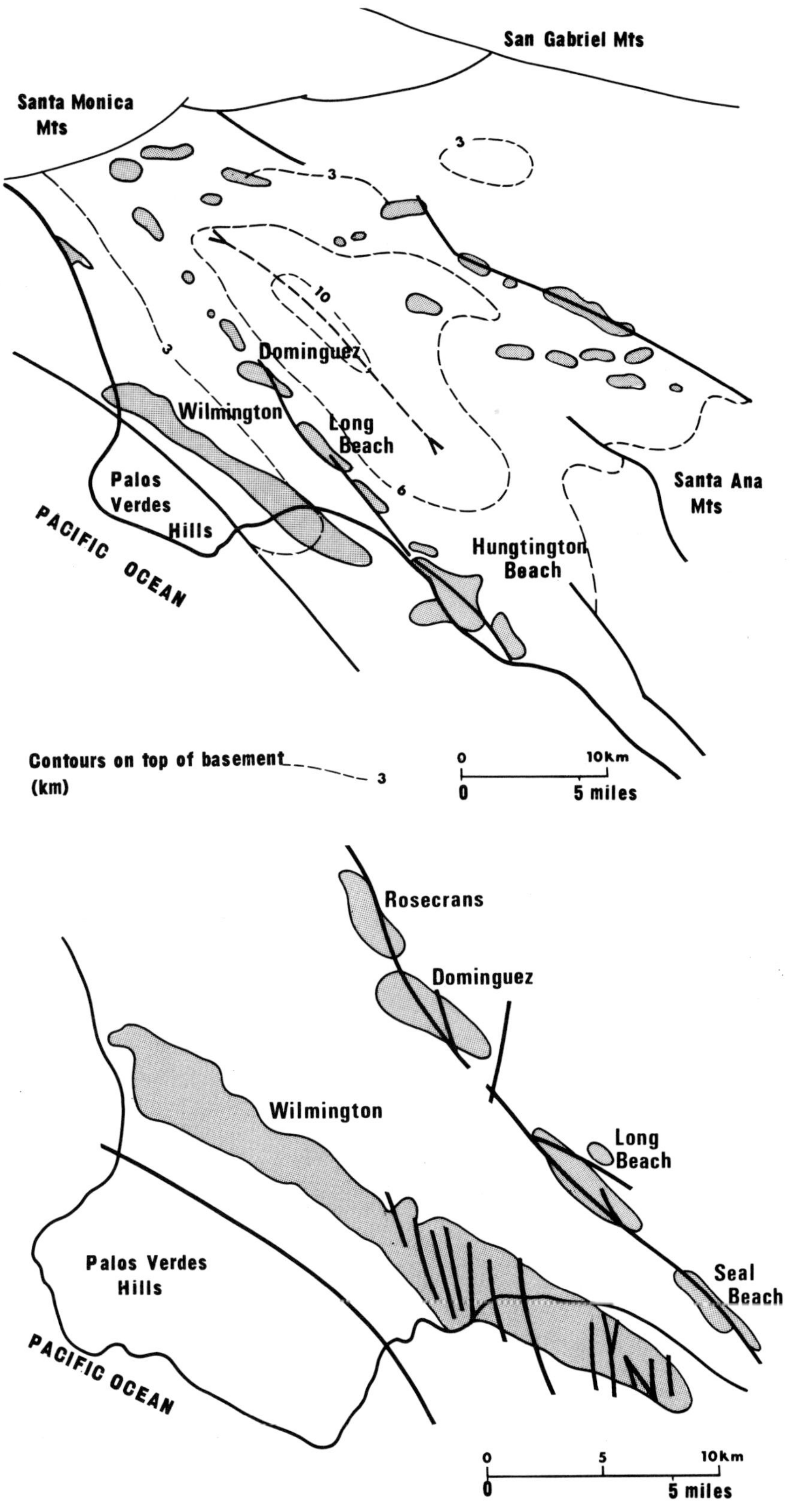

Fig. 162. — Structural map of western part of Los Angeles Basin and Wilmington Field (Mayuga, 1970).

2.3.3. **Provinces in convergent areas**

These provinces generally belong to mobile areas affected both by high subsidence and tangential movements, often associated with convergent plate movements. Depending on the share of continental and oceanic areas, two main families are distinguished :

— provinces of the foredeep or miogeosyncline type, by far the most important from the petroleum standpoint, with a special discussion of the highly unstable Chinese onshore basins,

— provinces associated with cordilleras or island arcs, generally of lesser petroleum interest.

2.3.3.1. ***Foredeep basins***

These basins generally consist of a platform area and a trough, subjected to compressive movements, quite often associated with subduction zones. They may also feature a broad predominantly monoclinal area, sometimes overthrusting, with thick and diversified series. The transition zones between these two areas, and especially the foothills zones with relatively moderate folds, fairly thick series and frequent facies variations, often appear to be the most favorable. But depending on the specific characteristics of each province, the fields may be concentrated in the folded zone (Ploesti, North Kalimantan), on the relatively stable platform or on the hinge Line (Tampico/Reforma, Prudhoe Bay), or on both, as in Alberta, the Orinoco, Volga-Urals and North Caucasus Basins, and especially in the Persian Gulf.

Some of the richest provinces are associated with this type, and especially those of the Middle East, West Canada, North Alaska, Arkoma, Tampico in North America, the Orinoco and Subandean Basins in South America, North Caucasus (USSR), Ploesti (Romania), and North Kalimantan (Southeast Asia). Among the lower potential provinces are those of the Molassic Basin in Germany and Austria, the Po Plain, and the Apennine margin in Italy. The Romanian province of Ploesti, a good example, is analyzed here. The next section contains a number of monographs of other provinces corresponding to giant provinces.

The Ploesti Province in Romania

This old and rich petroleum province corresponds to the central part of the Carpathian foredeep. It is an asymmetric basin covering over 20,000 km^2, very subsident, and the thickness of the sedimentary column exceeds 10,000 m in the deepest part (Figure 163). The sedimentation of the Cenozoic basin appears to be continuous in the Paleogene and Pliocene, more discontinuous and intersected by unconformities in the Miocene, and remarkable for a thick Saliferous episode in the Tortonian. The environment, which is marine at the base, becomes lacustrine in the Pliocene, as observed throughout the Neogene area of eastern Europe. Older Paleozoic and Mesozoic series have been identified, of which some facies are of significant petroleum interest.

The traps are mainly structural, highly tectonized anticlines, faulted and overturned, often with diapiric cores, in which the salt tectonics are added to compressive deformations. The shallowest structures are generally observed to be aquifers or producers of heavy oil. The main reservoirs are sand and are distributed in two major families. The Meotian (uppermost Miocene) sands, with a brackish facies, form the main reservoir of the Ploesti district. Their characteristics are excellent, but deteriorate gradually with depth : porosity 25 to 30 % and average permeability about 200 md up to a depth of about 2000 m, cumulative effective thickness 85 m. They are well sealed by Pontian marls, and contain an average of 1 Mt per km^2. In total, by early 1968, these reservoirs had supplied 60 % of the cumulative discoveries of the district. The Lower Dacian (Lower Pliocene) sands exhibit more favorable characteristics, especially in terms of thickness, but their distribution is more irregular. In the richest fields, they offer 4 Mt of recoverable reserves per km^2 and, overall, these sands have produced about 30 % of cumulative output.

Potential source rocks are numerous and their respective roles are still somewhat uncertain. Bituminous shales have been identified in the Paleozoic, Jurassic, Cretaceous, Oligocene and Tortonian, where they are often associated with evaporites and volcanic formations. The most interesting are probably the Meotian shales, with a brackish facies, which appear to have supplied the reservoirs of the same age with a paraffinic oil and with gas. The Dacian shales, often carbonaceous and with lacustrine facies, appear to have supplied the neighboring reservoirs with a naphtheno-aromatic crude, generally undersaturated and accompanied by CH_4 and CO_2. Gas becomes predominant in the younger series. Vertical migrations appear to be slight, although some diapirs may have facilitated transfers. Pressures are generally hydrostatic, with local geodynamic anomalies. Geothermal gradients are generally normal (30 °C per 1000 m), but higher, up to 40 to 50 °C, in the richer zones. The crudes vary in density, apparently in relation to biodegradation mechanisms. Mud volcanoes are frequent (Pl. 4).

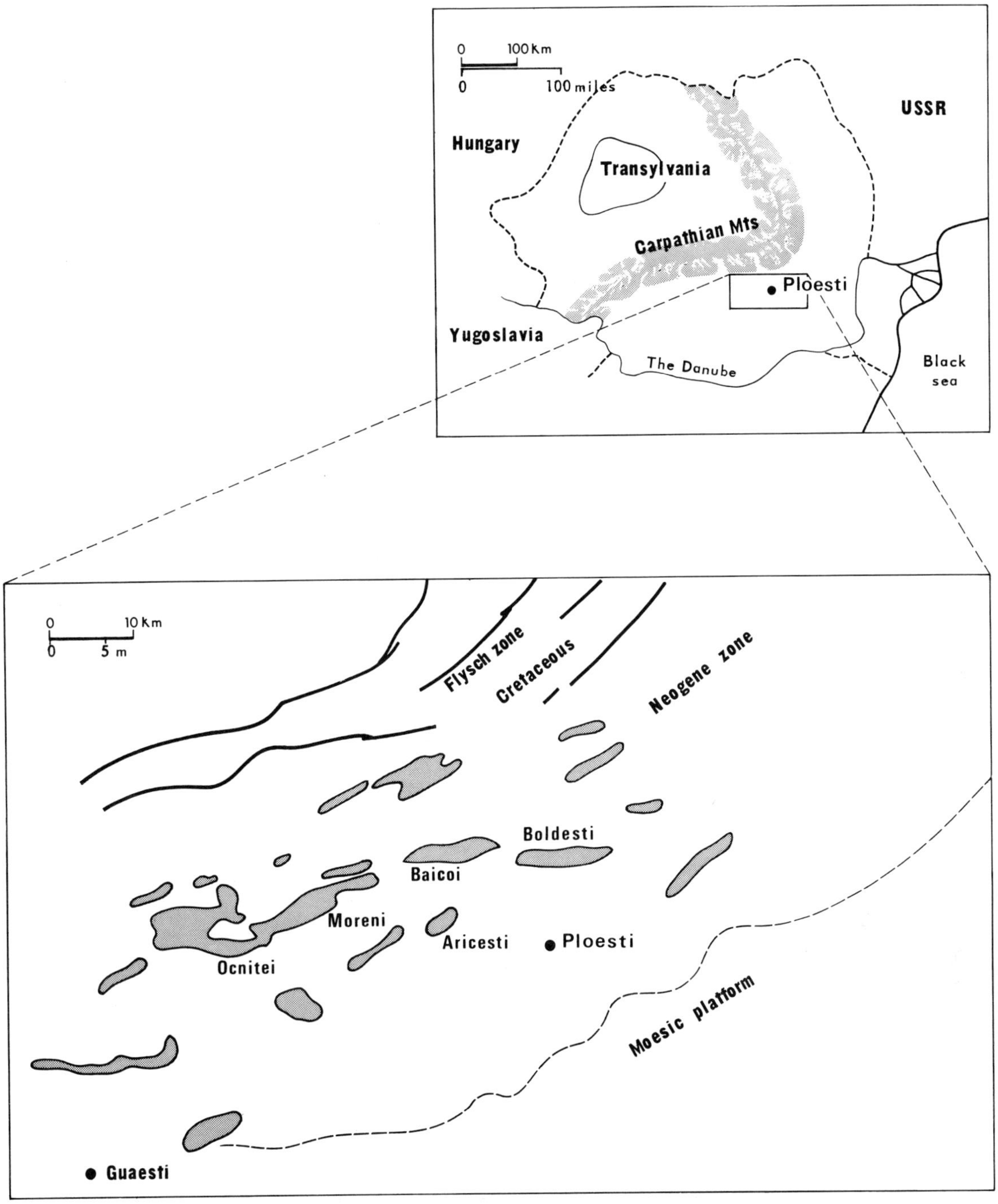

Fig. 163. — Petroleum map of Ploesti province (Romania).

Among the geological criteria that appear to have contributed most to the potential of this province are the following :

— narrow intercalations of sand and source rock shale, with a high proportion of porous and permeable sands,
— good seals, locally consisting of overthrusting nappes generally with a regional extension,
— rapid and recent sedimentation,
— the presence of some high zones and depressions and a large number of structural traps.

Mainly from the Ploesti zone, Romania has produced nearly 500 Mt of oil and substantially the same number of billions of cubic meters of gas, from some 300 fields, of which slightly over one-third are in the zone in question. Among the latter, several units appear to be giants. This is the case of the Moreni/Gura/Octinei trend, discovered in 1900, which, by early 1978, had produced 100 Mt or over 40 % of the cumulative

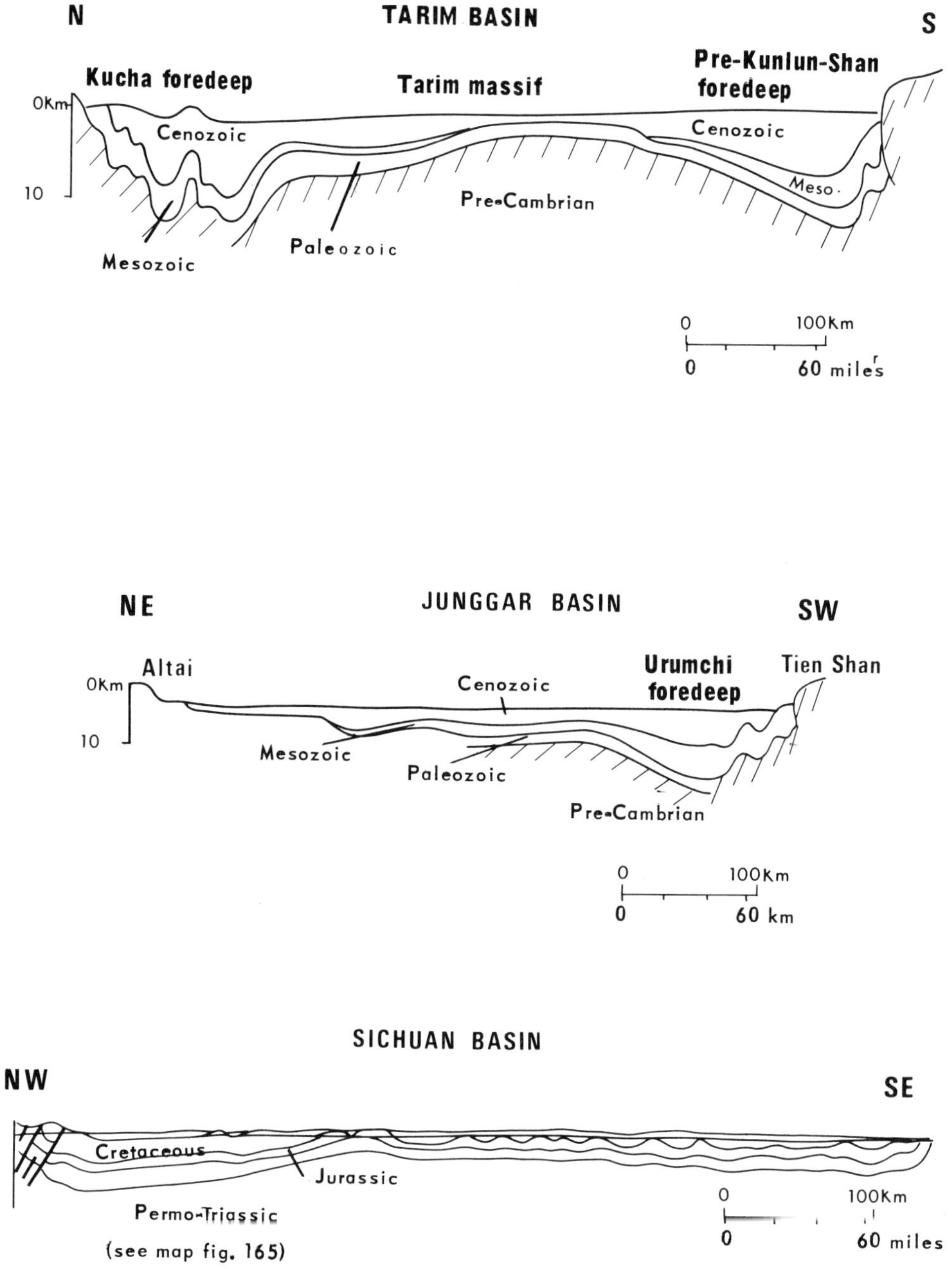

Fig. 164. — Sections across Chinese basins of Tarim, Junggar and Sichvan (Meyerhoff & Willums, 1976).

output of the Ploesti area, with initial reserves estimated at 110 Mt (Figure 84) (Paraschiv & Olteani, 1970; Paraschiv, 1975).

Chinese onshore provinces

The basins of central and western China correspond to vast, strongly folded intermontane troughs, separated by a central platform. These deeps, or foredeeps, may surround the more stable area, or stretch along one side only. The commonly old tectonically disturbed zones continued to be formed until very recently. These young ranges, partly formed of old orogenic materials, are referred to as 'tiwa'. Since Permian time, all the internal basins are filled with mainly non-marine formations, which supply the bulk of oil production (Figures 164 and 165). The coastal and offshore basins, as well as the Songliao Basin, are of the graben type, and also filled with Mesozoic and Tertiary non-marine formations, except in areas closest to the present sea, where marine beds are more common, especially in the Tertiary.

The Tarim Basin in Xinjiang is characteristic of this arrangement in two north and south deeps up to 7 to 8 km thick on the margins of uplifted folded ranges situated on either side of a more stable central area. It covers about 500,000 km^2 and exhibits a thick Jurassic through Tertiary sandstone/shale series with seams of coal and black shales with a lacustrine facies. Six fields (including three gas fields) were discovered in this basin, on surface anticlines, but exploration has not yet been significantly intense. Marine Cretaceous beds are common in the foredeep south of the central Tarim platform. The Jurassic rocks are believed to be the source beds for the oil and gas in the basin.

The Junggar Basin, farther north, is an asymmetric basin of the foredeep type covering 175,000 km^2, exhibiting a mainly continental sedimentary series stretching from the Permian to the Pliocene, and up to 10,000 m thick in the trough axis. The pools are in surface anticlines, commonly faulted, and productive in different zones. The main field is Karamay (1955), containing 100 Mt of oil, or about two-thirds of the reserves of the basin. Most production is from the Permian and Trias.

The Qaidam Basin, in the east central area, is an intermontane basin with an area of about 100,000 km^2, filled by a sedimentary series stretching from the Jurassic to the present, mainly continental and intensely folded. This basin, situated at an average altitude of 3000 m and bordered by reliefs of 7000 m, is highly promising. Nearly 240 Mt have been discovered there, distributed among some twenty fields. The largest, Lenghu (1958), is a dissymmetric anticline which is productive in the continental strata of the Upper Cretaceous, Eocene, Oligocene, Miocene, and Lower Pliocene.

The Sichuan or Red Basin in the south center, the oldest acknowledged productive zone in the world, covers about 160,000 km^2, with a sedimentary series extending from the Sinian to the Cretaceous, 4000 to 10,000 m thick. Two new producing zones are noted :
— Sinian carbonates with evaporites,
— Carboniferous carbonates.

It is characterized, in relation to the foregoing basin formations, by a marine Sinian, Carboniferous, Permian and Trias consisting of evaporites and carbonate formations, and by a reduced Tertiary. It is folded into many 'Appalachian' and recumbent anticlines up to 300 to 400 km long. The fractured limestones and dolomites of the Trias, Permian, Carboniferous and Sinian produce gas in the south and southeast, forming China's main gas province, while the Jurassic continental sandstones are oil producers in the north. The initial reserves of this basin, the second largest, amount to about 140 Mt of oil and 500 Gm^3 of gas, distributed among about ten oil fields and 29 gas fields. The largest gas fields are :

— Weiyuan, 55 Gm^3 probable,
— Shiyou, 50 Gm^3 probable,
— Huanguang, 20 Gm^3 probable,
— Wulunghou, 20 Gm^3 probable,
— Luzhou, 20 Gm^3 probable,
— Shengdenshan, 10 Gm^3 probable,
— Zigong, 10 Gm^3 probable,

and the main oil fields are :

— Longnüsu (1956), 20 Mt probable,
— Nanchong (1958), 15 Mt (?),
— Yingshan (1957 ?), 5 to 10 Mt (?).

The Erdous Basin in central China spreads over about 180,000 km^2, with a depth ranging from 3500 to 7000 m. The series are affected by acute post-Pliocene folds on both east and west margins. The reservoirs are formed by continental sandstones with coal seams of the Permian, Trias/Lower Jurassic and Cretaceous, and the source rock is represented by Lower Trias bituminous black shales. A dozen small fields have been identified in this basin, but only four produce commercial amounts of oil, about 6 million tons.

The chief remaining petroleum basins of China, located in the eastern part of the country, in Manchuria, and in the coastal zone, are at least partly associated with the graben type. The Songliao Basin in Manchuria, one of the richest, is a platform basin downwarped in a vast central graben, with an area of 200,000 km^2. It is formed by an overlapping sand/shale series, overlaid by Lower and Middle Cretaceous lacustrine bituminous shales, and sealed unconformably by continental green and red shales. The Daqing complex (1959), in the deep central part of this basin, is formed by a trend of anticlines with an area of about 1000 km^2, productive in the Lower and Middle Cretaceous sandstones. The initial reserves of these fields are estimated at about 800 to 1000 Mt.

The Bohai or North China province corresponds to a rift basin with an area of 270,000 km^2, including 100,000 km^2 offshore, possessing a lacustrine Meso-Cenozoic sedimentary series 3 to 5 km thick, exhibiting wide variations in facies. The reservoirs appear to be formed mainly by Tertiary (Eocene through Lower Pliocene) non-marine deltaic sandstones, and by buried fractured carbonates of Sinian, Cambrian and Ordovician ages. Some production is from fractured Permo-Carboniferous sandstones and Mesozoic volcanic rocks. Three sequences of source beds have been identified : continental Eocene and Lower to Middle

Oligocene bituminous shales, and Sinian to Ordovician carbonates. The main fields are those of the groups of :

— Shengli (Victory) (1960), 70 Mt,
— Dagang (1964),
— Renqiu (1975), a sub-unconformity field, 200 Mt,
— Liaohe (1967).

Renqiu may be 200 Mt alone (Meyerhoff & Willums, 1976; Meyerhoff, unpublished).

2.3.3.2. *Fore arc basins*

These provinces are situated in highly unstable areas and possess more or less developed sedimentary series. They correspond to narrow basins with steep slopes, partly filled by deep or slope deposits, rich in volcanic formations. However, basins of the graben type may also lie within these areas.

These geological characteristics are generally unfavorable from the petroleum standpoint, and these provinces are, in fact, poor and few in number. They may be illustrated by the small Santa Elena province, on the Pacific coast of Ecuador. This coastal province is formed by Upper Cretaceous and Paleogene formations, consisting of shales, silicified shales and sandstones, of the turbidite and olistrostrome type, and up to about 3000 m thick. The tectonics are highly complex and the traps are fractured and faulted. The oil from the small fields of this basin, Ancón, Santa Elena, El Tigre and Santa Paula, is found in silicified shale and fractured chert horizons. Ancón, the main accumulation, approaching depletion, will produce no more than 15 Mt.

The Peruvian coastal province, in the south prolongation of the Santa Elena province, exhibits certain analogies with the latter. It is more associated with the graben type in a highly faulted structural context. The main producing formation, the Eocene, is up to 5000 m thick, with many sand lenses. The traps consist of fault blocks and seeps are very abundant. The largest fields are :

— La Brea/Parinas/Talara (1868), the biggest, a vast mosaic of fault blocks containing some 140 Mt of oil, today practically exhausted by nearly 2000 wells,
— Lobitos, about 15 Mt.

The province also features a dozen smaller fields aligned more or less with the coastline.

Summary and conclusions

Petroleum provinces adapt to practically all types of sedimentary basin, and many similarities can be observed. However, the main structural features of the basins condition the characteristics of the provinces, whose potential is directly related to high subsidence, with relatively moderate tectonics.

Stable platform basins, for example, generally yield relatively poor provinces in the absence of reef structures, and a dispersed habitat. More mobile basins offer greater chances of success. A simple structure formed of broad arches or vast folds often means a concentrated habitat and hopes of a giant field, even in the absence of seeps. Highly tectonized and intensely faulted basins often turn out to be extremely poor, and this is often marked by the abundance of seeps. Paleozoic provinces normally belong to relatively stable platform basins, which succeeded in preserving their pools. On the other hand, Tertiary provinces, which are frequently in a hydrocarbon supply period, are found in orogenic basins.

3. – THE MAIN PETROLEUM PROVINCES

« Savoir pour savoir, mais peut-être encore plus, savoir pour pouvoir. »
'Knowledge can be an end in itself, but knowledge is also power.'
Teilhard de Chardin.

Some 600 sedimentary basins are already identified today throughout the world, onshore and offshore. About 160 out of nearly 400, more or less prospected, have yielded commercial hydrocarbon production, but only 25 represent cumulative discoveries of more than 1.4 billion tons, concentrating about 85 % of the total worldwide discoveries to date. Six of these superprovinces contain more than 5 Gt and account for nearly two-thirds of the world's total, while one of them by itself accounts for 40 % (Perrodon, 1978; Halbouty, 1978; Nehring, 1980) (Table II).

Petroleum potential is also concentrated on the scale of the field. Some 300 of the world's giant fields (reserves over 70 Mt or 500 10^6 bbl) which only account for about 1 % of the total number of fields, concentrate around three-quarters of cumulative discoveries, and 33 super-giant fields (over 700 Mt or 5 10^9 bbl), 23 of them in the Middle East, account for half (Nehring, 1978, 1980).

Gas appears to be slightly less concentrated, and nearly three-quarters of cumulative discoveries are found in some 15 giant provinces containing over 1000 Gm^3, two-thirds of this total in four provinces.

From the stratigraphic standpoint, post-Triassic series contain about 80 % of these hydrocarbon reserves, with the following distribution :

— Paleozoic, about 13 %,
— Mesozoic, about 63 %,
— Cenozoic, about 24 % (40 % outside the Middle East) (Halbouty *et al.*, 1970; Bois *et al.*, 1980).

Table II
Main oil and gas provinces

	Area 10^3 km^2	Cumulative discoveries and richness			
		10^9 t	10^3 t km^{-2} oil	10^{12} m^3 gas	10^6 m^3 km^{-2} t.o.e.
1 — Middle East	2,300	70	30	30	13
2 — W Siberia	3,200	10	3	22	7
3 — Gulf Coast	600	5.8	10	9	15
3 — Maracaibo	60	5.8	100	1.5	25
5 — Volga-Urals	500	5.6	11	6	12
6 — Reforma/Campeche	100	5	50	0.7	7
7 — Permian	300	4.2	14	2	7
8 — Sirte	200	4.2	21	0.9	5
9 — California	60	3.4	56	1	16
10 — Midcontinent	700	3.3	5	6	8
11 — North Sea	500	3	6	3	6
12 — Niger Delta	100	3	30	1.4	14
13 — West Canada	1,350	2.3	1.7	2.6	2
14 — Sumatra	210	2	10	—	—
15 — Orinoco/Trinidad	120	1.8	15	0.6	5
16 — North Caucasus	500	1.65	3	—	—
17 — South Caspian	150	1.65	11	0.4	—
18 — Tampico	100	1.5	15	0.25	2.5
19 — Triassic Sahara (Algeria-Tunisia)	330	1.5	4.5	3	1
20 — Rocky Moutains	1,500	1.5	1	1	0.7
21 — Arctic Slope	180	1.4	8	0.8	4.4
22 — Songliao	200	1.25	6	—	—

Fig. 165. — Map of the wo

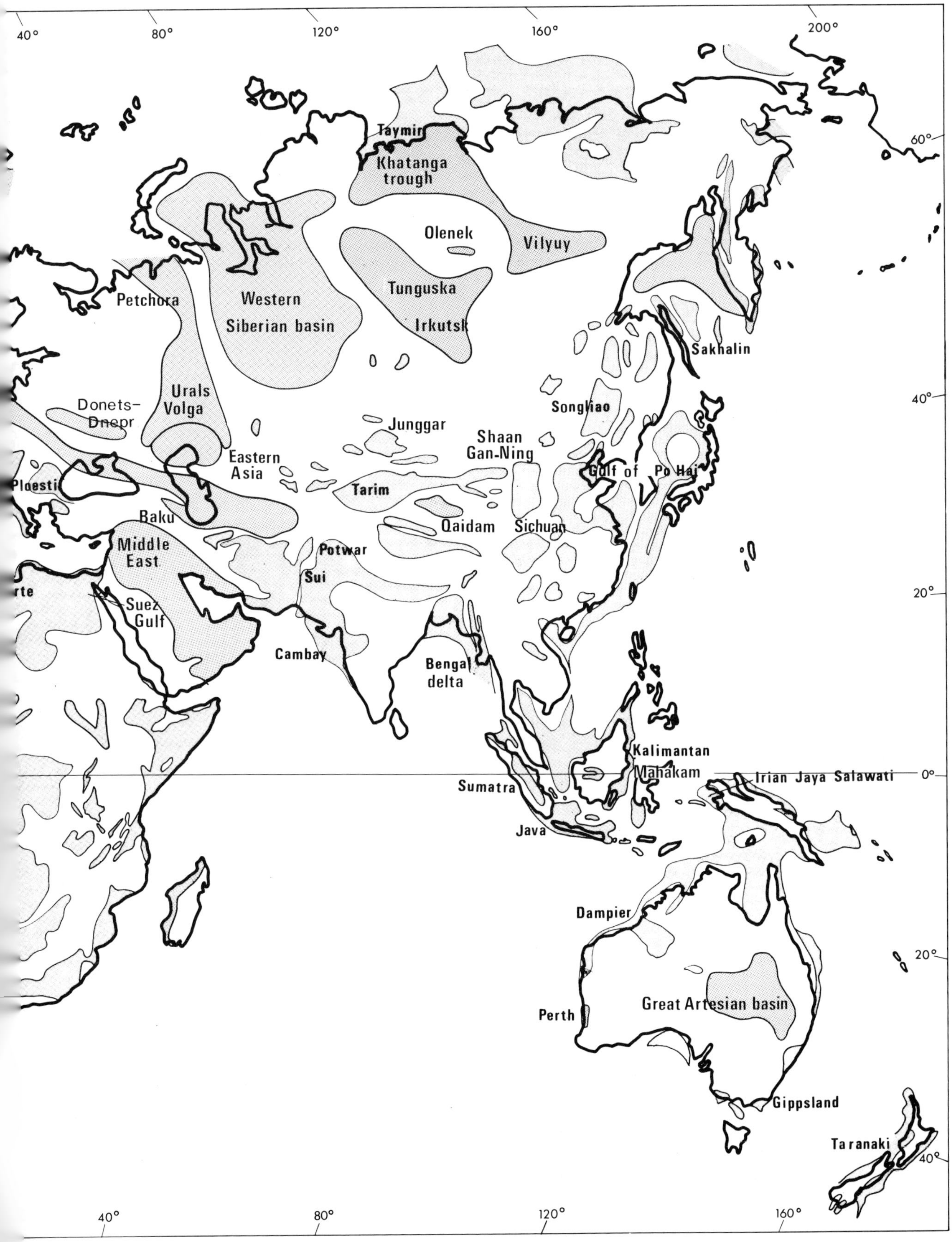

n sedimentary basins.

Structural traps, three-quarters of which are anticlines, contain 85 % of the giant fields and 95 % of oil and gas reserves (Klemme, 1975).

Global geotectonic sketch (Figure 165)

The oldest hydrocarbon accumulations are found in Eastern Siberia, in the Irkutsk and Tunguska Basins, in Upper Cambrian (Riphean, Vandean) and Cambrian series, aged between 925 and 500 million years. These productive formations are represented by alternating shales and carbonates with detrital, marine and non-metamorphosed horizons, overlaid by a Lower Cambrian series of carbonates and evaporites, which is also productive. The latter is covered slightly unconformably by largely continental Middle Cambrian and Ordovician deposits. These platform series, 500 to 2500 m thick, were only slightly or not at all tectonized.

This remarkable stability throughout the Phanerozoic, associated with the presence of anhydrite and halite seals, helped to preserve huge accumulations of wet gas and light, paraffinic crude oils. Some twenty pools have been identified to date, the two largest with reserves in the neighborhood of 300 and 170 Gm^3 (Kontorovich, 1979; Meyerhoff, 1981).

Paleozoic

The Paleozoic is characterized from the petroleum standpoint by vast platform basins distributed over the supercontinents of Laurasia and Gondwana. These basins are generally filled with shallow deposits, roughly distributed among the three major sedimentary cycles :

— Cambro-Silurian,
— Devonian,
— Carboniferous.

The subsidence and individuality of the basins tend to become gradually accentuated with time, at the approach of the Hercynian orogeny and, to begin with, the Breton or Acadian phase. Generally, the potential of the Paleozoic provinces thus increases fairly steadily with the increase in this tectonic instability.

This evolution is also observed on the North American continent with :

— the Cambrian/Mississippian Michigan, Illinois and Williston Basins, exhibiting initial reserves of 100, 500 and 650 Mt respectively, the third of which also exhibits a certain Mesozoic subsidence,
— the Devonian/Carboniferous West Canadian Basin, still remarkably stable, with about 1500 Mt reserves in the Paleozoic unit alone (in the United States, the prolongation of this Paleozoic and Mesozoic basin was deeply affected by the Laramide orogeny and subdivided into about ten tectonic basins),
— the complex Mid-Continent and West Texas Basins, located on the southern margin of the craton, characterized by many unconformities and capped by a thick Permian evaporitic series, with 3300 and 4400 Mt respectively, and the Arctic Slope province on the north margin of the same craton, relayed by a Mesozoic basin, the main hydrocarbon supply, with nearly 2 Gt today.

On the other hand, the Appalachian Basin, on the mobile eastern margin, only contains 0.5 Gt and 1.2 Tm^3 of gas, and the Devonian/Mesozoic Sverdrup Basin, which is highly subsident, has not yet demonstrated its potential.

In Eastern Europe, the Devonian/Permian Volga-Urals and Pechora Basins concentrate their richest zones, accounting for some 6 Gt of oil and substantially the same amount of gas, at the hinge of the Russian platform and the Urals fore-deep. In the Algerian Sahara, the Lower Paleozoic offers large oil and gas accumulations, about 2.3 Gt and 3.3 Tm^3, specifically associated with Hercynian and Mesozoic subsidence, but also with the presence of a Triassic evaporite seal.

Nearly 12 billion tons of oil and an equivalent amount of gas have already been found in the Paleozoic basins, distributed among sandstone and carbonate reservoirs. The Upper Devonian and Pennsylvanian represent the most favorable periods of this era. The largest provinces correspond to subsident basins, located in the neighborhood of prior troughs of the Hercynian orogeny, and to the presence of mainly Permian evaporitic cap rocks (Volga-Urals, Anglo-Dutch Basin, Mid-Continent, West Texas), though the latter may also be Triassic or Devonian.

In brief, the Paleozoic petroleum provinces appear to be characterized by great stability at the origin of a wide development of carbonates and, in certain regions, of reefs, by broad arches often sheltering giant fields suppiled by long lateral migrations. The provinces appear to be slightly 'fossilized', seeps are rare, and the hydrocarbons which are trapped therein, necessarily well protected to be there still today, are usually represented by light crude oils and gas. From this standpoint, while the Permian proves to be a mediocre oil generator, it nevertheless contributed strongly to the preservation of Paleozoic hydrocarbons.

Mesozoic

In the absence of major orogenic episodes, the Mesozoic appears to be characterized by the existence of vast platforms covered by the highest Phanerozoic onlaps, and by a progressive continental fragmentation which accompanied the birth of the great oceans, among which the Mesogean should be included.

The great Paleozoic platforms were partly covered by Cretaceous seas, after an initial attempt in the Jurassic. These high amplitude variations in sea level deeply mark the distribution of environments, and especially the combinations of source rock, reservoir and seal facies, as observed on the Arabian platform (Murris, 1980; Prestat & Riche, 1979). The vast platform basins, like those of Saudi Arabia, the internal Gulf Coast and West Canada, formed at this time, account for some 60 billion tons of oil and about 20 trillion m^3 of gas, distributed among 60 giant fields, to which must be added the 100 billion tons (in place) of heavy oil and tar in Northern Alberta.

Continental fragmentation is identified from the Triassic by the genesis of a number of rifts, generally offering interesting petroleum possibilities. Depending on the geotectonic context, these rifts evolve in accordance with three distinct scenarios to yield the following :

- aborted rifts which may pass progressively into basins with a broader extension, exhibiting obvious similarities with platform basins,
- rifts leading to passive margin basins,
- rifts subsequently transformed under the effect of subduction mechanisms into folded basins in an orogenic belt.

Aborted rifts, evolving through the widening of the basin with series largely overlapping on marginal platforms, account for at least four giant Mesozoic provinces, those of Western Siberia, the largest, Sirte, the North Sea and Songliao, which account for over 17 Gt and about 20 trillion m^3 of gas.

Mesozoic passive margin basins, well developed on the African and Brazilian coasts of the South Atlantic, on the west coast of India (Bombay High), and in Northwest Australia, without giant provinces, but with a number of productive zones in which some 1100 Mt of oil and 550 Gm^3 of gas have been discovered to date, the gas almost entirely concentrated in the Northwest Shelf.

Basins in mobile zones and fold belts, which generally extend into the Tertiary, are more specifically encountered in the Mesogean area, yielding a number of rich petroleum provinces, in Central Asia, the Middle East, the Gulf of Mexico and the Caribbean. It is necessary to stress the importance of the Mesogean and related basins, since this warm sea with East/West oceanic flows, hence devoid of polar currents, appears to be especially favorable to the development of source rocks. Most of the Mesozoic hydrocarbon accumulations are located in its immediate vicinity. A half dozen giant provinces are found there, accounting for nearly 30 billion tons, slightly over half in the Middle East, where they migrated vertically into Oligo-Miocene limestones, and about 20 trillion m^3 of gas. These reserves are mainly distributed among some 50 giant oil fields and 20 giant gas fields.

On the whole, at the current discovery level, the Mesozoic has yielded about 120 Gt of oil and 60 Tm^3 of gas in about 15 giant provinces. In this era, which lasted 165 million years, the Cretaceous, a period of maximum transgression and minimum oceanic flow, occupies an exceptional position, both for oil and for gas. On the whole, the Mesozoic provinces, divided among continental areas and orogenic belts, which began to be active in the Late Cretaceous, form the transition between the quiet and well sheltered basins of the Paleozoic and the mobile troughs of the Cenozoic.

Cenozoic

The Cenozoic basins, which usually succeed the subsident Cretaceous areas, are generally aligned with the orogenic belts of the Mesogean. Some of them, more specifically Tertiary, and often even Neogene, are aligned with the Pacific margins. These Cenozoic basins generally correspond to mobile, narrow and deep troughs, with a relatively limited area and duration, exhibiting sedimentary series, mainly sand/shale, frequently over 5000 m thick. Belonging mainly to the Alpine system, they are distributed among different families, depending on the predominance of tensile, shear or compressive forces.

In a tensile area, *rifts* rarely appear to be the pure products of the Tertiary period, such as the grabens of the Gulf of Suez and Cambay, but more often as the resumption of Mesozoic grabens. This is the case of the Viking and Central Grabens of the North Sea, and part of the Chinese basins (Qaidam, Tarim, Shaan-Gan-Ming, Junggar), noteworthy for the productivity of their non-marine source rocks. The most remarkable case of rifts in a passive margin basin situation is their conjunction with thick deltaic units, to yield rich deltaic provinces, the Niger, the Gulf Coast, the Mahakam and Mackenzie Deltas. The two former, the best explored, actually concentrate over 12 billion tons of oil and more than 3 billion m^3 of gas, in a typically dispersed habitat.

On the *Californian margin* of the Pacific, active and prolific Neogene basins were formed in the past 20 million years, mainly generated by shear movements. The major transcurrent faults of this Pacific margin induce high subsidence and continuous tectonic deformations, which contribute significantly to the potential of these provinces, where more than 3 Gt of oil and 1 Tm^3 of gas have been discovered to date. The remaining basins of these American margins have not yet yielded significant discoveries. Some of these failures may be related to the non-sialic nature of these basins, unlike the Californian area (North, 1971).

Tertiary basins belonging to *compressive zones* are the most numerous and most important from the petroleum standpoint. The following may be distinguished in this family, according to their geotectonic framework :

— back arc basins, like the provinces of Java, Sumatra, Sarawak/Kalimantan, which contain some 3 Gt of oil and over 1000 Gm^3 of gas,

— fore-deep basins, at least in one period of their development, in Venezuela, North Caucasus, the Carpathian Trough, and the Apennine zone, which contain about 10 Gt of oil and 2000 Gm^3 of gas, part of these hydrocarbons being of Mesozoic origin.

These Cenozoic basins are often affected by volcanic events, frequently beneficial for their source rocks, as well as vertical and horizontal tectonic movements, giving rise to upward migrations and seepage. Even if the often intense subsidence of these troughs failed to ensure the maturation of its own source rocks, it often contributed very actively to the genesis of hydrocarbons from older formations. The stable margin of some of these basins may cover large areas, facilitating long lateral migrations which may lead to the formation of large pools of heavy oils, like those of the Orinoco. At all events, these Cenozoic basins, normally in the course of formation and deformation, appear to be active and, in a way, living geological units.

*
**

To *summarize,* it may be considered that the *Paleozoic* basins, and exceptionally certain Proterozoic basins, constitute rich oil and gas provinces if they have not been affected by major tectonic events, and if they possess effective saliferous seals, which is often the case in the Permian, a period of great aridity.

The *Mesozoic* basins, generally more unstable than the foregoing, contain a majority of the giant provinces. The latter are more specifically associated with the formation of initial rifts, with the development

Table III

The leading oil-producing countries

	Production 10^6 t			Cumulative production (10 t)	Reserves
	1960	1970	1982	on 1 January 1983	on 1 Jan. 82
USSR	145.0	348.0	612	11.0	10.0
Saudi Arabia	62.0	175.0	325	7.0	26.0
USA	267.0	475.0	480	18.5	4.2
Mexico	13.0	21.0	148	1.2	7.0
Venezuela	140.0	183.0	100	5.2	3.0
China	1.7	20.0	102	1.2	2.7
UK	1.7	1.7	102	0.3	
Indonesia	20.0	42.0	65	1.3	1.3
Canada	26.0	62.0	73	1.4	1.0
Nigeria	1.0	53.0	64	1.3	2.5
Iran	53.0	164.0	98	4.3	8.0
Kuwait	80.0	135.0	42	3.1	9.0
Abu Dhabi	—	34.0	42	0.9	5.0
Libya	—	164.0	55	1.9	3.0
Iraq	50.0	77.0	48	2.3	5.0
Algeria	9.0	50.0	32	0.8	1.1
Egypt	3.0	16.0	34	0.3	0.5
Argentina	8.5	19.0	25	0.6	0.4
Norway	—	—	24	0.1	1.0
Qatar	8.0	18.0	16	0.4	0.5
WORLD TOTAL	630.0	2215.0	2755	~ 65.0	~ 100.0

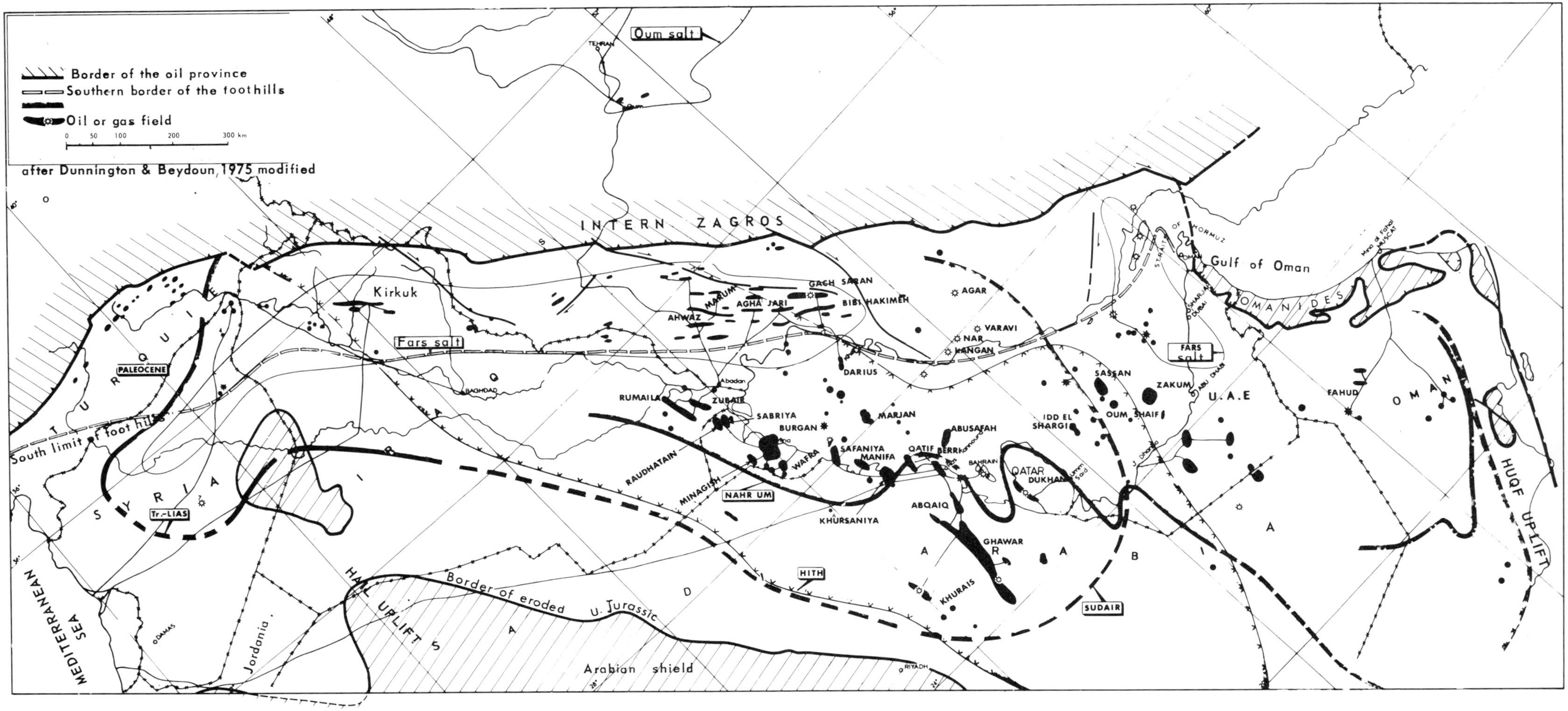

Fig. 166. — Oil map of the Middle East.

of the Mesogean, an ocean spreading in an East/West direction and sheltered from the cold oceanic currents. This era, and particularly the Upper Cretaceous, a period of exceptional, generalized transgression, thus appears to be especially favorable to the deposition of source rocks and reservoirs. The active subsident zones, which develop more specifically in relation to the preludes of the Alpine orogeny, often contribute effectively to the genesis of Mesozoic hydrocarbons and to their migration.

The *Cenozoic,* a period of basin formation 'par excellence', offers a number of provinces relatively small in area, but displaying all the perequisites for the genesis of pools, and the provinces of this age are among the richest, including those associated with oil shales (Halbouty, 1970; North, 1971 and 1972; Meyerhoff, 1975; Perrodon, 1978 a; Bois *et al.,* 1980; Bally & Snelson, 1980; Huff, 1980; Klemme, 1980).

Table III provides a list of the 20 main oil producing countries, based on their 1980 production; the chief provinces of Western Europe (apart from the North Sea) are discussed briefly at the end of the Section.

3.1. THE PROVINCES OF THE MIDDLE EAST

The Middle East corresponds to a vast and complex dissymmetric sedimentary basin covering 2.3 Mkm^2, including 600,000 km^2 of productive zones. From the geodynamic standpoint, this basin is located at the meeting point of the African plate which corresponds to the northeast advance of the African continent, and the Asiatic plate. The convergent zone, located on the old Mesogean oceanic area, is characterized today by thick, folded and faulted marine series. The Middle East province is thus bounded to the west by the great Levant flexure, to the southwest by the rise of the Arabian/Nubian shield, to the northeast by the Taurus/ Zagros overthrust/shear.

The following may be distinguished from the structural standpoint :

— a relatively stable platform area to the southwest corresponding to the countries of this margin of the Gulf, and characterized by a few major structural axes with very slight dip, towards the south,

— a compressive area to the northeast, marked by a trend of large pronounced folds, recently formed and striking NW/SE.

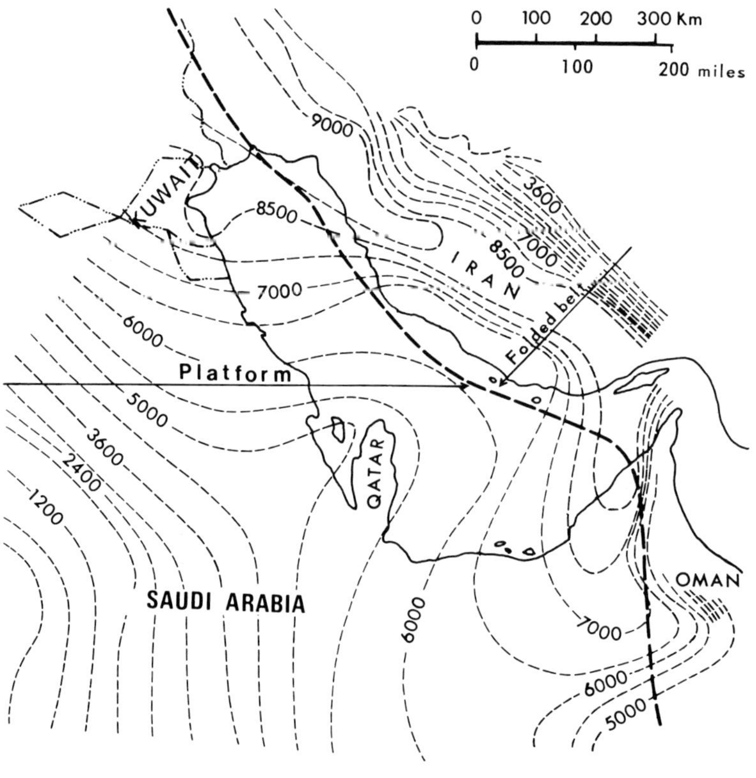

Fig. 167. — Isopach map of sedimentary series of the Middle East Basin (Kamen Kaye, 1970).

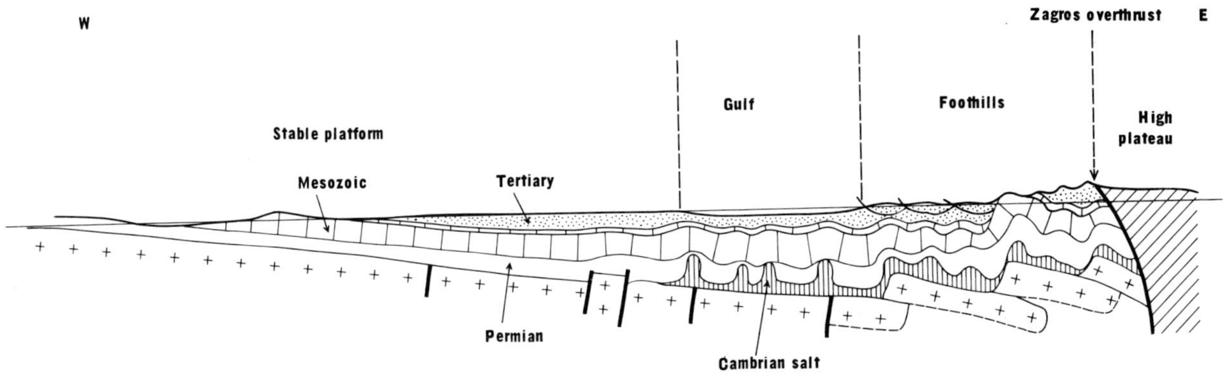

Fig. 168. — Section across the Middle East Basin.

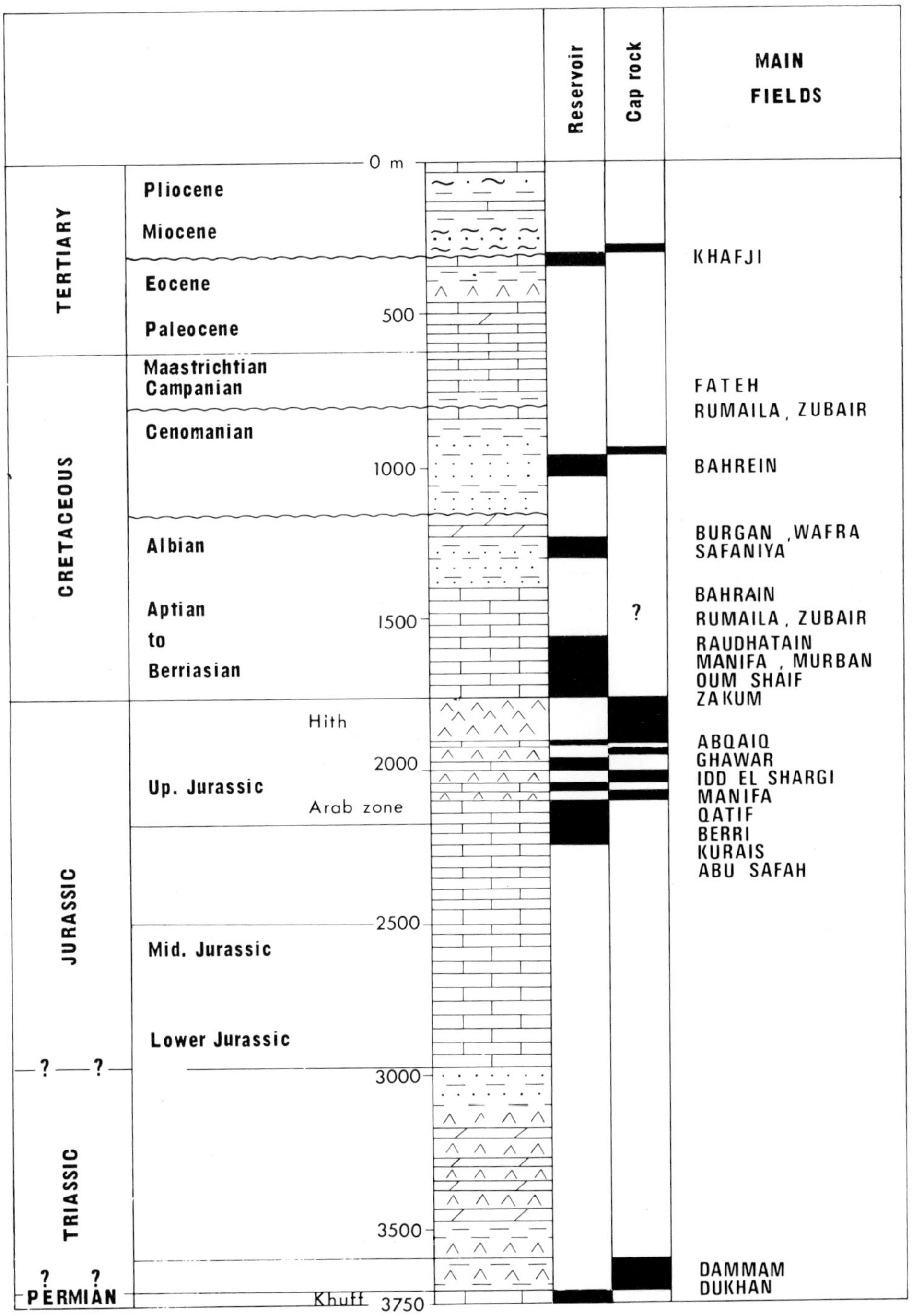

Fig. 169. — Typical log of stratigraphic series of the Arabian platform and situation of the main fields.

These different structures are often caused or complicated by halocinetic deformations, some of them deep and giving rise to simple domes, especially on the platform, and the others recent and developing in regional anomalies, such as the foothills of Iran and Iraq. Between the present coasts of the Mediterranean in the west and the Indian Ocean in the east, the shield is prolonged by the three regional highs of Hail in the north, Qatar in the center, and Hadramaut in the southeast, which delimit the main subsident areas of Mesopotamia and Rub al Khali (Figures 166 and 167). The richest area stretches between the highs of Hail in the nortwest and Hadramaut in the southeast. Areas located to the west of the former, in Syria and Turkey, are not dealt with here.

This structural pattern conditions the hydrodynamic framework, characterized by gravity drainage from the Zagros mountains and the Arabian platform converging towards the north central area of the Gulf, and creating vertical gradients that facilitate upward migrations in this region. A third pressure zone, caused by the undercompaction of the Upper Cretaceous shales of the Oman fore-deep, completes this pattern to the east. The geothermal gradients are relatively low in the subsident zones, higher on the margins and along the structural axes.

From the stratigraphic standpoint, the sedimentary series is more than 10,000 m thick, including some 3000 m of Cenozoic in the folded area on the northeast shore of the Gulf, and generally thins steadily and sharply in the direction of the shield and the Qatar arch (Figures 168 and 169). The stratigraphic series extends from the pre-Cambrian to the Present. It is substantially continuous from the Permian over a thickness of more than 7000 m, including about 3000 m in the Tertiary, reflecting the Alpine character of this province.

The Paleozoic displays a discontinuous series including the infra-Cambrian Hormuz saliferous formations, the source of major halocinetic movements, and Silurian organic shales, as well as a locally well developed marine Carboniferous. The Permian marks the start of a thick, relatively continuous series up to the Quaternary. It is represented by an essentially calcareous and evaporitic succession, with some sandstones and shales at the base. The Mesozoic consists of a sequence of detrital deposits, shales, carbonates and evaporites. The platform series are thinner, intersected by gaps and neritic facies, with broad detrital spreads on the platform. They appear to be thicker, more continuous and with a deeper facies in the subsident zone corresponding to the Mesogean trough.

However, the geological history of the productive province of southwestern Iran shows a certain diversity of detail which sometimes obscures the simplicity of the first order stratigraphic and structural features. In this series, 10 to 12 km thick, which initially appears to be continuous, slight unconformities are identified in addition to frequent variations in facies, emphasizing the number of sedimentary cycles. These pulsations locally mark the Permian/Triassic/Jurassic, Middle and Upper Jurassic, Jurassic/Cretaceous, Neocomian/Aptian, Aptian/Albian, Cenomanian/Turonian, Turonian/Senonian, and Cretaceous/Eocene boundaries. The isopachs of the different series delineate a number of mobile areas and more stable platforms, in which the depositional environments stretch from the shelf to the deep radiolarian troughs, especially in the Lower Cretaceous (Setudehnia, 1978). The Cenozoic series terminates in thick fill formations up to several thousand meters thick locally in synclinal areas. Sedimentation still continues in the Gulf.

Petroleum criteria

Some 70 Gt of oil have been discovered in the Middle East (recoverable reserves plus cumulative production), of which slightly more than 15 Gt have been produced, together with about 30 million m^3 of gas. The reserves in place have been estimated at 56 Gt in the Asmari limestones alone (Kent & Warman, 1972). These oil reserves, which correspond substantially to half of the world's cumulative discoveries, represent an average potential of 30,000 t km^{-2} or, more figuratively, an oil column of 3 cm over the entire region, more than 100,000 t km^{-2} considering only the proven zone. These reserves are distributed among some 60 giant oil fields, or one-third of all the world's giants, and in about ten gas giants. Despite this potential, large quantities of hydrocarbons have been lost and continue to be lost by vertical migration, yielding the vast asphalt seeps and gas emissions well known in Mesopotamia and Iraq.

It is customary to distinguish between two major petroleum provinces corresponding to the geological areas sketched out above :

— a platform province, or Arabian province, characterized by vast arches, generally aligned in a N/S direction, productive in the Jurassic and Lower Cretaceous,

— a folded province, the foothills of Iraq and Iran, productive in the Cretaceous and Miocene/Oligocene reservoirs fractured at the heart of 'Appalachian' anticlines, buried under a saliferous cap rock, the cause of deep disharmonies.

Considering the data available today, and due to the major role played by the seals in this basin, it is thus possible to define a number of petroleum systems characterized by the main producing horizons :

— the Khuff (Permian), producing gas, from Iran to Qatar, and even in Saudi Arabia and very recently Abu Dhabi (Umm Shaif),

— the 'Arab Zone' (Jurassic), producing oil on the platform,
— a more complex Cretaceous unit, sandstone or carbonate,
— the Asmari (Oligocene/Miocene) in the folded zone.

Broad overall homogeneity exists among the crude oils of this province, apart from the results of surface weathering (Law, 1957; Greig, 1958; Dunnington, 1967; Kamen-Kaye, 1970; Murris, 1980; Ayres *et al.*, 1982).

3.1.1. **The Khuff system**

The Khuff system (Permian) is characterized by a close association of limestone and dolomitic reservoirs, locally bioclastic and coralline (the Khuff), and dolomitic and anhydritic seals under high internal pressure (the Sudair Formation). The reservoir exhibits the following usual characteristics :

— total thickness of the Khuff : about 1000 m in the foothills of the Kangan region,
— highly variable porosity, up to 20 to 25 % in some horizons,
— fracture permeability.

The traps in the folded Iranian belt consist of magnificent elongated anticlines 8 to 12 km wide, 1.5 to 3 km high, perfectly sculpted at the surface by the Asmari limestones. Their height and closed area amount to over 1000 m and several hundred square kilometers respectively (Plate 3). However, these closures are affected by recent transverse strike slip faults that alter the seals. In the platform area, the traps are formed by broad domes.

The hydrocarbons are represented chiefly by wet gas (50 g natural gasoline per m^3). H_2S is present in very small quantities in the Kangan zone, but in larger amounts, and accompanied by CO_2 and N_2, on the Qatar Arch. Oil has been found in the sandstone reservoirs (Hauschi) at the base of the Permian, in Oman, with the Ghala North, Rarn Alam and Saih Nihayda pools. The main fields discovered to date but not yet produced are, in Iran :

— Pars, 1800 Gm^3,
— Kangan, 600 Gm^3 (Figure 74),
— Nar, 280 Gm^3,
— Agar,
— Varavi,

and the deep North West Dome Field offshore Qatar, whose reserves may be as high as 5000 Gm^3 of gas and 170 Mt of condensate and probably more (but with a high proportion of N_2 and H_2S).

3.1.2. **The Arab Zone system**

This Upper Jurassic system mainly consists of a repetition of sedimentary cycles beginning with high energy carbonate facies, including calcarenites (the four reservoirs of the Arab Zone) and terminating in anhydrite deposits, the entire unit sealed by the slightly overlapping anhydrite formations of the Hith, which covers most of the Arabian platform in the Late Jurassic. These reservoirs are supplied by the limestones and black shales of the Callovian/Oxfordian (Tuwaiq Formation) deposited in a transgressive period in a slightly subsident area, and brought to optimal maturation over an area of more than 100,000 km^2.

This unit is productive within vast domes and elongated arches with very slight dip (about one degree) and vast areas. The closed columns, often slightly more pronounced in the deeper levels, are up to 300 to 400 m. Many of these structures were created by the rise of deep salt domes, probably Cambrian. The reservoir characteristics, usually corresponding to vuggy dolomites, shelly limestones and calcarenites deposited in broad bands, are excellent, and the well outputs are outstanding, up to 200 t day^{-1} (14,000 bopd) at Ghawar, 2500 t day^{-1} (18,000 bopd) at Berri. The high sulfur crude oils display average densities of 31 to 38 °API).

The Arab Zone system has ten of the world's fifty biggest fields, including the No. 1, Ghawar, and contains nearly 25 Gt of total recoverable reserves. The Ghawar Field, discovered in 1948, has the following main characteristics :

— elongated anticlinorium 225 km long, 2275 km^2 closed area, with an oil column of 430 m,
— main reservoir formed by Arab Zone 'D', 80 m thick, at 2100 m depth,
— annual production about 265 Mt with 385 wells,
— cumulative production 3 Gt,
— initial reserves 11 Gt.

The Abqaiq Field (1940) is formed by a regular anticline with 350 km^2 productive area and a closure of 300 m, initially containing 1.75 Gt of oil. Manifa (1957) offshore, corresponds to an anticline which is productive in the three zones A, B and CD, containing 1.5 Gt. Qatif (1945), productive in zone D, has reserves of 1.2 Gt.

Among the remaining fields of Saudi Arabia are the following :
— Berri (1964), 1700 Mt,
— Khurais (1957), 1200 Mt,
— Abu Safah (1963), 900 Mt,
— Khursaniya (1956), 560 Mt,
— Marjan (Fereidon) (1967), 1400 Mt,

as well as Umm Shaif (1958) in Abu Dhabi, 700 Mt (reserves from St. John, 1980), and in Qatar, Idd el Shargi (1960), 300 Mt and Dukhan (1940), 400 Mt. This field, which corresponds to one of the outcropping anticlines of this province, stretching over 50 × 4 km, with average dips of 4 degrees, is productive in horizons C and D of the Arab Zone, respectively 25 and 56 m thick, separated only by 20 m of anhydrite. The upper horizon, a shelly limestone, contains a lighter saturated crude (Thralls & Hasson, 1956, and Steineke *et al.*, 1958).

3.1.3. **The Cretaceous system**

Pools belonging to this unit are divided between the platform and the folded zone, and are generally situated in Middle Cretaceous formations with platform facies, sands or carbonates. The oil originates in Jurassic source rocks and, in the northern part, where the Hith no longer provides an effective seal, in limestone or bituminous shale formations rich in sapropelic organic matter, corresponding to euxinic platform or deeper facies (Sulaiy (Berriasian/Valenginian) and Khazdumi (Albian)). The seals are mainly provided by overlapping Cenomanian and Senonian shales. The reservoirs are either detrital and formed by the Burgan (Albian) or Zubair (Barremian) sandstones in the region of Kuwait and the neighboring zones of southern Iraq and Arabia, or by carbonate deposits, often associated with reef environments containing Rudists (e.g. Mishrif). The traps are formed by gentle domes in a platform area, and by more pronounced anticlines in Iran.

Sandstone reservoirs

The reservoirs consists of broad continental facies sands and sandstones advancing westward on the platform. They display a column of 200 to 300 m with excellent characteristics. This sandstone unit contains about 15Gt of oil, and six giant fields among the 35 world's biggest belong to this family. Burgan, the largest, and the second ranking in all categories, discovered in 1938 in Kuwait, has reserves estimated at 10 Gt. It is a vast faulted anticline, ovoid and flattened, with an area of 90 km^2 and over 100 m closed height, with dips of 2 to 3 degrees, slightly sharper at greater depth, indicating the initiation of uplift in the Cenomanian. The reservoir is formed by the so-called Burgan sands, with a continental tendency, Albian/Lower Cenomanian, sealed by shales and black, brown and red marls which are only 30 m thick on the structure (Figure 170). The characteristics of these massive sands, formed of well rounded fine to medium grains, are rather exceptional :
— permeability up to 4 to 5 d,
— productive column 400 m, yielding unit outputs between 700 and 1200 t day^{-1}.

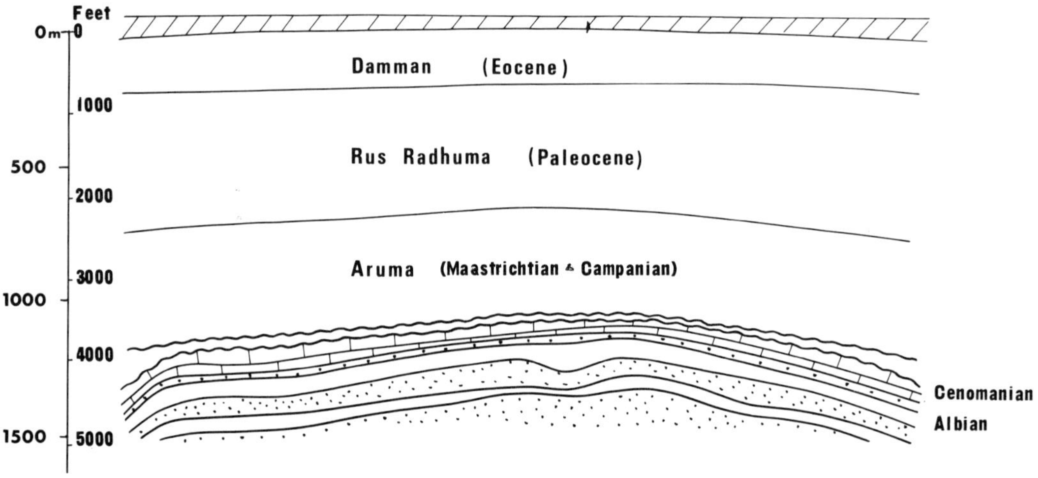

Fig. 170. — Section of Burgan Field (Kuwait) (see map Figure 166).

Large heavy oil reserves have been identified in the Eocene limestones, as well as asphalt seeps at the surface.

The No. 2 field in this family, Safaniya, discovered in 1953 on the coast of the Gulf, has a productive area of 500 km^2, and is the world's fourth largest field, the second of Saudi Arabia, and the world's biggest offshore pool. The oil column is 290 m and the sand permeability is 6 d. Initial reserves are estimated at 4.2 Gt of a crude oil with density 27 °API. Production reached 70 Mt in 1977 with 150 wells, or an average output of 1300 t day^{-1} per well, with some producing up to 3000 t day^{-1}.

The third giant of this zone is Rumaila in Iraq, discovered in 1953, also productive in the Middle Cretaceous limestones, displaying porosities of 25 % and permeabilities of 1 d at 3200 m. Its reserves are estimated at 2.8 Gt. Also worth mentioning are :

— Raudhatain (1955) in Kuwait, 1000 Mt,
— Rumaila North (1961) in Iraq, 700 Mt,
— Wafra (1953) in the Neutral Zone, 700 Mt, only partly in the Middle Cretaceous sands,
— Sabriya (1958) in Kuwait, 560 mt, also partly in the Middle Cretaceous sands,
— Minagish (1959) in Kuwait, 280 Mt, productive in a Lower Cretaceous oolith,
— Zubair (1948) in Iraq, 630 Mt,
— Nahr Umr in Iraq, 140 Mt,

(Iraq Petroleum Company, 1956; Thralls & Hasson, 1956; St. John, 1980).

Carbonate reservoirs

These reservoirs belong to different carbonate formations, often associated with reef environments, and especially bioclastic and rudistid limestones, stretching from the Neocomian (Thamama Formation) to the Eocene (Damman), with special mention of the Bangestan Group (Albian/Senonian) whose basin facies is reflected by a thick and relatively compact Oligostegine limestone, productive in certain fields of the folded

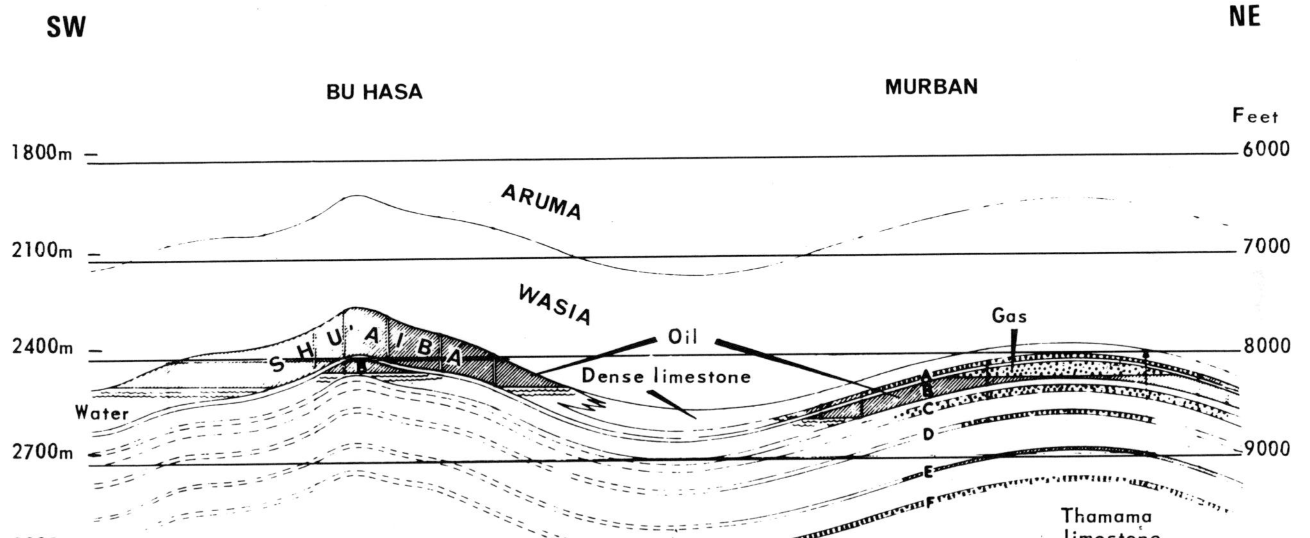

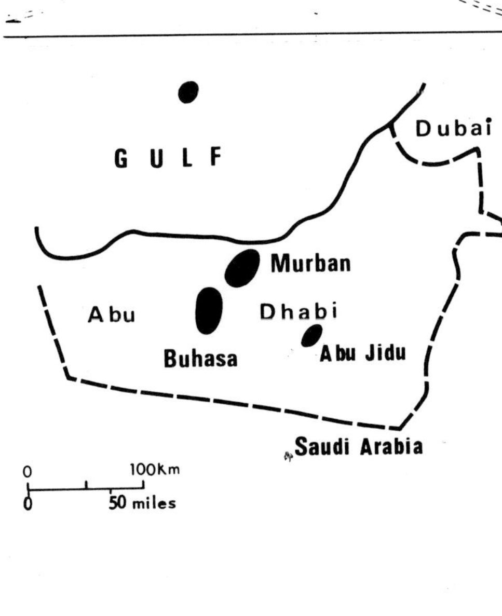

Fig. 171. — Section of Bu Hasa and Murban Fields (Abu Dhabi) (Hajash, 1967).

zone. On the basin margins, the complex and variable paleogeography of the synchronous reef environments offset, in time and place, the reservoir facies and source rocks from one field or zone to the next.

The Cretaceous carbonate formations contain a total of about 15 Gt of oil, including 5 Gt for the Thamama Formation. The productive fields in the Cretaceous carbonate reservoirs appear to be second order giants (for the Middle East). The largest consists of the twin Murban/Bu Hasa structure (1959/1962), in the Emirate of Abu Dhabi, with reserves of about 1200 Mt. It is productive in the Neocomian and Aptian Thamama and Shuaiba limestones (Figure 171). The Murban trap is a very flattened dome with a dip of 1 to 2 degrees, and a 240 m closure. The reservoir is represented by limestones with 27 % porosity and low permeability (5 to 250 md), and the gas cap consists of a high sulfur gas. On the neighboring Bu Hasa dome, substantially higher, the Thamama exhibits a reef facies, with better permeabilities. The crude oil (density 38 to 40 °API) is undersaturated (Hajash, 1967).

These fields are followed by :

— Zakum (1964) in Abu Dhabi, mainly productive in the Thamama Formation, with estimated reserves of about 1.6 Gt, including heavy crude oil,
— Sassan (1966), 200 Mt, Darius (1961), 200 Mt, Aga Jari (1938), 1.9 Gt, Marun (1963), 2.2 Gt, deep Masjid i Sulaiman (1908), 180 Mt, all in Iran,
— Fahud (1964) in Oman, 160 Mt, productive in an Albian chalky limestone (see Section 1.2 and Section 4 of Part 2),
— Rumaila, 2.8 Gt, and Buzurgan, 150 Mt, in Iraq,
— Awali (1932) in Bahrain, 140 Mt.

The Awali Field, the first discovered on this shore of the Gulf, thanks to the presence of a seep from an outcropping anticline, is productive at 600 m depth in a limestone horizon lying on the water bearing Wasia sands on this structure. The Arab Zone contains gas (Mina *et al.*, 1967; Riché & Prestat, 1979).

3.1.4. **The Asmari system**

This unit, which corresponds to the foothills province of Iran and northern Iraq, is characterized essentially by the following :

— from the structural standpoint, by 'Appalachian' anticlines, some of them probably overturned, very large, exhibiting depth differences up to 5000 m, and overlaid by disharmonic series,
— from the reservoir standpoint, by fractured Oligocene/Miocene limestones (Asmari) sealed by the evaporitic series of the Fars 1000 to 2000 m thick, Upper Miocene (Figure 172).

This nummulitic and locally dolomitized limestone reservoir, often with a micritic facies, generally exhibits low matrix porosity of 5 to 10 %, but high fracture permeabilities, yielding average outputs of 2000 to 3000 t day^{-1} over impregnated columns of several hundred meters. This reservoir merges into sands further northward (Ahwaz Formation).

The hydrocarbons of this system essentially originate in Lower Cretaceous (Albian) source rocks (Kazhdumi Formation), and to a lesser extent in the Pabdeh (Eocene) and Gurpi (Upper Cretaceous) Formations by vertical migrations during the formation of the structures, from the first accumulation in the Bangestan limestones. Both reservoirs have often remained in communication and have a single water contact (Ala *et al.*, 1980; Murris, 1980). The crude is generally an asphaltic high sulfur crude oil (at least 1 to 3.5 %) with density 30 to 38 °API. Its composition is homogeneous within any given pool. The GOR's are often high, and large gas Caps may occur. For instance, the gas reserves of Pazanun, considered for a long time as a gas field, are estimated at 1400 Gm^3. The traps, which are generally full, display a differential trapping scheme, with the gas more frequently found in the deeper structures. The crude oil becomes heavier in the shallower horizons.

This province contains some 15 Gt of oil and nine fields among the world's fifty biggest, including eight in Iran (Figure 173). Kirkuk, the oldest field, discovered in 1927 in Iraq, had original reserves of 2200 Mt, of which more than one billion tons have been extracted. It is formed by an anticline measuring 100 × 3 km, with a closed area of 420 km^2, productive in an altered limestone 300 m thick. Five fields have reserves exceeding 1 Gt :

— Gash Saran (1937), 2200 Mt, annual production (1977) 50 Mt (Figure 75),
— Marun (1963), 2200 Mt, annual production (1977) 67 Mt,
— Agha Jari (1936), 1900 Mt, annual production (1977) 35 Mt,
— Ahwaz (1958), 2400 Mt, annual production (1977) 50 Mt,
— Bibi Hakimeh (1961), 530 Mt, annual production (1977) 12 Mt.

Some of these fields have exceptional characteristics :

— 500 m reservoir thickness and 2200 m oil column at Gash Saran,
— outputs of 5750 t day^{-1} with only 7 m penetration into the reservoir at Agha Jari,

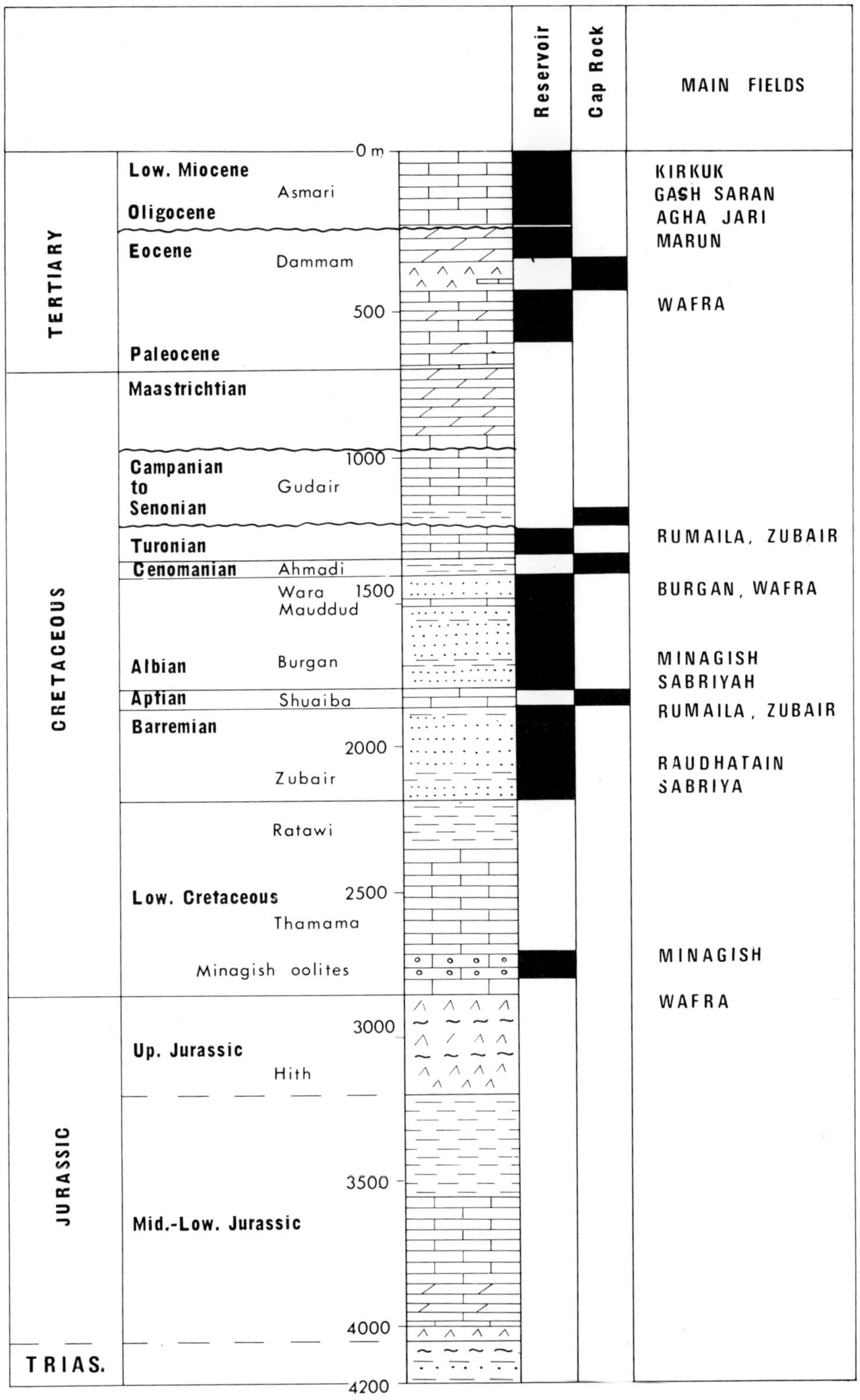

Fig. 172. — Typical log of stratigraphic series in the foothills zone and southern Iraq/Kuwait, and situation of main fields.

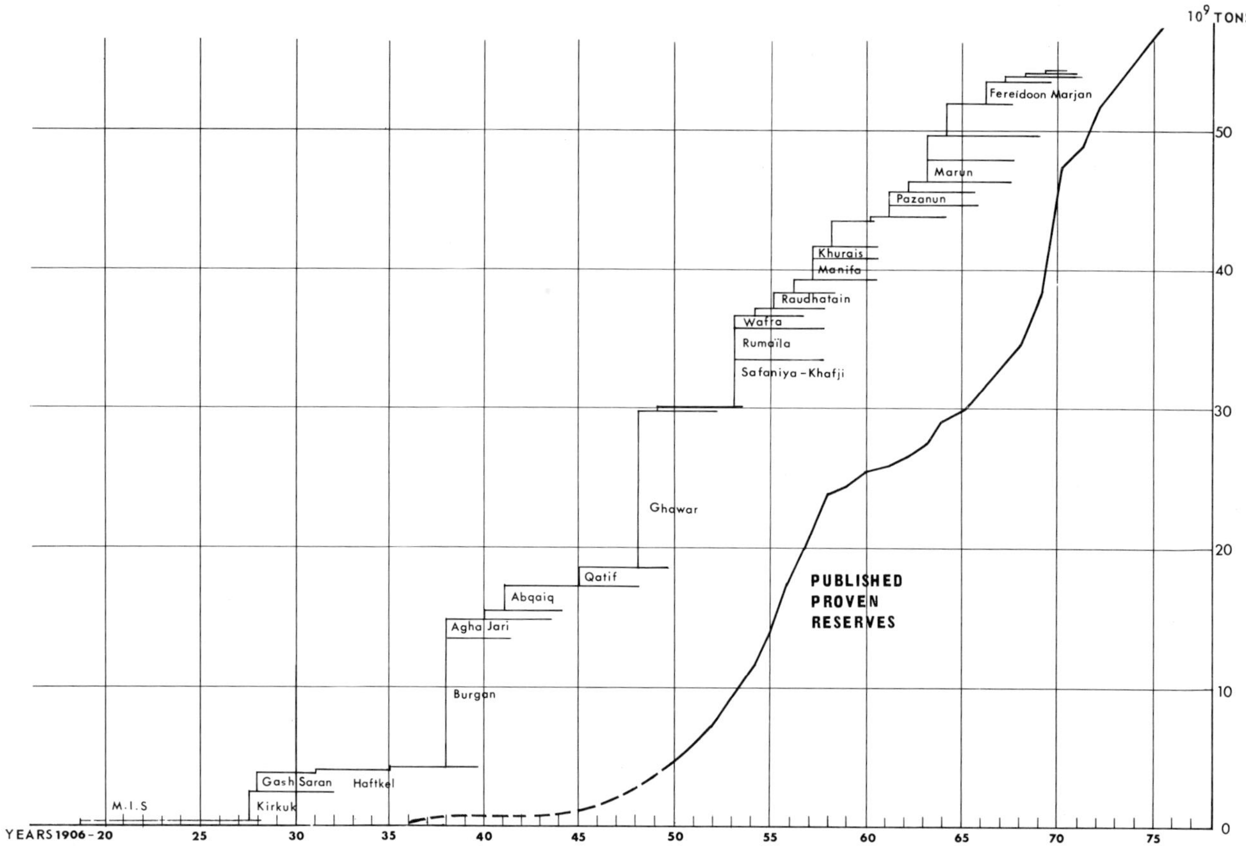

Fig. 173. — Graph showing the curve of oil discoveries in the Middle East since 1904 (Beydoun & Dunnington, 1975).

— 600 m gas column at Pazanun,

— significant production (4500 m^3 day^{-1}) from the eighth well only and after 49 years of various attempts at Ahwaz,

(British Petroleum Company, 1956; Dunnington, 1968; Falcon, 1968; Hull & Warman, 1970; Colman Saad, 1978; St. John, 1980).

Conclusions

Many geologists have wondered about the reasons for the petroleum potential of the Middle East. A certain number of causes are evident :

— large dimensions and filling percentages of the traps, all structural, whether formed by broad domes, or by long 'Appalachian' anticlines, both existing in large numbers,

— reservoir quality and variety stretching from the Permian to the Miocene, characterized by excellent porosities and permeabilities in a platform area, significantly improved by fractures in the folded zone, with porosities and permeabilities preserved thanks to early accumulation of the oil,

— effective and continuous seals, often evaporitic, which grow thicker as the risks of spillage are greater, despite considerable seepage,

— productivity of the source rocks, probably partly carbonate, close to the reservoirs and exhibiting excellent expulsion.

This testifies to outstanding qualities for the fundamental prerequisities for the formation of pools. Noteworthy is the repetition of these source rock/reservoir/seal units in the stratigraphic series, and their extension over large areas. Some authors have emphasized certain points concerning the relationships between these different criteria. Clark (1975) insisted on the probably exceptional yield of carbonate source rocks.

Also worth emphasizing is the extent of subsidence and Cenozoic sedimentation, which conveyed the source rocks to satisfactory maturation depths, and also the lateness of the deformations, which yielded high anticlinal structures still relatively unaffected by erosion. The geological history is also relatively simple, devoid of major orogenic phases before the Late Tertiary, and there is an absence of deep erosion. No major regional unconformities are observed, but nevertheless a significant number of low amplitude unconformities and non-deposition surfaces do occur, not only on the platform area, but also in the folded zones.

3.2. PROVINCES OF THE USSR

The Soviet Union exhibits a remarkable potential of large and rich sedimentary basins, substantially representing a third of the world's sedimentary area, and covering the entire sedimentary series. The Volga-Urals Basin contains most of the oil in Devonian and Carboniferous strata on the festooned margin of the Russian platform. However, huge Permian gas deposits are also known. The great basin of West Siberia contains a series of fields, and a large proportion of gas giants, in Jurassic/Cretaceous series. The small Caucasian basins, and especially those of Baku and Western Turkmenia, concentrate (or concentrated) their rich oil and gas reserves in narrow and subsident troughs. Added to these three giant provinces are those of Central Asia, especially rich in gas, the Dnepr/Donets Depression, provinces of the Vilyuy and Lena/Tunguska in East Siberia, and Sakhalin along the Pacific coastal zone.

3.2.1. The Volga-Urals Basin

The Volga-Urals Basin, 'the second Baku', extends over an area of about 500,000 km^2 between the Urals, the Russian platform and the pre-Caspian depression to the south. It corresponds to a Paleozoic platform type basin at the west, becoming a subsident trough in front of the Urals Range. It is prolonged northward by the Timan/Pechora province, and to the south by the North Caspian (Pericaspian) Basin (Figures 174 and 175).

Geological criteria

The sedimentary series begins from the east with the Givetian overlap on a folded and eroded basement. The detrital deposits of this stage and of the Lower Frasnian are overlaid by carbonate formations, passing into lagoonal and continental facies toward the west. Following a period of deformation, the Carboniferous marks a new sedimentary cycle, beginning with detrital deposits and continuing with carbonate facies. Finally, the Permian, well developed in the pre-Urals Trough, consists of two units of Artinskian limestones, commonly of the reef type, overlaid by the evaporitic Kungurian series. The entire sedimentary series thickens markedly from west to east, passing from 2000 to 3000 m on the margin of the Russian Platform to more than 6000 m at the east.

From a structural standpoint, this general flexure zone is characterized by alternating upwarps, promontories and depressions of large diameter. These large units are in turn affected by faults and more limited, low dip deformations. The geothermal gradients are generally normal, higher in the flexure zone and the productive domes.

Petroleum criteria

The reservoirs are generally sandstone in the Middle Devonian and the Visean, limestone in the Carboniferous and Artinskian, where reef developments are considerable. The Devonian sandstones, which form the reservoir of the biggest fields in the province, are generally fairly thin, around 10 m thick or more. They are generally interpreted as deltaic and littoral deposits, and they display good reservoir characteristics, porosity about 20 %, permeability 400 to 500 md, yielding natural flows of about 100 t/day and pumping 10 to 60 t/day. Most of the old wells (90 % of the wells) pump 22 t/day. The Permian limestones exhibit widely varying reservoir characteristics, depending in particular on their alteration and fractures. The huge Orenburg gas field produces from a very porous limestone zone. However, most Permian carbonates have poor reservoir characteristics.

The seals are provided by very continuous Devonian and Carboniferous shales, and also by the Kungurian salt and anhydrite. Their effectiveness appears to be excellent. Two main black shale horizons of the Eifelian and Frasnian form the chief source rocks of the region. They contain a high percentage of organic matter of the humic and sapropelic type, spread over vast areas. Other zones, especially in the Visean, may also have supplied gas and oil. Vertical migrations appear to have played an important role, and seeps have been found on the surface.

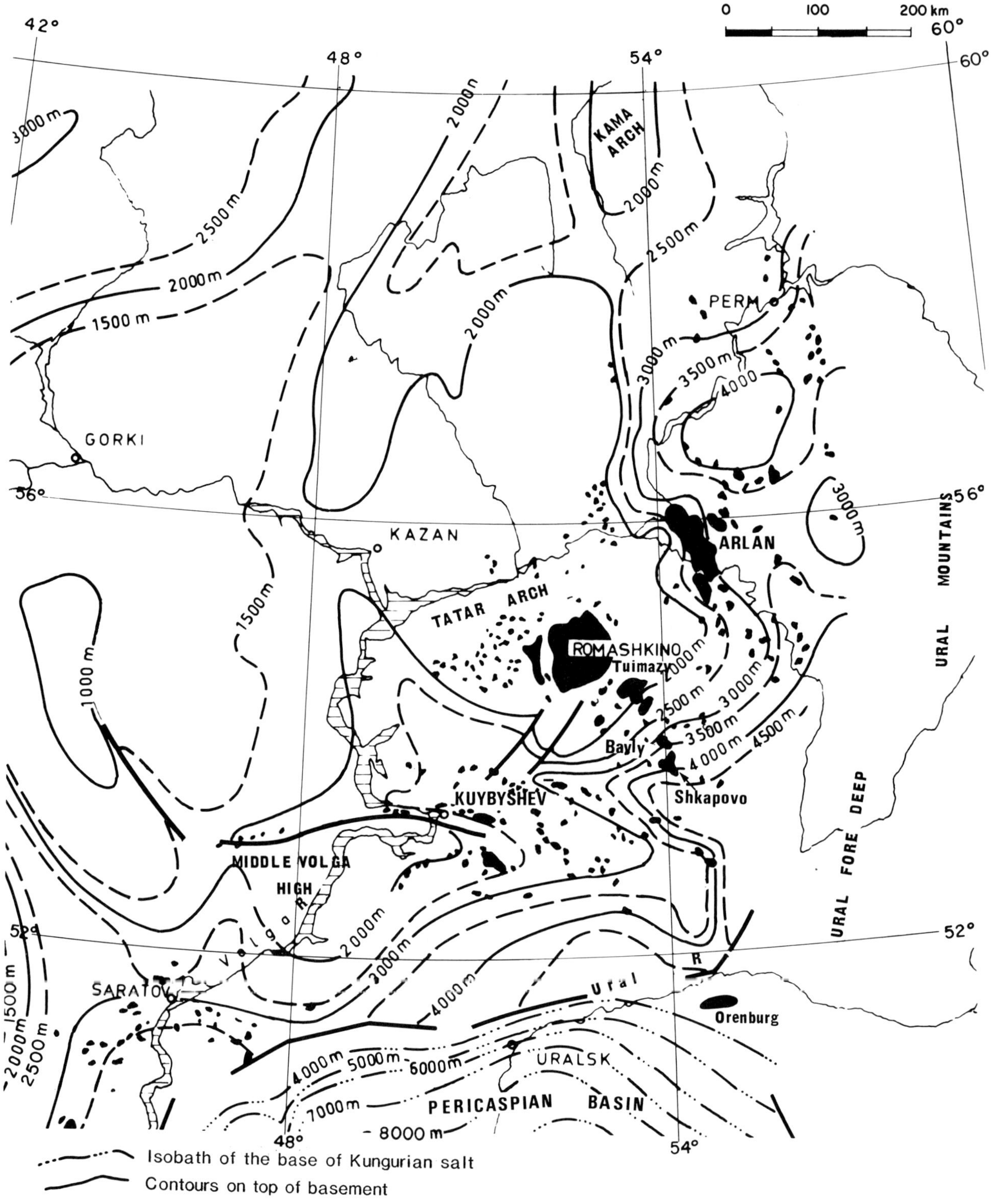

Fig. 174. — Structural map of Volga-Urals province with main oil and gas fields (from map of petroleum prospects of the USSR to 1 : 5,000,000 1967).

The Devonian crudes are of medium density (20 to 26 °API), medium sulfur (1 to 1.5 %), with GORs ranging from 40 to 63 t/day. The Carboniferous and Permian crude oils are heavier (specific gravity about 0.9) and high sulfur (up to 3.5 %). The Permian reef pools have gas caps containing nitrogen (15 to 90 %). The percentage of gas and light products increases with burial depth, to the detriment of sulfur compounds. The gas, especially that of the giant Orenburg Field, contains a high proportion of H_2S.

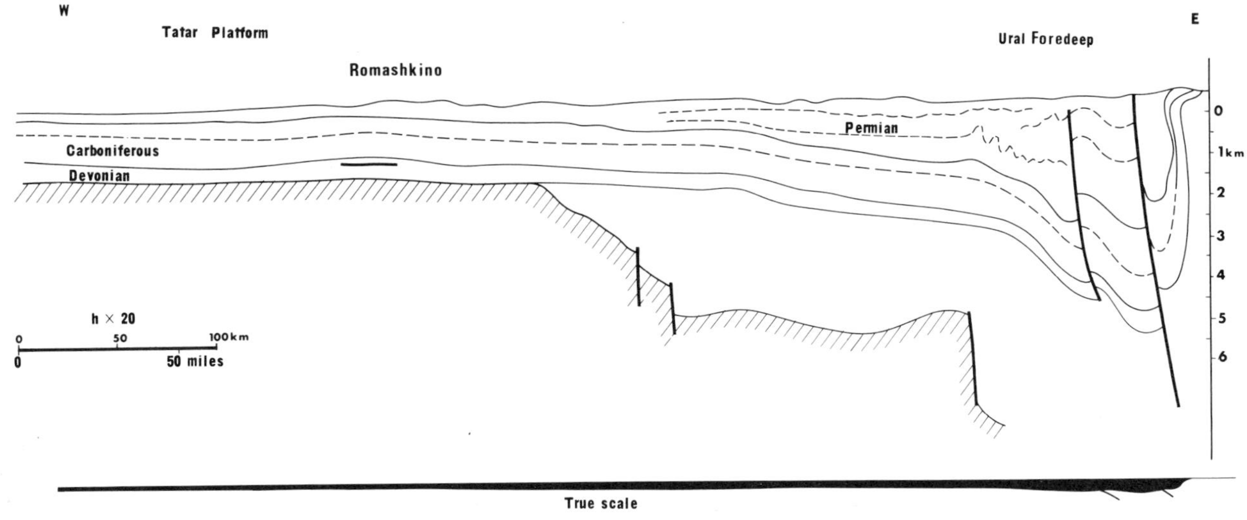

Fig. 175. — Section across Volga-Urals Basin passing through Romashkino Field (USSR).

The traps consist of domes and anticlines, usually with very slight dip (about 1 degree or less). The largest exhibit vast areas (3000 km^2 for Romashkino, the biggest). The pools are usually located on the south and southeast margins of the Tatar Arch and of the Middle Volga. The Permian reefs are stratigraphic traps of average potential, commonly aligned on the high axes.

Discoveries

The Volga-Urals Basin contains about 5 to 6 Gt of oil, of which more than a quarter has been extracted, distributed over more than 700 fields including a dozen giants. The biggest, Romashkino, has estimated reserves of 2 Gt, of which 1.4 Gt have been produced. The richest zone stretches on the margin of the Tatar and Bashkir Arches. This province is also rich in gas, with initial reserves estimated at more than 6000 Gm^3. The most important part is found on the Orenburg Field, of which the latest estimates amount to 2000 to 3000 Gm^3.

The Romashkino Field, discovered in 1948 and situated at the center of the large Tatar Arch, corresponds to a vast high zone of the platform type, with dips of about ten minutes (Figure 176). It produces mainly from five Middle Devonian deltaic sandstone layers at 1800 m. The pool temperature is exceptionally low, about 40 °C. The productive thickness ranges from 2 to 35 m with an average of 10 m. Mean production is about 40 t/day with wells at 125 t/day, and 8000 wells yielding annual production of 80 Mt. Cumulative production is more than 1.6 Gt. Water injection was developed very early on a large scale. Five other giant fields also produce in the Devonian : Novoyelkhov 400 Mt (productive in the Carboniferous), Tuymazy 300 Mt, Mukhanovo 200 Mt, Shkapovo 175 Mt, and Bavly 100 Mt. Another giant is the Arlan Field (Figure 177), productive in Moscovian limestones at 750 m and Tournaisian limestones at 1200 m, and also in Visean sandstones. Its reserves are estimated by different sources between 170 and 560 Mt (*).

By far the largest Permian pool is the Orenburg (originally called Krasnyy Kholm) gas field, with reserves estimated at more than 2000 Gm^3 (a figure of 3000 Gm^3 is sometimes advanced). This pool, which is productive in fractured Permian dolomites sealed by Kungurian salt and anhydrites, deformed by salt diapirism, contains a sour and very high sulfur gas (13 to 45 % H_2S). The trap is a faulted anticline with 700 m of closure, situated on a regional uplift on which the Permian exhibits considerable development. Production is expected to exceed 50 Gm^3 per year.

The Melekess Depression, between Kazan and Kuybyshev, contains about 18 Gt of heavy oil, mostly in the Permian sands on the south flank of the Tatar Arch. Most of the oil discoveries in this province are situated on the southeast flank of the vast Tatar and Bashkir Arches, on the margin of the Permian foredeep. The latter has not yet demonstrated the same oil potential, despite the presence of many reef developments. However, it contains the bulk of the gas reserves near Orenburg. The juxtaposition of elevated zones and regional depressions, marked by variations in geothermal gradients, appears to be one of the main reasons for the vast potential of this province, with deposits which are homogeneous and continuous (Trofimuk, 1955; Smirnov, 1958; Mirchink *et al.*, 1959, and Fohs, 1962).

(*) The unreliability of the reserve figures stems from various causes, especially the recovery rates, often optimistically estimated, and not always confirmed by the results of early water injections.

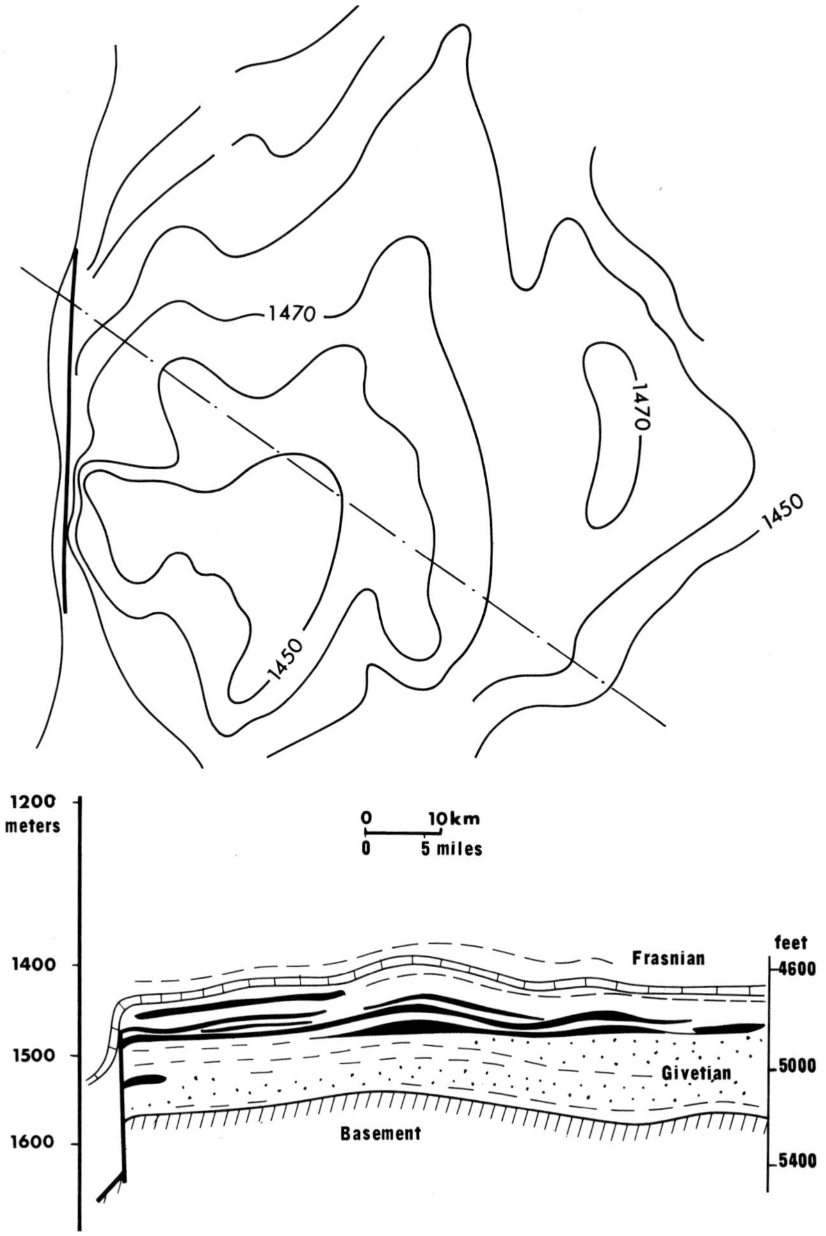

Fig. 176. — Structural map at base of Upper Devonian and section of Romashkino Field (USSR).

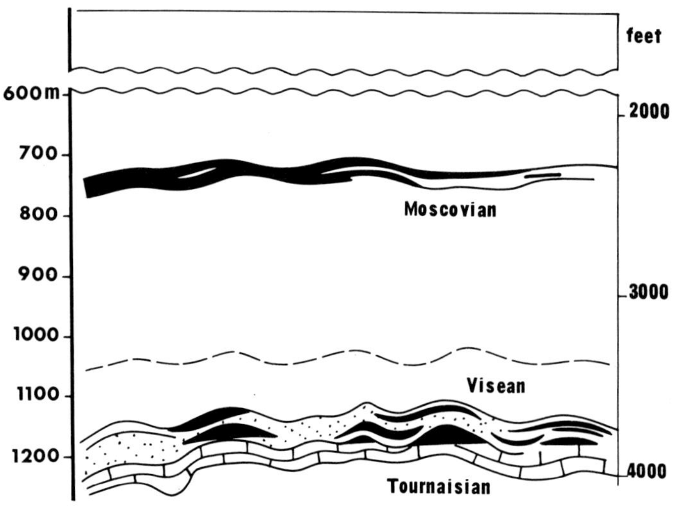

Fig. 177. — Cross-section of Arlan Field (USSR).

3.2.2. **The West Siberian Basin**

This province corresponds to the vast Meso-Cenozoic platform basin of West Siberia. Its area covers 3.8 Mkm^2, with 2.8 onshore and 0.4 on the Arctic continental shelf.

Geological criteria (Figure 178)

The geological history of this basin appears to be particularly simple, which is often a guarantee of high potential. From the structural standpoint, the basin consists of vast regional upwarps separated by depressions with gentle slopes, in which the basement is buried at a depth of 4 to 5 km, with dips of 2 to 3 degrees. The sedimentary series exhibits a thickness between 2 and 6 km. The Cenozoic formations are developed in the northern zone, where the total sedimentary column reaches 6 km, whereas a Paleozoic series unconformable beneath the Mesozoic cover has been identified in the southeast part. The Mesozoic and Cenozoic formations consist essentially of shales and sands corresponding to brackish or lacustrine littoral facies, and are subdivided into a number of major cycles.

The Lower and Middle Jurassic are represented by grey shales with sand and coal beds. The overlapping Upper Jurassic contains marine sands, black shales, bituminous at the base, passing into brackish shales with Valanginian/Hauterivian sand horizons. Some reefs are present in the west. A new Aptian/Cenomanian cycle displays a succession of shales, sand and coal beds. The Turonian marks a new onlap that continues until the Eocene. An offlap begins in the Oligocene towards the north, and continues throughout the Neogene.

The geothermal gradients are relatively high for this type of basin, especially on the high zones and in the central and western regions, where values of 30 to 60 °C km^{-1} have been recorded. From the hydrochemical standpoint, the presence of waters containing calcium chloride and sodium carbonate has been observed, with temperatures of 30 to 120 °C. The pressures are hydrostatic.

Petroleum criteria

The reservoirs, essentially sand, are numerous, broadly distributed over large areas, and display good characteristics. The Jurassic sands, except near the Urals, have the poorest reservoir qualities. Neocomian sands are of better quality. Cenomanian sands are the best. They are relatively non-cemented and exhibit high porosities and permeabilities (15 to 30 % and 150 to 4000 md). They are relatively thin, but multibedded. The well outputs are as high as 500 to 800 t/day. The main productive reservoirs belong to :

— the transgressive Upper Jurassic sealed by the Valanginian shales (sands and, in the west, calcarenite talus derived from reefs; oil and gas),
— the Valanginian/Hauterivian intercalated in shales of the same age (mainly oil),
— the Aptian/Albian/Cenomanian overlaid by overlapping Turonian shales (mainly gas).

Oil has also been produced from Silurian, Devonian and Lower Carboniferous carbonate in the area between Novosibirsk and the Middle Ob' at Surgut (ten fields : this is the Nyorol'ka Basin).

The seals are essentially shale, commonly thick and overlapping, especially in the Turonian. It appears that the permafrost also provides an effective seal in the northern zone. The source rocks appear to be :

— Upper Jurassic (Bazhenov Formation) black bituminous shales, silicious and radioactive, rich in organic matter (5 to 20 %), humic on the margins, sapropelic in the central portion, where they are 50 to 150 m thick (*),
— Lower Cretaceous shales (Georgiyev Formation), less rich in kerogen, but with 4 to 11 % values common,
— Middle Cretaceous (Aptian, Albian) coal measures in the northern part of the basin : the huge gas fields come mainly from Aptian/Albian/Lower Cenomanian coals.

The two first reach a maturation stage at depths of 2000 to 2800 m with temperatures around 80 °C occurring over vast areas.

The Jurassic crudes are generally of the naphtheno-aromatic type, fairly light (average specific gravity 0.84) and low sulfur. Productivity is 500 to 800 t/day. The Neocomian and Lower Aptian crudes, of the same type, are heavier (0.85 to 0.90). Well productivity is in the neighborhood of 10 to 15 t/day only. The oils become heavier towards the basin margins, where the waters are brackish (2 to 5 g/l). The depositional environments probably exert an influence on the composition of the oils, and some correlations have been established between littoral facies and low sulfur paraffinic hydrocarbons. The chemical composition appears

(*) These shales, often microfractured, especially along the bedding planes, in which the permeability varies with organic matter content, can yield oil, as in the Salym Field.

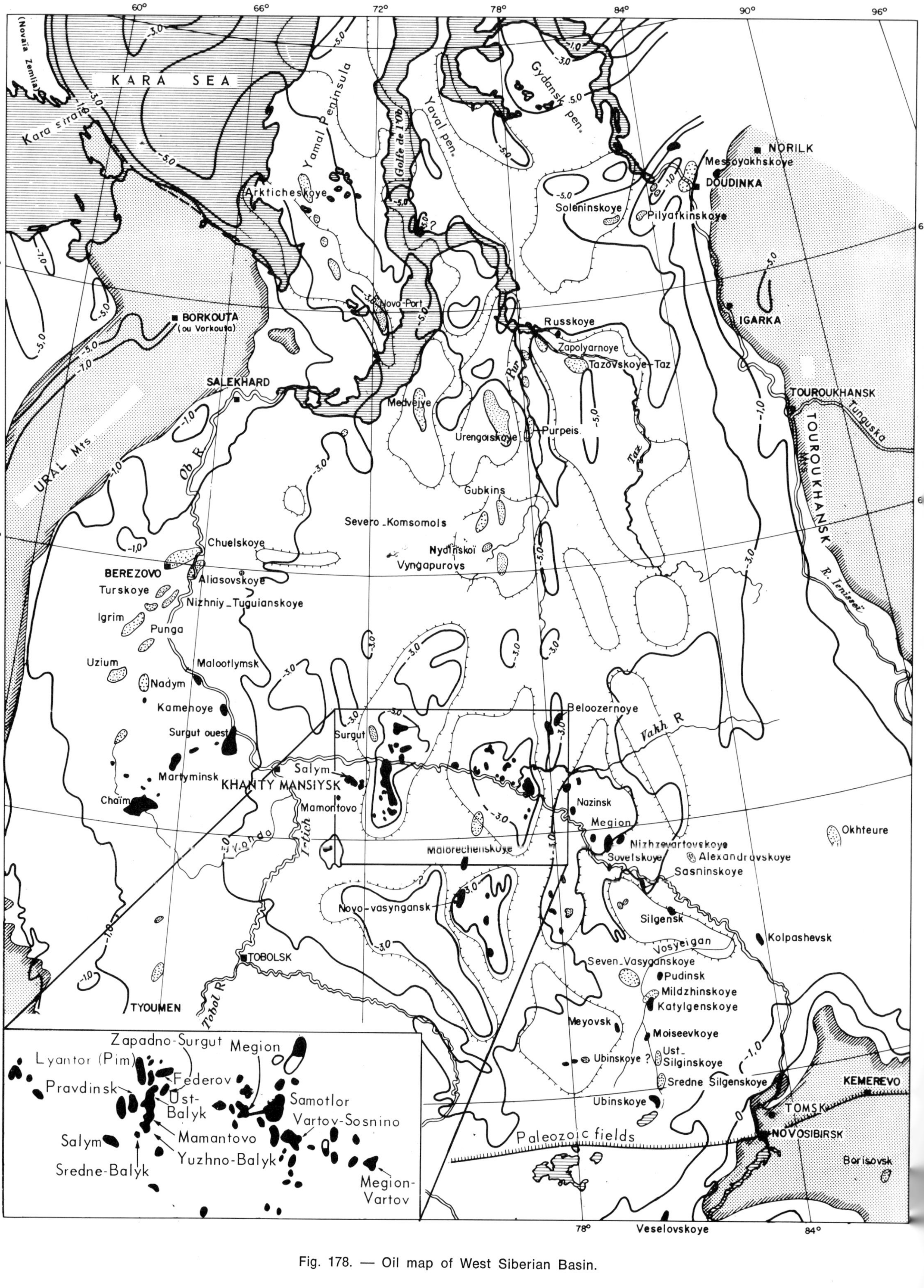

Fig. 178. — Oil map of West Siberian Basin.

to depend more on the stratigraphy than on depth. The gas is very rich in methane and generally in the higher homologues, whereas the Cenomanian gas is dry with a small percentage of nitrogen. The gas, especially in the Arctic zones, is mostly of shallow biogenic origin (Rice & Claypool, 1981).

The traps are structural, of the large radius anticline and dome type, and often appear to be associated with high zones of the granitic basement, formed in the Late Triassic and Early Jurassic. They exhibit average dimensions of 6 to 8 km by 10 to 12 km, more rarely 30 by 70 km, with closed areas up to 1500 km^2. The closure may be 300 to 500 m in the central zone, 1000 to 1500 m in the northern part. They are preferably located on vast regional upwarps that play the role of 'macrotraps', and are supplied from the neighboring syncline areas by lateral and partly vertical migration. The gas accumulates in arches of more recent formation (post-sedimentary). Some thousand structures have been identified.

This province of West Siberia appears to be one of the world's richest, especially in gas. Discoveries to date include 10 Gt of oil distributed among more than 300 fields, and more than 20 trillion m^3 of gas, while exploration is far from terminated. This basin contains a large number of giant fields, including the world's biggest gas fields :

— Urengoi	7500 Gm^3
— Yamburg	3000 Gm^3
— Yubileynyy	2000 Gm^3
— Arkticheskoye	1700 Gm^3
— Zapolyarnoye	1500 Gm^3
— Taz	1100 Gm^3
— Medvedzhiye	1000 Gm^3
— Kharasavey	1000 Gm^3.

There are 21 giant oil fields, the main ones being :

— Samotlor	(1965)	2600 Mt
— Fedorov	(1971)	400 Mt
— Ust'-Balyk	(1961)	170 Mt
— Vartov-Soviet	(1963)	150 Mt
— Mamontovo	(1964)	300 Mt
— Zapadno-Surgut	(1962)	100 Mt
— Pravdinsk	(1964)	130 Mt
— Salym	(1963)	50 Mt (?)
— Megion	(1961)	50 Mt
— Var'yegan	(1970)	200 Mt
— Russkoye	(1968)	100 Mt
— Lyantor	(1966)	150 Mt.

The Samotlor Field (Figure 69), discovered in 1965 in a particularly hostile lake and swamp region, is the USSR's biggest oil field and the fifth largest in the world. In 1979 it produced 150 Mt of oil. It is a vast very flattened anticline covering over 800 km^2. The reservoirs consist of Neocomian and Lower Aptian sands at 2400 m depth, stretching over a height of 20 m. Permeability is about 500 md, and the cumulative oil column is 50 m. The pressure is 245 bar and the temperature 88 °C. Average well production is 110 t/day, with a maximum of 1000 t. The oil is light, with density 30 to 39 °API. More than 3000 development wells are planned, and cumulative production reached 1 Gt in 1980 (Bazanov *et al.*, 1967; Dickey, 1972; Rovnin *et al.*, 1975; Meyerhoff, 1980).

3.2.3. **The Central Caspian Depression**

This province corresponds substantially to the Tertiary basin of the transverse type occupied by the southern half of the Caspian Sea and the adjacent coastal zones, particularly the zones of Baku and the Kura to the west (Republic of Azerbaijan) and Cheleken to the east (in Turkmenia). It corresponds to the eastern submersion of the Caucasus (Figure 179).

Geological criteria

The stratigraphic series extends from the Triassic to the present. The Tertiary occupies a predominant place both in terms of volume and petroleum value. The thicknesses are considerable in the southern part, corresponding to the North Caucasus Trough, where they reach about 20 km, and decrease progressively towards the north foreland. On the western shore of the Caspian Sea, Neogene detrital sand/shale series

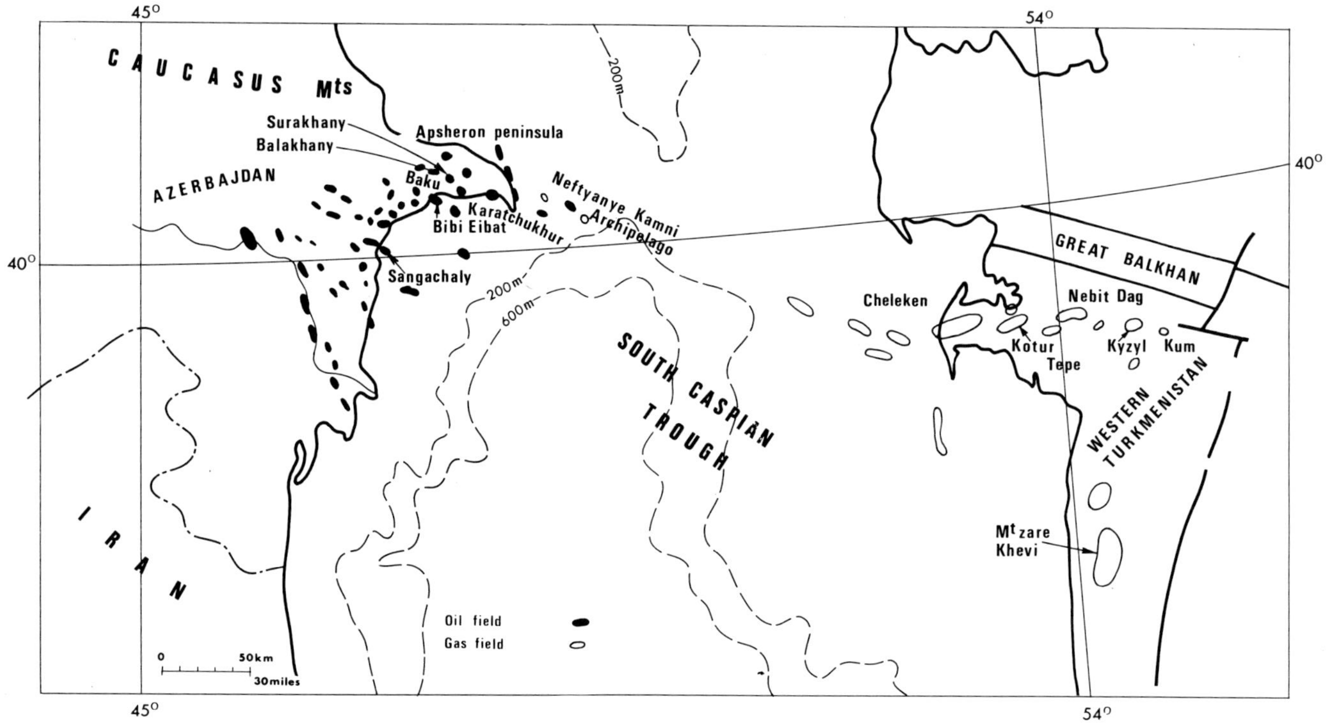

Fig. 179. — Map of the Central Caspian depression (regions of Baku, Kura, western Turkmenia).

from the paleodeltas of the Volga, the Kura and other rivers, form the bulk of the deposits. They correspond to massive deposits, rapidly settled in the thick series up to 4000 m thick, on limited, progressively foreset surfaces.

From the structural standpoint, this thick series is affected by sinuous and pronounced folds, separated by rounded synclines, commonly pierced in their axis by mud volcanoes, in other words, intrusions of under-compacted shales. In all these regions, vertical movements, many on a large scale, have played an important role. The geothermal gradients range from 25 to 35 °C km^{-1}.

Petroleum criteria

The traps, especially in the Apsheron Peninsula, consist of short "en échelon" anticlines, commonly diapiric, intensely faulted and fractured, and also of domes with a more subdued aspect. The closures, generally thicker than 100 m, may reach 1000 m, over lengths in the range of 10 km.

The reservoirs are formed of Pliocene deltaic sands, deposited in a current régime, well sorted and exhibiting good porosity (10 to 25 %) and permeability (100 to 700 md). The productive series, that corresponds to alternating sands and often red shales with fresh water facies, of the Middle Pliocene, extends over an interval of 2000 m. The lower part, well sealed by shale horizons, is the most prolific, and about fifty productive horizons have been identified. The wells have high potential, up to 10,000 t/day in exceptional cases. The potential per km^2 is also outstanding, up to 18 Mm^3/km^2 (450,000 barrels per acre).

The seals are largely provided by Upper Pliocene overlapping marine shales, and are up to 1000 m thick. The source rocks can be found in the deeper horizons, especially in the Eocene bituminous shales and in the fish and diatomaceous marls (Maykop beds) of the Oligocene. Vertical migrations play an important role and seeps are numerous. The crude oil displays variable characteristics (specific gravity 0.76 to 0.94), is often light in the upper beds, heavier and richer in paraffins deeper down. However, these grades commonly appear to be altered by the shallow depth and the continental environment of the pool. The eastern region is especially rich in gas condensate.

Main discoveries

Baku zone

The Baku zone, which is the richest, has produced about 1 Gt of oil over an area of slightly more than

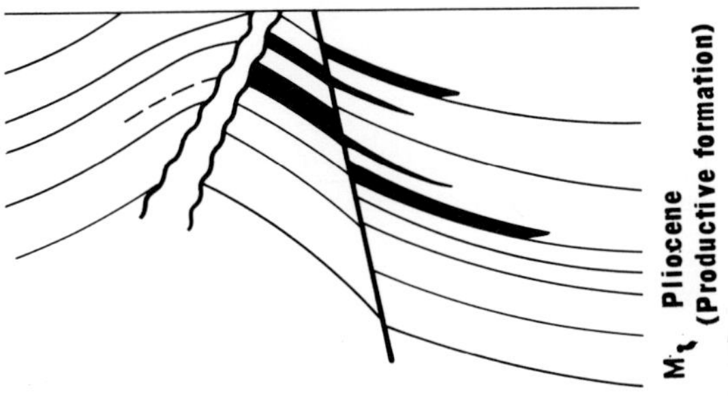

Fig. 180. — Cross-section of Neftyanye Kamni Field (USSR).

10,000 km^2. About thirty pools have been discovered there since the Balakhany gusher, found at 38 m depth in 1873 (Figure 179). The largest fields are :

— Balakhany 300 to 460 Mt (*),
— Karatchoukhour/Zykh 85 to 180 Mt,
— Surakhany (Pc 100 Mt in 1928),
— Bibi Eibat 150 to 300 Mt (Figure 85),
— Sangachaly 125 Mt,

and offshore :

— Neftyanye Kamni (oil rocks) 100 to 350 Mt (Figure 180),
— Bakhar 140 Gm^3.

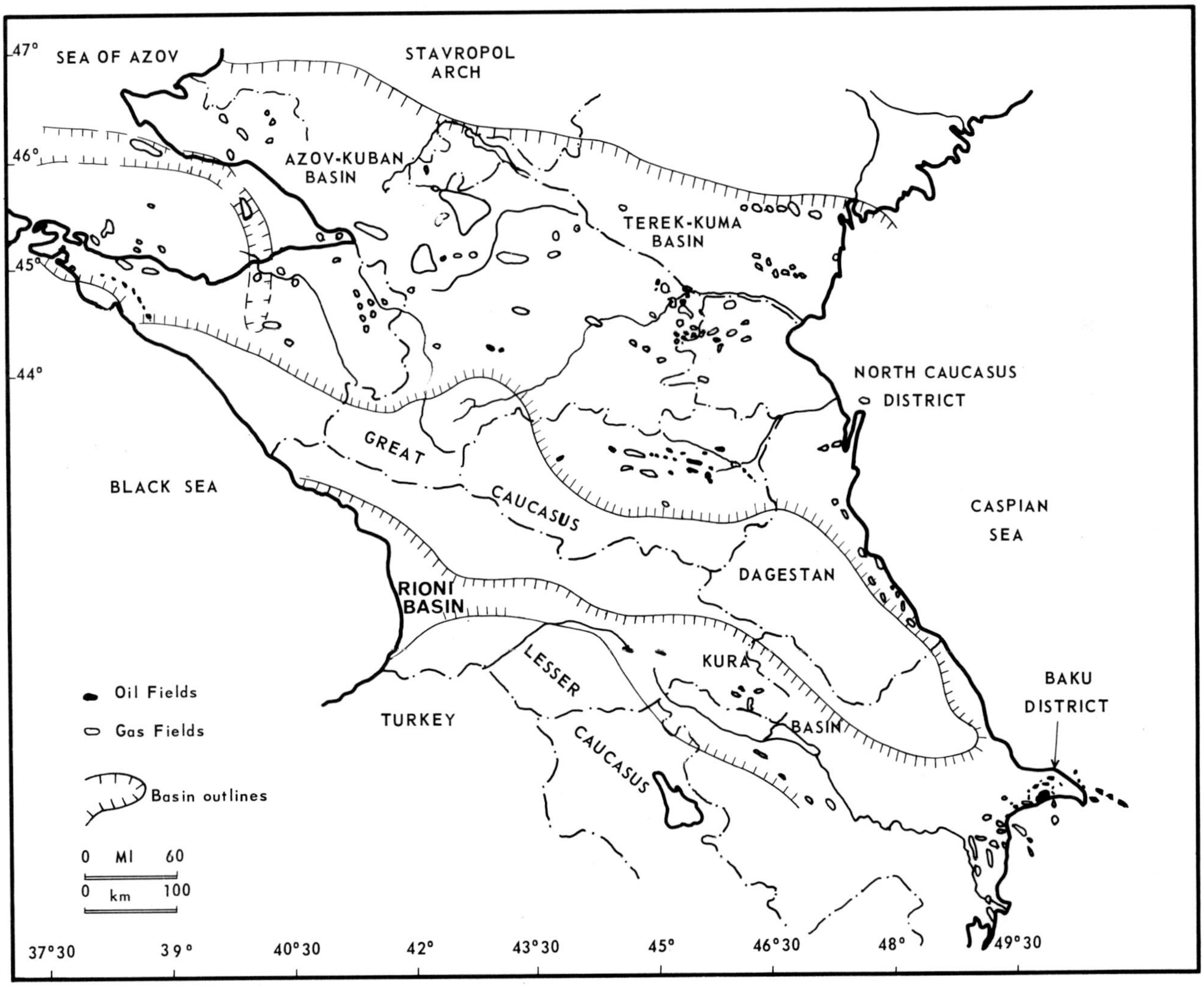

Fig. 181. — Map of the North Caucasus and Baku Districts.

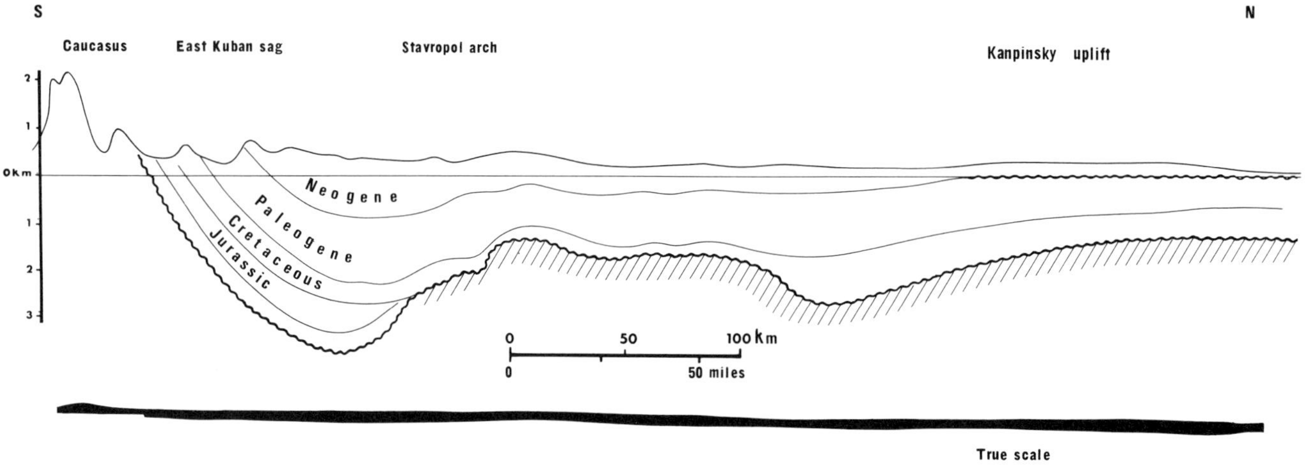

Fig. 182. — Cross-section of the North Caucasus Basins.

Western Turkmenia zone

On the eastern shore, the Western Turkmenia zone exhibits close analogies with the Baku zones. The structures are aligned with three main axes parallel to the Grand Balkhan Range, and the productive series, slightly older, belongs to the Upper Miocene (Figure 179). A dozen oil and gas condensate fields have been discovered. The Cheleken Field, whose initial reserves are evaluated at 85 Mt, has been produced since 1876, with the first productive zone at only 37 m. The Kotur Tepe pool, discovered in 1956, appears to be the largest, with reserves of about 200 to 500 Mt (110 Mt extracted by 1 January 1977) and 90 Gm^3 of gas. Also worth noting are the light oil and gas condensate fields of Nebit Dag, Kyzyl Kum and Mtsare Khevi.

***

These two zones are connected by a high structural and morphological axis that crosses the Caspian Sea in the prolongation of the Caucasus. This high zone is folded and displays a large number of structures, of which only those near the coasts have been drilled with success. However, it is worth noting that the Plio-Quaternary series thicken markedly offshore.

Other provinces

South of the Apsheron Peninsula, the intermontane basin of the Kura Delta has lower potential due to the invasion of the productive series by shale facies. The main gas pool is the Karadag, a faulted monocline with reserves of about 50 Gm^3. Farther westward, other basins in a foredeep position are aligned up to the Crimea (Figures 181 and 182). The most interesting from the petroleum standpoint is Groznyy, exploited since 1823. The productive pools in the Miocene sands and Upper Cretaceous fractured limestone with minor sandstone correspond to 'Appalachian' or overthrust anticlines aligned with two main anticlinal axes. The biggest and oldest fields are those of :

- — Malgobek (1915/1933) 300 to 430 Mt of initial reserves,
- — Stary Grozny (1893) 85 Mt,
- — Oktyabrskoye (1913) 75 Mt,
- — Novy Groznyy (1913) with cumulative production of more than 50 Mt.

The deeper fields in the basin mainly produce gas.

***

This Central Caspian Basin contains a total of nearly 2 Gt of oil. The Tertiary basins appear to owe their high potential to intense deltaic sedimentation, very rich in porous and permeable sands, well sealed, at least regionally, by overlapping Upper Pliocene shales, and supplied with hydrocarbons by source rocks rich in organic matter and recently buried. Vertical movements appear to have played a fundamental role in this genesis of hydrocarbons and in their migration. Many recent structures, present at the time of the

(*) These ranges are due not only to the lack of official figures, but also to the uncertainty of the field contours and to possible groupings.

hydrocarbon flows, have concentrated the hydrocarbons in pools, many of which are large, because a dozen giant fields have been identified (Gubkin, 1934; Halbouty *et al.,* 1970; and especially Ovanesov *et al.,* 1973 and 1975).

3.3. THE NORTH SEA PROVINCE

The North Sea province corresponds to many sedimentary basins stretching from the Devonian to the Carboniferous and the Permian at the south, to the Cenozoic of the Viking Graben in the north. The unit covers an area of about 500,000 km^2, with a sedimentary thickness often up to 5 to 6 km. The Hercynian, late Cimmerian and Laramide unconformities affect this sedimentary complex (Figure 183).

Geological criteria

The sedimentary series starts with the Devonian and then Carboniferous formations that constitute the Hercynian cycle. These formations were deposited in a late tectonic basin limited at the south of the North Sea, containing large coal formations spreading from north to south from the Visean to the Westphalian, but they invaded the entire sedimentary area in the latter era. After the Hercynian orogeny, a vast Permian basin, transgressing northward, was deposited with a positive basal sequence : the Rotliegendes with sandstones, red shales and salt; and a negative uppermost sequence, the Zechstein, predominantly carbonates and evaporites. These two sequences supply the reservoirs (sandstones and carbonates) and the saliferous seal of the gas province in the southern North Sea.

After the Triassic, which is mainly detrital, and still transgressing northward, comes the Precimmerian Jurassic, the Lias, Dogger and Lower Malm, which is only deposited in localized grabens in the south, with a negative megasequence (marine shales and rare sandstone horizons), whereas in the north, in the area of the Viking Graben, epicontinental sedimentation occurs, characterized by a positive megasequence in which sandstones with a fluviatile and then deltaic facies occupy a predominant place. This sedimentation is accompanied by considerable tectonic mobility which reaches its greatest intensity at the end of the Oxfordian, with a relief inversion mechanism in the southern grabens and rift tectonics in the north : the formation of faulted blocks directly against earlier normal synsedimentary faults.

The late Cimmerian compressive tectonics in the south, with a northern extension, is followed by erosion, and then by a general inversion of the relief. The Viking Graben collapses and is invaded by a deep northern sea, where the Kimmeridgian shales rich in organic matter are deposited, while in the south, the only occurrences in the Late Jurassic are localized deposits with paralic or evaporitic facies, on the margins of older grabens.

The Lower Cretaceous, represented in the south by black shales and transgressive sandstones, lies at the base of a regionally negative Cretaceous megasequence, with predominating carbonates in the south, and shales in the north in the Viking Graben. In the centre of the North Sea, the chalk facies of the Senonian extends up to the Danian. The regressive filling of the North Sea continues in the Tertiary. Subsidence reaches 3500 m directly below the central Ekofisk Basin. The series is essentially argillaceous, undercompacted up to the Miocene. In the north, at the periphery of the Shetlands platform, a large turbidite complex develops in the Paleocene and Lower Eocene, in which several major gas and oil fields are located, such as the Forties and Frigg Fields.

From the structural standpoint, a high east/west axis, the Mid North Sea High, occurs at the centre of the North Sea. It represents the earlier northern border of the Carboniferous basin. The North Sea is also affected in the south by a series of axial grabens, in relays, from the south to the north : the Hague Graben, the Dutch Graben, and the Ekofisk Graben. Also identified in an external position in relation to these grabens are basins with earlier mobility, essentially Triassic : the Sole Graben and the Dano-Norwegian Basin. North of the 58th parallel, the North Sea is affected by a structural system of the rift type, which schematically includes the Viking Graben, an anticlinorium with a north/south axis and a collapsed arch, and the Moray Firth Graben, an east/west trending synclinorium uplifted in a central horst. These two conjugate structures correspond to a relative displacement of the Shetlands platform towards the northwest (Figure 184) (especially Ziegler, 1980 b).

Petroleum criteria

Traps are associated with the general tectonics, very often antithetic fault blocks, either unconformable (Jurassic sandstones in the Viking Graben), or under the dysharmonic seal of Zechstein salt (Rotliegendes sandstone). However, the traps may also be turtleback structures, diapiric structures (piercement or not, Ekofisk chalk), or sedimentary traps. These include coastal sandstones (Lower Cretaceous in the Nether-

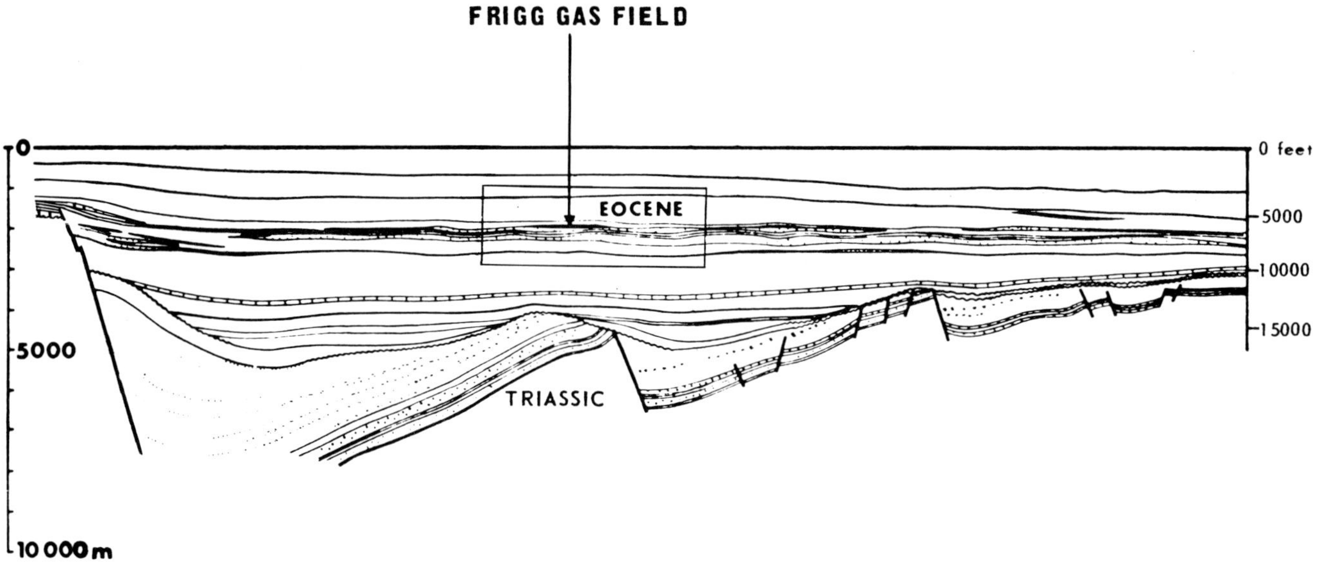

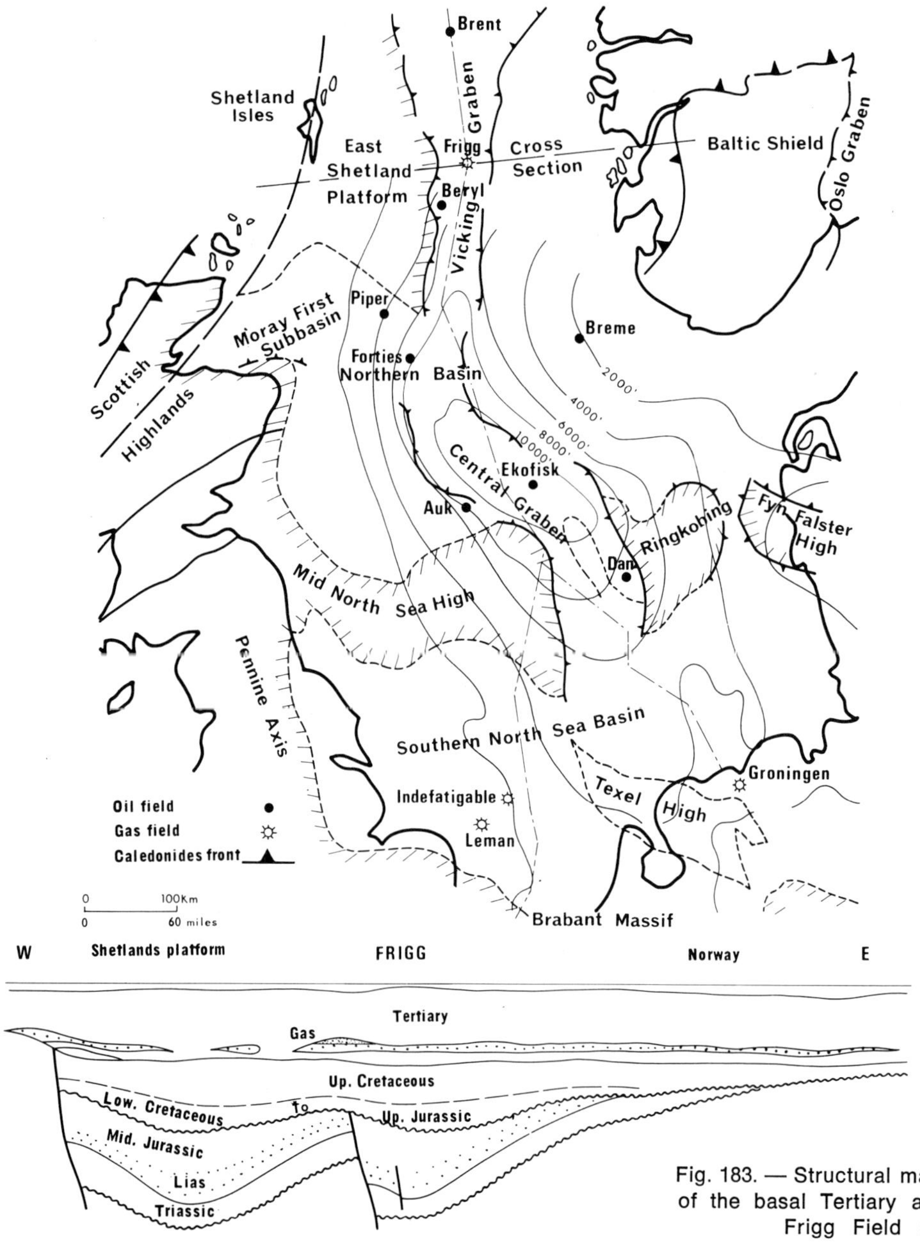

Fig. 183. — Structural map of the North Sea Basin. Isobaths of the basal Tertiary and cross-section passing through Frigg Field (Childs & Reed, 1975).

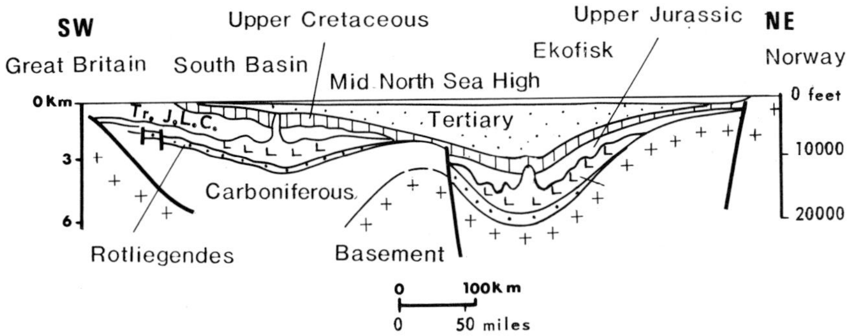

Fig. 184. — Cross-section of the North Sea Basins from London to Stavanger passing through Ekofisk (Dunn, 1975).

lands), deep sea fans (Frigg, Forties) under a shale seal. In the case of the Great Ekofisk fields, the oil column may exceed the present structural closure, as the oil was accumulated early in a still undercompacted chalk. The non-impregnated chalk was then compacted differentially into an impermeable mudstone, and subsequent structural deformations gave rise to accumulations with an irregular or tilted base. Similarly, an isolated reservoir or one of poor original quality, even at the top of a structure, may not necessarily have been impregnated. The Permian traps display areas and heights of 10 to 260 km^2 and 50 to 450 m respectively. The salt, which forms the cap rock, is subject to strong deformations capable of creating total disharmonies.

The reservoirs are highly varied, but mainly sandstones. This applies in particular to :
- Rotliegendes sandstones with a fluviatile and eolian facies, sealed by Zechstein salt,
- fluviatile to deltaic sandstones of the Lias and Dogger,
- sandstones and sands, attributable to the Eocene/Paleocene deep sea fans.

Also worth mentioning are the Danian and Maastrichtian chalk reservoirs in the Ekofisk zone (see Section 2 of Part 2) and the Zechstein carbonates (Auk, Argyll) made porous by early diagenesis of the 'breccia solution' type. These reservoirs generally display good characteristics. The Rotliegendes sandstones exhibit mean porosities of 12 to 13 % with useful thicknesses of 70 to 300 m at a depth of 1800 to 3200 m. The outputs range from 0.3 to 1.7 Mm^3/day. The Dogger sandstones show porosities of about 20 to 25 %, permeabilities from 0.5 to 1 darcy, which are preserved in the undercompacted zones, even at extreme depths. The Dogger's reservoir characteristics are less satisfactory in a hydrostatic pressure régime. The Paleocene and Lower Eocene sands (Frigg, Forties) display porosities of 25 to 30 % and permeabilities of up to 3 darcys.

The seals are formed by :
- Zechstein evaporitic series, ranging from 100 to 600 m in thickness,
- overlapping shales of the Kimmeridgian and the underlying Cretaceous (Viking Graben),
- Paleocene and Eocene shales, often undercompacted, in the Great Ekofisk Sector.

Their characteristics are excellent in each case. The gas in the Permian pools originates in the coal of the underlying Carboniferous, buried at extreme depths, often greater than 5 km.

The oil and gas in more recent series probably originate in the marine Upper Lias shales, the Dogger deltaic shales and, to a considerable extent, in the Kimmeridgian radioactive bituminous black shales containing algae and microplankton (Barnard & Cooper, 1981). The genesis of these hydrocarbons corresponds to sufficient burial since the Paleocene, in the Ekofisk region, but must have begun in more recent eras, the Eocene and Oligocene, in other cases. Vertical migrations appear to play an important role in the accumulation of the hydrocarbons, chiefly as regards the Zechstein salt horizon, which forms the seal of the Rotliegendes formation, but which is also a barrier preventing any migration towards the Triassic or Jurassic reservoirs. The crude oil exhibits medium to low density, 35 to 44 °API, and is low in sulfur (0.2 to 0.9 %).

The Rotliegendes gas contains 75 to 95 % methane, 3 to 6 % higher homologues, and, in the southern basin, variable amounts of CO_2 (up to 2 %) and nitrogen (1 to 9 %). The Frigg gas, the main Eocene field of the Viking Graben, contains 95.5 % CH_4, with traces of higher homologues, nitrogen and carbon dioxide. In accordance with the undercompaction and high pressures identified at great depths, the Jurassic oil pools evolve towards gas condensate accumulations that grow steadily smaller with increasing depth.

Main discoveries

Since 1964, the date of the first well, some 3 Gt of oil have been found in this province, distributed among 30 to 40 pools, including several giants : Brent, Statfjord, Ninian, Cormorant, Piper, Forties, and Great Ekofisk. More than 3000 Gm^3 of gas were also discovered, together with the 1700 Gm^3 of Groningen Field and neighboring pools.

3.3.1. **South Basin**

Slightly over 1000 Gm^3 of dry gas, 95 % of it in the Rotliegendes, have been found in this basin, distributed among some twenty pools, and around two-thirds concentrated in four giant fields, Leman, Indefatigable, Viking and West Sole. The remaining British fields are of lesser importance, together with the small accumulations found offshore the Netherlands. Leman (1966) is a faulted flattened dome measuring 30 × 8 km, and relatively simple. It is productive at 1800 m with a 240 m thick reservoir representing recoverable reserves of about 330 Gm^3. Indefatigable (1966), under a Mesozoic syncline, corresponds to two fault blocks containing recoverable reserves of 220 Gm^3 (Figure 76). Viking (1969) contains 100 Gm^3 (Gage, 1980). West Sole (1965), a faulted arch, corresponds to a surface syncline by variations in the salt thickness, and contains about 75 Gm^3 (Figure 76).

3.3.2. **North Basin**

The Jurassic contains about 1.7 Gt of oil and probably more than 1000 Gm^3 of gas, chiefly in the northern fields Statfjord, Brent, Ninian and Cormorant. The Danian and Maastrichtian of Great Ekofisk, a group of six diapiric anticlines, represent oil reserves of about 400 Mt and 260 Gm^3 of associated gas. The Paleocene (Forties, Frigg) contains some 350 Mt of oil and 600 Gm^3 of gas.

— *Oil fields*

• Statfjord (1974), in the Norwegian zone, containing 420 Mt of oil, is productive in the Dogger and Lower Jurassic/Upper Triassic sandstones, over an area of about 580 km^2 (Kirk, 1980).

• Brent (1971), in the British zone, with 280 Mt of oil and 100 Gm^3 of gas in the Dogger and Lias, is a typical Jurassic unconformity trap (Figures 96 and 185).

• Magnus (1974), containing 100 Mt of oil, produces from an Upper Oxfordian reservoir.

• Forties (1970), is a very low dip elongated dome producing from Paleocene turbidite sands at 2250 m, yielding outputs of about 2400 m^3/day and containing recoverable reserves estimated at 250 Mt (Hill & Wood, 1980).

• Ekofisk (1969), containing reserves of 170 Mt of oil and 125 Gm^3 of gas, at the center of a cluster of seven pools forming the Great Ekofisk Field (Figure 186) (Section 1 of Part 2) (Bark & Thomas, 1980) (*).

— *Gas field*

• Frigg (1971), straddling the Norwegian and British zones, is productive in a Paleocene sand mound deposited in relatively deep water, with reserves estimated at 200 Gm^3 (Figures 48 and 183) (Heritier *et al.*, 1980).

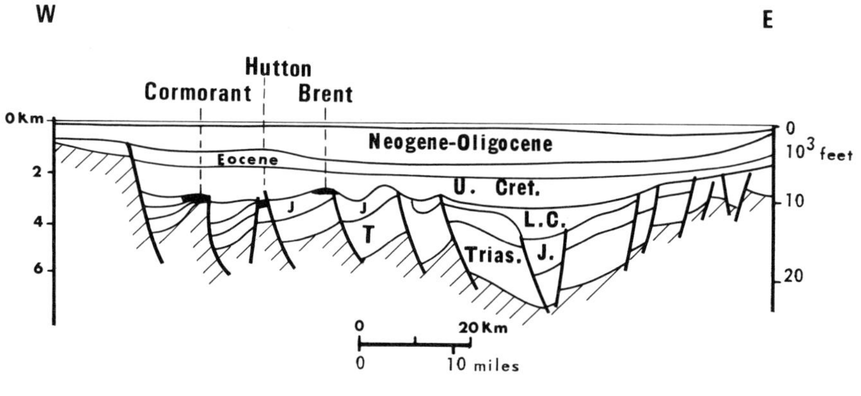

Fig. 185. — Section across the Viking Graben passing through Brent Field (Ziegler, 1975).

(*) Added to this list are the giants Ninian, with 170 Mt, Dunlin with 120 Mt, and Piper and Beryl with about 80 Mt each.

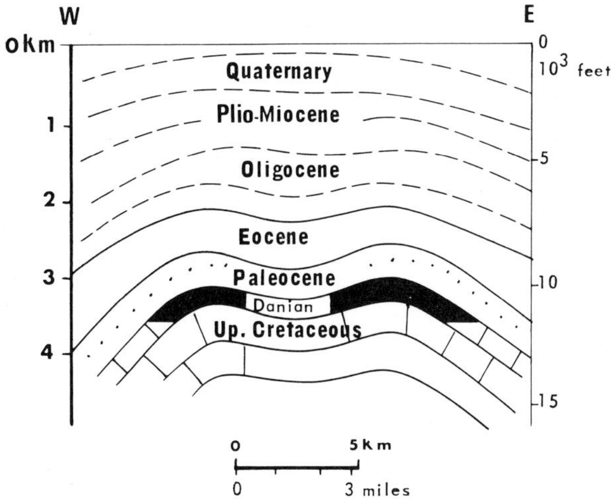

Fig. 186. — Section across the Ekofisk Field (Byrd, 1975).

*
**

It may be presumed that this potential, relatively concentrated in rather narrow grabens, aligned with the median line of the North Sea, especially with respect to the Mesozoic and Tertiary pools in the northern section, is more particularly related to :

— Jurassic shale source rocks, rich in sapropelic organic matter, raised to optimal maturation by burial which is often very deep and recent,
— a succession of detrital and carbonate reservoirs stretching from the Rotliegendes to the Eocene, sometimes corresponding to deep deposits,
— good evaporite and especially shale seals in succession from the Permian to the Eocene, with a special role played by Kimmeridgian radioactive shales, their effect accentuated by the under-compaction in the median Ekofisk Graben,
— regional unconformities, particularly the Hercynian unconformity, but above all the Late Cimmerian unconformity,
— a wide variety of structural traps, often complex, associated with extension tectonics and giving rise to faulted antithetic blocks, or with saliferous tectonics giving rise to non-piercement diapirs or turtlebacks, but apparently never associated with compressive tectonics, which is nevertheless present from the Late Cimmerian orogeny, directly below the southern grabens,
— recent subsidence causing the deposition of more than 2500 m of Tertiary deposits in the median grabens.

As for the South Basin, we have already noted (Section 1 of Part 3) the close superposition of deeply buried Westphalian gas source rocks, with reservoirs at the base of the Permian, and an effective evaporite seal, in a zone of strong catagenesis (Heritier, 1975; Pegrum *et al.,* 1975; Kent & Walmsley, 1970; Thomas, 1975; Watson & Swanson, 1975; Walmsley, 1975 and 1977; Kent, 1980; Ziegler, 1977 and 1980 a; Van Wijhe *et al.,* 1980).

3.4. THE NORTH AMERICAN PROVINCES

The great petroleum provinces of North America are characterized by their diversity, their size and the number of their fields, often resulting from intensive exploration. These provinces cover the entire stratigraphic scale from the Cambrian to the Pleistocene, from the Arctic Ocean to the Gulf of Mexico.

On the Arctic slope of Alaska, one province is practically characterized by a single field, which concentrates nearly all the reserves under Prudhoe Bay, with 1.4 Gt of oil and 800 Gm^3 of gas (see Section 4 of Part 2) (*). The onlap of a source rock over a series of eroded reservoirs, forming an uplifted hinge, appears to be the source of this potential (Figures 9 and 110). In Western Canada, the superposition of two vast Devonian/Mississippian and Mesozoic platform basins forms a vast province containing 2 Gt of oil and more than 2.5 Tm^3 of gas, with a tar belt containing more than 100 Gt of reserves in place. The Mid-Continent and the Permian Basin of West Texas, with reserves of nearly 7 Gt of oil and 6.5 Tm^3 of gas, represent the world's richest Paleozoic province.

On its western boundary, the Rocky Mountain complex, which is productive from the Cambrian to the Tertiary, contains cumulative discoveries of about 1 Gt of oil and 1.3 Tm^3 of gas in a dozen intermontane basins, as well as far greater reserves of oil shales. On the Pacific margin, the Californian Tertiary basins

(*) The Kuparuk field was recently discovered with about 140 Mt.

concentrate more than 3 Gt of oil and some 1000 Gm³ of gas in relatively narrow areas. The great crescent of the Gulf Coast provinces, with thick Mesozoic and Cenozoic detrital series, forming the southern margin of North America, also belongs to the rich area of the Gulf of Mexico and is discussed in the next section.

3.4.1. **Provinces of West Canada**

The West Canadian provinces cover about 1,350,000 km² in the provinces of Alberta and the adjacent zones of British Columbia and Saskatchewan. They include :

- a vast Upper Paleozoic/Mesozoic platform basin, remarkably stable, corresponding to the great inland plains,
- bounded at the west, without transition, by an intensely folded and eastward thrusting belt, forming the margin of the Rocky Mountains.

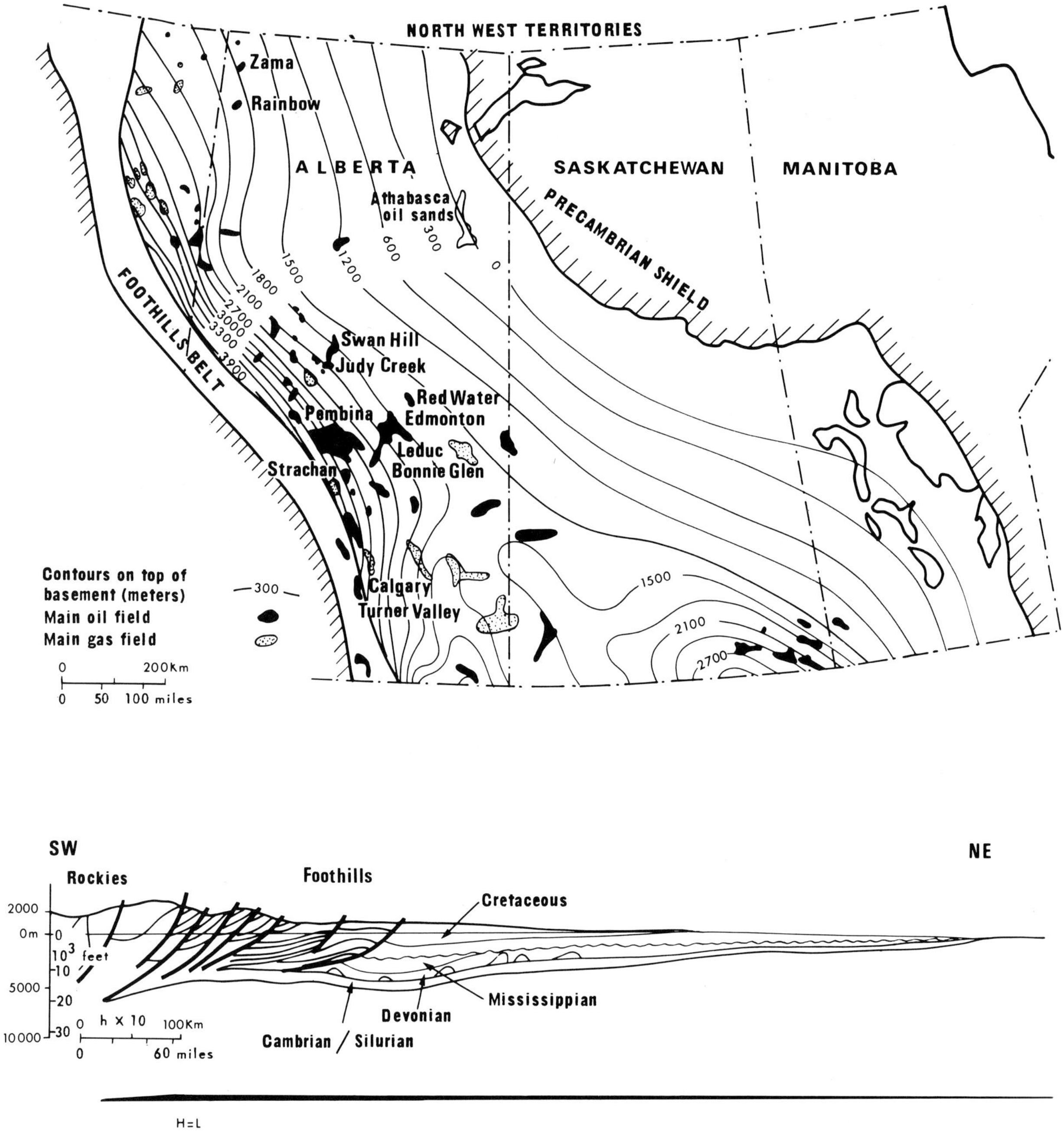

Fig. 187. — Map and section of Western Canada.

3.4.1.1. ***Platform province***

The sedimentary series of this province extends from the Cambrian to the Tertiary, with a number of hiatuses. It thickens steadily from the edge of the Canadian shield, where it thins very sharply, towards the base of the Rockies. It forms the 'Alberta syncline', where it reaches a thickness of more than 5000 m. The unit forms a very gentle monocline, tilted 0.5 to 1° towards the southwest, and marked only by a few regional very large radius arches (Figure 187).

In this context, the traps are essentially stratigraphic, reef traps in the Devonian and sand pinchouts in the Cretaceous. These two units lead to the distinction between two major themes or sub-provinces : Paleozoic and Mesozoic.

Paleozoic sub-province

The Middle and Upper Devonian is the principal producing system with 38 % of oil volume in place. However, some reserves are also found in Carboniferous series, especially gas (25 % of reserves) (Figure 188).

The reef traps, which spread from the Middle to the Upper Devonian, are grouped in three main districts. The Leduc District, between Edmonton and Calgary, the first discovery made in 1947, is productive in massive dolomites belonging to early Upper Devonian bioherms enclosed in black shales. The reserves of this group are estimated at 670 Mt of oil and 270 Gm³ of gas in place, distributed among about forty fields. The main fields are Redwater, containing about 110 Mt of recoverable oil, and Leduc/Woodbend, with 50 Mt of recoverable reserves. The entire Leduc Formation contains about 550 Mt of oil

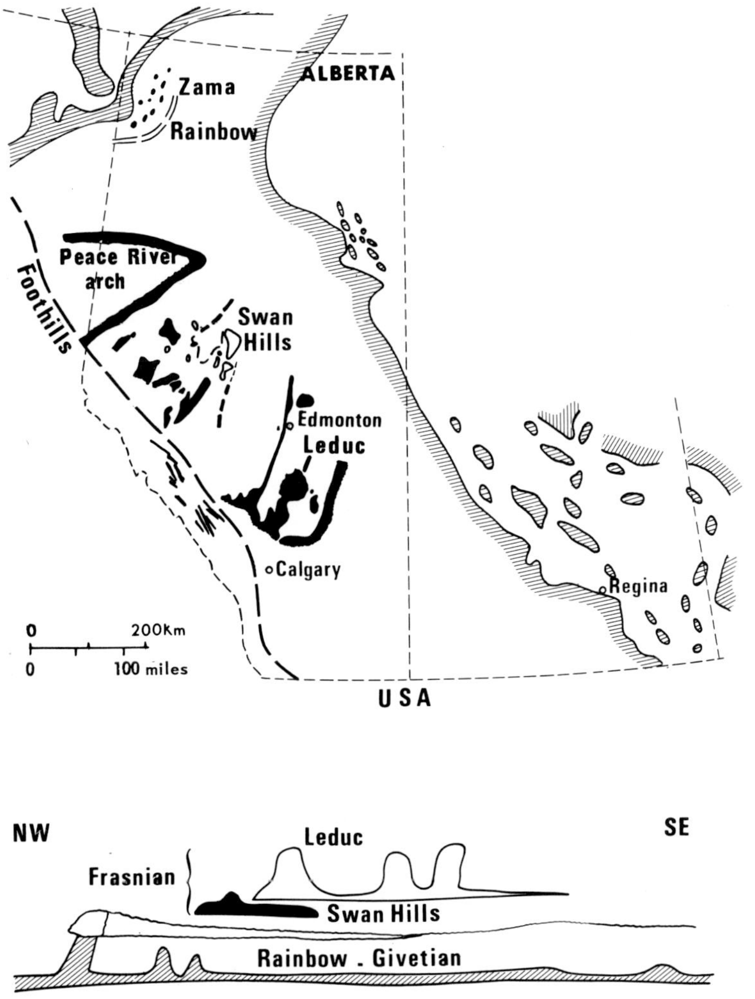

Fig. 188. — Map and stratigraphic section of main Devonian reef units in Alberta (Klovan, 1974).

and 500 Gm^3 of gas. Associated with the Leduc Formation are the two Strachan and Ricinus gas fields, discovered in 1967 and 1969 by seismics in the deep part of the basin, under the foothills and overthrust belt zone. These two reefs, located at 4000 and 4300 m depth respectively, are largely dolomitized and exhibit average porosities of 8 and 9 %, and matrix permeabilities of 15 to 20 md, considerably improved by fractures. The wells produce 1 to 2 Mm^3 day^{-1} for impregnated columns of 200 m. The reserves in place of these two fields, with an individual area of about 20 km^2, are estimated respectively at 46 and 50 Gm^3. The gas is sour, as throughout this region, and contains 80 % CH_4 and 9.5 % H_2S at Strachan, and 68 % CH_4 and 33 % H_2S at Ricinus (Bhat, 1975; Hriskevich *et al.,* 1980).

The Swanhills District, discovered in 1957, consists of a series of Lower Frasnian reef structures, grouped northwest of Edmonton on the southern flank of the Peace River Arch. The reefs, 100 to 150 m in height, consist chiefly of strotamoporoid limestones, enclosed in black shales and amphistegine marls. The productive areas range from a few km^2 up to 40 km^2 for Swanhills pools A and B, the largest, with reserves in place of 270 Mt. Next come the fields of Swanhills South, with 125 Mt in place (70 recoverable), Judy Creek A, with 113 Mt in place (70 recoverable), Swanhills C, with 77 Mt in place, and, with respect to gas, Kaybob South with 18 Gm^3 in place. The entire district contains about 900 Mt of oil in place (326 Mt recoverable), and 125 Gm^3 of gas in place (50 commercial) (Hemphill *et al.,* 1970).

The Rainbow/Zama District, discovered since 1965 in the northwest corner of Alberta, is productive in Middle Devonian reef formations. The reefs belong to a transgressive marine phase (Keg River Formation) starting with platform carbonate deposits and developing into bioherms. The latter, which are of different types (pinnacles, crescents, atolls), develop in front of a reef barrier located in the northwest. The cycle terminates in a regressive phase (Muskeg Formation), represented by evaporites, which bury and seal the reef traps. The latter are up to 180 to 200 m in height, with slopes of around 25 degrees. They consist chiefly of corals, stromatopores, bryozoa and crinoids. The evaporites appear to have contributed largely to the dolomitization of these carbonate formations. Porosities range from 3 to 13 % (mean 4.5 %), and permeabilities from 100 to 1000 md, at an average depth of 2000 m (Figure 67).

The oil originates in black fine-grained carbonate beds, rich in organic matter, corresponding to the beginning of the evaporite phase. Reserves in place vary widely from one pool to another, ranging from 0.5 to 38 Mt. The primary recovery factors are estimated between 20 and 50 %, and up to 80 to 95 % with the injection of miscible slugs. By the end of 1969, 72 pools had been discovered, for a total of some 200 Mt of oil in place. In the neighboring district of Zama, nearly 30 million tons of oil have been found in a series of small pools, generally of the pinnacle type. The overall Rainbow/Zama District contains about 220 Mt of oil and 30 Gm^3 of recoverable gas reserves, distributed among some 150 accumulations (Langton & Chin, 1968; McCamis & Griffith, 1968; Dumestre, 1969; Barss *et al.,* 1970; Evans, 1972; Klovan, 1974; Hriskevich, 1980).

Mesozoic sub-province

The transgressive Jurassic is represented by black shales with sandstone and limestone beds. In the Lower Cretaceous, the transgressions and regressions of the sea in a vast gulf opening westward are reflected by a series of variations in facies, and especially by large deltaic spreads, resulting in stratigraphic traps. The Eocene is represented by a sandstone/shale series up to 3000 m thick in the Alberta syncline, and, like the other sequences, thinning sharply towards the northeast.

In this entire sub-province, the traps are essentially stratigraphic, pinchouts and lenticular sand bodies, principally by facies variations and transitions to marls upstream, more rarely by transgressive pinchouts in the basal Cretaceous series. A few very gentle anticlines, generally containing gas, overlay early reliefs or Devonian reefs. The reservoirs are essentially detrital. The Upper Cretaceous sands, mainly the Cardium, contain about one-quarter of Canadian oil, the bulk of it concentrated in the giant Pembina Field. The Lower Cretaceous, especially the Viking, contains some 12 to 13 % of the oil and gas reserves. The seals consist essentially of shale. The hydrocarbons originate in Cretaceous shales. The oil density increases steadily from the deep areas of the basin towards the northeast margin, where it reaches values of about 1 in the tar sands. The source of this heavy oil appears to lie in the immaturity of the formations and the alteration of the migrated hydrocarbons.

The main Cretaceous field is the Pembina Field, discovered in 1953, after a reinterpretation of well logs. It is formed mainly by a vast pinchout of the Cardium sands, thinning and passing into shales updip, towards the east. The three sand bodies forming the reservoir are separated by two quite distinct unconformities. The reservoir, situated at 700 m, exhibits the following characteristics :

- area 3000 km^2 (755,000 acres),
- thickness 0 to 20 m, mean 4 m,
- mean porosity 12 %,
- mean permeability 25 md.

The formation waters are practically fresh waters, but the oil is nevertheless realtively light, with density 39 °API. The 2200 wells have low output, and produced 4 Mt in 1981 with 3400 wells, making an average of about 5 t/day. Initial recoverable reserves are estimated at 250 Mt, and cumulative production (1979) was 150 Mt.

The Mesozoic formations deserve our attention for two particular reasons :
— the existence of gas in the deep part of the syncline,
— the presence of very large accumulations of heavy oil in the shallow and even outcropping series.

Vast gas accumulations are found in the low porosity/permeability Cretaceous sandstones of the Alberta syncline. As we have shown (Section 4 of Part 2), this original deep synclinal trap appears to shelter resources of up to 12,000 Gm^3 of gas, including 250 Gm^3 in the Milk River Field. However, these volumes must be tempered by the low and locally variable well outputs.

Heavy oils

The Lower Cretaceous sandstones in the north and northeastern zones of Alberta contain very large volumes of heavy oil, estimated at some 750 Gbl (120 Gt) in place. These reserves are concentrated in the accumulations of Peace River at the west, Lloydminster and Cold Lake at the east, and largely at Wabasca and Athabasca in the northeast, the entire unit forming an arc about 1000 km long (Figure 189).

The characteristics of the oil deteriorate with decreasing reservoir depth, ranging from a density of 18 °API and more downstream, to conventional crude oils with a density of 10 °API at the outcrop. In these different accumulations, the oil is found in the basal Lower Cretaceous sandstones, mainly of the deltaic

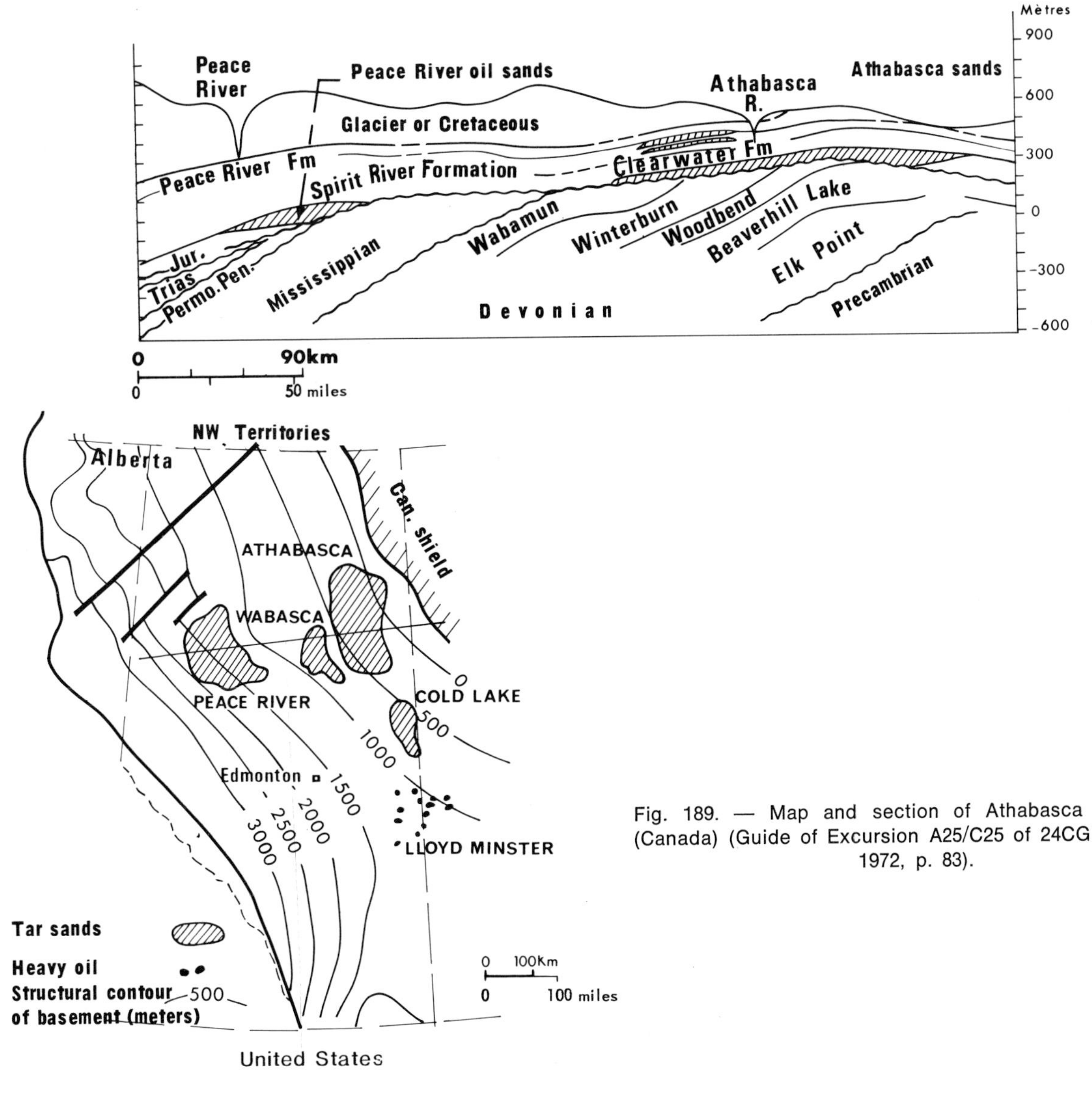

Fig. 189. — Map and section of Athabasca tar sands (Canada) (Guide of Excursion A25/C25 of 24CGI, Montreal 1972, p. 83).

and fluviatile type, overlapping the earlier Jurassic to Devonian formations, and forming an irregular relief. These sandstones are sealed by transgressive marine shales. Trapping is provided by a combination of structural and stratigraphic factors : shoals and regional sand pinchout towards the northeast.

The Peace River accumulation contains about 8 Gt (25 Gbl) of oil in place, in sands about 30 m thick at a depth of 500 to 800 m. The Cold Lake accumulation, situated above a Paleozoic uplift, contains about 11 Gt (75 Gbl) at 300 to 450 m, in a series of 'fields'. The fields in the Lloydminster zone, situated on noses and terraces in a pinchout zone, contain some 360 Mt (2 Gbl) of oil in place between 300 and 800 m depth, producible by conventional methods. The Wabasca accumulation represents about 8 Gt (50 Gbl) between 100 and 300 m. The largest by far is that of Athabasca, which contains between 90 and 100 Gt (560 to 626 Gbl) in place (Vigrass, 1968).

The Athabasca tar sands

The Athabasca tar sands, located in the northeast corner of Alberta, with some 90 to 100 Gt (560 to 626 Gbl) of oil in place, represents the world's largest resource of heavy oil. This accumulation, which covers about 30,000 km^2, is situated in Lower Cretaceous sands overlapping the Devonian, with an impregnated thickness up to 120 m. These sands outcrop in the valleys and are progressively buried under 600 m of overburden towards the southwest. Of the approximately 100 Gt of oil, about 12 Gt are producible by quarrying, and about 50 Gt could be produced by *in situ* thermal stimulation. Their average potential is about 4 Mt km^{-2} (100,000 barrels per acre) with a peak of 10 Mt (250,000 barrels).

From the geological standpoint, these sands correspond to a brackish to marine deltaic environment, with fluviatile facies. The bulk of this accumulation corresponds to a vast upwarp measuring 240 × 200 km, exhibiting a counterdip of about 60 m. The latter is closed by a slight syncline associated with the dissolution of the saline formations of the Elk Point group (Devonian). This unit forms a sort of giant field, probably the only outcropping one (Mossop, 1980).

3.4.1.2. **Rocky Mountains foothills province**

This province, which extends simultaneously in front of the Rockies over northeastern British Columbia and southeastern Alberta, is distinguished by its structure of faulted and overthrust folds, often complex, displaced eastward over the foreland of the platform during the orogeny. This tectonics appears to be the result of the subduction movements that generated the folded system of the Rockies. It collides with the especially stable area of the Prairies without any transition.

In this context, the traps consist essentially of acute folds, generally faulted and overthrust. The presence of the shaly Cretaceous between the rigid Carboniferous formations gives rise to a flexible and relatively simple pattern. If the Cretaceous is ejected, the Mississippian lies on the Mississippian, resulting in a complex network of slabs. In this zone, which has undergone an uplift of about 1000 m since the Miocene, the pools generally appear to be full. The reservoirs are chiefly represented by Mississippian limestones, more or less dolomitized, locally rather porous (6 to 7 %) and rendered permeable by intense fracturing. The Triassic carbonates and the Devonian and Cretaceous sandstone horizons also yield some production, especially gas. The seals are essentially shale, and usually Cretaceous. The source rocks appear to originate in the Mississippian (Ekshaw Formation) and Cretaceous (Fernie Formation) shales.

This region is extremely rich in gas, which is usually sour, and cumulative discoveries are in excess of 150 Gm^3. It is here that the first Canadian oil field, Turner Valley, was discovered, in 1914 for the Cretaceous horizons, and in 1936 for the more important Carboniferous reservoirs. This field, which is the only producer of oil in this sub-province, is practically exhausted today after having produced about 20 Mt of oil. The main gas fields are those of Waterton (100 Gm^3), the biggest (Figure 86) (see Section 4 of Part 2), Jumping Pound W (27 Gm^3), and Wildcat Hills (20 Gm^3) (Link, 1949; Fox, 1959).

Conclusion

About 2.3 Gt of conventional oil and 2600 Gm^3 of gas have been discovered to date in the great province of western Canada, with an area of 1.3 Mkm^2, corresponding to a total hydrocarbon potential of about 3000 t km^{-2}. These reserves appear to be distributed among more than 2700 pools, including half a dozen giant fields. Also worth noting, aside from these conventional discoveries, are the immense heavy oil and tar sand accumulations, with volumes in place estimated at more than 100 Gt.

Undeniably, one of the favorable factors in this province, which is on the whole not subsident and remarkably stable over most of its area, is a landscape of reefs and restricted settings in the Devonian, with a few temporary accelerations of subsidence, favoring the development of the reefs, broad deposits of evaporites forming excellent seals, and the varied landscapes of gulfs and deltaic complexes in the Cretaceous.

3.4.2. **The Mid-Continent Province**

The large Mid-Continent region, that covers nearly one million km^2, is distributed among half a dozen Paleozoic basins, including the West Texas basin, which is the centerpiece (Figure 190).

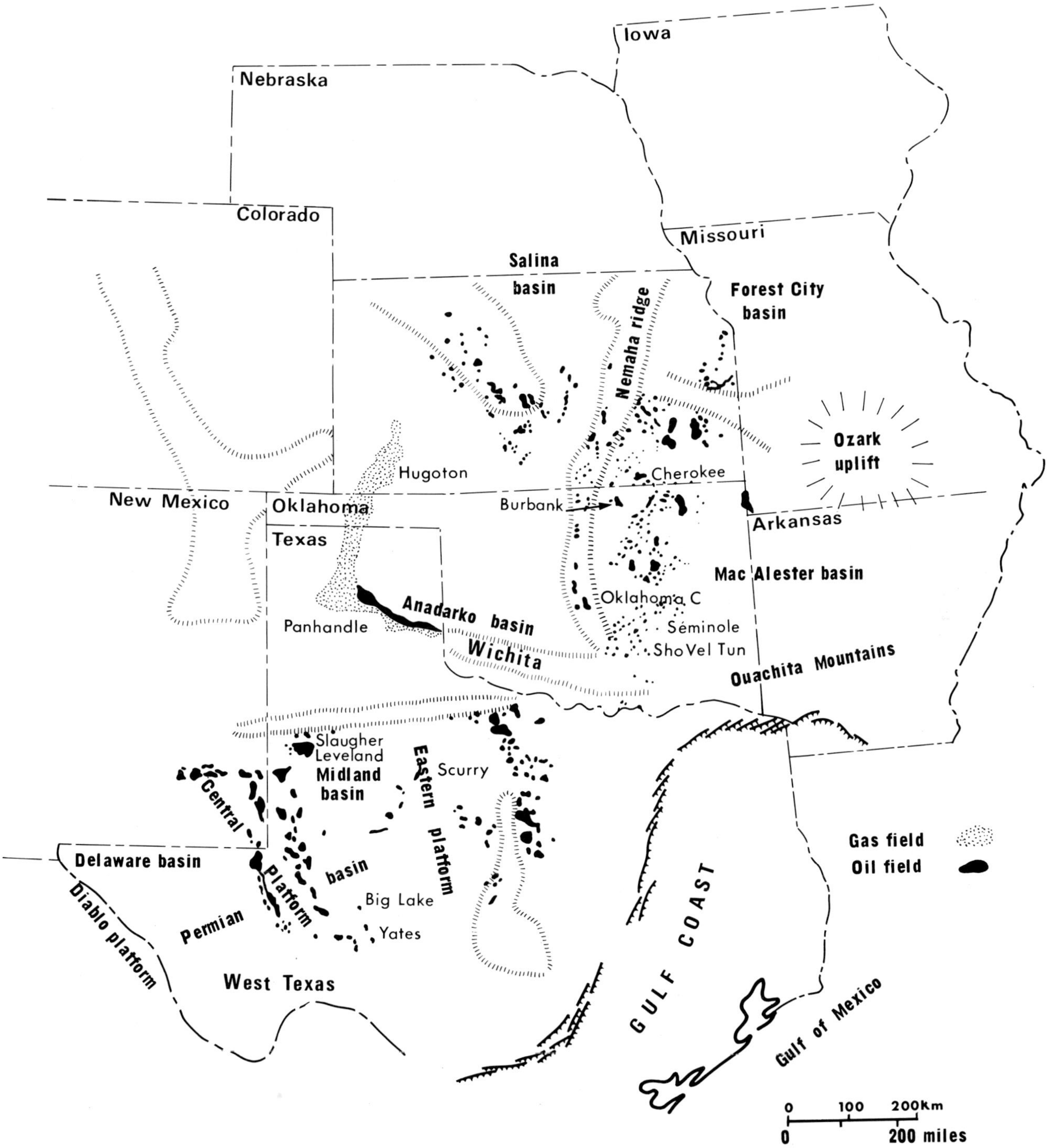

Fig. 190. — Map of main Mid-Continent provinces.

From the geological standpoint, this province consists of more or less stable platform basins, often structured with granitic ridges, and shoals between which large subsident areas are downwarped, the Delaware and Midland Basins in West Texas, and the miogeosyncline basins adjacent to the Ouachita and Wichita/Arbuckle Ranges in the southern prolongation of the Appalachians. From the stratigraphic standpoint, these basins are essentially filled with Paleozoic series, especially Carboniferous and Permian, themselves intersected by many hiatuses and unconformities, of which the regional transgression of the Pennsylvanian is the most important.

From the petroleum standpoint, these basins appear among the richest of North America, exhibiting a wide variety of traps, mainly structural, but also reefs, often very large, and formed early, as testified by the unconformities affecting them (such as the Oklahoma City anticline). The combination of stratigraphic and structural factors often appears to originate in vast and large accumulations, like the giant Panhandle/Hugoton Field. The sandstone and carbonate reservoirs are widely distributed in the stratigraphic series, often thick and showing good characteristics. The sandstone and carbonate formations of the Pennsylvanian and Permian deserve a special word, especially the reef developments of West Texas. The Guadeloupe series, for example, have yielded outputs of 16,000 m^3 per hectare (40,000 barrels per acre) and production of more than 10,000 tons per day (Yates Field).

On the whole, especially in the Delaware Basin, the sandstone reservoirs appear to contain a lighter and purer oil than the carbonate reservoirs, which correspond to depositional environments with a lagoonal tendency. The seals are mainly shales and evaporites covering the Permian limestones. Source rocks seem to be found at different locations in the stratigraphic series, especially in the form of Ordovician black shales, transgressive radioactive and bituminous black shales and limestones in the Devonian and Mississippian, Pennsylvanian phosphatic bituminous shales and limestones, and Permian black shales and limestones containing diatomites.

Main discoveries

In the overall Mid-Continent province, about 7.5 Gt of oil and more than 8 Tm^3 of gas have been discovered, including 4.2 Gt and 2 Tm^3 respectively in the West Texas Basin. These reserves appear to be distributed among more than 3600 fields, including about fifteen giants.

The Permian basin of West Texas, with an area of 300,000 km^2, thus appears to be one of the world's richest, with some 16,000 tons of oil and 15 Mm^3 of gas per km^2. Its main fields, which are productive in the Permian and sometimes Pennsylvanian limestones, are the following :

— Yates Field (1926), a broad dome on the central platform with 300 Mt,
— Wasson Field (1937) 260 Mt,
— Scurry Field (1948), on the Horseshoe Atoll (eastern platform) 250 Mt (see Section 4 of Part 2),
— Slaughter Leveland Field (1936) 215 Mt.

The pre-Mississippian and pre-Pennsylvanian unconformities appear to play a major role in hydrocarbon migrations, generated by intense Permian subsidence.

The Anadarko and McAlester Basins, extending mainly over Oklahoma, exhibit the same overall arrangement, with a vast platform in the northern part, an area of stratigraphic traps, especially the Grand Seminole (1936), 60 Mt (see Section 4 of Part 2), and at least for the second, the Ouachita foredeep at the south, charactorizod by overthrust folds, and strong catagenesis only allowing gas pools to remain. These two basins are separated by a major granitic ridge with a southerly trend, containing a series of structural pools, of which the largest are Oklahoma City (1928) with 120 Mt (see Section 4 of Part 2, and Figure 77), Burbank (1920) with 117 Mt, and Sho Vel Tun (1915) with 200 Mt. The western margin of the Anadarko Basin is formed by a right-angled dorsal that extends towards the north of the Wichita Spur. This vast shoal, that combines structural and stratigraphic features, shelters the biggest gas field of the United States, and one of the world's largest, the Panhandle/Hugoton unit.

The Panhandle/Hugoton Field extends over the States of Kansas, Oklahoma and Texas, covering a length of 440 km and a width of 13 to 30 km. The productive area covers 20,000 km^2 (5 million acres). The trap is formed by a combination of structural factors (basement ridge that became active again at Panhandle), and stratigraphic and hydrodynamic factors, especially in the western and northern closure (Figure 191). The wildcat was drilled in December 1918 in Texas on a surface dome, and produced at 2000 m with a gas flow of 140,000 m^3 per day. This well still showed a potential of 220,000 m^3 per day forty-eight years later. The northern part, Hugoton, in Kansas, was discovered in December 1922. Different producing wells were drilled in subsequent years, but the size of this field was only finally realized in 1927, nine years after the first wildcat. The oil pool was identified in May 1921, in Texas, with the first well delivering 200 bopd from a granitic wash. However, it was only in 1925 that the vast potential was recognized with deliveries of 1000 bopd in the dolomite. Since this discovery, more than 20,000 oil wells have been drilled on this field, with 11,827 wells in production at the beginning of 1967. On the whole, this field contained about 200 Mt of oil and 2000 Gm^3 of recoverable gas, with cumulative production of 190 Mt of oil and 1160 Gm^3 of gas on

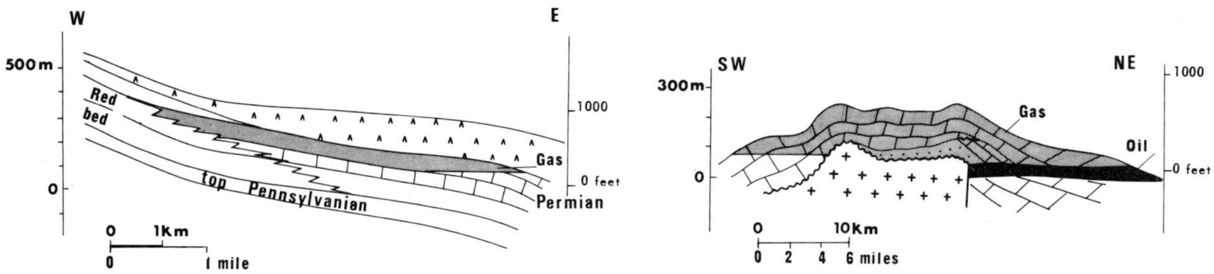

Cross-sections (Pippin, 1970).

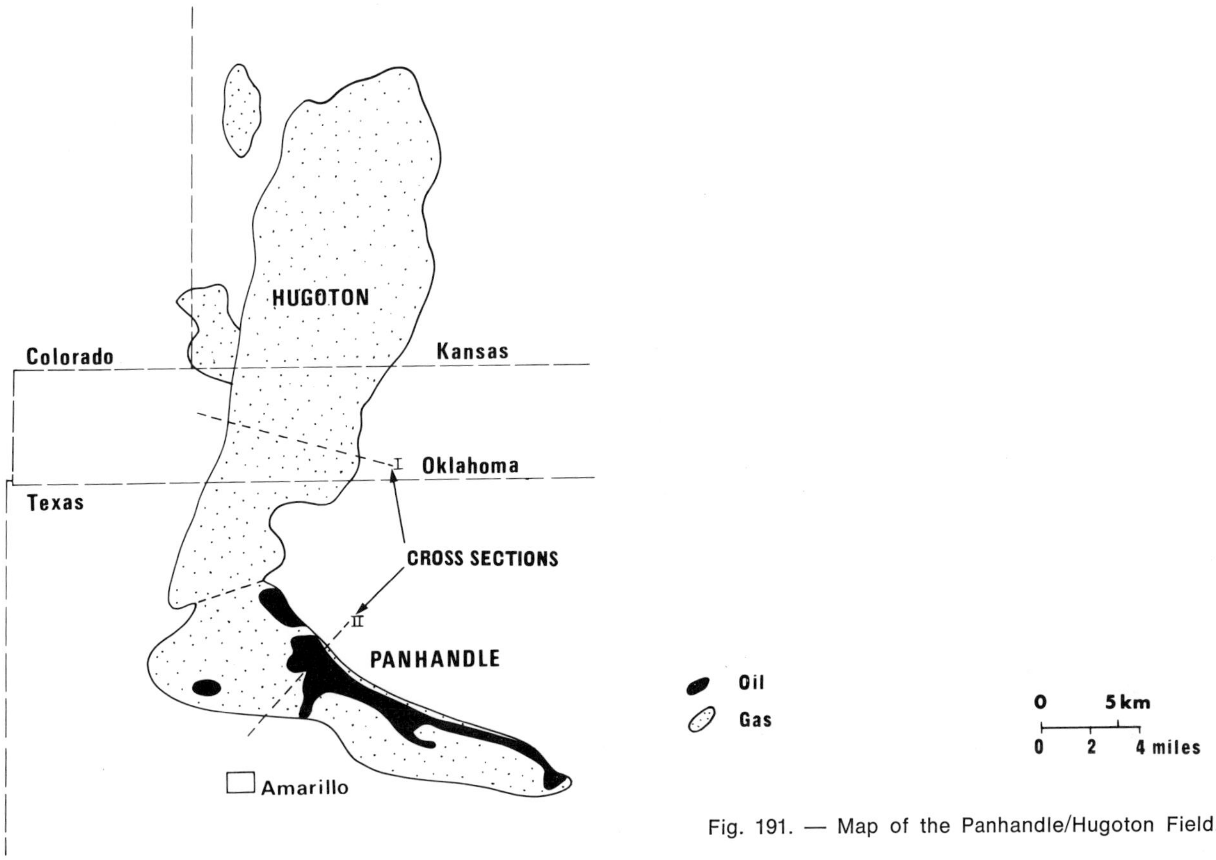

Fig. 191. — Map of the Panhandle/Hugoton Field.

1 January 1981. The gas contains an average of 0.5 % helium. A distinction is habitually drawn between the Panhandle and Hugoton Fields in this vast unit.

The trap of the Panhandle Field consists of a large ridge of Precambrian granites. Their altered area is directly covered by the folded Permo-Pennsylvanian carbonate formations. The reservoirs are formed by granitic wash and by coarse dolomitic and oolitic horizons, while the seal is provided by deposits of anhydrite and dense anhydritic dolomite. The oil is essentially found on the northwest flank of the field, and the water table is tilted westward. In the southeast, the gas rests directly on the aquifer, about 50 m above the water/oil interface on the other side. The flattening of the dips in the central part and a west/east hydrodynamic gradient lead to an extension of the productive area. The original pressure was 30 bar. The most productive reservoir, the oolitic brown dolomite zone, contains about 10,000 barrels per acre (400,000 m^3 km^{-2}), with the overall field yielding 1500 to 2000 barrels per acre (60,000 to 80,000 m^3 km^{-2}) in primary recovery. The oil and gas appear to originate in marine shales, from where they migrated into the granitic wash that played the role of a migratory path. Cumulative production by 1 January 1980 amounted to 180 Mt, with remaining recoverable reserves estimated at 14 Mt of oil.

The Hugoton Field is very similar to the Panhandle Field, and the main differences are :

— the exclusive presence of gas,

— the predominance of stratigraphic factors in the trapping pattern, with the updip closures to the west and north formed by a permeability pinchout of carbonate passing into red impermeable beds : the seal is also formed by the Wichita anhydritic dolomites that thin out westward.

The water table is tilted, sloping from + 50 m at the east to + 240 m at the west, over a distance of some 50 km, and permeability variations add their effects to those of the hydrodynamic gradient. At specific points, the level difference may be as great as 350 m, and the water table rises suddenly in the neighborhood of the pinchout.

***

Among the criteria that appear specifically to lie at the source of the potential of these basins, and especially of the Permian basin, are the following :

— a repetition of deeply buried source rocks,

— the juxtaposition of subsident troughs and shoals, and especially of vast reef formations exhibiting rich reservoirs,

— the intercalation of regional unconformity planes,

— excellent seals, especially of evaporites, regionally in an overlap position, at the top of the stratigraphic column,

— considerable Permian subsidence,

(Levorsen, 1934 and 1956; Galley, 1958; McNeal, 1965; Gatewood, 1970; Vest, 1970; Pippin, 1970; Adler, 1971).

3.4.3. **The Rocky Mountain Province**

It may seem artificial to speak of the Rocky Mountain province, considering its present breakup into a dozen productive basins, each displaying an individual personality. It is nevertheless necessary to consider this region before the Laramide orogeny, which cut out the vast Paleozoic and Mesozoic platform basins into independent intermontane basins, some of which continued their petroleum development (Figure 192).

In the Paleozoic, the basin, of which the general form is found in the northern part and in Canada, was in a situation closely comparable to that of the Volga-Urals Basin. The Jurassic and Cretaceous basins reflect this previous situation, especially in their eastern portions. This pattern was completely overthrown by the Laramide orogeny in the Late Cretaceous, which cut this early platform, except for its eastern margin which remains stable, into positive axes delimiting a number of deeply subsident basins during the Paleogene. This long and complex geological history lies at the source of a wide variety of petroleum zones and criteria.

The traps are mainly structural, with a large number of domes and anticlines of different types. Stratigraphic traps, reef structures, and especially sand pinchouts, are nevertheless numerous, particularly on the stable areas, and especially on the western margin. The reservoirs are also extremely varied, but mainly detrital, and extend from the Cambrian to the Tertiary. The seals are normally shale. The source rocks are distributed among the three stratigraphic systems, determining three major petroleum families. The Paleozoic series are rich in black shales functioning as a source rock, especially in the Mississippian and Pennsylvanian/ Permian. The crude oil generated appears mainly to be heavy, naphthenic and rich in sulfur. The Cretaceous shale series, with a marine to brackish facies, appear to give rise to a lighter oil, rich in gas and poor in sulfur. Finally, the Eocene deposits with a lacustrine facies, especially in the Green River and Uinta Basins, contain very rich beds of oil shales.

Main discoveries

The Williston and Paradox Basins, in the north and south respectively, have the bulk of their production from Paleozoic formations, and more precisely in the Mississippian for the former, and the Pennsylvanian for the latter. In the Williston Basin, which covers more than 200,000 km^2 in the United States and about 100,000 km^2 in Canada, the main pools, with one-third of the reserves, are aligned on the anticlinal axes of Nesson and Cedar Creek in the central, more subsident and better protected area (see Section 2 of Part 2). Its initial reserves were about 350 Mt, including nearly 80 % in the Mississippian. The Aneth Field, discovered in 1956 in the Paradox Basin (120,000 km^2), with reserves of around 50 Mt, concentrates about half of the discoveries of the basin. The trap and reservoir consist of a mound of algal limestones and calcarenites 60 m high, enclosed in black shales probably forming the source rock, and salt horizons of the same age.

The Cretaceous basins with a Paleozoic infrastructure are the most numerous. They can be divided into three main families, according to their position in the Cretaceous depositional basin :

— eastern basins, generally asymmetrical, in which the east flank corresponds to the stable margin of the depositional basin, the Denver and Powder River Basin,

— more central basins, completely remodeled by the Laramide orogeny : Big Horn, Wind River San Juan,

— western basins bounded by the great overthrust of the Rockies, the Green River and Uinta Basins.

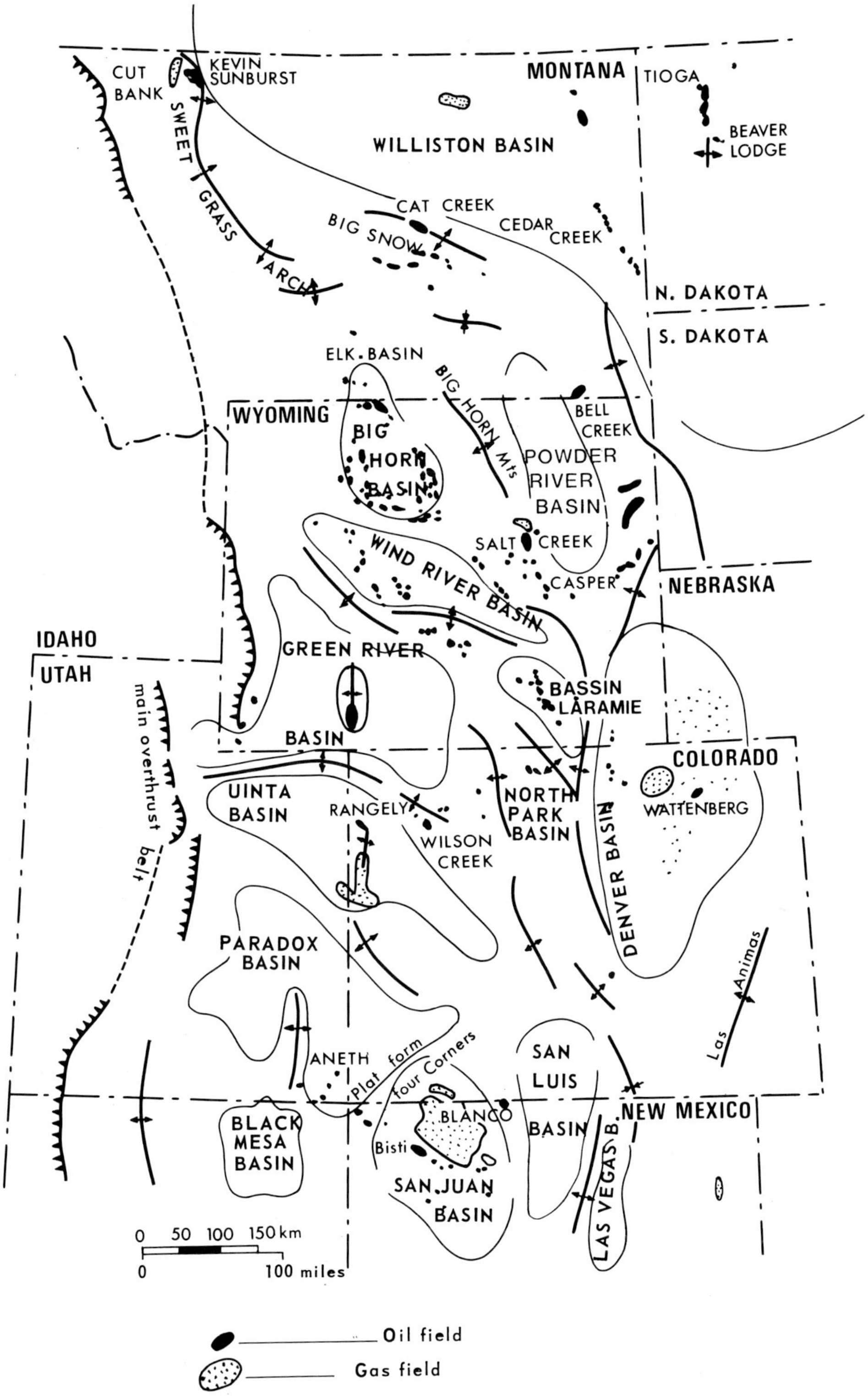

Fig. 192. — Framework of main Rocky Moutain provinces.

The Powder River Basin (100,000 km^2), straddling Wyoming and Montana, concentrates three-quarters of its discoveries in the southwest part, which is more subsident and folded. This is the location of the giant Salt Creek Field, discovered in 1889, an outstanding example of an anticline trap, exhibiting a closure of 500 m, productive in the Mississippian, then Pennsylvanian, the Jurassic and the Cretaceous, whose sand bodies, with a deltaic and offshore bar formation, are superimposed on the structure. This pool has produced more than 80 Mt, for recoverable reserves estimated at 100 Mt. The Bell Creek Field (1967) corresponds to a Cretaceous stratigraphic trap on the stable margin of the basin's eastern boundary (see Section 4 of Part 2).

The Denver Basin (150,000 km^2), formed mainly of a stable margin, is the seat of stratigraphic pools of modest size. Its cumulative discoveries amount to 100 Mt, distributed among over 2000 pools, of which the largest are Adena, with 10 Mt of reserves, and Wattenberg, with 36 Gm^3 of gas. The Big Horn Basin, covering 30,000 km^2, with a slightly asymmetrical form, contains slightly more than 350 Mt, half of which is concentrated in five fields, and distributed among a series of reservoirs stretching from the Silurian to the Cretaceous, with a preference for the Pennsylvanian. The biggest field was Elk Basin (1915) with 75 Mt. The Wind River Basin, with an area of 22,000 km^2, has a post-Laramide sedimentary series more than 3000 m thick, and contains about 200 Mt, distributed among around a hundred fields.

The Green River Basin, which extends over about 25,000 km^2 in southwestern Wyoming, is also productive from the Cambrian to the Eocene. The main fields are those of Lost Soldier (1916) and Wertz (1920), with reserves of 29 and 15 Mt respectively. The basin, in which cumulative discoveries are estimated at about 70 Mt, is chiefly remarkable for the size of its oil shale resources in the Eocene. The Uinta Basin, straddling Utah and Colorado, with an area of 40,000 km^2, displays close analogies with the Green River Basin, especially concerning its oil shale resources. Its eastern margin is the location of the Rangely Field, the biggest field in the Rocky Mountains, discovered in 1902, an outstanding anticline with 260 km^2 of closed area and a closure of 750 m, containing initial reserves of 110 Mt. The San Juan Basin, covering 50,000 km^2, produces oil in the Upper Cretaceous sand bodies of the submarine dune type (Bisti Field), as well as gas in the deep part of the basin, where entrapment is associated with hydrodynamism or capillary pressure (Blanco Field, 1927) containing 300 Gm^3 (see Section 4 of Part 2).

Conclusion

The Rocky Mountains group of basins, covering an area of about 1.5 Mkm^2, contains around 1.5 Gt of conventional oil and about 1000 Gm^3 of gas, distributed among some 800 pools. Added to these discoveries are about 300 Gt of oil, (in place), belonging to the lacustrine Green River Formatioin. This oustanding potential, concentrated in older Eocene lake deposits, illustrates the exceptional sedimentary characteristics that combine to form the biggest accumulations.

The very moderate potential (about 1000 t of oil per km^2) of these basins, however, reflects generally highly diversified but never exceptional geological conditions. The Paleozoic and Mesozoic platform basins do not offer particularly favorable depositional conditions. The Laramide orogeny, by setting the previous accumulations in motion, must have been a cause of considerable vertical migration. It is finally the 'small' and subsident Tertiary basins that satisfy the best conditions for the accumulation of organic matter, and for its transformation into free hydrocarbons (Davies, 1934; Pye, 1957; Haun & Kent, 1965; Haun *et al.*, 1969; Lyth, 1971; Curry, 1971; Mallory, 1972).

3.4.4. The Californian Provinces

The Californian provinces consist of the Upper Cretaceous and Tertiary intermontane basins spreading roughly parallel to the Pacific coast, along the great NW/SE trending shear faults, and separated by transverse ranges. These basins exhibit a number of common characteristics. Except for the Sacramento Basin, in the northwest part of the Great Valley, which essentially produces gas in the Eocene and Cretaceous, they are formed by a thick Neogene sand/shale series, often with a relatively deep facies, intersected by unconformities, folded and faulted, and especially prolific (Figure 193).

The geological history of this unstable area is complex and eventful. Roughly speaking, the following can be distinguished on the Franciscan series forming the petroleum basement :

- — a sandstone/marl Upper Cretaceous, locally very thick (more than 7500 m) deposited in a trough bordering an earlier island arc,
- — unconformably, after the major phase of the Laramide orogeny, an Eocene corresponding to a major sedimentary cycle,
- — after a new period of orogeny, a regressive Oligocene, largely continental,

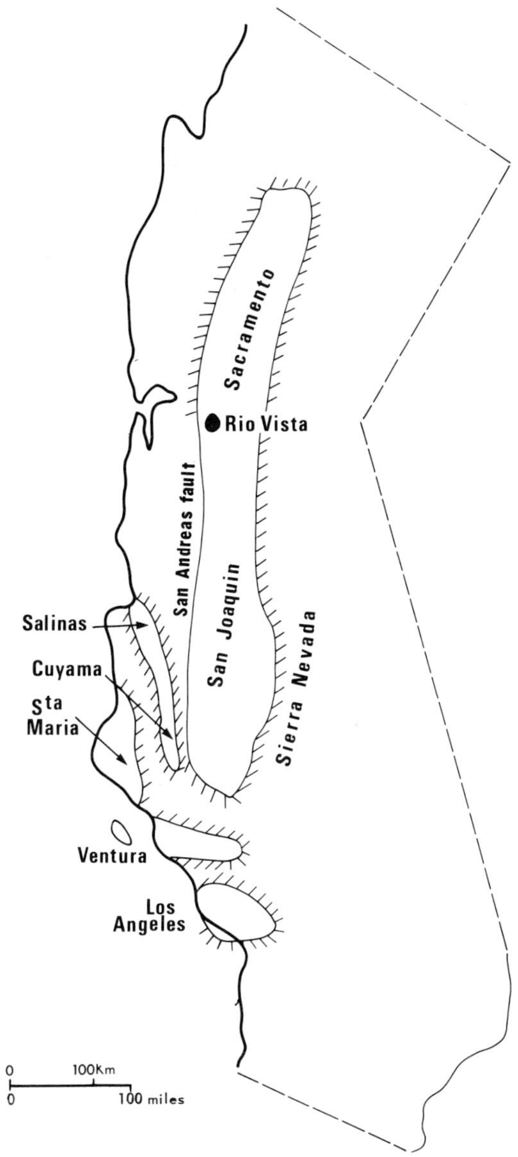

Fig. 193. — Map of Californian petroleum provinces.

— a highly developed Miocene in narrow and deep basins, reflecting intense volcanic activity,
— an unconformable sand/shale Pliocene, very thick in the open basins of the Pacific, and a Pleistocene which also testifies to great tectonic instability.

The bulk of the gas reserves is found in the Upper Cretaceous and Eocene, with 95 % of the oil in the Neogene, equally divided between the Miocene and the Pliocene. This intensely explored set of provinces, with an area of about 60,000 km^2, contains about 3.4 Gt of oil and 1000 Gm^3 of gas, distributed in slightly over three hundred fields, including a dozen giants (Figure 193). These discoveries are divided as follows :

— Sacramento Basin, about 200 Gm^3 of gas,
— San Joaquin Basin, 1.5 Gt of oil and 300 Gm^3 of gas,
— Los Angeles Basin, 1.2 Gt of oil and 200 Gm^3 of gas,
— Ventura/Santa Barbara Basin, 500 Mt of oil,
— Santa Maria Basin, 100 Mt of oil,
— Salinas Cuyama Basin, 85 Mt of oil.

Discussed below are the Great Valley Basins (Sacramento and San Joaquim). The Los Angeles Basin was covered in a previous section. These two provinces contain 80 % of California's oil.

Sacramento Basin

This Upper Cretaceous/Tertiary province has an area of 30,000 km^2. The sand/shale sedimentary series is very thick, up to 15,000 m in the deepest part, half of which lies in the Cretaceous. The reservoirs are essentially homogeneous and continuous Eocene sandstones, and more lenticular Upper Cretaceous sandstones, corresponding to deltaic and fluviatile deposits. Their characteristics are good, with porosity of 25 to 35 %, and permeability up to 1 darcy. This province is essentially a gas producer. The gas contains some nitrogen, especially in the eastern part, as well as CO_2. These gases originate in the Uppermost Cretaceous carbonaceous black shales, generally containing less than 1 % organic matter of the humic type, and subjected to temperatures ranging from 140 to over 200 °C.

The traps are formed by anticlines and fault blocks for the Eocene reservoirs, and by pinchouts and unconformities on the basin margins, especially for the Cretaceous sandstones. The province features a hundred gas fields containing more than 200 Gm^3 of gas. The main field is Rio Vista (1936), formed by a faulted

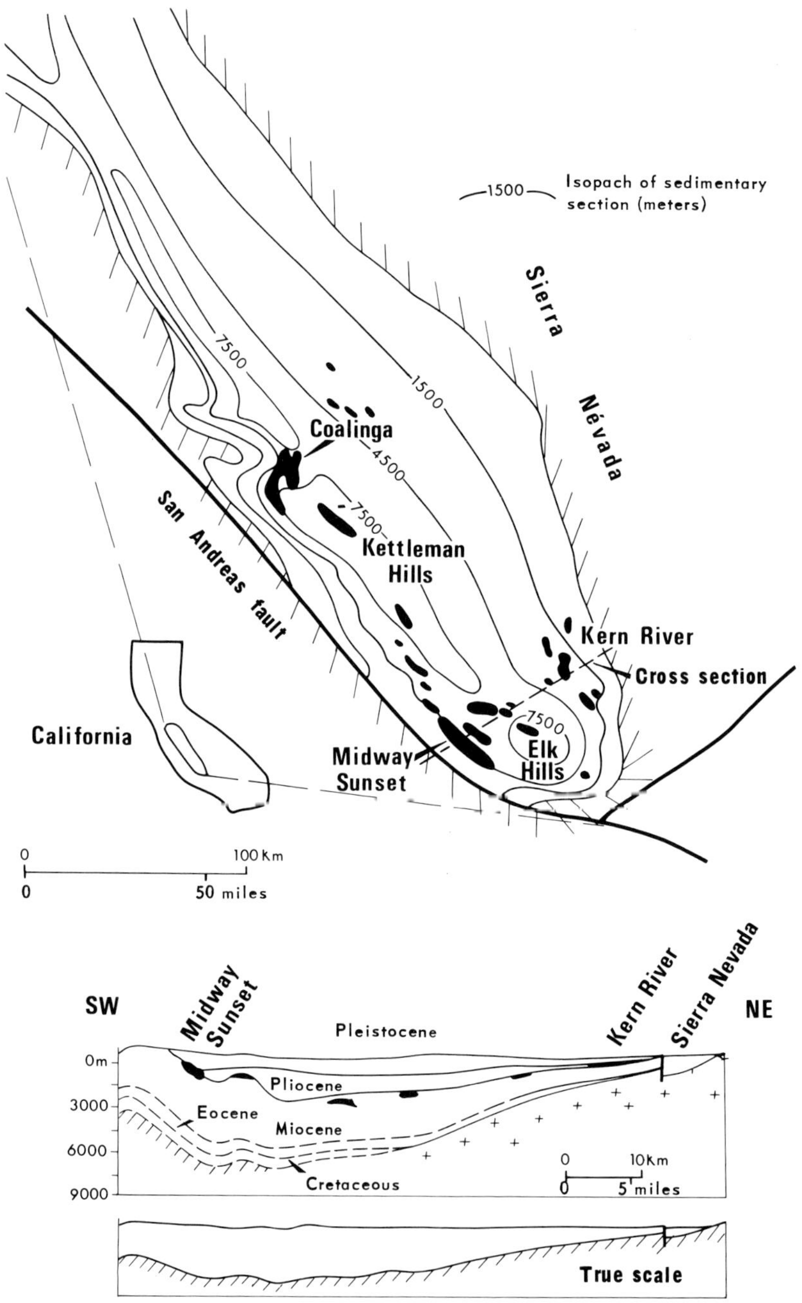

Fig. 194. — Map and cross-section of the San Joaquin Basin (California) (Simonson 1958 & Kilkenny 1971).

dome with a closed area of 90 km^2, productive in six Eocene/Paleocene sandstone horizons at 1100 to 1600 m. Its initial reserves are estimated at 100 Gm^3 (Morrison *et al.,* 1971; Zieglar & Spotts, 1978).

San Joaquin Basin

This basin, occupying the southern part of the Great Valley, covers an area of 40,000 km^2. Like the foregoing basin, it is of the elongated intermontane type and displays a clearly asymmetrical profile, in which the steeper west flank lies on the edge of the San Andreas Fault. The subsidence was gradually displaced southward, where the sedimentary series, up to more than 7000 m thick, is essentially Neogene (Figure 194).

The numerous, generally thick sand reservoirs have good characteristics. Their initial porosity, before burial, is 35 to 40 % irrespective of age, and decreases progressively with burial depth by 4.6 to 5.6 % per 1000 m. Permeability is as high as several darcys. As in many detrital reservoirs, the porosity/permeability relationships are fairly good, and permeability varies by a factor of 10 (for example, changing from 10 to 1 md) when porosity is divided by a factor of 0.6, decreasing from 18 to 11 %. The source rocks are represented by the Monterey shales (Upper Miocene), which are diatom shales and marls containing up to 5 % sapropelic organic matter. The formations were buried to more than 4000 m and carried to temperatures above 120 °C from the Early Pliocene during more than five million years (see Section 2 of Part 2) (Figure 30).

These reservoirs are covered by good shale seals, and seeps are frequently found on the basin margins. producing a few traps by asphalt plugging. The traps are many and varied. The anticlines, which are often pronounced and "en échelon", are the most numerous and account for nearly half of the pools. Fault traps make up about 20 %, and stratigraphic features, mainly transgressive onlaps, form the traps for nearly a third of the accumulations.

Main discoveries

About 1.5 Gt of oil and more than 300 Gm^3 of gas have been discovered in this basin, distributed among some 130 fields, including seven giants (six oil and one gas). The main fields on the western margin are :

— Midway Sunset (1894), an anticline under an unconformity with a neighboring pinchout on the edge of the basin, with the following characteristics : area 140 km^2, productive in 22 horizons between 200 and 1600 m, yielding a relatively heavy crude, with density 10 to 32 °API, initial reserves 270 Mt (Figure 109),
— Elk Hills (1919), a faulted anticline, containing 200 Mt,
— Coalinga, the first field discovered in 1887 in the northern area, where the Tertiary series is thin, and the reservoir is formed by an anticline and a neighboring onlap pinchout on the basin uplift : these two productive pools in the Miocene have estimated reserves of 190 Mt, of which 180 Mt have been produced,
— Kettleman Hills (1920), a gas producing anticline in Eocene and Miocene sandstones, with 100 Gm^3.

The largest field on the eastern margin is Kern River (1899), a stratigraphic pinchout which is productive in Pliocene sands and in the Jurassic basement, with reserves estimated at 175 Mt.

***

The reasons for the potential of these Californian basins can be found in the geological characteristics of the intermontane troughs, and, more specifically, in :

— shear tectonics giving rise to well-closed "en échelon" folds,
— strong and recent subsidence allowing the accumulation of thick sedimentary series, intersected by many transgressions and regressions,
— extensive deposits of effective source rocks, and high geothermal gradients,
— a high proportion of sands forming excellent reservoirs,

(Simonson, 1958; Callaway, 1971; Zieglar & Spotts, 1978).

3.5. THE PROVINCES OF THE GULF OF MEXICO

These provinces are divided between the United States and Mexico, between the vast Meso-Cenozoic sand/shale area of the Gulf Coast, terminating in the west beyond the Rio Bravo, and a very rich Mexican Cretaceous carbonate province, more specifically in the Faja de Oro, Reforma and Gulf of Campeche Districts, bearing a few similarities with the eastern part of the Gulf Coast, around Missouri, Alabama and Florida (Figure 195).

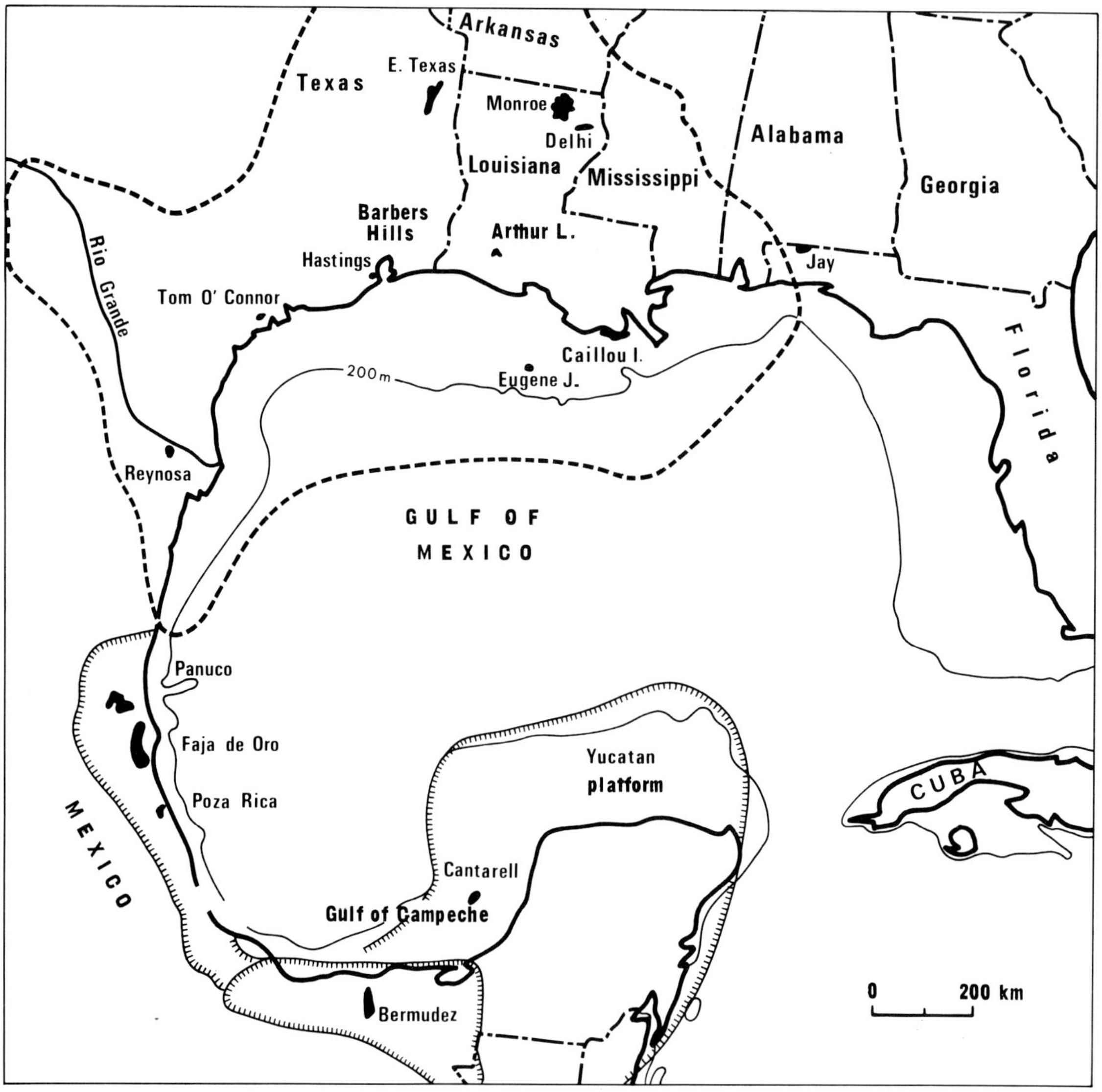

Fig. 195. — Map of main petroleum provinces of the Gulf of Mexico.

3.5.1. The Gulf Coast Province

This province corresponds to a vast Mesozoic and Cenozoic passive margin depositional basin, open on the present basin of the Gulf of Mexico. It covers an area of about 600,000 km^2.

Geological criteria

The sedimentary series plunges and thickens steadily towards the Gulf, at the edge of which it is up to 15 km thick. Due to the progressive stabilization of the margin, two major sub-provinces can be distinguished :

— an internal region in the north and northwest, Jurassic/Cretaceous, comparable to a platform province characterized by homogeneous and continuous facies, corresponding to vast transgressions and regressions of the sea, and on the structural level by an alternation of domes and broad depressions,

— an external region, corresponding to the alluvial plains and the continental shelf, with thick Cretaceous/Cenozoic paralic sedimentary prisms.

These units alternate in a number of sedimentary cycles, falling within a general regression framework, with a regularly spaced distribution of lacustrine, littoral, neritic and deep facies, which condition the distribution of the pools. The thick sand/shale series are intersected by synsedimentary faults and punctuated by deep salt domes. The mean geothermal gradient is 25 °C km^{-1}, slightly below average.

Petroleum criteria

The traps consist mainly of vast domes and broad pinchouts in the internal province, where a number of giant fields are found, such as East Texas and Monroe. In the coastal zone, the many average if not small traps are generally structural, mainly formed by salt domes, growth faults and roll overs. The salt domes offer an infinite variety of traps, from simple intumescences which undulate the Smackover in the north, to the slender columns 8 to 10 km high in the coastal regions. In the internal zone, deep domes give rise to more or less faulted anticlinal domes, at the source of large fields, like those of Hastings (120 Mt) (Figure 91), Conroe (120 Mt) and Webster (85 Mt). The roll overs and growth faults associated with them form a large number and a wide variety of pools, often compensating for the small enclosed volumes by an increased number of productive horizons. Thus some fields have large reserves, such as Seeligson (110 Mt), Tom O'Connor (125 Mt) (Figure 80), and Lake Arthur (75 Mt) (Figures 81, 111 and 195).

Early statistics compiled by Spillers (1965) for southern Louisiana show the following distribution of reserves and fields in the coastal zone :

— traps associated with salt domes : 62 % of oil reserves and only 18 % of the number of fields,
— anticlines and unfaulted domes, 19 % and 21 % respectively (and 47 % of gas reserves),
— faulted anticlines (roll overs) : 12 % and 20 %,
— fault traps 6 % and 31 %,
— stratigraphic traps, less than 1 % of reserves with 5 % of the fields.

Accumulations are found in both the upper and lower fault blocks.

The majority of the reservoirs are sands, and correspond to channels, and littoral and offshore bars. The pools are generally located in zones in which the proportion of sand is up to 30 to 50 % of the series. In the Upper Miocene in Louisiana, the productive facies correspond to deltaic plains for oil, and delta fronts for gas. The shale cap rocks often correspond to transgressive advances of the sea. The source rocks appear to be shales interbedded in the sands. Undercompacted formations may also play an important role in hydrocarbon genesis. It has been observed that the most productive zones are situated at the top of undercompacted shales, in the neighborhood of the 120 °C isotherm, and that the largest gas reserves are aligned at the south of the delta with a zone of thermal anomalies. Oil is most often found in littoral facies, and gas in deeper and more shaly neritic environments, generally more subsident. The presence of hydrocarbons in Pleistocene series argues in favor of rapid genesis, probably largely of biogenic origin (Rice & Claypool, 1981).

The crudes are generally light, paraffinic and low sulfur, and gas is abundant. Two major oil families may be distinguished :

— one unit exhibiting 70 % paraffins, 20 % naphthenes and 10 % aromatics, more or less characteristic of Mesozoic reservoirs,
— a Cenozoic group exhibiting an average of 43 % paraffins, 45 % naphthenes and 12 % aromatics.

The following substances have been identified :

— paraffinic oils, rich in waxes, and associated with deltaic environments,
— oil rich in aromatic compounds, corresponding to interdeltaic environments,
— Mesozoic high sulfur crudes in non-deltaic facies (Walters *et al.*, 1979).

The pressures are generally slightly higher than the hydrostatic pressure, and pressures in the productive horizons are generally below the pressures in the water bearing horizons, especially in the fault traps. High pressures are found at greater depths.

Main trends and discoveries

About 6 Gt of oil and 9 Tm^3 of gas, distributed among some 2500 pools, have been discovered in the Gulf Coast unit, which thus displays an average potential of about 10,000 t of oil and 15 Mm^3 of gas per square kilometer. The bulk of these hydrocarbons, or slightly over 3 Gt of oil, is found in the Mesozoic province, especially in the giant fields of East Texas (840 Mt) (Figure 105), Hastings (100 Mt), Conroe (100 Mt) for oil, and Monroe (200 Gm^3) and Carthage (170 Gm^3) for gas. The coastal province, consisting of a series of closely interlinked deltaic units, distributes its fields in accordance with the paleogeographic logic of sand bodies.

The Cotton Valley trend (Uppermost Jurassic) groups some hundred fields with a total of about 75 Mt, among the sand bars of the continental platform. The alignments of Comanche and Coahuila (Lower Cretaceous) contain about 300 pools with 230 Mt of oil and 300 Gm^3 of gas. The Wilcox (Lower Eocene) aligns some 540 fields with cumulative production of nearly 200 Mt along older shorelines. The Yegua (Middle Eocene), distributed in the framework of a vast deltaic unit, includes 700 fields with cumulative production of about 220 Mt. The Frio trend (Lower Miocene) spreads slightly more than 200 accumulations over a belt measuring 1100 km by 60 km. The Upper Miocene contains more than 500 gas pools with a total of nearly 3 Tm^3, and the Pliocene has about sixty fields, the biggest being Eugene Island 330 with 30 to 70 Mt (225 to 500 Mbl) and 26 Gm^3 of gas (950 billion m^3) (Holland *et al.*, 1980, St. John, 1980).

On the whole, the habitat in this coastal zone is praticularly dispersed, and the average field yields 2 to 3 Mt of oil and 3 to 4 Gm^3 of gas only. However, at the cost of an especially intensive exploration effort, this province appears to be remarkably rich. Among the reasons for this potential are the following :

— rapid subsidence,
— intimate imbrication of sands and shales with a relatively high percentage of sand, about 30 to 50 % in the richest zones.
— the repetition, at the base of the sedimentary sequences, of transgressive marine shale seams performing the role of a seal and perhaps also source rock,
— a large number of complex structural traps, roughly outlined if not formed during sedimentation,

(Grimes, 1970; Meyerhoff *et al.*, 1968; Mills, 1970; Myers, 1968; Spillers, 1965; Stanley, 1970; Tipsword *et al.*, 1971).

3.5.2. **The Mexican carbonate province**

Mexico is an old petroleum country which rediscovered a new and brilliant future in recent years. Production is obtained from the following (Figure 196) :

— Tertiary sand/shale series situated in the northeast, in the prolongation of the Gulf Coast Basin, and in the Chicontepec and Macuspana foredeeps, on the edge of the Gulf of Mexico,
— essentially in Cretaceous carbonate series, locally Jurassic and Paleocene, extending more or less parallel to the coast, from the Tampico area to offshore Yucatan, which are examined here.

Recent discoveries have also been reported :

— in the folded Mesozoic (Upper Jurassic sandstones and Lower Cretaceous dolomites) of the region of Coahuila (Sabinas Basin) in the north center, for gas,
— in Baja California, in Cretaceous/Tertiary series.

Mexico, which has been producing since the turn of the century, now boasts :

— cumulative production of around 1000 Mt,
— proven reserves of 10 Gt of oil and 2200 Gm^3 of gas,
— ultimate potential reserves of about 30 billion tons.

The carbonate province is divided into five main units :

— three old zones in the north : Ebano Panuco, Faja de Oro and Poza Rica, forming the Tampico Province,
— two new zones in the south, Reforma and the Gulf of Campeche.

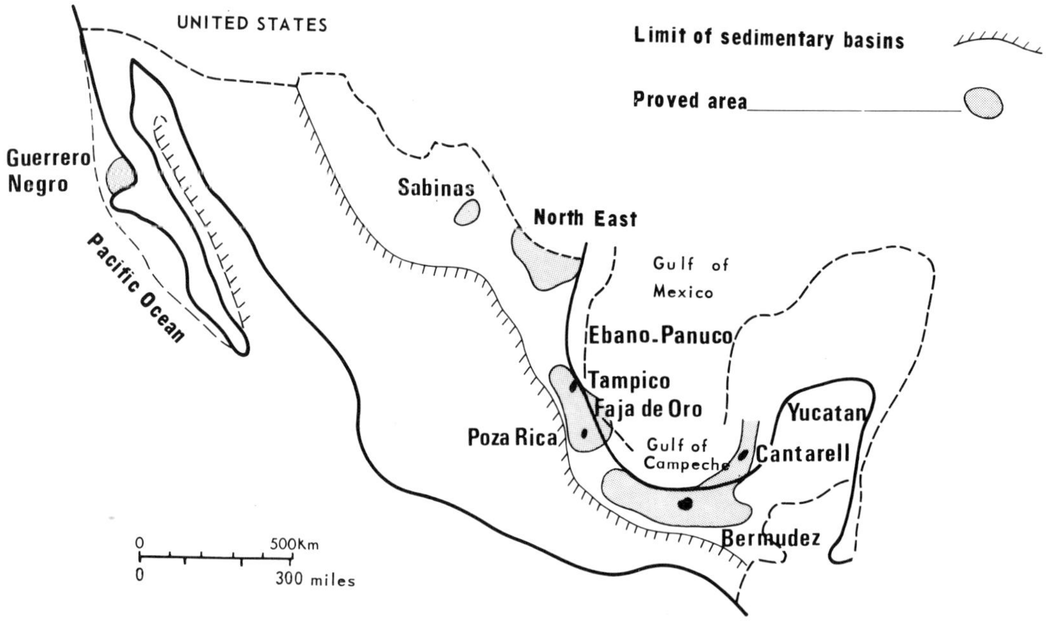

Fig. 196. — Map of petroleum provinces of Mexico.

3.5.2.1. *Tampico Province*

This region, which covers about 100,000 km^2, corresponds to a Mesozoic basin in which several regional highs are individualized in the Cretaceous, conditioning the distribution of the pools. In the northwest, in the burial of the Tamaulipas Siérra, compact micritic limestones are deposited (Tamaulipas facies) with bituminous beds, fractured by folding. This is the Ebano Panuco District, with initial reserves of around 140 to 180 Mt disseminated in some fifty fields (Figure 196).

At the periphery of a shoal (Tuxpan Island), the limestones thicken substantially, with a rudistid biostrome facies, recalling the present Bahamas. This is the El Abra facies and the Faja de Oro District. The succession of highs in the form of an arc along the coast, closing offshore in an oval some 200 km long, forms a series of pools with initial reserves estimated at about 250 Mt (Figure 197). Aligned slightly to the west and at the edge of this belt are rudistid limestones and algal biodetrital limestones with reef affinities known by the name of the Tamabra facies. This second barrier forms the Poza Rica trend, which contains the biggest field of the province, with slightly over 400 Mt. Oil is also found in this province, particularly in Jurassic calcarenites in the Constitutiones (23 Mt) and neighbouring Arenque (about 100 Mt) Fields, which are often linked together, and also in the San Andres District, near Poza Rica.

Faja de Oro area

This zone corresponds to a shallow platform, subsident in the Lower and Middle Cretaceous, during which miliolitic and rudistid limestones were deposited to a thickness over 1000 m. Emergence and erosion affected these series in the Late Cretaceous, and the karstified limestones were only covered by marine shales in the Eocene to Oligocene, and then buried during the Neogene (Figure 198). The reservoirs have excellent characteristics (mean porosity 35 %, very high permeability, producing outputs up to 38,000 t day^{-1}) resulting from the conditions of deposition, alteration and fracturation. Their depth ranges from about 400 m in the north to 2400 m in the south. The seals are provided by Upper Cretaceous to Oligocene shales several hundred meters thick.

The crude oil is of the intermediate type, with relatively high density, 13 to 30 °API, rich in sulfur (3 to 4 %). The oil appears to come from Upper Jurassic black shales by vertical migration. The geothermal gradient is high, 65 °C km^{-1}, and seeps (called Chapopoteras) are numerous and large. The traps are formed by the topographic highs of a narrow belt 0.3 to 7 km wide, continuing in the form of an arc over 200 km, of structural origin and modeled by erosion. The slopes are gentle, 1 to 5 degrees towards the east, steeper and up to 45 degrees towards the west. The closures of these highs are about 100 m on the average, and up to 250 m in Cerro Azul. The productive areas range from 1 to 18 km^2.

The initial reserves of Faja de Oro are estimated at about 250 Mt of oil of which 220 Mt have been identified onshore, and about a dozen offshore. The main pools are onshore, in the southern part or Nueva Faja de Oro :

— Santa Agueda, which has supplied nearly half of cumulative production,
— Ezechiel Ordonez, next to it in the southeast, from which nearly a quarter of total production of the zone has been extracted,
— Acuatempa/Mozutla,

and further to the north, in the older Faja :

— Cerro Azul 45 Mt,
— Los Naranjos 45 Mt.

A dozen fields of a comparable type, forming a symmetrical arch, have been found offshore, and Atun, the biggest, has reserves of about 55 Mt.

Poza Rica zone

The trend of Poza Rica, the name of the largest field, extends parallel to and on the west of Faja de Oro. The reservoirs are formed by Tamabra limestone with an average thickness of 260 m at a depth of about 2000 m. This formation contains calcarenites, bioclastic and rudistid limestones, which are more or less dolomitized and fractured. The porosity ranges from 8 to 25 %, permeability from 5 to over 1000 md. The initial productivity of the wells was 120 t/day. The seal is provided by Upper Cretaceous shale, marl and carbonate series, 200 m and thicker. The intermediate crude displays average density, 21 to 31 °API, and appears to originate in Portlandian black shales and limestones, commonly bituminous.

The traps are of the combination type, like that of Poza Rica, whose closure is provided updip by a transition to compact limestone facies. The main field, Poza Rica, discovered in 1930, has estimated reserves of more than 300 Mt. This overall zone contains about 430 Mt of recoverable oil, distributed among a dozen fields. Apart from Poza Rica, the main fields are San Andrés/Hallazgo (1957), with 70 Mt, a stratigraphic trap, and Tres Hermanos (1959) with 20 Mt.

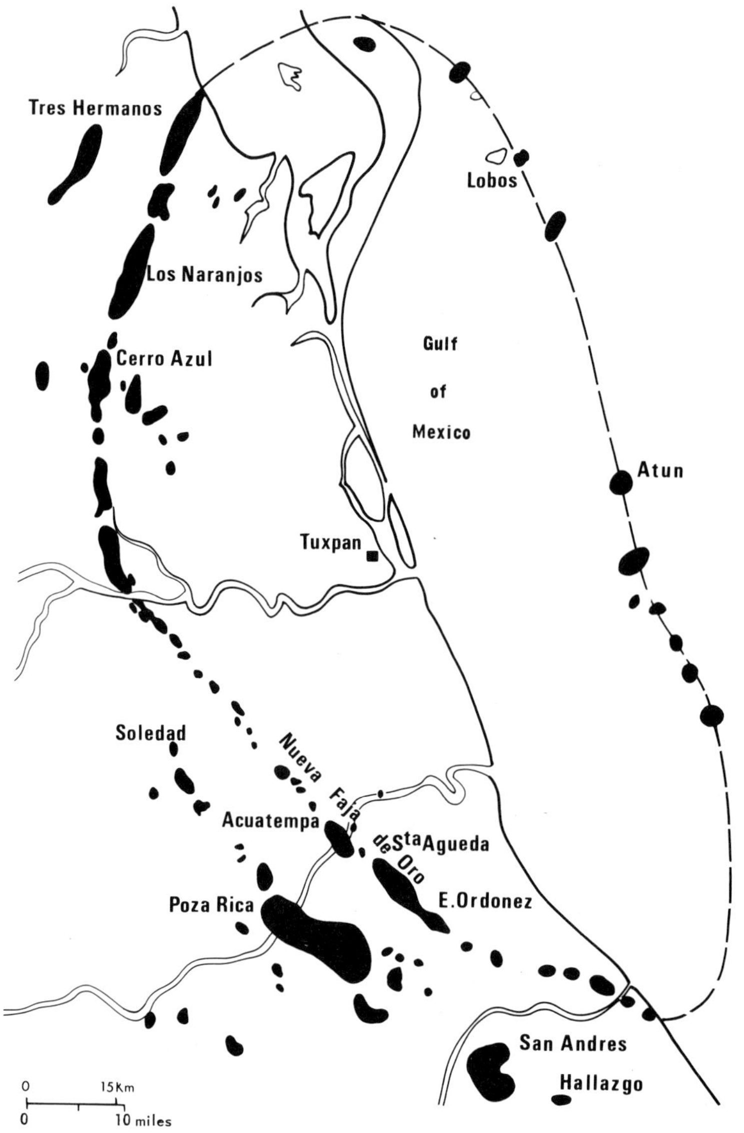

Fig. 197. — Map of fields in the Faja de Oro and Poza Rica zones (Mexico) (Coogan *et al.*, 1972).

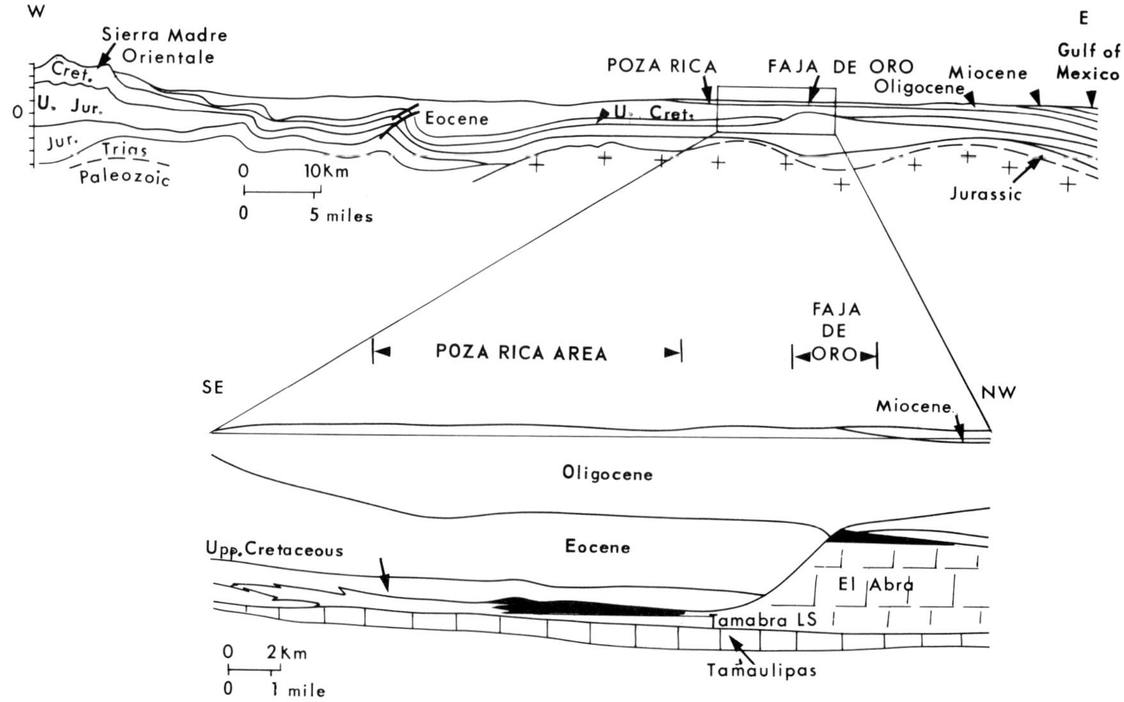

Fig. 198. — Section across Tampico province (Mexico) (Lopez Ramos, 1959).

3.5.2.2. *Reforma/Gulf of Campeche province*

This second carbonate province, which exhibits close analogies with the former, extends in the southern part of the Gulf of Mexico, and is divided into two sub-provinces :

— the Reforma zone onshore,
— the second zone on the continental shelf, in the Gulf of Campeche, off the Yucatan Peninsula (Figure 199).

The onshore zone is located in the States of Tabasco and Chiapas, around Villahermosa. It represents a regional arch corresponding to the plunning nose of the Chiapas anticlinorum and separating the two Miocene basins (Isthmian or Saline on the west, Macuspana on the east), exploited since the turn of the century (initial reserves about 170 Mt and 215 Gm^3 distributed among about forty pools) (Acevedo, 1980).

The known stratigraphic series begins with the Jurassic salt deposits, giving rise to a number of domes, and black shales and limestones, which appear to form the main source rock. A thick carbonate series is developed above, extending substantially from the Callovian to the Cenomanian, under three large types of facies recalling the series of the Tampico province :

— towards the Gulf, under the slope and outer shelf, fine-grained limestones of the deep type, intersected by black shales, comparable to Tamaulipas facies,
— to the south and east of the latter, detrital carbonates and calcarenites of bank edge talus, with reef facies, usually dolomitized, closely comparable to the Tamabra facies, forming the main reservoir,
— farther east and south, towards the Yucatan Peninsula, shallow water and lagoonal carbonates of back reef, with compact limestone and anhydrite seals, recalling the El Abra facies.

The stratigraphic column begins in this area with salt series of the Late Jurassic, which are overlaid by Tithonian black shales. During the Neocomian/Aptian era, the period of the north/south opening of the Atlantic Ocean, a marine transgression covered these regions, with carbonates and evaporites of very shallow water in the Yucatan Peninsula, talus facies highly dolomitized in the Reforma area, and carbonate of deeper water deposition westward. The Albian/Cenomanian had a very similar history and can be characterized as follows : large areas of the present region of Mexico were covered by a shallow sea; reef banks grew with the El Abra facies on the Yucatan Peninsula, bordered westward on the bank edge talus by dolomitized bioclastic and calcarenite fans (Tamabra), which constitute the principal reservoir of the Reforma trend.

After this maximum transgression, the region began to emerge and certain large carbonate banks were at least partially exposed and altered.

The Paleocene consists mainly of marine transgressive shaly and sandy sediments, generally of very deep water origin, some being deposited in fore-deeps that formed during the Laramide orogeny.

Post-Paleocene strata are represented by a thick sand/shale series, more calcareous in the offshore zone, reaching a thickness of 3000 to 4000 m. Unconformities are at the top of the Cenomanian, and also in the rest of the Upper Cretaceous, the Paleocene and the Lower Eocene.

The reservoirs are formed by the thick bioclastic and calcarenite carbonate series, commonly dolomitized, fractured and karstified, of the Callovian through Lower Cenomanian, in the areas where the section is most complete, and from the Callovian through the highest Cretaceous unit, where the section is not complete, especially offshore. The oil-bearing columns are continuous on an average of about 500 m and can be as high as 1080 m. The Paleocene limestones are also productive in the offshore zone. The reservoir unit ranges from 200 to 300 m at depths of 3200 to 5100 m in the Reforma section, and is 1260 to 3500 m offshore. The reservoir characteristics are excellent, with permeability up to 7 darcys, yielding average outputs of 700 tons per day with peaks of 4500 t/day (32,700 bopd) (Stewart-Gordon, 1981).

The seals are formed by Tertiary shales up to several thousand meters thick, and especially Late Paleocene overpressured shales, locally Cretaceous shales. The oil originates in deep Jurassic and Lowest Cretaceous source rocks, by vertical migration, or perhaps laterally from Cretaceous shale facies. The Reforma zone produces a light crude oil, density 25 to 45 °API, with high sulfur (1.6 %). The offshore zone produces a heavier oil, 20 to 33 °API (average 22.5 °API), with higher sulfur (1 to 3 %). The GORs are said to be typically 1400/1 to 1500/1. However, several gas condensate fields have since been found onshore with much higher GORs. Commonly there is a gradation from the highest GOR values at the top of the producing column to the lowest GOR values at the base of the producing column (Viniegra, 1981). The average pool temperature is about 120 °C. The traps are largely structural, associated with salt pillowing, basement horsts, and Laramide tectonics, superimposed on an original and regional stratigraphic facies variation. The dips are low, and normal and wrench faults abundant.

It is not possible to provide an idea of the potential of a province undergoing rapid development, whose continually changing physiognomy is identified by new wells. One fact is certainly true : this province henceforth appears to be extremely prolific. In early 1981 its reserves were estimated at about 6 Gt, distributed among a few giant fields, formed by the coalescence of pools initially considered distinct, as often occurs in

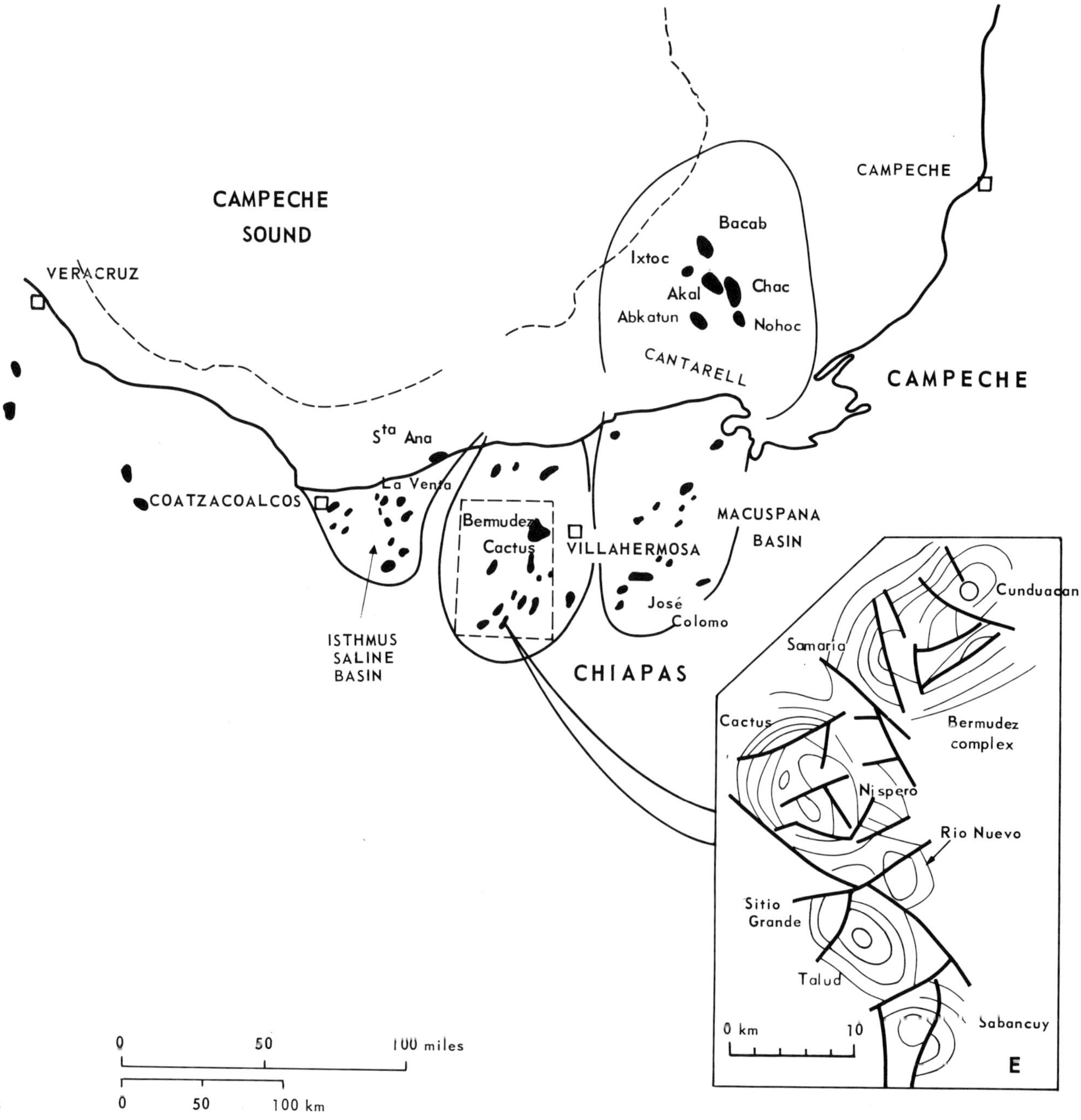

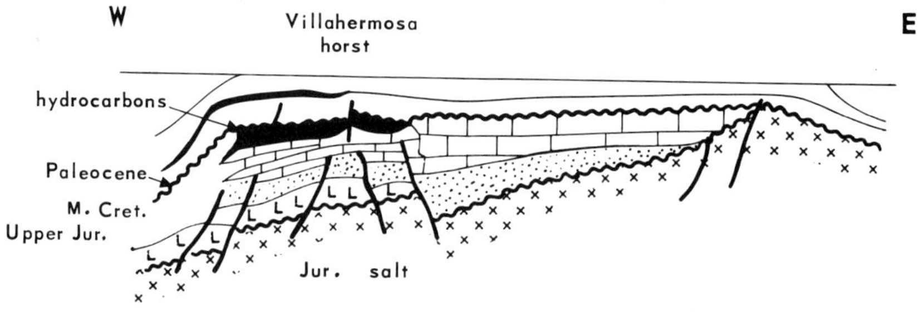

Fig. 199. — Maps and stratigraphic cross-section of the southern province of Mexico (Meyerhoff, 1980; Meneses de Gyves *et al.*, 1980).

this case. The Reforma zone includes the giant A.J. Bermúdez Field with an area of 150 km^2, and reserves estimated in early 1981 at 900 Mt (6.5 million barrels). Among the other fields are Sitio Grande (1972), Cactus (1972), Agave and Sabancuy, containing around 100 Mt. Offshore, the first discovery was Chac (1976). A total of eleven discoveries have since been made and infill-drilling is showing that some of these are in fact single giant field complexes. The best-known example to date is the Cantarell Complex, which includes three separate discoveries, Chac, Akal and Nohoch (Viniegra, 1981).

In January 1979, Pemex estimated the proved crude oil and gas liquids at about 12.3 Gt (16,500 billion barrels) and nearly 420 Gm^3 (15 Tcf) of gas in the Reforma/Campeche shelf trend. The proved plus probable reserves exceed 4.75 Gm^3 (34 billion barrels) of liquids at the same date (Viniegra, 1981).

The Tampico/Campeche carbonate province, that contains cumulative discoveries of about 7 Gt to date, with a considerably higher potential, exhibits the characteristic of containing the bulk of its hydrocarbons in several reef type carbonate units, mainly Upper Jurassic through Cenomanian. The good reservoir characteristics stem from favorable depositional conditions, and also karstic mechanisms and fracturing. Trapping appears to be associated both with sedimentation conditions of a sub-reef character and with vertical movements in the Late Cretaceous. The oil appears to originate in black Jurassic and Lowest Cretaceous shales and limestones corresponding to the first marine invasions, immediately after the evaporitic period. It seems to have been expelled during the considerable Cenozoic subsidence (Cervera, 1967; Viniegra *et al.,* 1970; Coogan *et al.,* 1972; Bishop, 1980; Meyerhoff, 1980; Meneses de Gyves *et al.,* 1980; Acevedo, 1980; Viniegra, 1981).

Briefly, the exceptional potential of the Mexican Cretaceous formations is related in particular to :

— a porous volume of carbonates, favored by depositional conditions on the sides of broad reef zones, by emersions and by intense fracturing,
— a thick Cenozoic shale series offering an effective seal and deep burial,
— vast structural traps uplifted and fractured progressively by salt movements,
— effective source rocks at the base of the series.

3.6. THE VENEZUELAN PROVINCES

Venezuela is occupied by a complex area of troughs and cordilleras corresponding to the prolongation and northeastern ramifications of the Andean Cordillera, surrounding the Guyana Shield to the south. Proceeding from south to north, the following major units can be distinguished (Figure 200) :

— a vast foredeep basin, the prolongation of the great sub-Andean basin, and including, on both sides of the El Baul swell, the Barinas Apure Basin to the west, and the Orinoco or Oriental Basin to the east, with its prolongation up to Trinidad,
— bordering this miogeosyncline on its northern limb, the Merida Cordillera and the Caribbean Andes, with a geosynclinal character and overthrusting southward,
— to the northwest of the Merida Cordillera, the Falcon Basin and the Maracaibo Basin, of the intermontane type, bordered on the northwest by the Sierra de Perija, of very recent formation.

From the petroleum standpoint, two major provinces are individualized in this unit, the Orinoco/Trinidad province to the southeast, and the Maracaibo Basin to the northwest, accounting for a total of 8 Gt (0.2 Gt on Trinidad) distributed among nearly 300 fields.

3.6.1. The Orinoco/Trinidad province

This basin, of the miogeosyncline type, with an area of 120,000 km^2, is filled with Cretaceous/Cenozoic deposits, with more recent formations developing eastward. It includes a platform zone to the south, resting on the Guyana Shield, and a subsident trough to the north, on the edge of the Caribbean Andes overthrusting towards the south. These two units are linked by a faulted hinge zone that plays a major role in the trapping of the oil, especially in the median high. The evolution of the basin and the Caribbean Cordillera is reflected by transgressions and regressions of the sea on the southern platform, also contributing to the trapping of hydrocarbons. The deposits are generally of the platform type in the Cretaceous, with paralic facies in the Tertiary. They are nearly 10,000 m thick in the northern trough and pinchout towards the shield.

The sandstone reservoirs stretch from the Eocene to the Upper Miocene. In the Oficina zone, the largest, the Oligocene and Lower Miocene sands, often corresponding to channels in a transgressive series, account for about 30 % of the series. They exhibit an average porosity of 20 to 25 %, permeability of 300 to 1000 md, and a thickness of about 10 m. Well productivities vary from 40 to 400 t day^{-1}. On the north limb,

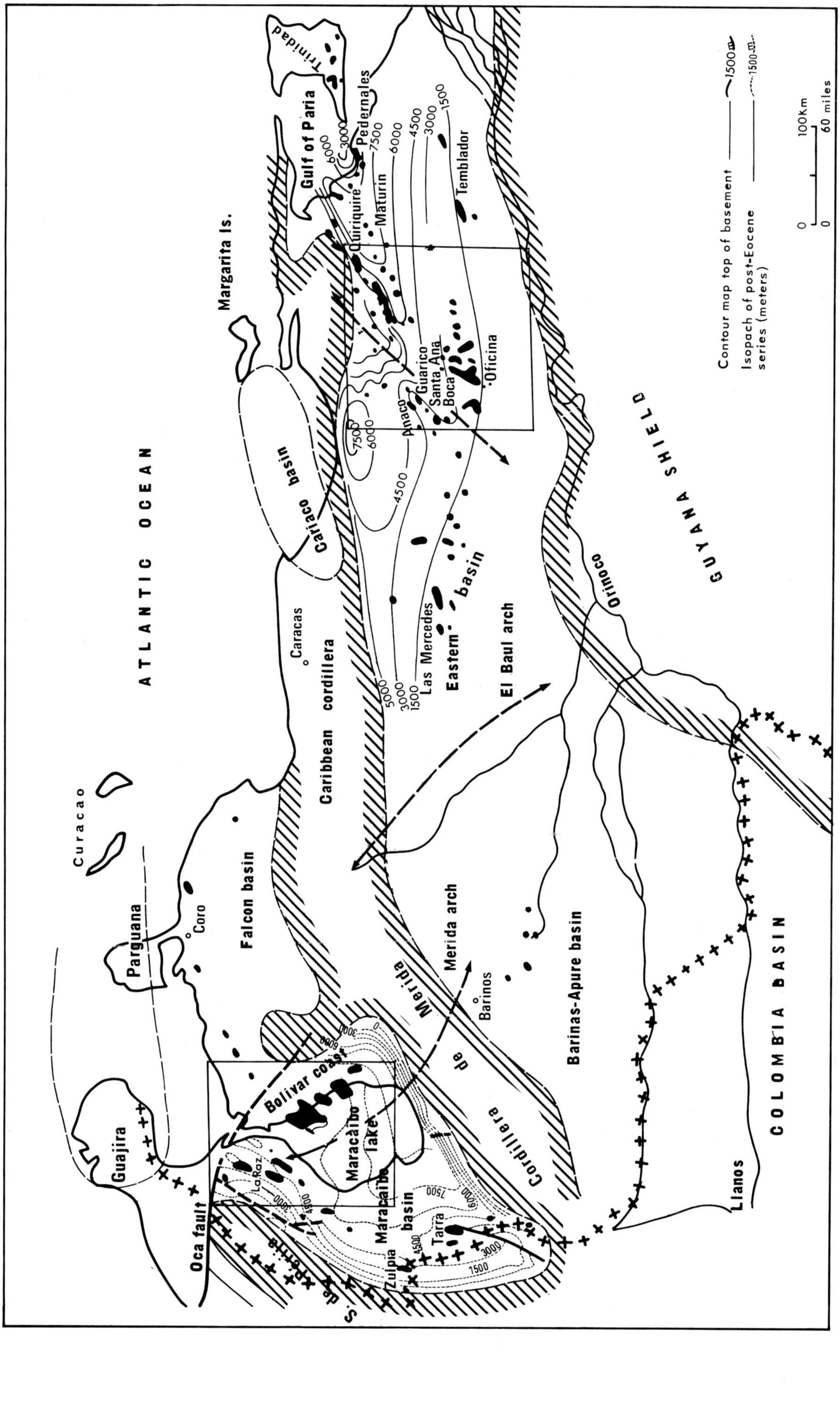

Fig. 200. — Map of sedimentary basins of Venezuela and Trinidad.

the Miocene sands correspond to turbidites, belonging to a second sedimentary cycle specific to this region. The seals are provided by shale horizons about 10 m thick.

The hydrocarbons, chiefly liquids, originate in mainly Oligocene marine and brackish shales intercalated with the reservoirs, or in deeper horizons. In the Oficina zone, the geothermal gradient is 36 °C per km. The naphthenic crude is generally rich in paraffin, and its density, which varies widely, is often high, especially when close to outcrops. The sulfur content may be up to 2 %. The traps are generally formed by faults, or are of the combination type, with faults parallel to the basin margin and oblique faults, pinchouts or noses (Figures 201 and 92). On the northern limb, the pools are formed by anticlines, intra-Miocene unconformity pinchouts (Maturin zone) and transgressive overlaps (Quiriquire), by permeability barriers, and by tar plugs, especially large and yielding considerable seeps in this zone in the process of rapid erosion. One of these remarkable seeps is the Guanoco tar lake, situated at the pinchout of onlapping Mio-Pliocene sands on the Cretaceous.

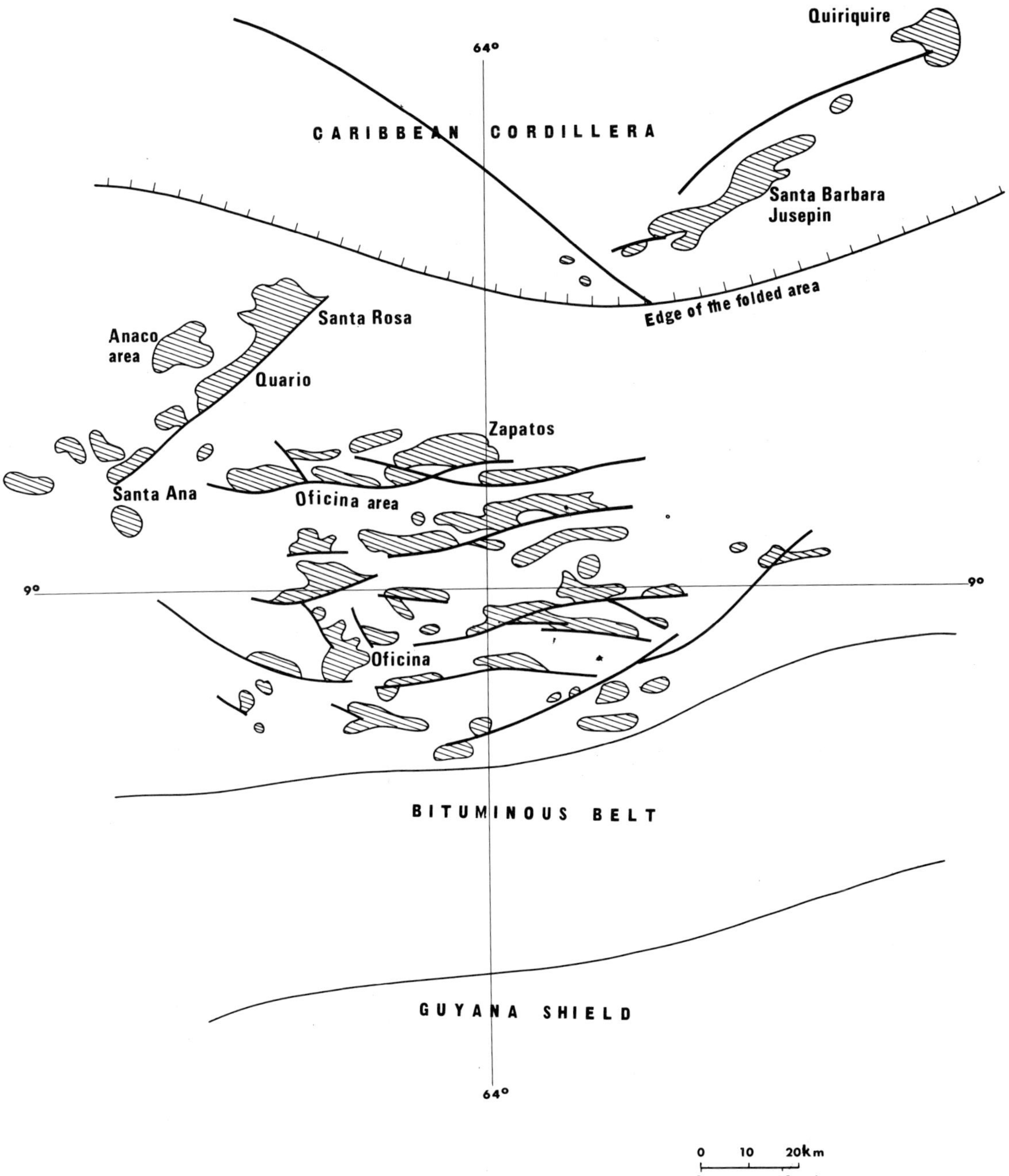

Fig. 201. — Map of fields in the central part of the Orinoco Basin (Venezuela).

Main discoveries

The eastern basin has initial recoverable reserves estimated at about 1.8 Gt (including 1.2 Gt in the Oficina/Anaco District) distributed among some 210 fields. The main productive zones are clustered on the northern and southern limbs in the central part, crossed by a transverse high.

South margin

On the southern limb, the richest zone is the Oficina/Anaco District, which corresponds to a transverse arch, with 140 Mt distributed in some 160 fields. The biggest are those of Oficina, Santa Rosa/Guario (80 to 90 Mt). Productive areas are generally small (not more than about 10 km^2) and larger in the slightly folded Anaco zone.

The Orinoco tar belt (Faja Petrolifera)

On the southern uplift of the Guyana Shield, the Cretaceous and Tertiary sandstones concentrate what is probably the world's largest oil accumulation in a deltaic environment. This belt of tar sands stretches over nearly 500 km from east to west, with a width of about 50 km over an area of around 30,000 km^2. It passes progressively towards the north and the Oficina zone via heavy oil accumulations.

This tar belt covers a monocline plunging 2 to 4 degrees northward containing a few low relief shoals and many synthetic or antithetic normal faults. The impregnated sands, with an average porosity of 25 %, have a cumulative thickness between 50 to 100 m, at depths from 0 to 1000 m. Trapping is associated with pinchouts on the flanks of the highs, facies changes and faults, and also with tar plugs updip.

The heavy asphaltic crude, with a density of 1 to 10 or 12 °API, rich in sulfur (up to 5 %), nickel and vanadium (maximum 800 ppm), originates in the Oligocene source rocks and pools of the Oficina zone. Reserves in place are estimated between 100 and 160 Gt, and up to 300 Gt by some geologists.

North margin

On the northern limb, the two zones of Maturin and Quiriquire contain slightly more than 300 Mt, two-thirds of which are concentrated in the Quiriquire and Santa Barbara/Jusepin Fields (1938) with 30 Mt. The Quiriquire Field, discovered in 1928, is a specific example of a very shallow suboutcropping pinchout trap. Its cumulative production is 100 Mt, out of initial reserves estimated at 140 Mt (see Section 4 of Part 2, and Figure 115). Two fairly large fields are situated outside this central area :

— Las Mercedes (1941) in the eastern part, with estimated reserves of 25 Mt,

— Pedernales (1933) at the eastern end of the north limb, in the Orinoco Delta, formed by a diapiric anticline with shale intrusions, with reserves of 16 Mt, a type of formation found further west on Trinidad.

Trinidad

The southern part of this island, which belongs to the northern flank of the East Venezuela Basin, with an area of 2000 km^2 (10,000 km^2 with the neighboring offshore zones) is characterized by a thick Cretaceous/Cenozoic series more than 10,000 m thick, intensely folded and faulted. The series stretching from the Paleocene to the Miocene are represented by relatively deep marine shales, fairly rich in organic matter and considered as source rocks, intersected by turbidite sand bodies. These shales are often undercompacted and form diapirs and mud volcanoes. The geothermal gradient is low, around 20 to 24 °C per km.

The reservoirs are represented essentially by Upper Miocene sands originating in the Guyana Shield and belonging to a vast deltaic unit, up to 5000 m thick, formed by three sedimentary cycles. The sand/shale ratio is especially high, often greater than 1. About 85 % of the oil discoveries are located in the Miocene delta front sands, while the gas is found preferentially in the thicker sands at the center of the deposit.

The traps are formed mainly by synsedimentary faulted anticlines (like the Guayaguayare Field) (see Section 4 of Part 2 and Figure 102) pierced by shale diapirs and distributed around the flanks of the basin. The main productive diapiric folds are ailgned substantially east-west. Traps, mainly gas-bearing, are formed at the south of the foregoing zone by fault blocks (Michelson, 1976). Cumulative discoveries amount to about 200 Mt of oil, and 100 to 200 Gm^3 of gas. The main fields are Forest Reserve (1913) with 80 Mt, Soldado (1955) offshore with 75 Mt, Fyzabad (1918) with 100 Mt, and Teak (1969) with 14 Mt of cumulative production (1979) (Bane and Chanpong).

*
**

The eastern basin of Venezuela and its prolongation offshore and on Trinidad have a potential, apart from the tar belt, of about 16,000 t km^{-2}, five to six times higher if one counts a tenth of the reserves in place of tar sands. This potential appears to be related to :

— zones with particularly faulted flexures and a transverse upwarp of a mobile basin which was highly subsident in the Cenozoic,

— for the southern margin, pinchouts on a monocline covered on various occasions by transgressions of the sea,

— thick Miocene deltaic sedimentation, rich in sand on the northeast front,

(Renz *et al.*, 1958; Young, 1971; Michelson, 1976).

3.6.2. **Maracaibo province**

This province, with an area of 60,000 km^2, may be compared to an intermontane basin in a foredeep position in relation to the ranges surrounding it (Figure 202). Its history begins with the progressive trans-

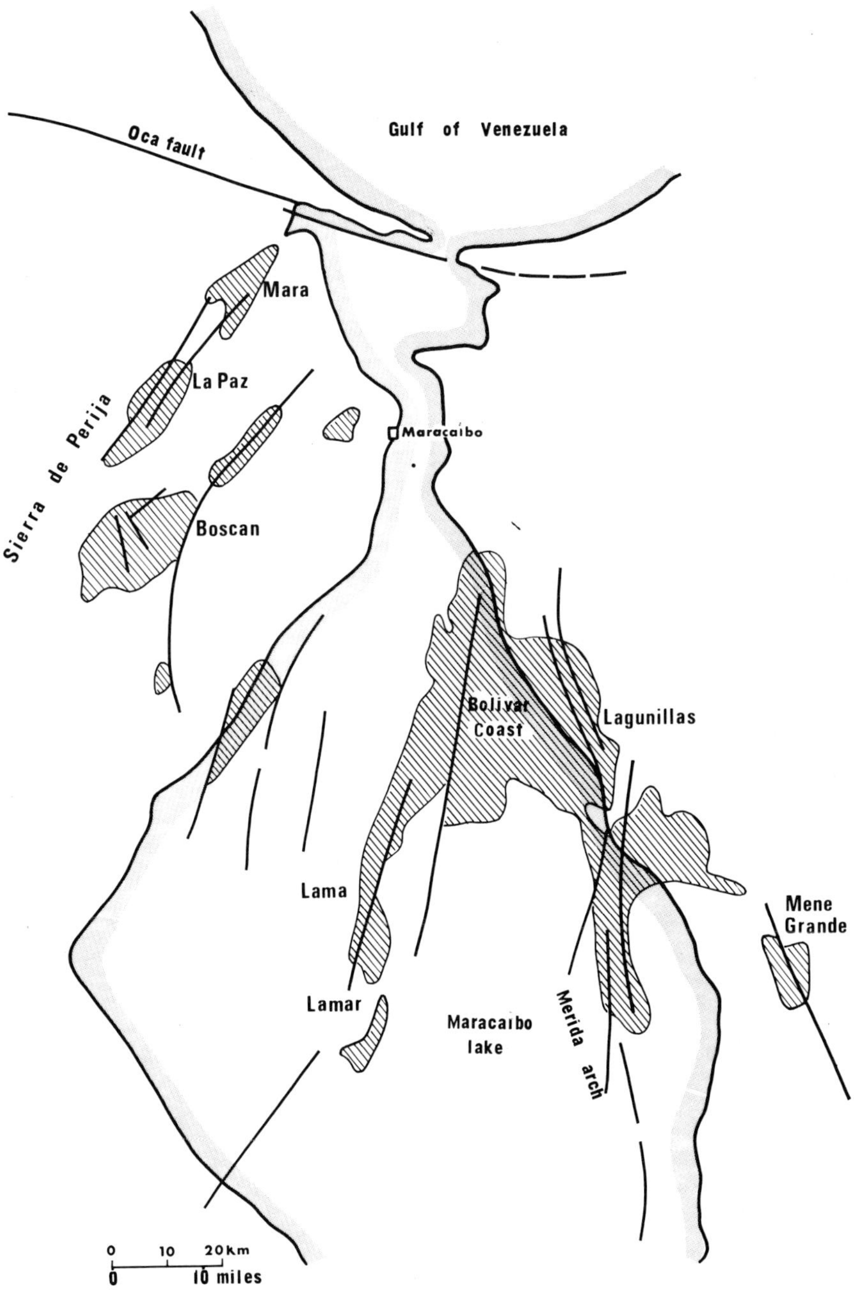

Fig. 202. — Map of fields of the Maracaibo Basin (Venezuela).

gression of the Cretaceous sea which deposited platform series on an eroded basement : detrital in the Albian, carbonate and shale with pelagic and euxinic facies in the Coniacian, shale and open sea facies in the Maastrichtian. The first Laramide movements in the southwest altered the general framework and gave rise to irregular deltaic and paralic sedimentation in the Paleocene and Eocene, with intercalations of sands, shales and coal, passing to marine shales in the east. The series reaches its maximum thickness (7000 m) on the northeastern margin of the basin. After the Upper Eocene movements, which folded this region in a north/south direction. the sea returned in the Oligocene, and subsequently in the Miocene and Pliocene, depositing alternating sands and shales. These series exhibit a maximum thickness at the edge of the Merida Cordillera, which was formed in the Late Pliocene.

The reservoirs of the main fields consist of the following :

— in the Mara zone at the northwest, Cretaceous limestones, often compact and fractured, and fissured crystalline rocks, yielding mean productivities of 700 t/day,

— in the central and eastern zones, Eocene, Oligocene and transgressive Miocene sands, generally porous (25 to 30 %) and permeable, with an effective thickness of several dozen meters, yielding mean productivities of 90 to 450 t/day.

The seals consists of 400 m thick Upper Cretaceous shales in the west, where the tectonics are intense, and thinner Eocene shales (a few tens of meters), for the Tertiary reservoirs, in the center and east, where the folds are less pronounced.

The Cretaceous crudes are paraffin base, relatively light, and fairly rich in sulfur (up to 2 %). They appear to originate in the Luna limestones (Upper Cretaceous), fine grained limestones, rich in sapropelic organic matter, and possibly in marls of the same age. The Tertiary oils, with a widely varying density (10 to 42 °API) depending on depth, are of the intermediate type, and appear to originate in Eocene and possibly Cretaceous shales and marls. The traps are formed :

— at the northwest by fractured fault blocks,

— at the center and east by clearly marked faulted anticlines under the Eocene/Oligocene unconformity plane (closed areas about 30 km^2) in an Oligocene/Miocene regional pinchout zone, as illustrated by the giant fields of the Bolivar Coast : closures may also be provided by tar plugs at the surface.

Main discoveries

About 6 Gt of oil have been found in the Maracaibo Basin, distributed among some thirty fields, together with 1500 m^3 of gas. By far the largest field is the Bolivar Coast unit, discovered since 1917, with estimated reserves of 4200 Mt. The trap is formed by Eocene fault blocks closed by the transgressive Oligocene and Miocene, and by the transgressive pinchouts of the latter. The Lamar Field (1958) with 180 Mt is located under Lake Maracaibo.

The third largest field, La Paz (1922), is the biggest in the northwest zone, with reserves of 125 Mt, located in Cretaceous limestones and in the fractured basement. The Boscan Field (1946), in the same zone, contains about 140 Mt of heavy crude (density 10 °API) with high viscosity, relatively poor in hydrocarbons, trapped in Eocene sandstones pinched out on an anticline. In the eastern part, the Mene Grande Field (1914) corresponds to the combination of a transgressive pinchout of Miocene sands on a faulted Eocene nose which is also productive. The Miocene sands outcrop at the surface and are impregnated with tars, with initial reserves estimated at 85 Mt.

***

The very high potential of the Maracaibo Basin, amounting to about 100,000 t km^{-2}, appears to be specifically associated with :

— pelagic facies with a euxinic tendency in the Upper Cretaceous, and a succession of paralic environments particularly rich in sands in the Eocene,

— a set of vast transgressions, especially post-Eocene, followed by active subsidence troughs,

— the existence of a hinge arch between the Eocene and the Miocene, in the northeast part of the basin, where the Bolivar Coast Fields are concentrated,

(Miller *et al.*, 1958; Zambrano *et al.*, 1970; Martinez, 1970).

3.7. THE SAHARAN BASINS

The Saharan basins stretch from the Cambrian to the Tertiary, from the Tassilian margins of Algeria to the Saharan Atlas and from the Eastern Sand desert to the Gulf of Sirte in Libya. The essentially Paleozoic basins of the Tassilis, Ahnet and Illizi, contain dry gas and oil in moderate quantities. The saliferous basin

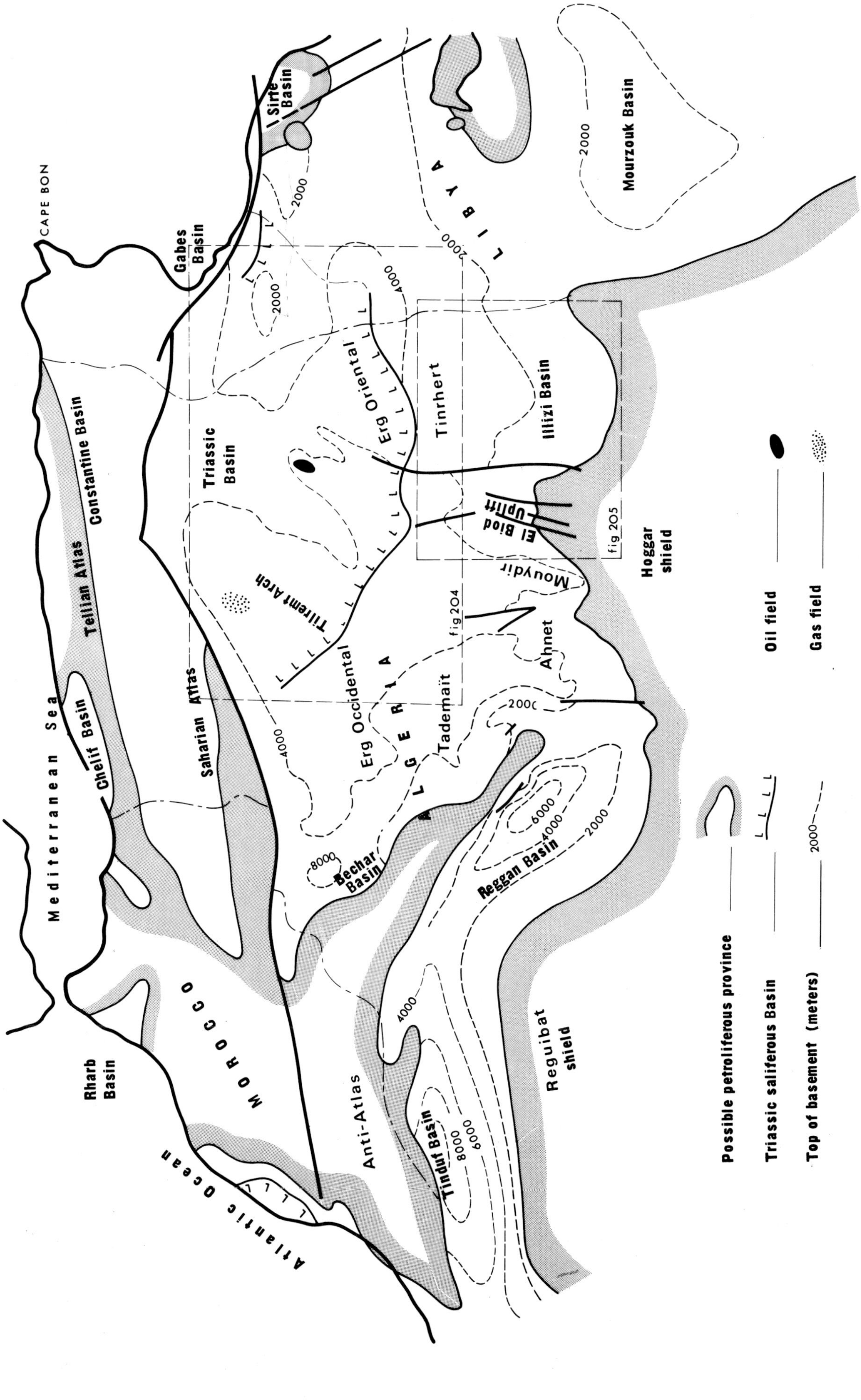

Fig. 203. — Map of main sedimentary basins of North Africa.

of the northwest Sahara owes its potential to the combination of abundant Paleozoic source rocks and a thick saliferous Triassic seal. The Sirte Basin, essentially post-Jurassic, originates in a deep splitting into vast horsts and grabens, allowing a fruitful juxtaposition of source rock and reservoir facies.

3.7.1. **Provinces of the Algerian Sahara**

The Algerian Sahara includes three large Paleozoic and Mesozoic oil and gas basins, covering a total area of about 600,000 km^2, and containing around 1500 Mt of oil and some 3000 Gm^3 of gas (Figure 203). These are the Triassic Basin in the northeast and the Illizi and Ahnet Basins in the south.

3.7.1.1. ***Northeast Triassic Basin***

Geological criteria

This Paleozoic and Mesozoic basin covers an area of about 330,000 km^2 in the northeastern part of the Algerian Sahara and southern Tunisia. It extends in the northeast into Tunisia, by the development of Carboniferous and Permian formations with a Mesogean facies (Figure 204).

This province corresponds to a vast platform basin characterized by two large regional arches, separated by a subsident trough, and to the southeast by an alignment of horsts. The sedimentary series contains two major units separated by a long hiatus :

— a Lower Paleozoic series, formed mainly of the Cambrian and Silurian, with predominating shales and sandstones,

— a Mesozoic series, characterized by a thick Triassic and partly Liassic saliferous series, overlaid by the "Continental Intercalaire" sand/shale complex, and by carbonate series of the Upper Cretaceous,

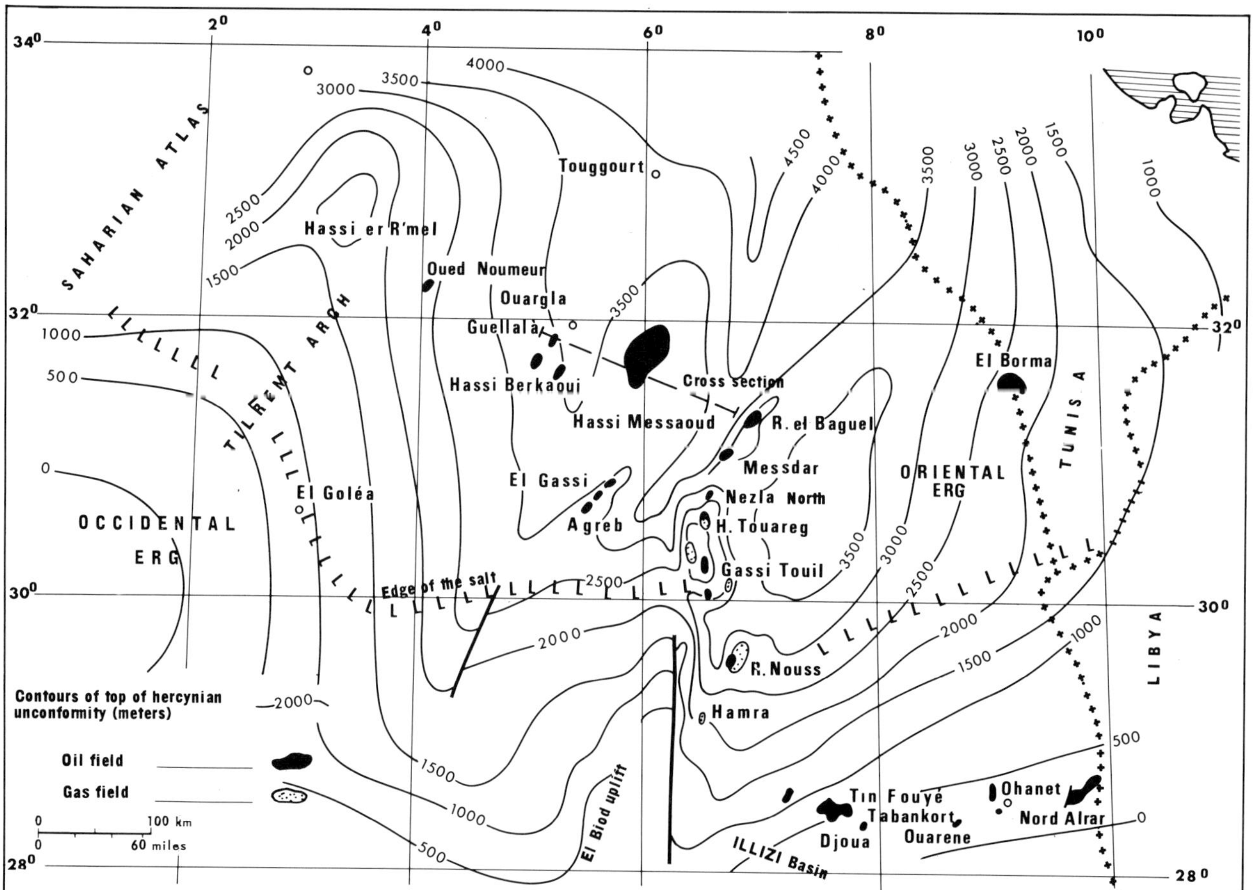

Fig. 204. — Map of the Northeast Sahara Basin.

with a few Tertiary components : this Mesozoic series is cut in the horst zone by an 'Austrian' unconformity.

Petroleum criteria

The traps consist of broad domes, eroded below the Triassic transgression (Hassi Messaoud), locally exhibiting evidence of transgressive pinchouts (Hassi R'Mel), and also of more pronounced anticlines, often faulted, folded in the Lower Cretaceous and covered unconformably by the Albian (Hassi Tuareg zone). The non-structural traps are few in number, and the main traps are situated on earlier regional uplifts.

The reservoirs, which are always detrital, are formed by Triassic basal sandstones and Cambrian and Ordovician quartzitic sandstone horizons. The only cap rock of the province is provided by the Triassic red shale and salt series, which often overlaps the underlying sandstones on the southern margin of the basin. The source rocks very probably consist of Silurian and more rarely Devonian black shales, preserved in the synclinal areas and buried at about 4000 m under the Mesozoic deposits, which guaranteed the burial necessary for hydrocarbon genesis and expulsion. The hydrocarbons are light sulfur-free crudes, density 45 to 35 °API, and condensate-rich gases. The aquifers are very saline, sometimes saturated, and the pressures are high, approaching geostatic pressures.

Main discoveries

In the northeastern basin, straddling Algeria and Tunisia, slightly over one billion tons of oil and some 3000 billion m^3 of gas have been discovered. The largest quantities are found in the pools of Hassi Messaoud (1956), with about 800 million recoverable tons (3500 Mt in place) and Hassi R'Mel (1956), with over 1.5 trillion m^3 (Figure 104). In addition to these giants are the following :

— El Borma (1964), most of it in Tunisia, about 100 Mt,
— Rhourde Nouss (1962), about 150 Gm^3 of a condensate-rich gas,
— Gassi Touil (1961) and Rhourde Baguel (1962), 50 to 70 Mt each (see Perrodon, 1956, sections),
— El Agreb (1960) and Haoud Berkaoui, about 35 Mt each (Hamouda, 1980).

Three geological criteria appear to play a major role in the hydrocarbon distribution of this province :

— the Triassic saliferous seal,
— the Hercynian unconformity,
— earlier regional arches.

The latter merit consideration, especially owing to the differences in their geological history.

The Mzab arch, whose northern burial contains the Hassi R'Mel gas pool, is a permanent uplift forming the eastern margin of the Paleozoic basins to the west, and the western margin of the Mesozoic basin to the east. The promontory of Hassi Messaoud is a Paleozoic arch that collapsed in the Mesozoic and is bordered by two subsident troughs. The eastern Hassi Tuareg/Gassi Touil Ridge, separated from the foregoing by the Dorbane Trough, appears to be a Paleozoic high axis, uplifted and folded in the Lower Cretaceous. The two axes, which remained positive in the Mesozoic, are characterized by the presence of gas, while the Messaoud Ridge, subsident during this period, exhibits undersaturated pools.

The two main fields, Hassi Messaoud and Hassi R'Mel, were discovered in 1956, very soon after exploration was begun. Subsequent discoveries were steadily smaller. At the end of 1970, recoverable reserves in the northeastern Triassic basin were estimated at more than 600 million tons of oil.

3.7.1.2. ***Illizi Basin***

Geological criteria

This slightly folded cratonic basin, with an area of 110,000 km^2, is filled by a Paleozoic series stretching from the Cambrian to the Carboniferous, represented by sandstones, shales and carbonates. The formations are buried deeper to the north, under the Tinrhert Hammada, below a Mesozoic series of increasing thickness. At the Libyan border, the ancient Tihemboka Arch is characterized by a regional transgression of the Visean on the different formations of the Primary series. This basin continues eastward, in Libya, with the Rhadamès Basin, which is more stable and has a thinner series (Figure 205).

Petroleum criteria

The traps are formed mainly by elongated anticlines with a southerly trend, moderate dip, sometimes faulted, formed in the Late Carboniferous. The age and paleotectonic situation of the structures appear to have played a major role. The largest pools are found in the vicinity of earlier regional arches. Hydrodynamic mechanisms play an important role in the trapping of the oil, and may offset the weakness of certain closures (Tin Fouye/Tabankort). The aquifers are widely varied, and sometimes brackish. The reservoirs are formed

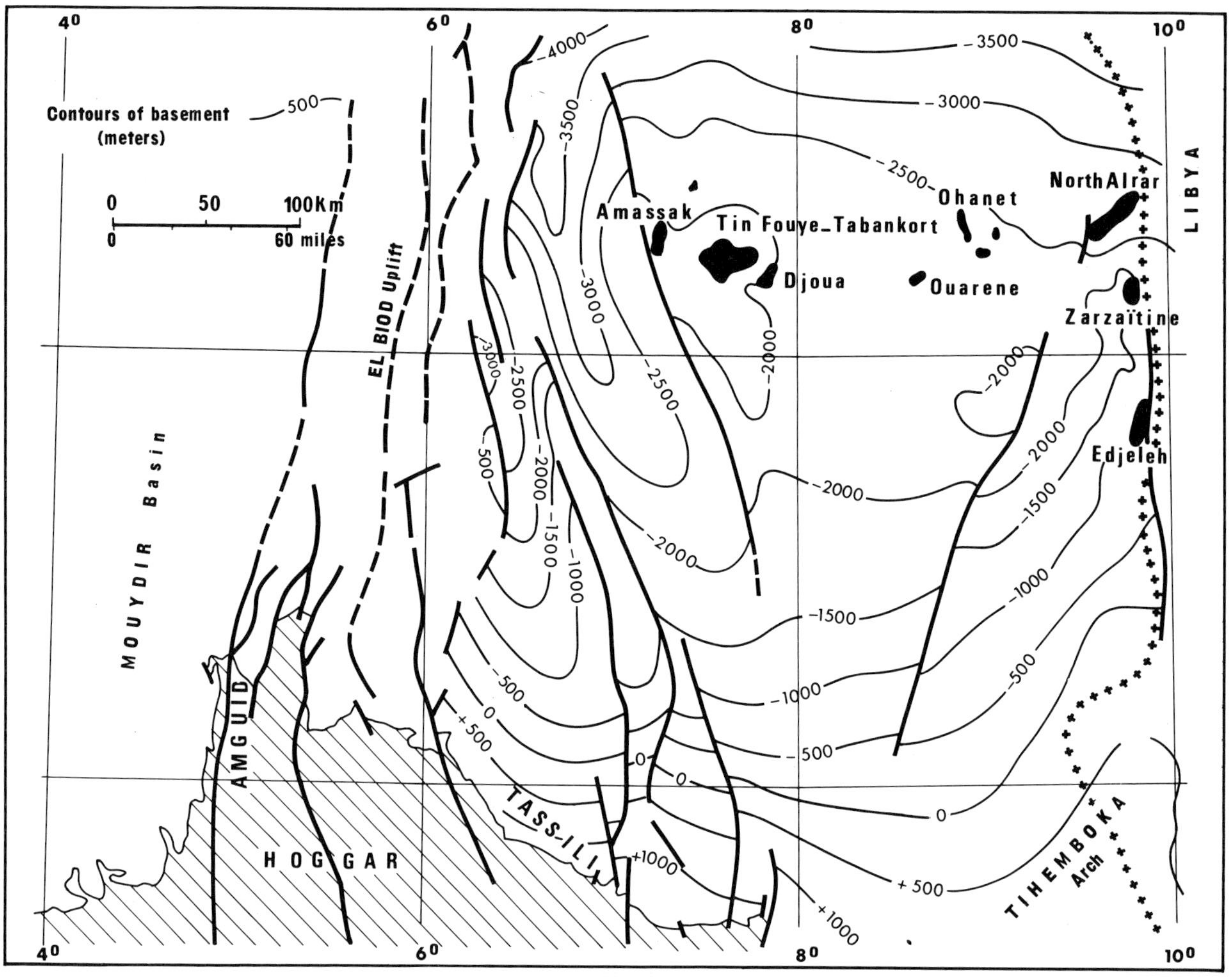

Fig. 205. — Structural map of the Illizi Basin (Algeria).

essentially by Lower Devonian sandstone horizons, more rarely Cambro-Ordovician, Upper Devonian and Carboniferous. The latter horizons only appear to be impregnated if supplied through a fault by the Upper Devonian reservoirs. The seals are essentially shale. The source rocks most probably consist of Graptolitic Silurian black shales with radioactive beds and Middle Devonian shales. The hydrocarbons are light sulfur-free oils, density 45 to 35 °API, and wet gas. The pressures are normal

Main discoveries

About 400 Mt of oil and more than 400 Gm^3 of gas have been discovered in this basin. The main fields are :

— Edjeleh (1956), a faulted anticline, with initial reserves of about 40 Mt,
— Zarzaitine (1959), a faulted nose with reserves of about 150 Mt,
— Tin Fouye/Tabankort (1960), a faulted anticline affected by a high hydrodynamic gradient, causing a tilt in the water table of 5 to 15 m km^{-1}, with reserves around 100 Mt of oil and 40 Gm^3 of gas.

3.7.1.3. ***Ahnet Basin***

Geological criteria

The stratigraphic series stretches from the Cambrian to the Visean. It consists essentially of sandstones (Cambro-Silurian, Lower Devonian especially), carbonates (chiefly Middle Devonian) and shales. The sandstones are sometimes coarse-grained, and catagenesis appears to have been fairly strong. This basin is folded in a N/S direction, connected at the west to the SSE/NNW faults of the Ougarta Ranges.

Gas criteria

The basin contains only dry gas, essentially methane, in Cambro-Ordovician and Lower Devonian sandstone reservoirs, and Devonian/Carboniferous transition series. The traps are formed by often well pronounced anticlines, many of them faulted. A dozen accumulations have been detected, but not proven, with reserves of around 50 Gm^3 and perhaps more. The main pools are those of Djebel Berga, the first one discovered in 1954, and In Salah (Balducchi & Pommier, 1970; Chanut *et al.*, 1971; Chiarelli, 1978; Claret & Tempere, 1967; Magloire, 1970; Perrodon, 1966; Tenaille *et al.*, 1975).

3.7.2. **The Sirte province, Libya**

The Sirte Basin covers an area of about 200,000 km^2 in central Libya.

Geological criteria

This rich province corresponds to a vast Mesozoic and Cenozoic rift basin, succeeding a Paleozoic regional uplift (Figures 206 and 207). The stratigraphic series began in the Albian/Cenomanian with sandstone deposits, and, especially on the shoals, with carbonates. The shaly Cenomanian transgression progressively buried, an irregular surface cut by active faults, and some horsts were covered in the Maastrichtian. The Upper Cretaceous corresponds to two cycles :
— a unit of marls, carbonate and locally Cenomano-Turonian anhydrite,
— a thick transgressive series of Senonian black marls, sometimes bituminous, lying on the previous cycle.

The sedimentation continued in the Danian/Paleocene with black Globigerina shales and carbonates.

The sedimentation reached its maximum extension in the Eocene, with the deposition of marls and limestones. The Oligocene is represented by fine limestones with shale and sandstone intercalations, and the Miocene principally by shales. This series is up to 5000 m thick in synclinal areas, but is characterized by many rapid variations in thickness and facies, associated with horst and graben tectonics. From the structural standpoint, the basin is intersected by major faults trending NNW/SSE, delimiting four troughs separating two major positive zones :
— the central platform, with the Dahra and Raguba axes,
— the Zelten/Defa high.

Volcanic occurrences, often considerable, locally accompany the renewed activity of the major faults.

Petroleum criteria

The traps, chiefly faulted anticlines and blocks, are generally associated with the horst tectonics conditioning the paleomorphology of the basement, and also with variations in facies in the Cretaceous and Tertiary. These regional uplifts thus provide the source of the productive trends. The closed areas of some traps cover up to 400 km^2. Paleocene/Eocene reefs form excellent stratigraphic traps (Intissar), reaching areas of 20 km^2 and a height of 300 m.

The many reservoirs show wide variations in facies and are generally of good quality. The chief reservoirs are formed by the following :
— calcarenites and carbonates, mainly Paleocene and Eocene, yielding well productivities of 200 to 2000 t/day,
— sandstones of the basal formations of the Cretaceous transgression, with thickness up to 200 to 300 m and permeabilities up to 2 to 3 darcys, and also by Upper Cretaceous and Oligocene sands and sandstones,
— Cambrian quartzites, and exceptionally granite.

The cap rocks are provided by shale and marls, mainly Senonian, Paleocene, Eocene and Oligocene, and locally by anhydrite. The source rocks consist of transgressive Senonian and Paleocene marls, and locally of Eocene and Oligocene shales. The crude oil, mainly Cretaceous, is generally paraffinic, undersaturated, of medium density (32-36 °API) and low in sulfur (less than 0.25 %). The highly paraffinic Sarir crude contains 15 to 20 % waxes and the same amount of asphaltenes. It 'congeals' between 13 and 24 °C.

Main discoveries

The Sirte province, which until now accounts for the total production of Libya, has initial reserves estimated at 5 Gt of oil and 500 Gm^3 of gas, partly associated, corresponding to a mean potential of 30,000 tons of hydrocarbons per km^2. The province contains twelve giant oil fields out of a total of around 100.

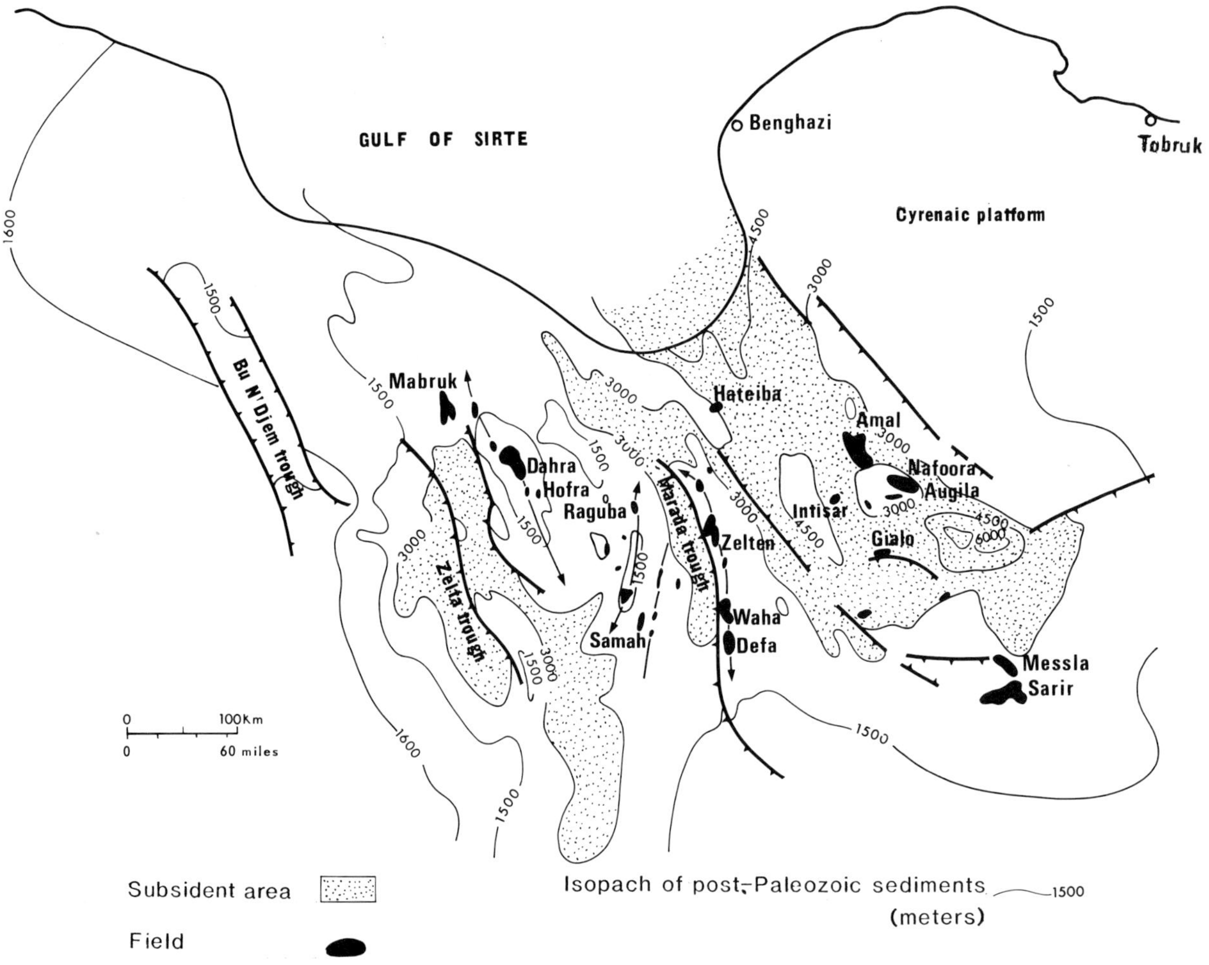

Fig. 206. — Structural map of the Sirte Basin (Libya).

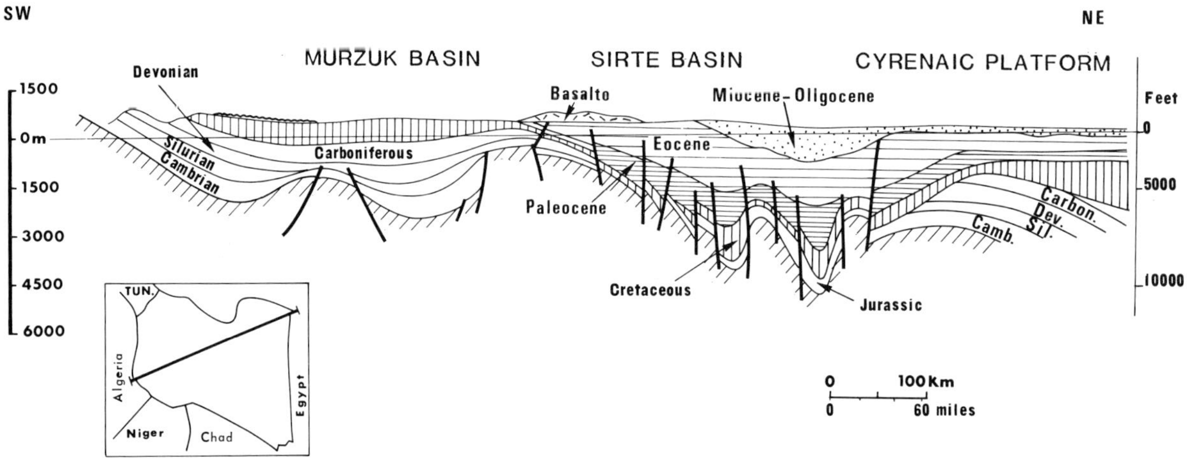

Fig. 207. — Geological cross-section of the Sirte and Murzuk Basins (Libya).

The biggest field is Sarir (Figure 78), discovered in 1961, credited with one billion tons of recoverable reserves. The reservoir consists of Upper Cretaceous basal sandstones, deposited in the depressed zones of the basement, sealed and supplied by Senonian marine shales. The trap is formed by a group of fault blocks on the flank of an uplift, under the Senonian unconformity. The productive area covers 380 km^2 for a closed area estimated at 400 km^2, with an oil column of 100 m for a closure of 120 m. A tar mat, up to 18 m thick, extends in contact with the aquifer. The best wells yield productiviites up to 4000 t/day (Gillespie, 1967; Sanford, 1970).

The Messla Field, discovered in 1971, 40 km north of the Sarir pool, also consists of an Upper Cretaceous basal sandstone pinchout, on a broad structural flexure, sealed by Senonian shales. The mean porosity is 16 % and permeability is 450 md, yielding well productivities of 14,000 t/day (100,000 bopd), of an oil with density 40 °API. The closed area covers more than 200 km^2, with an oil column of about 30 m, at a depth of 2600 m. Recoverable reserves are estimated at 210 Mt (Clifford *et al.,* 1979, 1980).

After Sarir, three fields of a quite different type offer recoverable reserves of about 550 to 600 Mt : Nasser (Zelten), Amal and Gialo. The Nasser (Zelten) Field was discovered among the earliest, in 1959. It is productive in limestone with debris of reef materials, more or less dolomitized, of the Lower Eocene and, to a lesser extent, the Maastrichtian, about 100 m thick, exhibiting very good porosities and permeabilities. The wildcat produced 2400 t /day^{-1}. The trap, with a productive area of 140 km^2, is a gentle anticline initiated in the Cretaceous, which conditioned the development of the reservoir facies (Fraser, 1967). The Amal Field, also discovered in 1959 on the east flank of the basin, consists of a faulted horst, with pinchouts on its north side. The impregnated area spreads over more than 400 km^2. The reservoirs are represented by basal Cretaceous sands, and largely by Paleozoic feldspathic quartzites (probably Cambro-Ordovician), with mediocre characteristics, but very extensive (Roberts, 1970).

The Gialo Field (1961) is a faulted anticline with an impregnated area of 125 km^2, productive in nine distinct horizons :

— four Oligocene pay sands,
— one Upper Eocene sandstone horizon,
— Middle Eocene limestones and calcarenites, the most important at a depth of 850 to 870 m,
— Paleocene limestones.

In the 170 to 250 Mt range, the Defa, Waha, Samah (70 Mt), Nafoora/Augila, and the Intissar A, B and D Fields are worth mentioning. The Intissar A (Idriss A) Field, discovered in 1967, consists of a reef structure of the coral type, built on an oval shoal extending over 4 × 5 km, and containing algae and foraminifera. Its height is between 200 and 365 m. The porosity varies from 20 to 25 % on the average, over an impregnated height of 300 m. The wells yielded an initial output of 7000 t/day (50,000 bopd) and reserves in place were estimated initially at 280 Mt, but are probably higher (160 Mt recoverable?) (Terry and Williams, 1969). The Intissar D Field contains about 140 Mt (Brady *et al.,* 1980). The main gas field is Hateiba Field, with estimated reserves of 340 Gm^3.

*
**

Among the major causes of the potential of this province are the following :

— the concentration and extension of the Santonian source rock, overlapping a large variety of reservoirs, and Paleocene, Eocene and Oligocene black shales,
— the quality of certain reservoirs, especially carbonates, well developed on the positive zones and passing laterally into marls with pelagic facies,
— the action of horsts and grabens throughout the Cretaceous and Tertiary, and especially the considerable subsidence in syncline areas,

(Parsons *et al.,* 1980).

3.8. THE NIGER DELTA PROVINCE

This province corresponds to the Cenozoic basin presently covered by the Niger Delta and its prolongation on the continental shelf. It supplies the entire production of Nigeria, Africa's leading producer.

Geological criteria

From the geodynamic standpoint, this region corresponds to the meeting of the Cretaceous rift valley or aulacogene, trending NNE/SSW, which marks a zone of major weakness of the African continent, and the oceanic fracture zones generally trending north/south and east/west, materialized today by the right-angled passive margins of this part of Africa. Its area is about 100,000 km^2, with around one-third situated

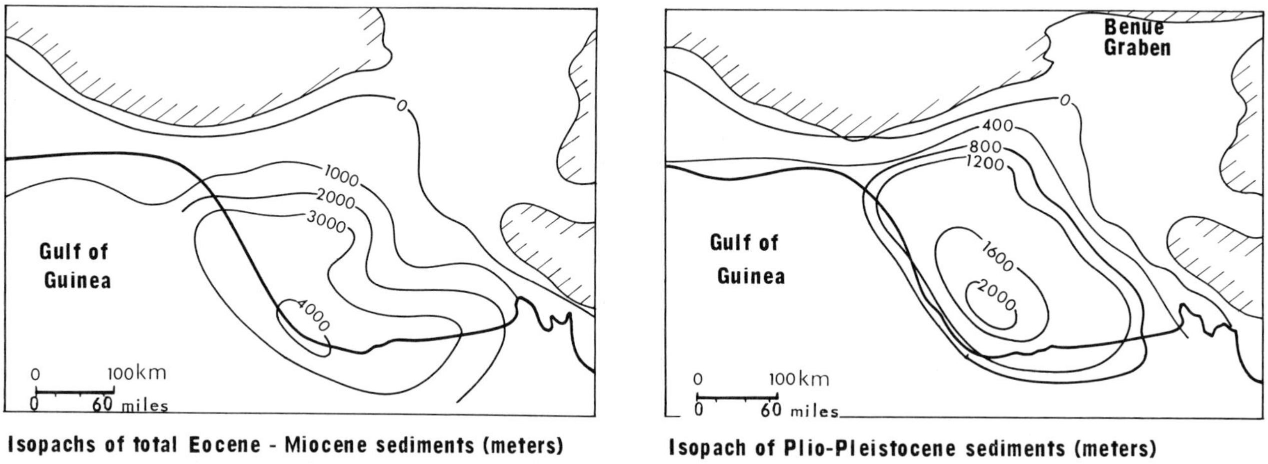

Fig. 208. — Isopach map of Tertiary series of the Niger Delta Basin (Avbouro & Ogbe, 1978).

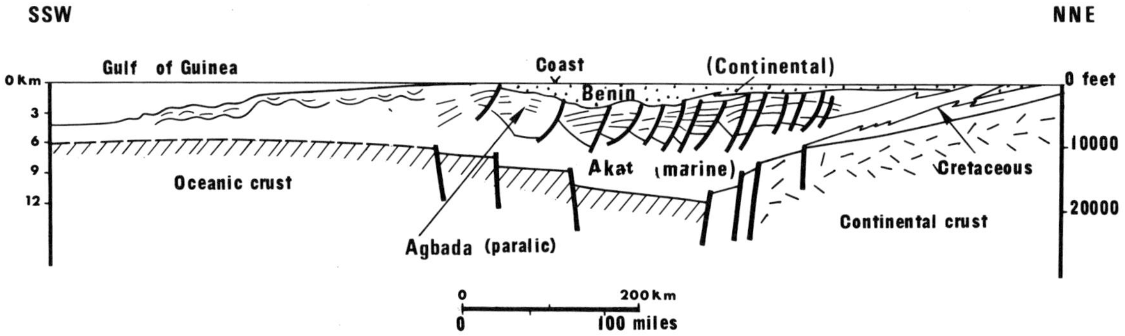

Fig. 209. — Cross-section of the Niger Delta Basin (Evamy *et al.*, 1978).

on the continental shelf. The sedimentary series, continuous from the Lower Cretaceous to the Quaternary, reaches a thickness of 12 km in the central part of the basin, situated directly above the coastal zone of the present delta (Figures 208 and 209). It is an essentially sand/shale series, of which the major sequences cover the stratigraphic succession, following the construction of the deltaic unit, throughout the Cenozoic. Hence three major units can be distinguished, proceeding upward in the series, and from north to south :

— a continental formation of fresh water sands, the Benin, at the top,
— an alternating sand/shale zone, with brackish to neritic facies, the Agbada, which forms the main productive series below,
— marine shales, partly undercompacted, the Akata Shales, at the bottom : these shales give rise to a series of deep intumescences, especially in the southern part, where they are shallower.

From the structural standpoint, this thick progradating series is cut by synsedimentary faults, consisting of growth faults and compensation faults. The faults, which exhibit a curved form in plan view and section, concave towards the sea and less marked with increasing depth, are generally synthetic, but they are often supplemented by a network of antithetic faults. The network of these faults is more complex with increasing subsidence. The Akata Shales also form diapirs, especially in the downdip part of the delta (Figures 88 and 89). From the hydrodynamic standpoint, the bulk of the basin, except for the surface horizons which are devoid of seals, is subjected to centrifugal hydrodynamism of geostatic origin, with a clear tendency to vertical migrations. On the geothermal level, this basin, of the 'cold' type, is characterized by low gradients ranging from 25 °C km^{-1} on the margins to 16 °C km^{-1} in the central zone. They are lower in the zones that contain predominantly sands, and are up to 54 °C km^{-1} in the shale series. In the central part, 116 °C is only reached at 4200 m (Figure 210).

Petroleum criteria

The reservoirs are essentially sands, Eocene in the north in the updip part, Mio-Pliocene in the south. The bulk of the reserves (about three-quarters) lies in the Oligo-Miocene series in the median portion. The

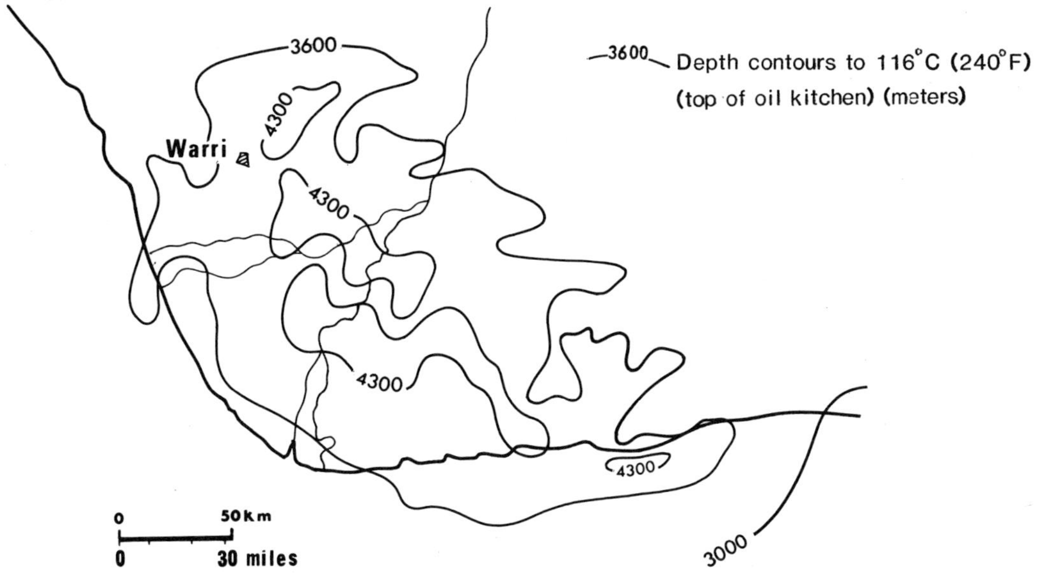

Fig. 210. — Isobath map of 116 °C isotherm of the Niger Delta Basin (in meters) (Evamy *et al.*, 1978). Subsurface depth to 116 °C (240 °F) temperature.

sands correspond largely to fluviatile and marine deposits situated between the fresh water continental sand environments and the marine shale facies. However, production has been obtained on the shelf in fresh water sands. The percentage of sands is generally high, and the reservoirs are multibedded. Their mean thickness is about 15 m, but some productive horizons are up to 100 m thick. The characteristics of these reservoirs are generally good, and average well productivity ranges from 400 to 500 tons per day.

The cap and source rocks are essentially shale. The Akata Shales appear to form the main source rock of the basin. They correspond to marine deposits of the continental slope in front of a delta rich in planktonic microfauna. Other shale horizons, with a paralic facies containing humic organic matter, probably contributed to a varying extent to the genesis of oil and gas. On the whole, source rocks are abundant but of average quality. They appear to have generated their oil in the 'kitchen' between 115 and 150 °C, temperatures which these rocks, at least in the eastern area, only reached at their present state of burial, or very recently.

The first migrations occurred from the high pressured Akata source rocks towards the Agbada sand horizons, when the latter were placed in contact by a fault. This would appear to be the major source of accumulation in the downthrown blocks (Figure 89). The faults appear to have played a major role in vertical migration and in supplying of traps. They only appear to have performed a barrier function when their throw exceeded 150 m. They appear to be tighter as the shale percentage increases, in other words as the chances of setting the shales against the sand horizons are greater. Very often, these faults do not ensure the closure of the pools, which are filled in this case in the roll over only up to the fault contact.

The traps consist mainly of roll overs bounded by one or more faults. Out of the 150 fields identified, only a fifth are not faulted. Stratigraphic traps, especially unconformity traps, are well known, particularly on the continental shelf. The closed areas and columns are modest on the average, and only 5 % of the impregnated columns exceed 45 m and 70 % are less than 15 m. The crude oils are generally light, of the paraffinic type and wax-rich, with high pour points (around — 7 to + 32 °C). Gas is frequently found, and its proportion increases with depth and distance offshore. This oil, which may be qualified as indigenous, is typical of paralic series rich in humic organic matter. A second, heavier category (density over 15 °API), of the naphthenic type, poor in waxes, and with a lower pour point, is found in the shallower series, and particularly in the beds in which the temperature is lower than 65 to 80 °C. It is a bacterial degradation product of the lighter crude.

Main discoveries

The Niger Delta is a rich oil and gas province. Since 1952, the date of the first deep well, more than 3 Gt of oil and 1400 Gm^3 of gas have been discovered, distributed among around 150 fields. Cumulative production is 1150 Mt (end 1980), and annual output is about 100 Mt. The habitat is typically dispersed with an average field of about 20 Mt. The main fields are :

— Jones Creek (1967) 110 Mt,

— Imo River (1959), 100 Mt,

— Nembe Creek (1973), 100 Mt,
— Meren (1965), 90 Mt,
— Bomu (1968), 70 Mt,
— Okan (1969), 70 Mt,
— Foro Foro Yoka (1971), 70 Mt,
— Obagi (1964), 70 Mt.

More than 700 exploration wells have been drilled with a success rate of 20 to 25 % and an average yield of 4 to 5 Mt per wildcat (Frankel, 1967; Weber, 1971; Weber & Daukoru, 1975, Evamy *et al.*, 1978; Weber *et al.*, 1978; Avbovdo & Ogbe, 1978; Nelson, 1980; St. John, 1980).

3.9. THE PROVINCES OF SOUTHEAST ASIA

The province of Sumatra falls within the framework of the great Southeast Asian province, divided mainly between the islands of Java, Sumatra, Kalimantan, Irian Jaya and the surrounding seas. This unit contains about thirty sedimentary basins, with more than 3 Gt of oil distributed among some 150 fields. The main province, Sumatra, has three onshore zones, North, Center and South, that contain two-thirds of the total discoveries (Figure 211).

Fig. 211. — Map of petroleum basins of Southeast Asia (Soeparjadi *et al.*, 1975).

Geological framework

The essentially Neogene oil series belongs chiefly to narrow, folded troughs of different types, among which the foredeep basins appear to be the most prolific (Figure 212). The sedimentary series is predominantly sand/shale, locally limestone, with volcanic horizons. It starts in Sumatra with a mainly detrital Paleogene with coal horizons. The Miocene corresponds to a major sedimentary cycle, beginning with sandstones and conglomerates overlaid by shales and Globigerina marls interbedded with tuff (Telissa

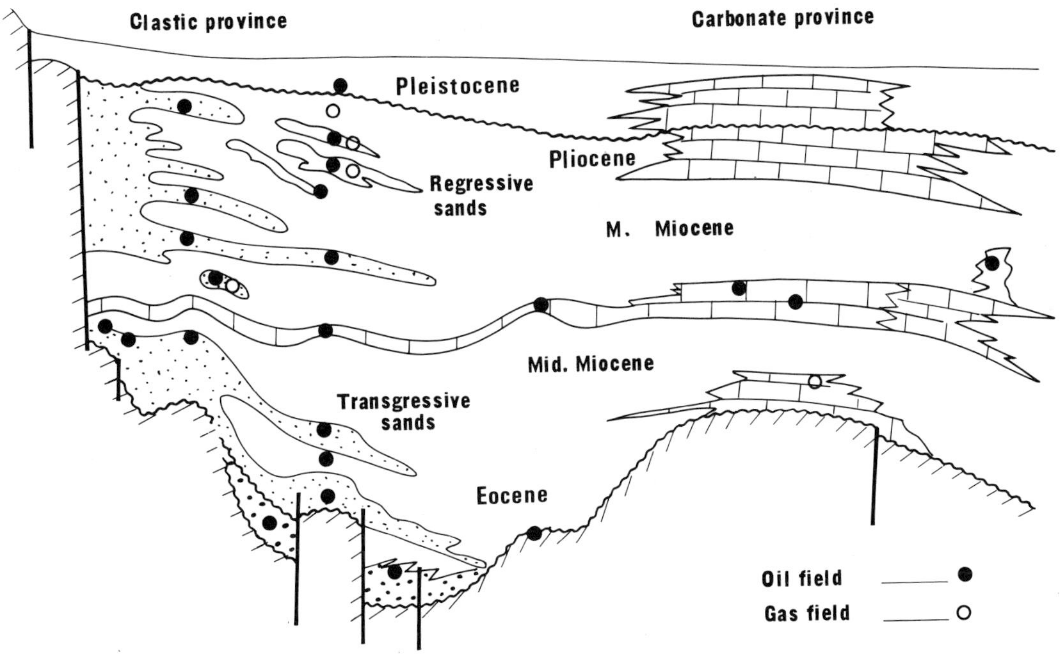

Fig. 212. — Cross-section of main producing horizons of Indonesia (Soeparjadi *et al.*, 1975).

Formation) of the Lower Miocene. The Middle Miocene mainly consists of black shales interbedded with sandstone and limestone, with a relatively deep facies. The Upper Miocene, regionally slightly unconformable, is mainly sandstone. The Pliocene is predominantly detrital and continental. This Neogene series is up to 8000 m thick in the North Basin of Sumatra, and only 3500 m in the Central province. It has been affected by many tectonic phases, especially the end of the Lower Miocene, in the Late Miocene, and the Pleistocene, which gave rise to folds, often acute, faulted and overlapping (Wennekers, 1958; Sappenfield, 1974; Pulungonno, 1976).

Petroleum criteria

With 20 % of the total volume of sediments of the different Indonesian basins, the Lower and Middle Miocene contain about 90 % of the hydrocarbons discovered to date (Fletcher & Soeparjadi, 1977). Many seeps have been found. All the discoveries of these provinces are distributed among half a dozen petroleum systems, falling within a major geological cycle. Most of the reserves are concentrated in the basal transgressive and uppermost regressive series of this megasequence. In declining order of importance, they are the following :

- transgressive basal sands, Eocene to Lower Miocene, especially well developed in the Sumatra Basins, and particularly in the Central Basin, relatively shallow,
- regressive sands from the Middle Miocene to the Pliocene, represented in particular by the North and South Basins of Sumatra, highly subsident,
- deltaic complexes of the Upper Miocene and Pliocene, the most important being that of the Mahakam Delta, in East Kalimantan (see Section 2 of Part 2),
- carbonate platforms with a reef tendency, Upper Oligocene/Lower Miocene, especially in northern Sumatra,
- reef formations of the pinnacle type, like those of Salawati, Irian Jaya (see Section 2 of Part 2),
- volcanic reservoirs and fractured tuff, illustrated by the Jatibarang Field in Java, with reserves of about 25 Mt of a highly paraffinic oil, in a petroleum zone containing about 100 Mt in six pools (Koesoemadinata, 1969; Soeparjadi *et al.*, 1975).

3.9.1. **The province of northeast Kalimantan** (Sabah, Brunei, Sarawak)

This corresponds to a Cenozoic fore-deep basin, partly offshore, covering more than 200,000 km². It has led to the discovery of more than 600 Mt of oil and 500 Gm³ of gas, distributed among about fifteen fields, mainly formed by narrow and faulted anticlines, productive in Neogene multibedded sands, supplied by

Oligocene/Miocene black shales rich in gas prone organic matter. The main fields are :
— Seria (1929) onshore, a faulted anticline with reserves estimated between 150 and 250 Mt,
— Ampa SW (1963) offshore, estimated at 150 Mt,
(Scherer, 1980).

3.9.2. **Sumatra province**

The Sumatra province is divided into three North, Central and South foredeep Basins, covering an area of 210,000 km^2, essentially showing the development of prospects associated with transgressive basal sands (Center and South Basins) and uppermost regressive sands (North and South Basins).

System of transgressive sands

The basal sands, which 'rise' in the stratigraphic series from the Eocene to the Lower Miocene, correspond to fluviatile and littoral deposits or to turbidites. The reservoirs are sealed by Middle Miocene marine shales, and supplied with hydrocarbons by contemporaneous or older marine or paralic shales. The geothermal gradients are generally high. The traps are formed by broad Plio-Pleistocene anticlines. The oil is paraffinic, with average densities 17 to 36 °API, low in sulfur and with a low GOR. This system has provided the bulk of Sumatra's oil, and has supplied some fifty fields, including Minas, the biggest.

System of regressive sands

The reservoirs of this prospect consist of generally fine, multibedded sands deposited at the end of the cycle, in a regression period, and in a shallow environment. They are spread in the stratigraphic series from the Middle Miocene to the Upper Pliocene. They appear to be sealed and supplied by interbedded marine shales. The traps are the result of folds that affected the basin in the Late Pliocene and the Pleistocene. The oil is light, density 37 to 57 °API, paraffinic, with a low GOR, although it is often found at shallower depths than the deeper transgressive horizons. The system includes about forty fields, which have produced slightly over 200 Mt to date.

Main discoveries in the Sumatra province

More than 2 Gt of oil have been discovered in the three onshore basins of Sumatra, distributed as follows :
— North Basin, area 85,000 km^2, including 25,000 km^2 onshore, about 200 Mt, slightly more than a quarter in the Rantau Field (1929),
— Central Basin, 50,000 km^2, the richest and the shallowest, 1700 Mt, including 1 Gt at Minas and 400 Mt at Duri,
— South or Palembang Basin, 75,000 km^2, 300 Mt including 80 Mt at Talang Akar (1922).

The Minas Field (1944) corresponds to an elongated anticline with an area of 230 km^2, productive in five Miocene sand horizons between 600 and 670 m depth. Production in 1966 amounted to 18 Mt with 235 wells. The crude oil has a density of 37 °API. Initial reserves are estimated at 1 Gt, of which 300 Mt have already been produced. The Duri Field (1941), located 70 km northwest of Minas, has initial reserves estimated at 300 Mt. In 1977 it produced 1.7 Mt with 426 wells, yielding a crude with density 22 °API from five horizons located between 90 and 300 m depth.

*
**

The potential of these Indonesian provinces, and particularly that of Sumatra, which amounts to about 10,000 t of oil per km^2, appears to be associated in particular with :
— rapid and recent subsidence, allowing thick sand/shale sedimentation, rich in reservoirs,
— high geothermal gradients, contributing to the maturation of organic matter which is mainly gas prone,
— varied structural traps, formed late but contemporaneous with the hydrocarbon genesis, associated with vertical migrations.

3.10. THE MAIN PETROLEUM PROVINCES OF WESTERN EUROPE (not including the North Sea)

The sedimentary basins of the North Sea are by far the chief petroleum province of Western Europe. Other basins, particularly onshore, nevertheless offer interesting oil and gas zones.

3.10.1. Permian province

The largest is the Permo-Triassic gas province of north Central Europe, that spills westward over the southern sector of the North Sea, and extends deep into the Netherlands and North Germany. This gas system is formed by the following stratigraphic trilogy :

— Westphalian coal deposits, deeply buried, especially in the Mesozoic and Cenozoic, forming the source rock,
— Rotliegendes fluviatile and eolian sands, preserving excellent reservoir qualities (porosity 17 to 20 %) up to 4000 m depth,
— Zechstein shales and especially evaporites, testifying to the extremley arid climate, forming the cap rock.

The gas reserves of this system amount to about 4000 Gm^3, of which nearly three-quarters are situated onshore. Variable but sometimes high percentages of nitrogen and carbon dioxide are frequently found in this basin, particularly in Germany, where the dolomitic Zechstein also contains slightly over 200 Gm^3 of a gas which is often also rich in N_2 and CO_2. The main onshore fields are those of Groningen (or Slochteren) (1959) with 1830 Gm^3 (80.5 Tcf) (see Section 4.1 and Figure 70), Wustrow/Salzwedel, at the frontier between the two German, and those of the South Oldenburg and East regions, East Hanover (Van Wijhe *et al.*, 1980 ; Ziegler, 1981).

3.10.2. Mesozoic provinces

The Jurassic and Cretaceous seas largely covered the vast platforms of Europe where, during more or less long periods, certain more unstable areas formed different intracratonic basins, regionally grabens, yielding several medium-scale petroleum provinces. The top formations of the Lias, represented by black shales rich in organic matter, especially Posidonidae beds, the main source rocks of these different basins, constitute the common feature of these provinces. The principal petroleum province is that of Lower Saxony, in which the following can be individualized :

— the *Gifhorn trough,* to the east, mainly productive in the Liassic series, overlaid unconformably by the Dogger Formation, with the Hankensbüttel Field (1954), 15 Mt (1980),
— the *Emsland province,* on the German/Dutch border, which is the richest, productive in Lower Cretaceous sands, with the following fields : Schoonebeek (1943) in the Netherlands with cumulative production of 27 Mt (1980), Ruhlermoor (1949) with 25 Mt (1980), Georgsdorf (1944) with 20 Mt (1980).

A total of about 300 Mt of oil, today largely extracted, have been discovered in this vast province covering an area of 70,000 km^2, distributed among slightly over 100 fields, of which about forty are exhausted.

The chief petroleum components consist of the combination of the following :

— Liassic shale source rocks and, to a lesser extent, Late Jurassic/Early Cretaceous shales, carried to maturity by high Cretaceous and Tertiary subsidence,
— predominantly sand reservoirs, chiefly Liassic in the Gifhorn trough, Neocomian in the Emsland province,
— Jurassic and Cretaceous shale cap rocks,
— structural deformations due to folding and fracturing, or caused by salt tectonics.

In the Gifhorn Trough, the accumulations contain an average of 0.14 to 2.8 Mt (1 to 20 Mbbl), with a peak of about 7 to 8 Mt (50 to 60 Mbbl). The accumulations in the Emsland province contain 1.4 to 7 Mt (10 to 50 Mbbl), with the largest containing 14 to 35 Mt (100 to 250 Mbbl) (Ziegler, 1981).

In the Netherlands, the small *Hague Basin,* with a proven area of about 2000 km^2, is productive in the Upper Jurassic and Neocomian transgressive sands, supplied as in Lower Saxony by the Toarcian Posidonidae shales. The traps are formed by faulted anticlines created by an Uppermost Cretaceous compressive phase, which uplifted and eroded the northern part of the basin. About 30 Mt have been discovered in this basin, of which two-thirds have been extracted, and the main fields are those of Ijsselmonde (1956) with 13 Mt, and Wassenaar (1956) with 7 Mt (Bodenhausen & Ott, 1981).

In France, two basins merit special attention :

— the *Paris Basin,* productive in the Lias, the Dogger and the Barremian, with total discoveries of about 13 Mt distributed among some twenty pools,

— the *Aquitaine Basin,* the largest (90,000 km^2), itself divided into two provinces, the Parentis province in the northwest, and a complex trough in the south, the Lacq/Saint Marcet province.

The *Parentis Basin,* which essentially produces oil, contains nearly 50 Mt distributed among eight fields, on an area of slightly over 5000 km^2, in which the largest fields are the Parentis Field (1954) with 34 Mt, and the Cazaux Field (1959) with 10 Mt, productive in Lower Cretaceous carbonates and sands. The *Lacq/Saint Marcet province,* which mainly contains gas, has about 300 Gm^3, added to which are about 50 million tons of sulfur on the southern margin occupied by the Upper Cretaceous flysch trough. The main fields, productive in the Upper Jurassic and Lower Cretaceous carbonate reservoirs, are those of Lacq (1954) with 200 Gm^3, Meillon/Saint Faust (1959) with 50 Gm^3, and Saint Marcet (1939) with 7 Gm^3, already extracted (see Section 4.1 and Figures 93 and 101). Half a dozen small oil pools, with a total of about 15 Mt, were discovered in particular in the north of the basin, marked by a tar belt. The chief accumulations are those of Upper Lacq (1949) with 4 Mt, and Vic Bihl (1979) with 6 to 8 Mt recoverable (Coustau, 1980).

In Italy, the Jurassic and Triassic series are known to be productive, particularly in the Po Plain and in southeastern Sicily. In the *middle Po Valley,* near Milan, *Malossa,* Italy's largest gas condensate field, is productive at more than 5000 m depth in a Triassic dolomite, sealed by Cretaceous limestones and by Oligocene high-pressure marls. Since the Tertiary unit is more than 4000 m thick, the main source rock appears to be formed by Rhetian black shales. The reservoir exhibits high pressures and temperatures, in the range of 100 kPa (15,000 psi) and 150 °C (300 °F). The trap is formed by a fault block in the form of a wedge, and reserves are estimated at 50 Gm^3 and 40 Mt of condensate (Chierici *et al.,* 1979; Groppi *et al.,* 1981). In *southeastern Sicily,* the Triassic dolomites that pass laterally into and are overlaid by black shales rich in organic matter, form the reservoir of the two main fields of this province : Ragusa (1954) with 18 Mt, and Gela (1956) with 18 Mt. The *Marches Basins,* along the Adriatic margin, appear to be an important petroleum province which may contain, in the platform carbonate formations, often karstic or formed of Cretaceous and Paleocene basin type calcarenites, nearly one billion tons in place of a heavy and viscous oil, of which it is expected to recover 70 to 140 Mt. The main accumulations, with an order of magnitude of 200 to 300 Mt in place, are Emma, Emilio, David, Rospo Mare and Sarago Mare. The Santa Maria a Mare pool (reserves of 3 Mt) produces a lighter crude (see Section 4.1 and Figure 87).

3.10.3. **Cenozoic provinces**

It is often difficult to establish the boundaries between Mesozoic and Cenozoic provinces. While petroleum Tertiary basins certainly exist, especially troughs and grabens, which contain their own source rocks and reservoirs, many post Mesozoic subsident zones give rise to the burial and maturation of older source rocks. Numerous examples are available in the North Sea, where the largest European Tertiary basin is situated, but also in North Germany, the Netherlands, Aquitaine, and the Po Plain. There are excellent grounds for believing that the potential of these provinces is often directly associated with the intensity of this Cenozoic subsidence. From the petroleum standpoint, apart from this major orogeny phase, Alpine Europe is characterized by two series of basins :

— marginal basins of the great folded ranges,

— rift valleys, roughly perpendicular to the former.

Among the first are the following :

— the extra-Alpine *Molassic basin,* in southern Germany and Austria, slightly productive in the Paleogene (about 10 Mt of cumulative discoveries and 16 billion m^3 of gas, distributed among about sixty pools),

— the Neogene basins of the *Po Plain,* the *Adriatic margin,* the *Marches,* and *Bradano.*

The latter, which collapsed between the Alpine and Appenine Ranges, folded and bristling with faulted horsts and a Mesozoic limestone platform, are essentially productive in the Lower Pliocene sands. About 300 Gm^3 of gas have been found, including 240 Gm^3 in the Po Plain alone, distributed among about 75 fields, of which the largest are Cortemaggiore (1949) with 20 Gm^3, Ravenna (1952) with 20 Gm^3, Minerbio (1956) and Caviaga (1944); and offshore Porto Garibaldi (1968) with 18 Gm^3, Porto Corsini Mare (1967) with 17 Gm^3, and Cervia Mare (1966) with 15 Gm^3. Also situated in the Tertiary are the fields of Luna, in the Gulf of Taranto, with about 50 Gm^3, and Gagliano, in central Sicily, with about 20 Gm^3.

Other productive rift basins are found on the Mediterranean margin of Spain. The largest basin lies offshore the *Ebro Delta.* On this part of the Spanish continental shelf, a small petroleum province was identified in the 1970s, so far containing about 30 million tons distributed among half a dozen pools. The three largest pools are those of Amposta Marina (1970), with recoverable reserves of about 8 Mt, Dorada (1975), an accumulation with 6 Mt of oil in place, and Casablanca (1975) with nearly 40 million tons of oil in place. The petroleum characteristics of this zone may be summarized as follows :

— traps formed by a fault block, a monocline, or a buried hill under an unconformity,
— reservoirs consisting mainly of altered, fractured and karstified Jurassic carbonates,
— supplied and sealed by transgressive Miocene black shales and limestones.

The crude oil is generally heavy, as in the Italian Adriatic province, which exhibits closely comparable characteristics (Siñeriz, 1979).

In north and central Europe, two grabens merit particular attention :

— the *Rhine Graben* which produced oil in the Paleogene, with the Pechelbronn (3 Mt) field in France, and the Landau (3 Mt) accumulation in Germany, with gas in the northern part, with total cumulative production under 10 Mt,
— the *intra-Alpine Vienna Basin* in Austria (about 100 Gt and 100 Gm^3) (see Section 2.2).

To complete this short review of Western Europe, it is important to point out recent discoveries, which are still isolated, but nevertheless likely to open new oil and gas zones. Among these are the following :

— West Shetlands (1977), a large heavy oil accumulation (25 °API),
— Wytch Farm (1973), in the southern United Kingdom, productive in Triassic sandstones (30 Mt),
— Morecambe (1974) in a small Triassic basin in the Irish Sea. 150 Gm^3 (5 Tcf) at low pressure,
— Kinsale Head (1971) in the Irish Sea, with 30 Gm^3 of gas in the Albian/Aptian sands (1 Tcf),
— oil in the offshore Porcupine Basin (Ireland),
— gas in the Bay of Biscay (Spain),
— small gas accumulations in the Gulf of Cadiz (about 3 Gm^3), and onshore in the south Pyrenees/ Serablo zone, in the Jaca Basin (see especially Illing & Hobson, 1981; Tissot & Bessereau, 1982).

Continued new successes and even major discoveries in sedimentary basins, often actively prospected for more than half a century, are an encouraging factor for the intensification of exploration efforts.

Summary and conclusions

Twenty giant provinces, each containing at least one billion tons of oil or gas equivalent, concentrate about 90 % of the cumulative oil discoveries and nearly 80 % of all gas discoveries. These provinces, which are situated approximately according to a log-normal distribution, are found in the northern hemisphere, one of them near the equator. From the geological standpoint, they exhibit a strong predominance of structural traps, regional unconformities, and high and generally recent subsidence. They correspond chiefly to large Mesozoic and Cenozoic basins, half of them covering more than 400,000 km^2, and two of them over 2 Mkm^2. The habitat is generally of the concentrated type. The history of their discovery and exploration reveals a certain acceleration which appeared to culminate in the 1950s.

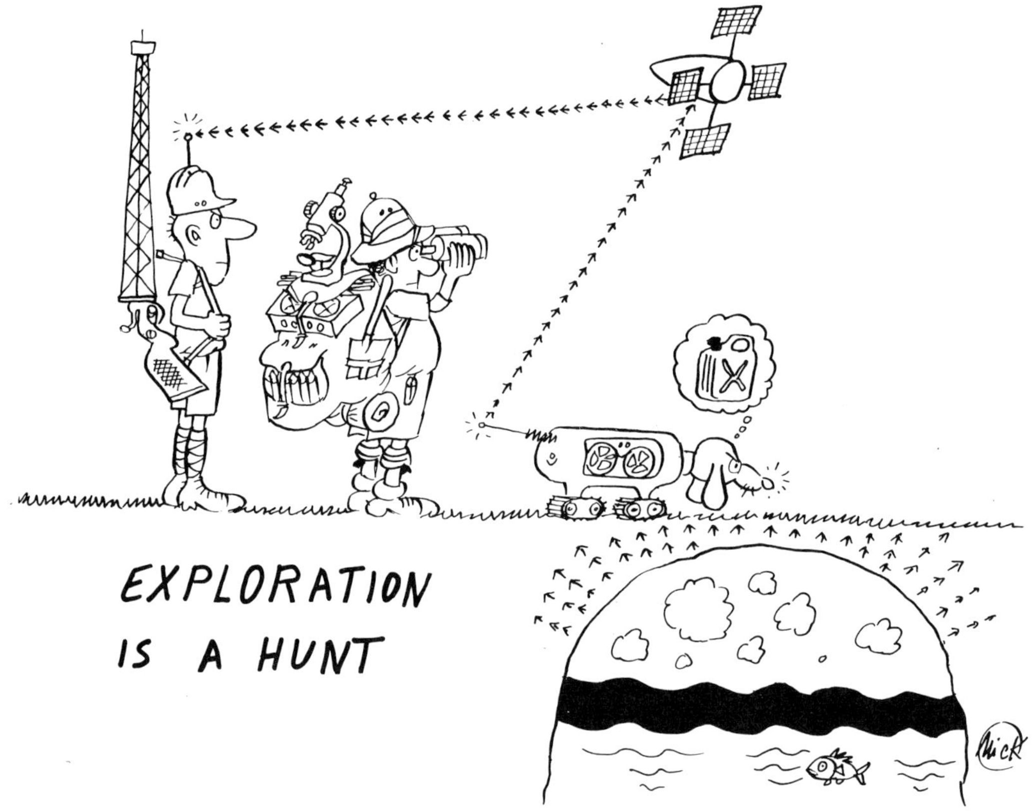

Fig. 213. — Exploration is a hunt !

4. – PETROLEUM EXPLORATION

'We usually find oil in new places with old ideas. Sometimes, also, we find oil in an old place with a new idea, but we seldom find much oil in an old place with an old idea. Several times in the past we have thought that we were running out of oil, whereas actually we were only running out of ideas.'

P.A. Dickey.

« Se mettre en face de la nature et l'interroger à l'aide de moyens d'investigation de plus en plus parfaits. »

'Come face to face with Nature and interrogate her with better and better investigative tools.'

Claude Bernard.

« Devine si tu peux, et choisis si tu l'oses. »

'Guess if you can, and choose if you dare.'

Corneille.

History

Petroleum exploration is a hunt. It has been going on for more than a century, during which it entered the industrial phase, by different stages which are still superimposed, depending on the country and degree of sophistication in prospecting techniques (Figure 213). To begin with, petroleum prospecting may be considered as a direct or semi-direct method, with the drilling of boreholes in hydrocarbon seeps, or in their immediate vicinity. Large reserves have thus been discovered in a number of basins, and especially in the provinces of Baku, Sumatra, Romania, Trinidad, Venezuela, California, the Rockies and the Appalachians.

Since some prospectors noticed that discoveries were more frequent on seeps situated on anticlines, the anticlinal theory was proposed, which is still valid today, though not as exclusive as in the past. It is basically a method involving an indirect approach, because it is aimed at finding closed zones in which the oil or gas has accumulated, assuming that one exists in this part of the basin. With the help of seismics, it has led to spectacular discoveries, even in basins like those of the Arabian and Saharan platforms. The large majority of today's giant fields correspond to anticlines.

The uncertainties of this method, together with progress achieved in our knowledge of petroleum geology and the geopetroleum characteristics of the basins, have given rise to what may be called the 'global approach'.

After having gathered a body of geological data concerning the reservoirs, seals, source rocks and fluids from the first boreholes, generally drilled on structural objectives, in other words, once in possession of data making it possible to make certain assessments concerning the oil generating and petroleum potential, exploration is gradually focused on finding traps in the zones recognized to be the most favorable. This approach is practically implemented through intensive exploration that develops, geological and economic conditions permitting, after the discovery of the major accumulations, aimed to find objectives smaller in area and involving a more subtle approach. It is curious to note that this logical path, which involves all our geological and technical knowledge, is sometimes run down by the partisans of a blind approach, of boreholes drilled in a regular grid. Logically speaking, however, the meshes of this grid become steadily smaller, and investments rise steeply as the traps looked for become increasingly rare.

It seems more interesting to try to enhance the effectiveness of exploration by developing tools and techniques designed to return to the direct approach from the surface. Advances in seismics and seismic interpretation have already led, in some series, and up to medium depth by recordings and specific processing techniques, to the identification of 'bright spots' corresponding to gas accumulations. In these matters, it is important to keep a cool head, while remaining resolutely optimistic. There is every reason to believe that progress will continue, and that more efficient and precise techniques will some day be available.

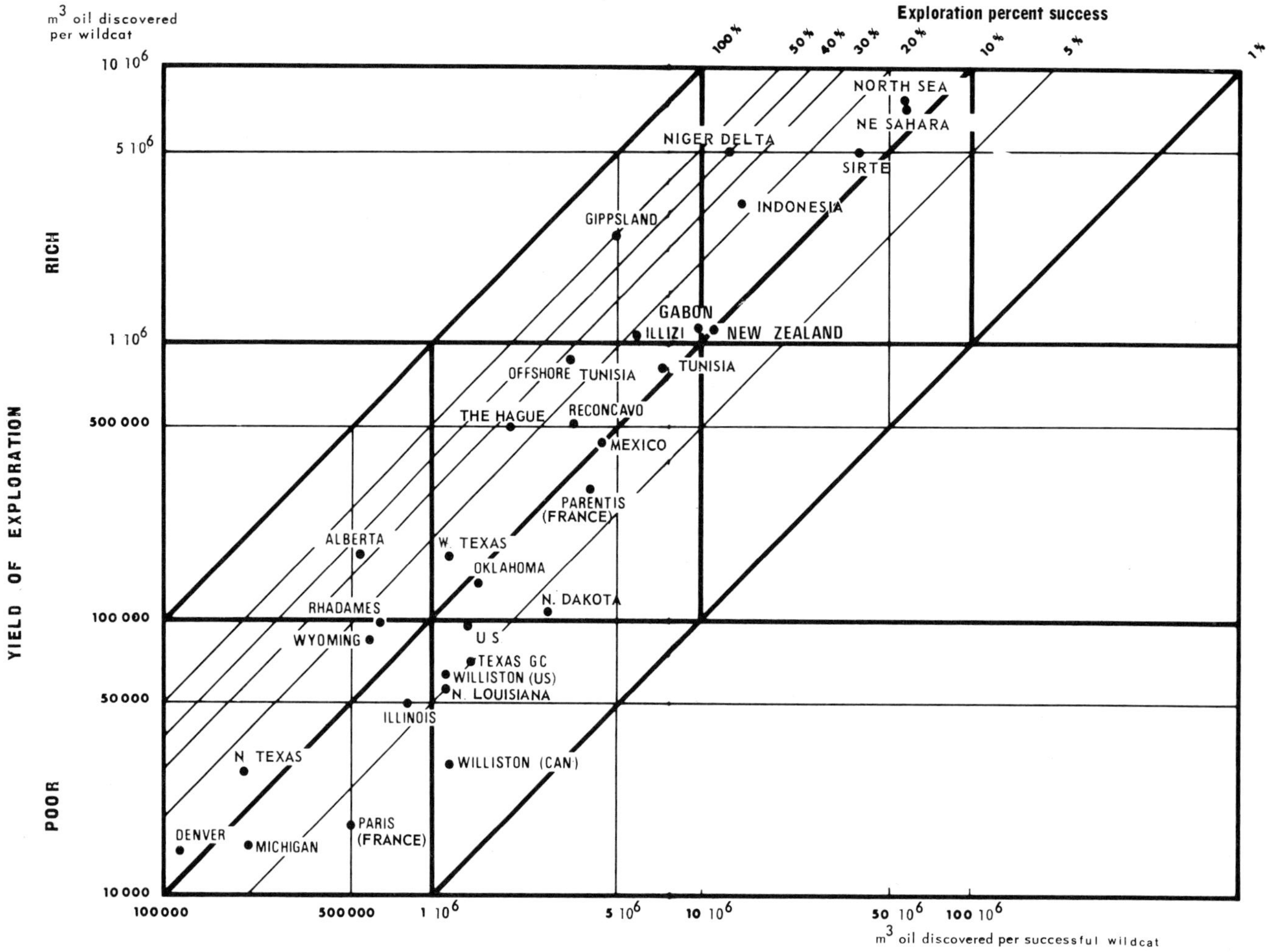

Fig. 214. — Diagram showing exploration yield versus potential and exploration intensity.

However, one must keep in mind the complexity of the sedimentary environment, the mass of noises that always tend to confuse the small signal that we are trying to identify. Furthermore, the ever more detailed knowledge of geological processes will continue for a long time to assist the interpretation of new measurements.

*
* *

An exposition of the techniques used for petroleum exploration would easily fill a book of this size. Hence there is no question of dealing with this methodology and with exploration tactics here. However, the results of prospecting control the knowledge of petroleum provinces and fields, of their anatomy and their physiology. This strategy is guided on the technical level by geological data about the sedimentary basin. Keeping this in mind, this chapter offers some aspects of these relationships between geological knowledge and exploration results, and more specifically :

1 — some examples of exploration yield and strategy,

2 — some considerations of the American example,

3 — a rough idea of ultimate hydrocarbon resources.

4.1. EXPLORATION YIELD AND STRATEGY

Main criteria and definitions

The analysis of exploration strategy and yield must take the following into account :

— the history of previous work, chiefly the means implemented, expressed by the number of wildcats and exploration investments, or in a relative manner, the number of these wells and the amount of these investments per 10,000 km^2, for example,

— in addition, the results obtained by these means, namely cumulative oil and possibly gas discoveries, in gross figures or by km^2, and the number of discoveries per 10,000 km^2.

The areas considered are interesting sedimentary areas and, in some cases, within the limits of the permit area. The concept of discovered reserves or discoveries must be clarified, and must be qualified, as a first approximation, as gross or all-categories (it is relatively homogeneous in this case), net or profitable, in which case it depends closely on economic conditions.

The relative result, or exploration yield, is conventionally and logically expressed by reference to drilled units in :

— cumulative discoveries per wildcat (total of dry and positive wells),

— cumulative discoveries per 'significant' wildcat (average pool or discovery),

— success ratio in %, gross if all successes are considered, net if only successes exploitable within a given context are considered.

The two first indications more closely reflect the economic value, while the third reflects the risk factor. The lower the risk, or the higher the success ratio, the smaller the economic discovery and the lower the commercial value, irrespective of the type of province or habitat.

The essential factor is to compare comparable data only : average discovery in all categories and gross success ratio, or, even better, average significant discovery and net success ratio. In Nigeria, for example, some 740 wildcats were drilled from 1950 to 1975, distributed as follows :

— 412 dry wells,

— 123 non-significant 'discoveries',

— 205 significant discoveries, for a total of about 2500 Mt.

Thus the average significant discovery is 12.5 Mt (2500/206) and the net success ratio is 0.16, while the average discovery and the gross success ratio are 7.6 Mt (2500/330) and 0.35. The economic (technical) result is expressed as technical discovery cost to make a valid approximation of exploration investments, as the latter partly depend on the yield of the means employed. These different data are expressed globally over a certain period, and hence depend on two main parameters, exploration effort and the potential of the basin or type of habitat.

Main factors governing exploration yield

The exploration yield obviously primarily depends on the geological conditions, which themselves condition the potential and type of habitat of the hydrocarbons. However, it is also determined (and discovery costs especially) by geographic conditions, by the state of technology, and by the degree of exploration of the province, not to mention political and economic factors.

4.1.1. Geological factors

The petroleum characteristics of a province are essentially expressed in terms of potential and type of habitat, and these data are closely dependent on the characteristics of the sedimentary basin (see Chapter 1 of Part 3).

Potential

A rather weak correlation can be observed (Figure 215) between potential (the quantity of oil discovered per km^2) and exploration effort or exploration yield, expressed in tons of hydrocarbons discovered per wildcat. Very rich provinces (over 10,000 t km^{-2}), but in different exploration stages, stretch over yield scales ranging from 100 Mt per well for the Persian Gulf to less than 100,000 t for North Texas. However, neither

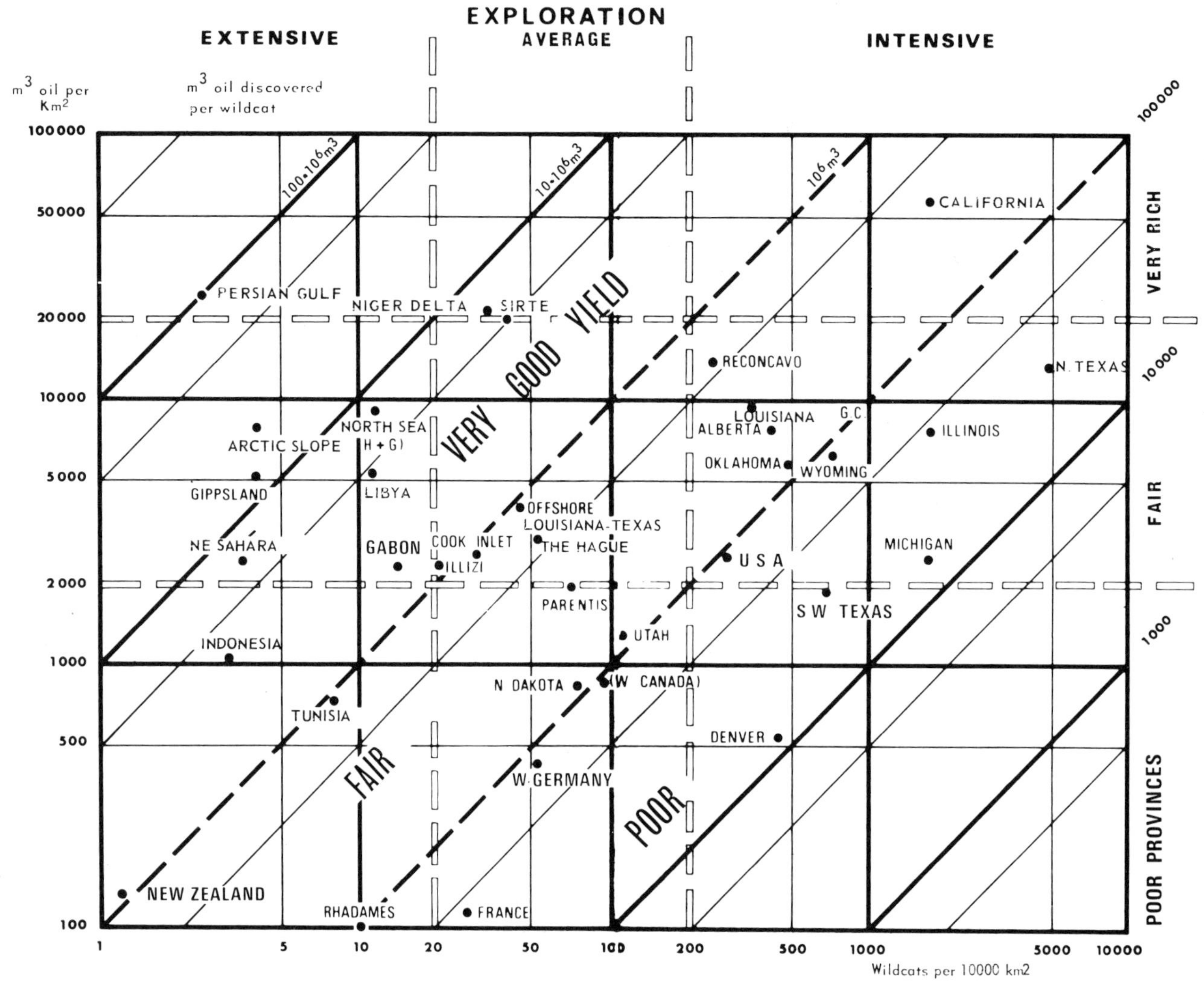

Fig. 215. — Diagram showing success ratio versus exploration yield and average discovery.

of these provinces, any more than tho rich provinces, offers yields less than 30,000 or 40,000 tons per wildcat. On the other hand, poor provinces only rarely offer yields exceeding 10,000 tons per well. The potential appears to have a more direct influence on the success rate. Most rich provinces generally offer high success rates, often above 20 %, while poor provinces display rates generally below 10 % (Figure 216).

Type of habitat

The type of habitat has a more direct and generally important influence on exploration yield and discovery cost. Provinces with a concentrated habitat, characterized for instance by less than ten discoveries per 10,000 km^2, very often exhibit high yields (more than 10 Mt per wildcat) while basins with a dispersed habitat generally fall below this value.

The latter, when they correspond to rich provinces, display high success rates, while the average discoveries are obviously smaller than in provinces of comparable richness with a concentrated habitat. Relationships have been demonstrated between these two parameters. For example, by using the Statsid (*) program, as confirmed by exploration results, Coustau has shown that the concentrated habitat predominates in rich provinces. On the other hand, it has been observed that the poorer the basin, the greater the tendency for the habitat to be dispersed, and this characteristic is often itself associated with a type of stable platform basin or certain intra-continental rift basins.

(*) Statsid : Petroleum Statistics Data Bank of Elf Aquitaine's Information and Documentation Department (SID, Service Information Documentation).

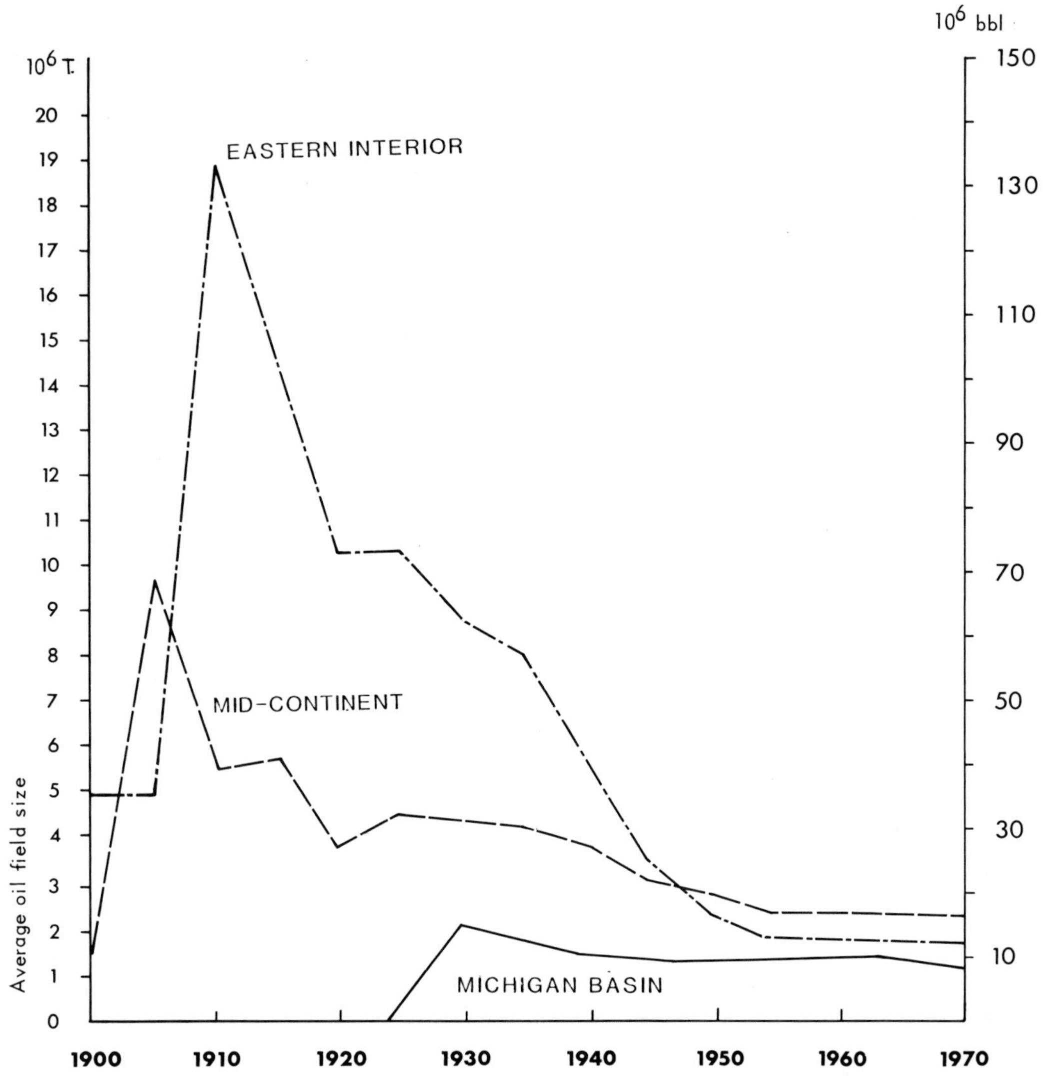

Fig. 216. — Variation in average size of fields discovered in the Michigan, Mid-Continent and Eastern Interior provinces (Illinois and neighboring zones) from 1900 to 1970 (Vinkovetsky, 1979).

Importance of basin type

On the average, every sedimentary basin has one chance out of two or three of containing a commercial hydrocarbon accumulation, but only one out of ten offers a field with more than 80 Mt (Tatsch, 1974). It is well known that these possibilities are related to the geological characteristics, as the type of basin closely conditions the criteria which themselves determine the potential and type of habitat. However, this relationship appears to be complex and indirect, and the predictions that may be made always retain a random character. As a rule, the large unstable platform basins with vast undulations provide the majority of the giant provinces. On the average, two basins of this type out of three contain giant fields, often concentrating half of the reserves. These giant fields are generally discovered fairly early, but small fields may continue to be found for a long time after the first successes (Klemme, 1971).

On the other hand, since stable platform basins do not exhibit regional arches, they generally offer a far more random distribution of the accumulations, obeying a sedimentary and paleogeographic logic in a number of cases, always longer and more difficult to decipher. This means low exploration yields and high discovery costs, a sequence of successes with frequent gaps, associated with the exploitation of a reef trend or the prolongation of a shoreline, for example. The basins of this type are often among the poorest, and among those offering the smallest yields.

In West Canada, the sucess rate of all categories has been 12.6 %, but that for 'significant' pools (1 million barrels at least) is half, and that of Class A pools (50 million barrels) is 0.5 % only. The average discovery is 2.75 Mt counting all discoveries, 5.5 Mt if discoveries less than 1 Mbl, accounting for 1 % of reserves, are ignored. These results, which are more favorable than those of the other stable basins, are related to the presence of reefs, themselves created by a certain intensification in subsidence.

The Roma province in southeastern Queensland (Australia) offers an example of the prospecting of stratigraphic traps. It covers a substantially square area of about 10,000 km^2, which corresponds to a stable platform basin, exhibiting a relatively thin sedimentary series stretching from the Permian to the Tertiary, about 1500 to 2500 m deep. These formations, essentially marls, silts and sandstones, are intersected by half a dozen regional unconformities. The beds progressively bury the shoals of the Devonian basement and overlap the basin margins, especially in the Upper Jurassic (Evergreen Formation), which could contain the main source rocks. The sandstone reservoirs, generally thin and discontinuous, correspond to fluviatile or littoral deposits. They have medium porosity, and permeability may be several darcys locally. About ten formations are productive, but the bulk of the hydrocarbons originates in the fluviatile sandstones of the Upper Jurassic Precipice Formation. These sandstones are closed on the uplifts of the basement, which were only buried in the Uppermost Jurassic. The traps, in this only slightly structured basin, are formed by stratigraphic pinchouts, often associated with structural components. All these traps exhibit small areas and closed heights.

The first gas discovery was made in 1900. In 1970, around 200 wells had been drilled, identifying 36 pools with a total of 7 Gm^3 of gas only. The yield is low : while the success ratio is 30 %, the average discovery is only 200 Mm^3 of gas, with an average of 26 Mm^3 per wildcat. Production in 1970 was 250 Mm^3 per year. This type of basin and traps clearly illustrates the mediocrity of stable basins and the difficulties of intensive exploration. Only at the cost of many boreholes can one discover the small reserves existing in this type of basin. On the other hand, the wells can be rapidly and cheaply drilled : in this particular case, the time required to drill a well was reduced from six months to six days (Trades, 1971).

Rift basins could be highly attractive hunting grounds. Their strong point is the presence of favorable conditions for the formation of source rocks, intense diagenesis, and often good protection. Reservoirs are their weak point. A relatively specific but exemplary case for judging the prospects of deep offshore basins is the Los Angeles Basin, which is exceptionally rich, whose abundant and top-quality reservoirs are represented by turbidite sands, deposited at a depth of a few hundred meters. The exploration of this province, rich in seeps and anticlines, was pursued actively from 1880, in accordance with surface considerations and the search for the 'trend'. At the end of 1949, only 41 wildcats were productive out of 1377, but this 3 % success rate meant the discovery of more than 1200 Mt of oil, or a yield of about 1 Mt per wildcat and 30 Mt per productive wildcat. About 1 Gm^3 of gas was discovered from locations situated by surface geology (Owen, 1975).

When these rift basins are productive, one out of three shelters giant fields, that may contain 60 to 90 % of the basin's reserves (Klemme, 1971). Hence from the strategic standpoint, it is important to be stringent in the choice of prospects, and to carry out their prospection first. In these conditions, the exploration yield is often high, especially in offshore basins, where systematic seismic exploration rapidly serves to find the major objectives (examples are the Cook Inlet, Gippsland and Dampier Basins). The pull apart basins, in particular, thus offer considerable assets, but also undeniable risks. It should be recalled that the Ottway Basin lies next to the Gippsland Basin, the Cuanza Basin lies next to that of Cabinda, the Barrow Basin next to the Dampier Basin, and that, in each case, a few divergences in the geological framework subsequently provided a satisfactory explanation.

The Cenozoic deltaic basins usually contain a high porportion of gas and few giant fields, despite their generally high potential. Hence their exploration is a long-term affair. Success ratios are usually high, ranging from 25 to 50 %, while average discoveries per wildcat are relatively low, about 10 Mt in Nigeria, 0.3 Mt in coastal Louisiana, 0.5 Gm^3 of gas in the Mexican Burgos Basin in northeastern Mexico, where the average pools contain around 2 Gm^3, and 1 Gm^3 not counting the giant of Reynosa, which alone contains nearly half of total discoveries (1938/1954 period). Also in this area, subsidence appears to be a prerequisite for high potential, if one merely compares the data of these Tertiary provinces to deltaic platforms such as the Cretaceous in West Canada or the Denver Basin.

In view of these considerations, it may be stated roughly that :

— yields are generally high (more than 1 Mt per well) and discovery costs minimal in rich provinces with a concentrated habitat, which often correspond to provinces containing giants, and, in many cases, to giant provinces,

— yields are low (less than one million tons per well on the average) and costs high in poor provinces with a dispersed habitat.

4.1.2. **Geographic factors**

Contrary to what may be believed, geographic conditions have little effect on onshore exploration yield. Using a broad range of examples, Klemme (1971) points out that exploration rates vary extremely little between the jungle and the deserts. However, these conditions have a considerable influence on discovery cost. On the other hand, the difference is very significant, both from the standpoints of strategy and yields, between onshore and offshore exploration, mainly due to the ease, speed and efficiency of offshore seismics.

High exploration performance, at least in provinces with a concentrated habitat, like the North Sea, Gippsland and Cook Inlet Basins, may offset the high drilling expenditures and finally produce relatively low discovery costs. Thus exploration in the North Sea, thanks to a high exploration yield, about 4.5 Mt per wildcat, and a success ratio of only 12 %, produces discovery costs of 3 F per m^3 (or 11 cents per barrel) (Héritier, 1975), with 15 F/t or $0.50 per barrel for Cook Inlet (Klemme, 1971).

The exploration of the Gippsland Basin was especially easy and fast. It began with 18,000 miles of airmag and 5320 miles of seismics, representing a pre-drilling investment of 5 cents per acre, or about 6 F/km^2 (Weeks, 1969). The drilling campaign, which began in late 1964, was immediately positive :

— 1st wildcat gas discovery (Barracouta) followed by an appraisal well.

— 2nd wildcat dry,

— 3rd wildcat gas discovery (Marlin), followed by two appraisal wells,

— 4th and 5th wildcats, two major oil discoveries, Kingfish and Halibut (Owen, 1975). On the whole, out of six wildcats covering the period from December 1964 to September 1967, this represents four major discoveries with a total of 250 Mt of oil and 270 Gm^3 of gas, or a success rate of 80 % and a yield of 25 Mt of oil per new field wildcat. From 1967 to 1976, 34 wildcats, including ten step out wells, were drilled, culminating in more modest discoveries. Subsequently, the exploration yield gradually decreased, while the systematic character of geophysical reconnaissance made it possible to find the main targets very quickly. However, the discovery costs remained quite satisfactory, from $0.005 to 0.02 per barrel (Klemme, 1971).

4.1.3. **The time and technical progress factors**

Half of the oil discovered worldwide has been found since 1964. An ever-increasing number of productive basins are being discovered, but the success of exploration is progressively declining, essentially due to the rarefaction of easily discoverable traps. On the whole, however, in a given basin, the exploration yield decreases with increasing effort. All the statistics prove this, with a few discrepancies related to the discovery of a new play and/or the implementation of new techniques. The evolution in the size of fields discovered in three provinces of the United States (Figure 217) between 1900 and 1970 clearly illustrates the history of the

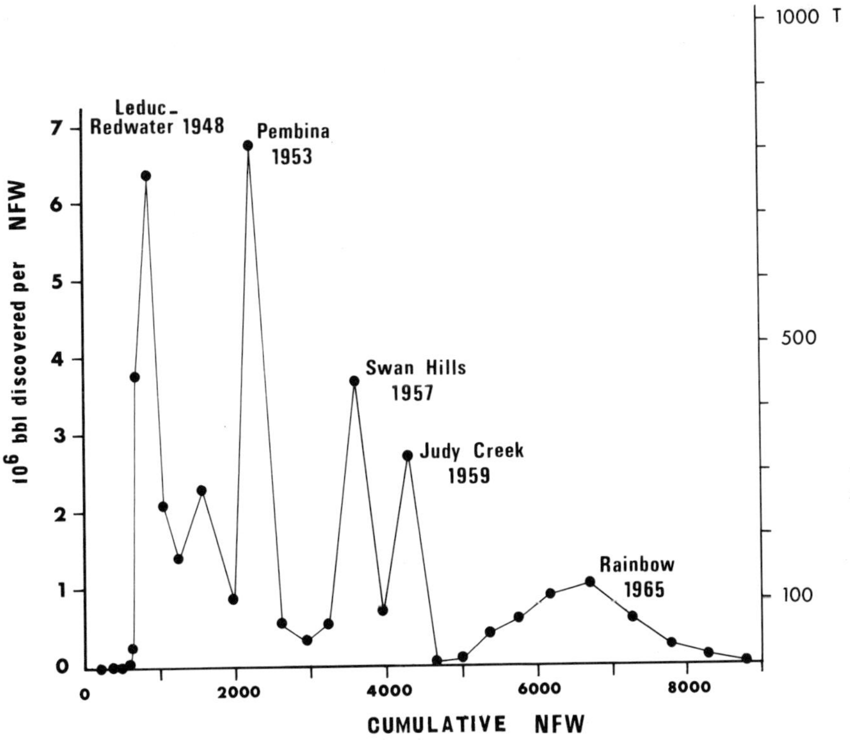

Fig. 217. — Variation in average discoveries per NFW in Alberta (Ryan, 1973).

exploration of a basin. In the Mid-Continent and Illinois provinces, characterized by the presence of a few 'major' fields, the most important discoveries were made in the early years of exploration, after which the size of the discovered fields rapidly declined.

In the Michigan Basin, which exhibits a dispersed habitat and modest accumulations, the size of the discoveries remained substantially constant with the passage of time. A closer analysis of the distribution of the 95 fields of this basin at different stages of its exploration reveals a steady increase in the number of small discoveries, offering a few million barrels, and near-stagnation in the number of larger fields (Figure 129). A closely comparable pattern is found in the Powder River Basin (Vinkovetsky, 1979). The same historical pattern is found in the case of a larger and more complex area, such as the 48 States of the United States. Another clear example of this decline is offered by the variation in size of discoveries as a function of the number of NFW in Alberta (Figure 218).

The statistics for the overall United States (not including Alaska) reveal a relatively constant gross success rate (about 10 %) which then climbs back to 16 % in 1976. But there is a steady and significant drop in

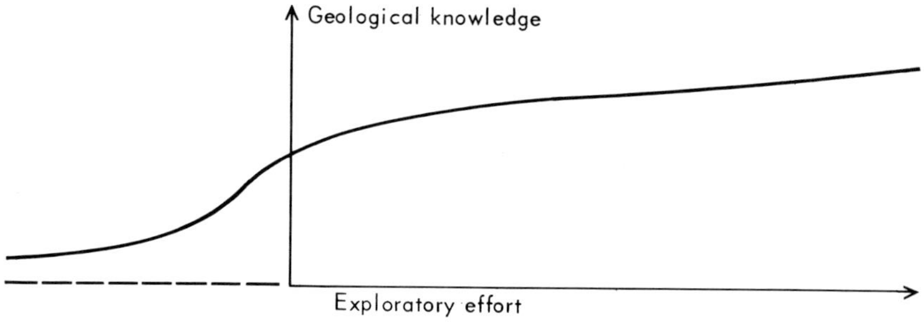

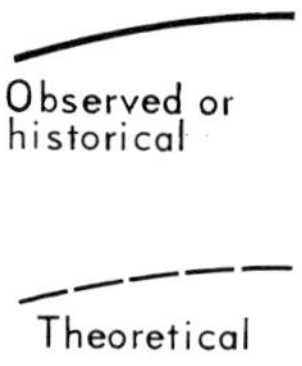

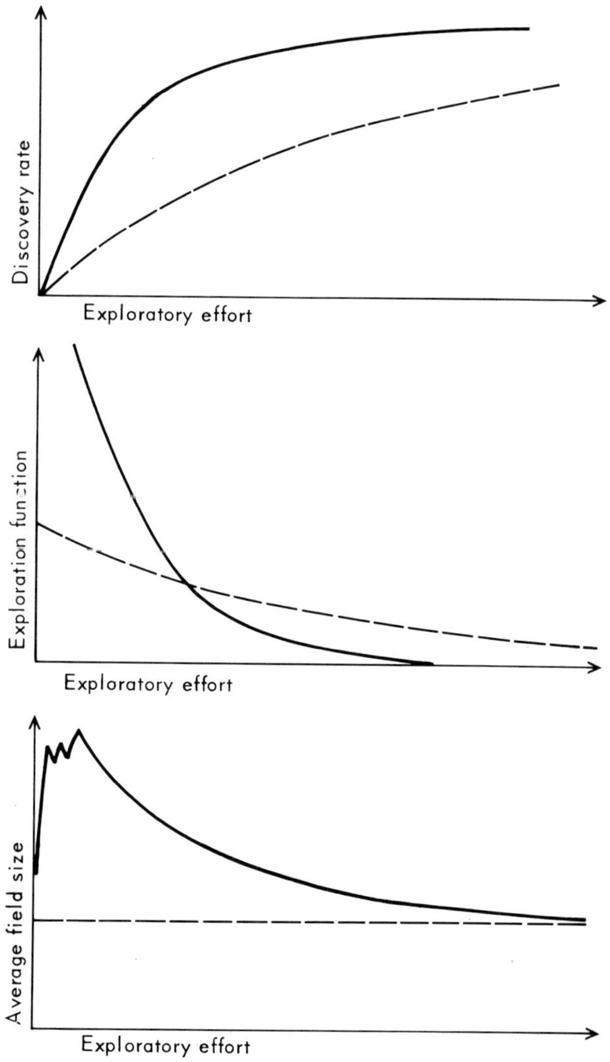

Fig. 218. — Diagrams showing theoretical variation (models of wells and history of different exploration parameters) as a function of exploratory effort (Vinkovetsky, 1979).

the net success rate or the number of significant successes, which falls from 3.5 % in 1945 to 1.5 % in 1972 (Owen, 1975). This would imply that technical progress offsets the progressive exhaustion of the 'prizes' to be won. Simultaneously, discovery costs show a steadily upward tendency. These 'replacement costs' rose, for example, from $ 0.43 in 1942 to $ 1.73 in 1957 in the Rocky Mountains, and from $ 0.73 to $ 3.76 during the same period for Illinois and Michigan (Lovejoy & Homan, 1963) (*).

Also worth noting is the fact that the exploration rate is gradually accelerated and that the interval between the initiation of prospecting and the first discovery is increasingly shorter, whatever the environment, and this can be ascribed chiefly to the benefits of technical progress (Klemme, 1971). However, the author adds, 'More basins are being explored. More producing basins are being found. However, the industry is experiencing a lower and lower succes ratio in this search !' This interval largely depends on the geological characteristics of the basin, and on technical progress. Concerning the first point, by compiling a statistical analysis of the volume of exploration performed before the first discovery, Coustau (1975) found that this effort generally remains low for provinces subsequently recognized to be extremely rich (1.2 wildcats per 10,000 km^2 on the average), high and often dispersed in poor provinces (8.76 wildcats per 10,000 km^2 on the average, or 7.5 times more).

One can only make a valid assessment of the degree of exploration of a basin in relation to the sedimentary volume really and effectively explored, in other words, explored throughly by boreholes based on valid seismics at the depth considered. Anything beyond largely remains in the area of the unknown. It is here that technical progress, and specifically advances in seismic techniques, are opening new zones in areas that have already been partly explored. On the other hand, the same standards cannot be applied to all basins, and assessment criteria must be adapted to the characteristics of the basin.

The northwest Australian shelf offers an example of the difficulties gradually overcome by technology. On this vast continental shelf, seismic exploration began in 1964, after an aeromagnetic survey that revealed the existence of a sufficient sedimentary volume. The considerable difficulties in positioning logistics and seismic processing were gradually overcome. By the end of 1971, 28,000 miles were selected from the 380,000 km^2 licence area. The six wildcats drilled in this initial period proved disappointing. However, the next eleven, drilled between May 1971 and June 1972, yielded six commercial discoveries with a total of some 350 Gm^3 of wet gas, and three other 'successes', which had to be checked by appraisal wells (Owen, 1975). Following this, no results were obtained offshore on the Exmouth shelf, in water depths up to 1000 m, after the drilling of a dozen wildcats.

Some illustrations

All prospecting is a pitting of technique against Nature, and to attack effectively it is necessary to know the defense system, in other words to understand the geology and the geodynamics of hydrocarbons. In a new province, it is always difficult to conceive accurately the prospection pattern to serve as the exploration guideline. This formulation is very often inferred by experience, or by an apparent analogy with a neighboring or recently explored province. But this analogy, which is based on fragmentary and sometimes erroneous geological knowledge of the province to be explored, is always dubious, often false. In many cases, the definition of the pattern is directly and strongly influenced by the first indications, generally by the first shows, and experience indicates that these early signals often point in the wrong direction, as the exact meaning of the seeps is always difficult to decipher.

This is why guidelines must not be adopted prematurely. It is better to proceed by sampling an overall zone to be prospected by a wide mesh grid, by seismics and boreholes, which permits a better comparison with known models. Examples of staggering failures, followed a few years later by success of equivalent magnitude, have been described time and time again. A typical example was the early exploration in Venezuela, especially in the Maracaibo Basin, where only three companies out of seventy-three succeeded from 1922 to 1943, in a province that finally proved to be exceptionally rich (Owen, 1975).

It is often forgotten that in the northern Algerian Sahara, twenty dry wildcats were drilled before the discovery of Hassi Messaoud. Naturally, technical difficulties, and particularly the inability of seismics to define the structure of the traps, and the difficulties of drilling Triassic shales, were largely responsible for these failures. But added to this was the fascination with the western area resulting from the presence of shows in the Devonian, in formations which turned out to be devoid of reservoirs.

Reference to a model is always a delicate matter, often haphazard, sometimes the cause of striking failures. The history of exploration on the Alaskan North Slope is well known : 37 wells were drilled on 18 structures in the 'foothills' from 1944 to 1953, culminating in the discovery of a mere 10 million tons of

(*) A strict definition of discovery cost is not universally accepted, and it is often difficult to compare the figures given by different authors. However, the definition is perfectly valid within the framework of any single author's work.

reserves. This was followed by the drilling of nine dry wells from 1963 to 1967, and finally Prudhoe Bay and its nearly 2 Gt of oil and gas equivalent in a very specific zone of the Basin. Thanks to this super giant, the exploration yield remained high, and discovery costs were brought down (Owen, 1975).

The changing fortunes of exploration are also well known in West Canada, where twenty years of effort and 33 wildcats were required to discover Leduc, the first large field after the initial discovery of Turner Valley. In this case the problem was one of imagining the target, the stratigraphic trap, and defining it by a still imprecise seismic technique. In a completely different geological context, exploration in Nigeria also experienced early problems for similar reasons.

In the Oriente province of Peru, from 1937 to 1950, 3000 km of seismic surveys were conducted and eight dry wells were drilled, of which the last, Tiputini 1, was located only 7 km from the discovery well of the field of the same name, drilled a few years later. Most of the structures drilled correspond to faulted anticlines of the foothills zone, leached by active hydrodynamism. From 1961 to 1972, a new partnership drilled twenty wildcats that led to the discovery of eleven fields of more than 450 Mt. The corresponding exploration investment was $ 140 million or a discovery cost of $ 0.04 per barrel for this group, with a success ratio of 0.55 and an average yield of more than 35 Mt per wildcat. The remaining groups operating during the same period in neighboring sectors were less lucky, discovering barely three pools with about fifteen wildcats, since several discoveries were not commercially viable (Duffaud, 1970 ; Owen, 1975).

The history of the petroleum conquest of Western Siberia is perhaps less well known, but was recently summarized by Trofimuk (1974). Since 1932, the great petroleum geologist Gubkin advanced the hypothesis that the Mesozoic coal facies of the Urals margin passed towards the east, at the center of the basin, to petroleum facies. The first prospection, just before the war, proved negative. In 1948, a summary report, sponsored by Kudriatsev, used the Volga-Urals basin as a model. The first wells drilled in the southern part of the basin were dry. At the end of 1953, however, a small gas pool was discovered in the Jurassic on a structure on the western margin. This success inspired a vigorous exploration effort in this sector, with no results. Only seven years later, in 1961, was the first great oil discovery made in the Lower Cretaceous in the middle Ob region, in the central part of the basin. Subsequently, major successes were achieved every year, and today this province is ranked the world's second or third biggest.

If a good guideline exists, discoveries have good chances of succeeding as soon as the field distribution logic is identified, and success rates are then high. This logic is often structural : it may be stratigraphic or sedimentological, with the extension of a sand, a reef facies, or a cap rock. The discovery of reef trends, for example, obeys this law. In Michigan and Alberta, exploration hesitated for years, and then a guideline emerged, a 'play' took shape, and for a few years the success ratio was as high as 30 to 40 %, and even 57 % as at Rainbow, after the development of seismic techniques (Owen, 1975).

The exploration of the Salawati Basin, in West Irian, also illustrates this arduous quest. The initial exploration phase stretched from 1935 to 1960. The first well was drilled in 1936 on a surface anticline with many seeps. It discovered heavy oil with density 19 °API at 130 m depth in very porous Miocene limestones, identifying the Klamono Field. The reservoir was actually the top of a 420 m high reef. The field produced slightly under 5 Mt. Following this success, 25 wells were drilled, all dry, apart from two non-commercial discoveries. Thanks to a detailed reinterpretation of the data from all the boreholes, and following geomorphological analyses, a new campaign was initiated in 1970 by another group. The first well drilled in 1972 discovered the Kasim Field, where the trap was a Miocene reef. In the following year, 13 wells were drilled on six prospects, yielding seven oil wells. Three fields, Jana, Walio and Kasim Utara, were discovered subsequently on other reefs, and the outputs reached 20,000 to 32,000 bopd. Once the guideline is discovered, the exploration yield is high, but, in this case, the guideline took a long time to unravel, and this was only achieved by a thorough reinterpretation of previous findings (Vincelette & Soepardaji, 1976).

The hydrocarbon distribution logic may be clear or unclear, easy or difficult to identify. Sometimes it may be practically absent, or very limited in space, which is generally the case of stable platform basins. It may also result from a sequence of complex stratigraphic events, such as sudden variations in facies, or structural events at the source of vertical migrations, raising a particularly difficult problem. This was observed in the Aquitaine Basin (apart from the Parentis sub-basin) where nearly every success appeared to be a unique case, and where the reservoir facies, seals and source rocks appeared to be numerous, but very often discontinuous. Obviously, the poorest provinces generally mean the most difficult exploration conditions.

Conclusion

To the extent that the characteristics of a basin result in certain sedimentary, geochemical and structural conditions, the type of basin induces its petroleum properties. This emerges very clearly for the giant provinces and a few models of poor provinces, particularly the stable platform basins. However, the presence of pools obeys too many variables for clear rules to be set forth based on the geology of the basin alone, and in all cases the specific characteristics of the geological history play the leading role. A few geo-petroleum models are thus identified, in relatively small numbers, quite independent of the framework or type of basin involved.

Relationships between the exploration results and petroleum characteristics of a province are rarely very close. A definite but weak correlation is found between exploration yield and potential, a clearer relationship with type of habitat. In particular, the average discovery appears to be higher if the habitat is more concentrated, and vice versa, smaller if the habitat is more dispersed.

As a rule, a certain correlation is found between the three parameters that express the exploration yield : success ratio, average quantity discovered per wildcat, average discovery. The relationships between the various terms are determined in particular by the type of habitat, and hence to a certain extent by the type of basin. On the other hand, hardly any relationship is found within a given period, between exploration yield and exploration effort, except that following a trial and error period, a general declining tendency emerges with the passage of time. More than the intensity of exploration, the geological characteristics must nevertheless guide our strategy. It would be pointless, for example, to measure the degree of knowledge of a basin by the mere density of the wildcats drilled. In fact, it is obvious that a basin with complex structures and rapid facies variations requires intensive exploration, but this cannot achieve significant additional success in cratonic basins with broad and quiet anticline arches. Since the ratio of the number of wildcats per unit area is currently 1 to 100 between the Middle East and the United States, it is hard to conceive that the countries of the Persian Gulf should be the target of exploration as intensive as in North America to discover the bulk of their potential. On the other hand, the current effort of 1/50 for Europe in comparison with the United States can and must certainly be significantly intensified. In fact, it appears evident that in a group of basins, intensive exploration helps to discover significant additional reserves. Thus in the United States, the 'major' fields (over 100 Mbl) only produced 50 % of total discoveries, while for the world taken as a whole, the 'giants' (over 500 Mbl) contain more than three-quarters of cumulative discoveries (Owen, 1975).

A dispersed habitat implies a natural distribution of the chances of success, and if the area explored is not too narrow, the operator can work with a high share of the venture. On the contrary, in the case of a concentrated habitat, and particularly if the province is not very rich, or offers few chances of being so, it would be advisable to compensate for the poor natural distribution of the target by a smaller share of the venture, possibly over a larger area.

These general considerations must obviously be supplemented by the degree of exploration of the basin : the large units are generally discovered in the early years of exploration, which is easier and more effective with a simpler though fairly differentiated basin structure, as happens to be the case in platform provinces with broad arches. Faced with the diversity, originality and, one is tempted to add, the fantasy of petroleum provinces, it is difficult to formulate more specific recommendations in the current state of our knowledge. Experience is undeniably the most invaluable asset in this area, and very often only the variety of cases that have been analyzed and understood, indeed lived, can guide the imagination. In this field, uncertainties in assumptions often prove more costly than inaccuracies in analyses and measurements.

In fact, every geologist reasons more or less consciously in a similar manner, and the effectiveness of this argument is directly related to the degree of accuracy in the knowledge of the geological and petroleum data about the prospected zone, on the one hand, and the example taken as a reference on the other. This applies particularly to offshore areas, thanks to the possibilities afforded by seismics. It implies that at any moment of exploration, both in the most elementary stages, in which information is rare and scattered, and when a large body of data is available, it is necessary to recapitulate all the material at hand, to sketch out and to detaïl the portrait of the province investigated, to achieve a better comparison. It is often very useful to repeat the procedure, to reinterpret a basin with a fresh outlook. In fact, quite frequently, a particular fact, a relationship in space and/or time, determines the presence of an accumulation. It is very often the original idea plus geometric analysis that brings success.

Whatever the uncertainties and inadequacies of geological and geophysical methods, and despite the low effectiveness of our exploration (very often technical advances merely compensate for the exhaustion and diminution of prospects), the approach which they allow still very largely surpasses any exploration carried out at random.

The comparison of observed or historical data with values calculated from theoretical models, based on random drilling scenarios, yields a certain number of observations. The curves in Figure 219, plotted by Vinkovetsky (1979), initially reveal an overall convergence of exploration yield and results. It may be observed that the volume of discoveries grows more rapidly and remains at a higher level in reality, namely with due consideration of geological knowledge, than in the random drilling scenario. Simultaneously, the exploration yield is in fact much higher in the early stages of exploration, but decreases more rapidly than in the theoretical model.

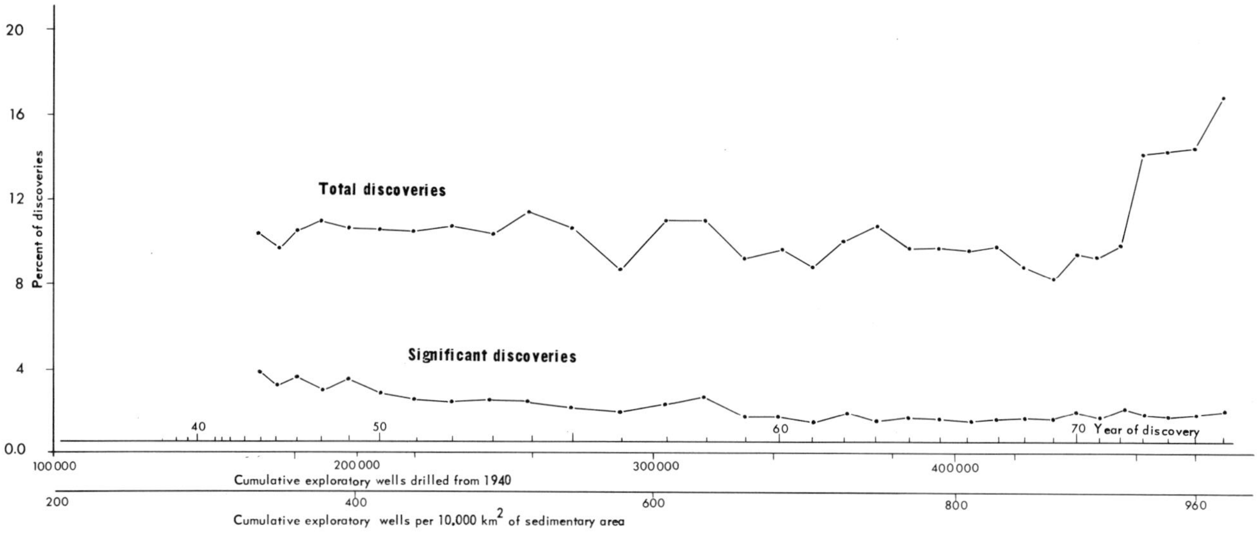

Fig. 219. — Variation in percent success of new field wildcat (NFW) exploratory wells in the United States from 1945 to 1976 (Coustau, 1979 unpublished).

On the whole, we tend to underestimate the quality and quantity of geological data provided by boreholes, both dry and productive, and advances in concepts and techniques. Geological information offers an invaluable asset, the utilization of which is not always as optimal as it should be. The interpretation of seismic sections, for example, does not always take into account all the geological data provided, and these data are not always sufficiently well rearranged, within the paleogeographic and historical framework of the basin.

On countless occasions, exploration has been resumed due to the advent of a new technique that explores a deeper zone of the basin. This also applies to the formulation of a new concept, *considering that this brings success closer when it is difficult to escape from standard thought patterns and to discover what is poorly imagined.* No technical progress can be fully exploited without an effort of the imagination. It would be outright pessimism to claim that a method based exclusively on random prospection could be more effective than an approach that implements all the resources of the human spirit. This way of thinking is contradicted by the entire history of civilization from time immemorial.

4.2. THE AMERICAN EXAMPLE

The knowledge of a sedimentary basin and a petroleum province depends on the exploration effort, and especially on the number of wells drilled, with the clear understanding that this effort is itself influenced by the basin characteristics, and by the types of oil and gas habitat. Few provinces in the world have been so intensely drilled as the United States and Canada. In this respect, the North American basins offer a series of vitally useful models for planning exploration strategies in other regions. The 48 contiguous States of the United States and the Canadian 'provinces' obviously appear to be the most exemplary.

4.2.1. Geological and petroleum data

In the United States, sedimentary basins of petroleum value cover a total of slightly over 8 Mkm^2, 6.6 Mkm^2 onshore and 1.4 Mkm^2 offshore (up to 200 m). The basins of the 48 contiguous States cover about 6.5 Mkm^2, including six onshore, and Alaska about 1.5 Mkm^2. These figures represent approximately 10 % of the world's sedimentary space. In actual fact, only about 2.6 Mkm^2 have proved to be sufficiently rich to merit intensive exploration. The Canadian basins occupy an area of about 7 Mkm^2 onshore and offshore (continental shelf and slope), including 1.7 Mkm^2 for the 'classic' areas, mainly the Provinces. The corresponding sedimentary volumes amount to nearly 20 and 4 Mkm^3 respectively.

These sedimentary areas are remarkable for the number and variety of their petroleum provinces. In the United States, for example, broad stable platform basins are encountered, like those of Michigan and

Williston, undulated and folded basins as in the Mid-Continent and West Texas, thick sedimentary prisms in the Gulf Coast, intermontane troughs in the Rockies, and active margin basins in California. The productive stratigraphic series stretch from the Cambrian to the Pleistocene, practically uninterrupted.

The fields of North America cover the entire range of petroleum traps : 'textbook' anticlines in the Rockies, Appalachians and California, anticlines under unconformities in the Mid-Continent, overthrust folds in the Rockies, traps against faults and salt domes in the Gulf Coast, reefs in West Texas, Michigan and Alberta, sand body pinchouts in the Mid-Continent, East Texas and the Eastern Rockies, turbidites in California and in the Delaware Basin.

Results

— *United States*

About 21 Gt of oil have been discovered since inception, 16 Gt in cumulative production and 5 Gt in remaining reserves, with about 20,000 Gm^3 of gas, including 13,400 Gm^3 already produced and accounting for one-third of cumulative world production, with 6,600 Gm^3 in reserve. Oil discoveries, corresponding to an average recovery rate of about 32 % (13.5 % average in Ohio, 65 % in East Texas), represent a mean potential density of 2500 t km^{-2}, 3250 t for the 48 States, to which 3 Mm^3 of gas should be added, making a gas/oil ratio of 0.9 m^3/m^3 (in recoverable quantities).

These averages are obviously the result of widely differing values for the basins, ranging from more than 8000 t km^{-2} for the Pacific Coast, West Texas and Gulf Coast provinces, to only 700 t km^{-2} for the Western Rockies. These 21 Gt of oil are distributed among some 26,000 'significant' fields. Considering all the oil and gas fields in the 48 States together, their distribution, which obeys a substantially log-normal scheme, is the following (Ivanhoe, 1976) :

2	>	560		Mt of oil or gas equivalent,
45	between	560	and 70	Mt of oil or gas equivalent,
216	between	70	and 14	— —
600	between	14	and 7	— —
1500	between	7	and 3.5	— —
4000	between	3.5	and 1.4	— —
about 20,000	<	1.4		Mt of oil or gas equivalent.

According to the annual tables in Oil and Gas Journal, 270 fields of over 100 Mbl or 14 Mt account for about two-thirds of cumulative discoveries, and all fields of more than 200 Mbl or 28 Mt account for around half. This simply means that half of the reserves are distributed among pools of less than 28 Mt. According to the tables of the AAPG (Bull. August, 1976, p. 1177), only 20 % of the pools discovered between 1945 and 1969 have reserves above 1 Mbl (0.14 Mt), and 80 % obviously have smaller reserves. In other words, 40 % of the reserves discovered are situated in the United States in giant fields (70 Mt), against 70 % in the world taken as a whole, where very often only rather large fields have been prospected and catalogued.

— *Canada*

In the 1946-1970 period, about 2250 Mt of oil and 2400 Gm^3 of gas have been discovered in the Provinces, corresponding to an average potential of about 1600 t and 1.75 Mm^3 of oil and gas respectively per km^2.

These discoveries are distributed as follows :

81 Class A pools (> 7 Mt),
54 Class B pools (7 to 3.5 Mt),
121 Class C pools (3.5 to 1.4 Mt),
637 Class D pools (1.4 to 0.14 Mt),

making a total of 893 pools of more than 1 Mbl, added to 952 pools of smaller size. In other words, half of the number of pools contain approximately one-hundredth of the reserves, and 4 % (the biggest) contain 72 % of reserves.

4.2.2. **Means implemented and exploration yield**

4.2.2.1. ***Means implemented***

United States

The total number of wells drilled for oil since inception is estimated at about 2,425,000, including about 650,000 wildcats, accounting for three-quarters of all wells drilled worldwide. About 482,000 wildcats, including

174.000 New Field Wildcats, have been drilled in this country since 1945 (Grossling, 1976). This total represents an exploration effort of about 750 wildcats per 10,000 km^2, including 300 since 1945, in the 48 States, or roughly an average of seven wells per 10,000 km^2 and per year. The average depth of the wildcats rose from 4000 ft in 1946 to 6000 ft in 1975.

It is well known that the geophysical effort has never been sustained as intensely as the drilling rate. Hence during the 1945-1958 period, about 350,000 seismic crew-weeks of geophysical prospecting were carried out, approximately 630 per 10,000 km^2, compared with 157,000 wildcats, or an average of 2.2 crew-weeks per wildcat. This ratio is a mean between the values ranging from 7 for the Rockies, to 4.5 for Louisiana, with only 0.1 for the Appalachians, Michigan, Illinois and California.

Canada

From 1946 to 1970, 14,500 New Field Wildcats were drilled in West Canada, or an average of 108 per 10,000 km^2.

4.2.2.2. *Exploration yield*

The following discoveries were made during the 1945-1975 period :

in the 48 States :

12 Gt of oil
13.4 Tm^3 of gas,

in Canada :

2.2 Gt of oil and 2.4 Tm^3 of gas.
The average exploration yield, expressed in reserves discovered per NFW, amounts in the United States to :
12 Gt with 174,000 wells, or 70,000 t of oil,
13.4 Tm^3 with 174,000 wells, or 77 Mm^3 of gas,
and in Canada to :
156,000 t of oil,
163 Mm^3 of gas.
These values correspond to a mean, between yields of around 200,000 t per well in the years 1946-1948, and 40,000 to 50,000 t in the 1960 s.

Arguing in terms of success rate, the United States, still for the 1945-1975 period, show 11,258 and 4983 NFW which yielded oil and gas shows respectively during tests. The gross 'success' rate thus amounts to 0.06 for oil, and 0.03 for gas, making a total of 0.09. This corresponds to average 'discoveries' or pools of 1 Mt of oil and 2.7 Gm^3 of gas. If one only considers the wells corresponding to a discovery larger than 1 Mbl or gas equivalent, or 'significant' discoveries, the success ratios fall to 0.01 for oil, and 0.008 for gas, or slightly under 2 % for oil and gas combined, with average discoveries of about 7 Mt and 10 Gm^3 of gas. If one considers only Class A discoveries (> 7 Mt of oil or gas equivalent), the success rate falls to 3 per 10,000 for oil and 6 per 10,000 for gas, making a total of 1 per 1000 approximately.

If one examines the pattern as a function of time, it may be seen that the wildcat success ratio remains less than 18 % between 1938 and 1943 (10 to 18 %), but that from 1944 onwards, it generally remains above 18 %. However, one must not be misled by this relatively high rate : for the New Field Wildcats, the 'significant' success ratio, namely corresponding to a discovery greater than 140,000 t of oil or recoverable gas equivalent, was only 3.75 % in 1945. This rate declined steadily and fell to 1.7 in 1975 (Figure 220). The cumulative ratio (i.e. yearly average) fell from 3.75 to 2 %. However, the decline in this significant success ratio from 1945 to 1975 is not at all random. In fact, the following may be observed :

- a certain stabilization of the number of Class A and B discoveries, and a smaller number of Class C and D oil discoveries,
- a slower decline in gas discoveries of Class A (> 8.5 Gm^3), C (4 to 1.6 Gm^3) and especially D (1.6 to 0.16 Gm^3), than the oil discoveries of equivalent dimensions,
- a slight increase in non-significant oil discoveries,
- a considerable increase in non-significant gas discoveries.

In other words, this overall pattern impiles that the intensification of the drilling effort undertaken since 1945/1950, added to the improvement in techniques and the substantial deepening of the wells, resulted in slowing down the decline in the number of significant gas discoveries, from 1955 stopped the decline in large oil discoveries (pools undoubtedly situated in an unconventional position), and considerably increased the number of non-significant gas discoveries. One interpretation among others may be suggested for this increase in non-significant gas discoveries : the deepening of the wells may have led to penetration into earlier, diagenetically evolved basins, where a rapid drop in permeability occurs below 1800 to 2000 m. On the other hand, the sedimentary series in zones less affected by diagenesis, by offering considerable gas discovery prospects below 3500 m, may exist, but has not yet been substantially drilled due to the high cost of drilling (Coustau, 1979 unpublished).

Some data concerning the methods used to discover wells larger than 100 Mbl for the 1877/1945 period are also interesting (Owen, 1975).

Method	No. of fields	%	Reserves Mm^3	%	Average pool Mm^3
Random drilling	17	7.7	1560	15.5	89
Seeps	38	17.3	1685	17.3	44
Trends	17	7.7	556	5.7	33
Surface structure	50	22.7	607	26.8	52
Subsurface structure	22	10.0	525	5.4	24
Geophysics	76	34.6	2280	28.7	36
	220	100 %	9200	100 %	44

This shows that one-third of all discoveries of pools with more than 14 Mt and 38 % of the oil were made by random drilling or on superficial seeps, and that over half of all pools and two-thirds of reserves were obtained by drilling surface structures.

It took 66 years (1859/1925) to discover the first 10 billion m^3 of oil and gas, exactly 9.3, with less than 20 % of the total drilling effort and only 8 % of the total footage drilled, considering the shallow wells sunk at the time. The corresponding exploration yield amounts to approximately 100,000 m^3 per well. In the next twelve years (1926 to 1937) another 10 billion m^3 were discovered, with slightly under 10 % of the drilling effort. This period represented a considerable advance thanks to the advent of geophysics, with an average yield of nearly 230,000 m^3 per well. The next to 10 billion m^3 were discovered in thirteen years (1938 to 1950), but with one-third of the total drilling effort, and the yield fell to 140,000 m^3 per well. This was followed by the ageing period of decline (1951 to 1974), with nearly three-quarters of the footage drilled to discover another 10 Gm^3, with a yield of less than only 38,000 m^3 per well (Coustau, unpublished report).

4.2.2.3. *The future of exploration*

Despite the density of drilling already achieved, reserves which remain to be discovered in the 48 States (and also offshore up to a water depth of 200 m) are estimated, with 95 % probability, at about 7000 billion m^3 of gas and around 4 billion m^3 of oil. The upper limit of these reserves, with 5 % probability only, is estimated at about 13 billion m^3 of oil and 16 trillion m^3 of gas. These figures can be summarized by stating that another 10 billion m^3 of hydrocarbons certainly remain to be discovered, and possibly another 30 billion m^3 (including volumes extracted by increasing the recovery rate).

These new reserves, added to the 15 Gt of cumulative production and the 5 Gt of proven reserves, point to ultimate discoveries of more than 30 Gt of oil and 34 Tm^3 of gas, figures corresponding to average oil and gas potentials of about 4000 t and slightly more than 4 Mm^3 per km^2. This ultimate potential is about 800 t km^{-2} for the provinces of the Appalachians, Michigan and the Western Rockies, and 10,000 to 11,000 t km^{-2} for the California, Gulf Coast and West Texas Basins. It is noteworthy that the anticipated discoveries could be of the same order of magnitude, if not greater, than the cumulative discoveries in provinces as intensely explored as Michigan and the Northern Rockies, around 40 % in the Appalachians and the Pacific Coast Basins, 20 % in the Permian basin (USGS Circular 725-1975) (Table IV).

In the United States the authors show a tendency to expect more from proven areas, chiefly by enhancing recovery rates, despite the drilling density, rather than from unexplored or little explored areas. Considering the effort that was required to discover the ten last billion m^3 from 1951 to 1974, one can well imagine the investments required, both in terms of money and manpower, to discover the remaining 10 to 30 billion m^3.

In Canada, a recent estimate of conventional oil resources of the onshore basins and continental shelves (1977) gives a range below the estimates of a 1973 study, lying between 3.5 Gt with a probability of 90 %, 6 Gt with a probability of 10 % only, and 4.15 Gt with a probability of 50 %. As for gas, anticipated discoveries are estimated at :

— 6400 Gm^3 with 90 % probability,

— 7750 Gm^3 with 50 % probability,

— 10,600 Gm^3 with 10 % probability.

Table IV. — Oil discoveries in different provinces of the United States (Mt) (source : USGS Circular 725)

	Cumul. prod.	Demonstrated reserves	Total	Area (1000 km²)	Total per km²	Undiscovered recoverable reserves	Ultimate reserves	Ultimate reserves per km²
ONSHORE								
Alaska	21	1394	1415	650	9/4	1700	3115	4800
Pacific Coast States	2135	530	2666	320	2177	1000	3600	11,250
Western Rocky Mountains	156	70	227	855	8331	500	700	820
Northern Rocky Mountains	843	240	1083	930	265	1000	2000	2150
West Texas and North-eastern New Mexico	3000	1260	4261	500	1164	1120	5300	10,600
Western Gulf basin	4390	1072	5462	620	8522	1120	6600	10,650
Mid-Continent	2408	282	2690	1150	8809	840	3500	3040
Michigan basin	90	13	103	315	2340	140	250	800
Eastern Interior	608	42	650	430	327	140	800	1860
Appalachians	355	31	386	530	1512	140	420	800
48 States	13,985	3556	17,451	5950	728	6160	23,700	4000
plus Alaska	14,006	4950	18,956	6600	2948	7840	26,800	4000
OFFSHORE								
Alaska	63	22	85	820	2872	2200	2200	2700
Pacific Coast States	210	156	366	46	7956	420	800	17,400
Gulf of Mexico	580	315	895	290	3086	700	1600	5500
Total Lower 48 offshore	789	473	1262	603	2092	1540	2800	4660
Total offshore US	853	494	1346	1427	943	3640	5000	3500
Total Lower 48 States	14,770	4022	18,802	6555	2868	7700	26,500	4040
Total United States	14,860	5444	20,304	8030	2490	11,500	31,500	3900

Most of the expected discoveries will lie with a probability of 50 % in the three most difficult 'pioneer' regions, the Mackenzie Delta, the Beaufort Sea/Arctic Islands, and the Labrador continental shelf, which colud contain 1.9 Gt of oil (1.08 to 3.55 Gt with high and low probabilities) and 3855 Gm^3 of gas (2240 to 7000 Gm^3).

Western Canada, where 2.35 Gt of oil and 2400 Gm^3 of gas have already been discovered, distributed among some 3000 pools, could still yield discoveries of about 0.5 Gt (range 0.4 to 0.8 Gt) and 330 Gm^3 (range 110 to 600 Gm^3). It seems improbable that all the resources of the pioneer regions could ever exceed those of Western Canada, where the average potential will rise from 1800 to 2200 t km^{-2}. These estimates of traditional hydrocarbon resources do not account for what the Canadians are classifying as unconventional products : oils from tar sands, heavy oils, and also hydrocarbons beyond the continental shelf. The tar sands are known to be present in vast quantities in Canada (9.4 to 24 Gt of presumably recoverable synthetic crudes, including 3.8 Gt by ore extraction) but 'this potential is not considered likely to contribute significantly to our short to medium-term supplies' notes the author (Energy, Mines and Resources Canada, 1977).

Summary and conclusions

The sedimentary basins of North America, and especially those of the 48 States and the Canadian Provinces, represent a good average sampling of poor to rich petroleum provinces, but, unlike those of the Middle East, never exceptionally prolific. These basins have been subjected to intense exploration, mainly by drilling, ten to a hundred times more dense than on the other continents. Despite the low average yield, this intensive exploration served to find large quantities of economically exploitable oil and gas. The reasons for these successes are known :

— an 'incentive' mining law,

— most often, shallow and inexpensive wells,

— a particularly dense industrial infrastructure,

— a climate of competition and a highly developed sense of enterprise, sometimes resulting in waste.

America does not offer us miracles, but an example, that of drilling to find new pools, an example which cannot be blindly aped, but which must be adapted to circumstances. To drill on good objectives, many locations are necessary, implying detailed subsurface surveys, a thorough knowledge of the targets, tenacity in the search for a prospect, and imagination in making new assumptions. Also necessary is the ability to question accepted ideas, and to remain open to new ones, to avoid ignoring series which are often considered as 'overburden', and to remain constantly vigilant and openminded. The American example shows us that when the economic and technical framework is favorable, an intense exploration effort normally succeeds in discovering new reserves in large quantities, and not only marginal pools. In other words, low exploration yields in a favorable economic context can yield economically valid successes.

4.3. ULTIMATE HYDROCARBON RESOURCES

One of the major problems of the day, and one besetting geologists in particular, is the estimation of the quantities of oil and gas likely to be produced in the forthcoming decades. Such attempts to quantify undiscovered reserves are necessary to anticipate the future share of hydrocarbons in the overall balance of requirements, and also to orient and guide the exploration effort towards zones that offer or appear to offer the best hopes of success. This endeavor, which involves the routine activities of the geologist on another scale, is obviously full of uncertainties and unknowns :

- — of a geological nature to begin with, concerning the characteristics of the basins and zones to be explored,
- — of a technical order, for everything associated with advances in methods and prospecting and production tools,
- — of an economic order, and more specifically associated with crude oil prices, fiscal factors, and political considerations.

4.3.1. Review of some definitions of reserves and resources

The definition of the different categories of reserves is determined essentially by geological considerations and by technical and economic criteria.

According to the degree of geological knowledge, and proceeding from the known to the unknown, it is customary to distinguish three main categories (Table V).

- — Proven reserves are defined qualitatively and measured quantitatively, to within 20 %, by interpolation between boreholes and limited extrapolation, based on reliable seismic data. These are the 'measured' and 'indicated' reserves of the English-speaking world, and Categories A and B in the Soviet Union. In fact, they can be defined in terms of probability, as quantities offering at least 80 % chances of being reached or exceeded (Table V).

Table V. — Different categories of reserves and resources

<table>
<tr><td rowspan="4">Exploitable (economic)</td><td colspan="3">Discoveries</td><td colspan="2">Undiscovered</td></tr>
<tr><td rowspan="3">Produced</td><td colspan="2">Reserves</td><td colspan="2">In areas</td></tr>
<tr><td rowspan="2">demonstrated</td><td rowspan="2">inferred</td><td>Known</td><td>Unknown</td></tr>
<tr><td>hypothetical</td><td>speculative</td></tr>
<tr><td>Potential (subeconomic)</td><td colspan="5">RESOURCES</td></tr>
</table>

Feasability: 100 % (top) — 0 (bottom)

100 % ← Increasing degree of geological assurance

— Probable reserves are estimated by extrapolation from a well and the geophysical survey of a structure, and possibly of one or more neighboring structures well known from the geological standpoint. They correspond to the 'inferred' categories in America, and C1 and partly C2 for the Soviet Union. This category is progressively clarified by extension wells or outposts and by advances in recovery. The very term 'probable' clearly indicates the probabilistic nature of the assessment of this category, which may be characterized as quantities offering 40 to 80 % chances of being discovered.

— Reserves, or even better resources, which are possible or potential, in other words hypothetical, predictable or speculative, designate undiscovered volumes, globally and reasonably estimated on the basis of regional geological criteria, and by comparison with known proven provinces (these are the so-called 'hypothetical' categories in the known zones, 'speculative' in the unknown zones in America, and D1 and D2 reserves in the Soviet Union). As a rule, the degree of probability of their discovery remains under 40 %, and often falls in the estimates to 10 and even 5 %.

In accordance with technical and economic criteria, as a rule, a distinction is made between reserves in place and recoverable reserves.

Reserves in place, also called 'accumulations' (in French), before the start of production, represent the total volume of hydrocarbons existing in the subsurface and, in general, likely to be at least partly exploited. This estimate essentially depends on the geology, and more specifically the geometry of the trap, its impregnated volume, and the petrophysical characteristics of the reservoir; in short, the concentration of hydrocarbons in the rock, and the physical and chemical characteristics of the fluids (pressure, GOR). It is obtained from measurements that remain extremely limited and dispersed, even in the most favorable cases. In fact, reserves in place and recoverable reserves are closely related, and one often speaks of the former as 'mobilizable' quantities, in other words, likely to be partly produced. A minimum concentration is in fact necessary when speaking of a pool.

Recoverable reserves represent quantities which are really likely to be produced within a given period. In other words, they represent the sum of foreseeable production over a period of 10 to 20 years. This concept of 'exploitability' is obviously dependent on :

— the geology, and more specifically the petrophysical characteristics of the reservoirs, especially permeability, and fluid characteristics,
— the state of production techniques, or in other words the 'feasibilty' of production,
— the economic context in the broad sense of the term (tax, market, policy considerations), in other words, criteria which are valid from the geographic standpoint (physical, political), and in the long term.

These technical and economic factors are closely interrelated. In particular, it is well known that the increase in the value of a raw material (and any rarefaction pushes up prices) is reflected within certain limits by an increase in extractable quantities. But it is also obvious, at the extreme, that these economically recoverable reserves, especially with respect to energy sources, are themselves limited by the amount of energy required for their extraction.

The two major categories of reserves, in place and recoverable, are actually closely interrelated and both correspond to variable degrees of probability, especially with respect to probable and possible reserves. In fact, the latter depend on the degree of geological knowledge and the evolution of techniques and of the economy. However, in the case of recoverable reserves, added to the probabilities of a geological order are the probabilities of technical 'feasibility' and economic implementation. The accuracy of estimates increases as the work proceeds in the field, and finally the volume of recoverable reserves coincides with final cumulative production.

Within the framework of the production of a pool, it is also customary to define initial, or original reserves, namely those known at the start of production, or recompiled during extraction by the addition of cumulative production to remaining reserves. The sum of cumulative production and remaining reserves is also qualified as cumulative discovery, more specifically when a group of fields is involved. The total of cumulative discoveries and recoverable reserves likely to be discovered is qualified as ultimate reserves.

Estimates of recoverable reserves are especially difficult to make and are very uncertain, for products such as sand oils, more generally for heavy oils, and shale oils, of which quantities in place are often accurately known, but with a very sketchy and vague idea of actual possibilities of technical and economic production. The product of these oil shales is often a complex fuel, partly solid, which it is equally easy (or difficult) to compare with coals as well as oils.

Hence the term resource is more suitable. In the case of these 'unconventional' oils, and specifically oil shales, one actually speaks of quantities of fuels likely to be extracted from the rock in the most favorable conditions, and this figure is multiplied by the volume of rocks presenting the same percentage of hydrocarbons. This brings us closer to mining data than to pretroleum standards. Hence a certain convergence exists between oil in place and recoverable oils. But apart from the fact that the entire impregnated volume cannot be produced other than by open cast mining, it is not generally known how to accomplish this operation in profitable conditions, and this makes it risky to speak of recoverable reserves.

4.3.2. **Estimate of conventional oil resources**

Despite these difficulties, and considering a number of hypotheses, following a growth in the value of estimates comparable to that of proven resources, a certain convergence of estimates is observed in recent years (Moody, 1975 ; Halbouty, 1978 ; Desprairies, 1978 ; Nehring, 1978 ; Halbouty & Moody, 1979 ; Meyerhoff, 1979 ; Klemme, 1980 ; Colitti, 1981). Until now, about 160 Gt of oil have been discovered, of which some 60 Gt have been extracted. Proven reserves tend to level off in recent years at the level of 90 to 100 Gt (depending on sources and accepted definitions). These discoveries were made on a 'useful' sedimentary area of about 75 Mkm^2, 50 to 55 Mkm^2 onshore, 20 to 25 Mkm^2 on the continental shelf, adjacent zones of the slope, and inland seas, up to a water depth of 1500 m (Meyerhoff, 1979). Speaking of basins, about 400 entities have been explored out of a total of 600, but as we have have seen (Chapter 2 of Part 3), only 25 of the 160 productive basins contain approximately 85 % of all the oil discovered.

The usual estimates, and those presented below, are primarily concerned with so-called conventional resources, in other words excluding deep offshore oil, polar regions in general and at least the Antarctic, and also the 'synthetic' crude from tar sands and oil shales. Hence an attempt is made to distinguish between products resulting from exploration and production using techniques currently considered to be fully developed, as opposed to the hydrocarbons requiring substantially different technical and economic conditions. Possible discoveries are usually concerned with the year 2000 and the following few decades, and refer (for the author quoted) to a crude oil price of US $ 20 to 25 per barrel (1978 value) in this period.

In a closely interrelated manner, these estimates account for the oil belonging to yet undiscovered accumulations, and also oil likely to be produced from known pools, thanks to enhancement of the recovery rate. The share between these two sources is generally poorly defined, especially with respect to future discoveries, but a tendency has emerged to raise the recovery rate more easily than to conceive new productive zones. In the context thus defined, reserves in place proven today are estimated at about 400 Gt, or 100 Gt of recoverable reserves with a recovery factor of 25 %, which may be considered as the present world average. They could supply 50 % more, or about 150 Gt, if the factor is raised to 40 %, which is far more than can be hoped for, considering all the world's fields taken together (Figure 221). The covery rate certainly rises steadily with technical progress and the application of enhanced recovery processes. However, increases in this percentage often come up against growing difficulties and rising costs. The increase in crude oil prices, which contributes to their slow but steady development, nevertheless barely appears to increase the rate of return, and intensified exploration seems necessary. New enhanced recovery techniques are certainly promising, but due to the geology involved and to their costs, they are still very often reserved for a relatively limited number of pools.

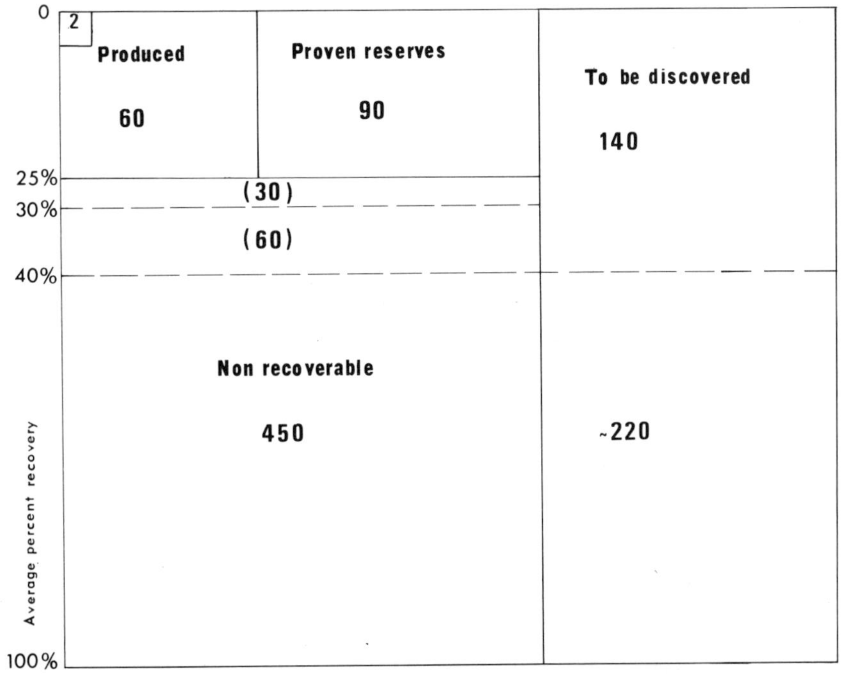

Fig. 220. — Ultimate conventional oil resources in billions of tons.

In the United States, for example, which presents the most favorable technical,fiscal and economic context, the average recovery coefficient rose from 30 to 32 % between 1967 and 1976.

For sedimentary basins as a whole, the experts generally agree that, as far as conventional oil is concerned, one can hope to discover and produce substantially the same amount that has already been produced, or a quantity of about 150 ± 50 Gt. More accurately speaking, these predictions are advanced within a probability range spreading from a 10 % hope of discovery (optimistic) for the highest values, to 90 % (pessimistic) for the lowest but most probable figures, with an average value with 50 % chances of success.

It is in these terms that the results of the survey of the World Energy Conference, held in Istanbul in 1977, are expressed. According to the experts consulted, reserves remaining to be discovered amount on the average to 100 to 400 Gt, but nearly two-thirds of the experts estimate these hopes at less than 200 Gt. If we tighten the range of answers of the central, most uniform group, representing two-thirds of opinions, the figures obtained for reserves to be discovered range from 40 Gt with 90 % chances of success, to 300 Gt with 10 %. More recent analyses give a range of 137 to 144 Gt for this category (Mackay and North, 1975) and 100 to 180 Gt (Nehring, 1978). Halbouty (1978) estimates that half of the 600 known sedimentary basins could finally yield commercial production. However, he continues, no more than a dozen new basins could contain over 1.4 Gt of oil or equivalent, and two or three only more than 8 Gt. Moody (1979) considers that there are 90 % chances of discovering another 140 Gt and 10 % of discovering 345 Gt.

Colitti (1981), Vice-Chairman of AGIP, estimates the volume of reserves remaining to be discovered, including all offshore discoveries, but excluding the highly sophisticated enhanced recovery processes, at 1017 billion barrels (140 Gt), which he divides as follows :
- — $450 \cdot 10^9$ bl (63 Gt), or 45 % to be discovered with present technologies,
- — $316 \cdot 10^9$ bl (50 Gt), or 85 % with enhanced recovery processes,
- — $205 \cdot 10^9$ bl (29 Gt), or 20 % by 're-evaluations'.

On the continent/continental shelf level, possibilities are more or less evenly divided, with a slight advantage to the former, where the favorable zones are about three times larger in area. With respect to the degree of knowledge of the basins, some experts tend to divide the hydrocarbon volumes into three substantially equal parts :
- — a small third from well known productive basins, both in traditional objectives, and in new targets or types of trap, and by improved recovery methods,
- — a large third from explored productive basins, only slightly explored up to now,
- — the remainder in new zones, in frontier zones, and especially in deep waters (Halbouty & Moody, 1979).

The distribution by continents or by major regions reveals a number of divergences. The United States and Canada combined have witnessed the discovery of 25 Gt to date, of which 17 Gt have already been produced. Hopes of discoveries lie in the range of 4 to 36 Gt (for 90 and 10 % probabilities), with an average value of 20 Gt (14 to 26 Gt, Nehring, 1978). It should be remembered that, at the present time, this area has an oil density per km^2 of about 2000 t, achieved at the cost of an exploration effort comprising nearly 500 wells per 10,000 km^2, out of all proportion to the means implemented on the other continents (in the United States alone, over an area representing only 7 % of the total of the sedimentary basins contains three-quarters of all the wells ever drilled worldwide). However, the average yield is very low, around 50,000 t per well only. It is worth noting that a large share of these future reserves is expected to be derived from advances in recovery techniques, and by re-evaluations (64 % of reserves remaining to be discovered for Colitti, 1981).

The 'Communist world' including the USSR, Eastern Europe and China, has produced some 8 Gt to date, and is estimated to have nearly 22 Gt of proven reserves. Its possibilities of future discoveries appear to be high and lie within the range 15 to 100 Gt, with an average of about 50 Gt (9 to 17 Gt for the USSR and Eastern Europe according to Nehring, 1978). The average density of present discoveries is about 1900 t per km^2, in basin which are mostly attractive though as yet relatively little explored (around 95 wildcats per 10,000 km^2, a figure slightly under the world 'average' of about 100). The exploration yield is about the same as the world average, about 200,000 t per wildcat. However, the production of the fields does not appear to have met initial hopes in a number of cases, especially owing to the sharp increases in the percentage of water.

As we all know, the Middle East offers outstanding potential. Around 72 Gt of oil, including 60 Gt of reserves, have already been discovered in this region, with an exploration effort of only five wells per 10,000 km^2, and an unrivalled yield of over 35 Mt per well. It is still hoped to discover at least 30 billion tons there, in the range 7 to 70 Gt (48 to 86 Gt according to Nehring, 1978, 55 Gt including 35 Gt by re-evaluation and enhanced recovery according to Colitti, 1981).

The remaining continents display more modest results and hopes :
- — Western Europe has seen the discovery of nearly 4 Gt, essentially in the North Sea, with an average but irregularly distributed exploration effort of about 20 wells per 10,000 km^2, five times less than the world average (particularly low when compared to North America) and an average yield of

300,000 t per well. Hopes for discoveries lie in the range 2 to 12 Gt, with an average of 5 to 6 Gt, and perhaps a quantity of the same order of magnitude from the known fields (Colitti, 1981).

— Around 12 Gt have been discovered in Africa, with the drilling of about five wells per 10,000 km^2 and a fairly high average yield of 1.8 Mt per well. Hopes for discoveries are estimated at an average value of about 12 billion tons, 17 billion tons for Colitti (1981).
— Latin America has witnessed the discovery of nearly 20 billion tons, essentially concentrated in Venezuela and Mexico, with an exploration effort of around 12 wells per 10,000 km^2, and a yield, prior to the latest Mexican discoveries, of slightly under one million tons per well. Hopes for discoveries are high and appear to lie in the range of about 40 billion tons, largely concentrated in the Mexican province.
— The Far East (Southeast Asia, Australia) has barely yielded 5 Gt to date, with a relatively modest exploration effort of about 12 wells per 10,000 km^2 and a yield of 800,000 t per well. Hopes for discoveries lie within a fairly broad range, from 2 to 20 Gt, with an average of about 10 Gt.

Natural gas

These hydrocarbons have not been investigated as intensively as their liquid counterparts. The current situation, which is still unclear owing to the quantities of gas produced but not marketed, is the following (excluding associated gas) :

— cumulative production estimated at 20,000 Gm^3,
— proven recoverable reserves of about 70,000 Gm^3,
— ultimate reserves of 200,000 to 300,000 Gm^3.

Gas resources are often closely related to oil, but their recovery rate is about 70 % of the quantities in place, while these materials may include fluids other than hydrocarbons, and sometimes incombustible gases. Some provinces, however, like the Middle East and the southern basin of the North Sea (Permian), the Algerian Sahara (mainly Triassic), Western Siberia (Cretaceous), the Canadian Arctic, Australia and, in general, the polar and sub-polar provinces, have a higher potential of gas than oil.

Known resources today are distributed as follows :

— USSR	40 %,
— Middle East	26 % (including 16 % Iran),
— United States	7 %,
— Western Europe	5 %,
— Algeria	5 %,

Resources remaining to be discovered are as follows, according to Colitti (1981) :

— Asia, USSR, Oceania	$51 \cdot 10^{12}$ m^3 or 38 %,
— Middle East	$38 \cdot 10^{12}$ m^3 or 28 %,
— The Americas	$21 \cdot 10^{12}$ m^3 or 15 %,
— Africa	$15 \cdot 10^{12}$ m^3 or 11 %,
— Europe	$11 \cdot 10^{12}$ m^3 or 8 %.

While little or nothing can be expected from secondary recovery processes, the quantities subjected to re-evaluation of current discoveries could amount to more than two-thirds of the total according to Colitti (1981), considering the frequent misappreciation of a number of discoveries.

4.3.3. **Unconventional oil resources**

Oil from unconventional zones, and in particular the oil in deep offshore basins and the Antarctic, is the subject of far more open evaluations, varying according to the experts from a few units to about 100 billion tons, in other words, not very meaningful. Additional geological data are required, especially borehole data, to provide a clearer picture. However, the similarity of the geological criteria of a number of these deep offshore basins, with petroleum provinces known onshore or under the continental shelf, provide grounds to postulate the existence of a few rich provinces. Difficult technical and economic conditions will restrict the exploitation to the largest and most productive formation. We know that while the giant fields contain the bulk of the discoveries, their discovery rate, due to their limited number, remains very low. This could lead to a series of failures and an increasingly pessimistic outlook.

Between conventional oil and the oil from tar sands lies the large category of heavy oils (density in the order of 15 to 20 °API) of which the poorly defined resources are very large. Their exploitation, which normally precedes that of sand oils in place, chiefly relies on thermal processes.

Sand and shale oil represent considerable resources which are very difficult to assess at the present time in terms of reserves in place, in the absence of the economic and technical ability to exploit most of these resources. Presumably the quantities extracted will be very large, but at relatively low production rates, spread over long periods. Unconventional oil could prolong world supply by several decades. However, the forecasts compiled, for example, for the Athabasca Sands, probably the richest and the best known, show that despite their volume, their exploitation could not account for more than 20 % of the needs of Canada in the 1990s.

*
* *

It may be felt that, until the end of the century, barring a 'political accident', the energy balances must, from the technical standpoint, be normally satisfied by oil, which is capable of supplying 35 to 40 % of world energy consumption in this period. In fact, the world's physical petroleum resources are still considerable, but their location and exploitation are faced by increasingly difficult technical, financial and especially political conditions. Only a major exploration effort, particularly in politically stable geographic zones, will serve to limit the scarcity at the end of the century, while awaiting the development of other energy sources. This intensified exploration will only bear fruit if it goes hand in hand with a vigorous scientific and technical effort, the only way to overcome the obstacles that we have failed to surmount so far.

Furthermore, this success can only be achieved if the companies and States agree to make the required investments, the former by increasingly considering their vocation as a public service, the latter by no longer regarding the petroleum industry as a sure and easy source of income, while the satisfaction of our energy requirements, especially oil, will require ever greater financial resources. As for ourselves, rest assured that nothing can replace the creative imagination of geologists to conceive new prospects and to design better tools, with an unshakable determination to try and succeed.

Summary and conclusion

Petroleum exploration is the key not only to the discovery of pools, but also to the knowledge of petroleum provinces. A seismic profile and borehole are not only the instruments of final success, but primarily a source of information about the deep, hidden zones of the sedimentary basins, and the geological mechanisms that gave rise to the formation of hydrocarbon accumulations.

Geological knowledge thus serves to diminish the risks of prospecting, but, as the latter proceeds more or less at the rate that discoveries are made (and vice versa), the number and size of the pools remaining to be discovered inevitably tend to decline as exploration makes further headway. The American example perfectly illustrates this rule of exploration. However, it also shows us that technical progress and imaginative endeavor fortunately succeed in periodically opening new frontiers. If we are determined, and we find the means, quantities of oil and gas, probably greater than those that we know today, remain to be discovered.

CONCLUSION

'There are more things in heaven and earth, Horatio,
Than are dreamt of in our philosophy.'
Shakespeare (Hamlet)

In brief

Petroleum strategy, and its underlying considerations, have long been dominated by the concept of the structural trap and the priority assigned to high zones. This approach has led to a large number of major successes. The need to discover pools which have not yet been found, and a better knowledge of hydrocarbon geology and geodynamics, lead us today to rethink certain concepts and to test other approaches. Thus it appears of fundamental importance to develop coherence between the knowledge of the mechanisms culminating in the concentration of oil and gas and the strategy of exploration. It is not enough to start with the source rock and try to find the pool, by somehow tracking the hydrocarbons. However, this pattern could be followed in another way, by a logical integration of the different geological data playing a part in this history.

Thus a good starting point is the identification of the facies, and particularly of environments that are favorable to deposits rich in organic matter, as well as detrital horizons and biogenic buildups. The reconstruction of paleogeographies, of which the coherent sequence constitutes the stratigraphy, as Maurice Gignoux declared, thus represents a fundamental basis for petroleum prospecting. Among the different depositional areas, special attention is thus paid to subsident areas, where the two basic dimensions of geology, space and time, are found intimately linked. In fact, it is in these mobile zones that the sediments accumulate, some of which play the role of source rocks and seals, while others are reservoirs. It is here, during periods of subsidence which are more or less long, more or less intense, that temperature and pressure perform their role, necessary for the genesis of hydrocarbons and their migrations.

Petroleum geology thus assigns great attention to geothermal gradients and paleogradients, to heat flows and to the thermal conductivity of rocks. In effect, these thermal criteria largely determine the limits of the formation and destruction of hydrocarbons. But the latter can only concentrate if they migrate from the rocks where they originated to concentration points. These flows can only occur through the continuity of aquifer systems, depending on continuous spaces with matrix permeability or associated with fracture networks. Thus the study of oil pools includes the data of hydrogeology, hydrodynamism and hydrochemistry on the one hand, and geodynamics on the scale of the basin on the other.

Hydrocarbons are normally trapped by the first barriers that stop them in their migrations and, consequently, whenever the seals and closures permit, they accumulate in relatively deep traps, near the oil and gas source, and well protected. In view of this, it is logical for relatively deep closed zones to be impregnated, and for more distant and consequently shallower zones to be aquifers, as so often occurs. While priority should not be assigned to the base of syncline areas, at least equal interest should be shown in the relatively deep margins of subsident zones as in the uplifts, which were the priority objectives of yesterday.

Hydrocarbons are fugitive and fragile materials, and unless very effectively protected, the permanence of their accumulations is liable to be of limited duration. In these conditions, the continuous or at least recent supply of the traps with oil and gas is the best way to combat alterations and vertical migrations. The marginal areas of recent and present troughs, generally in a tensile regime, thus have better chances of being supplied and productive. The age of the traps should only be considered in relation to the flow periods of the hydrocarbons likely to supply them.

Added to the standard and always necessary conditions of finding a trap, a reservoir and a source rock, is the progress in time of the different acts in the play. Movement intimately involves the time factor, the fourth dimension, the historical perspective, which is the true key to geological events. Consideration of all these factors on the scale of the basin enables the geologist to penetrate beyond this apparent and complex heterogeneity, and to discover the basic logic of the mechanism and of the geological events. Observation of the facts, compared with the analysis of the processes involved, the setting up of rules and the testing of models, helps to compile a representation as faithful as possible to reality, proceeding step by step, patiently, judiciously, and sometimes haphazardly.

Petroleum geology, and the exploration which applies its conclusions, thus assume a dynamic and global perspective, that contrasts with the often static character of older concepts. They form a closely integrated chapter of all the Earth sciences, side by side with sedimentology, geochemistry, stratigraphy, and geodynamics, with which they are linked in many ways.

Petroleum geology's return contribution to the Earth Sciences

With this new image, and offering its various branches, petroleum exploration contributes to the enrichment of the different geological disciplines. It is well known that boreholes and geophysical surveys have made an inestimable contribution to the knowledge of the deep zones of the sedimentary basins. Entire basins, unknown sedimentary series on the margins, and especially salt and undercompacted shale deposits, have thus been revealed on the continents and more often on their margins.

High resolution seismics today offers an overall view of the structure of a basin, of the geometry of its strata and its facies, allowing the reconstruction of the main stages of its history. The latest advances in seismics thus add a new dimension to stratigraphic and sedimentological analyses. Well logs, electrical, nuclear and sonic logs, offer a rigorously continuous analysis of the sequence of deposits. The combination of the different recorded physical parameters serves to obtain accurate lithologic interpretations, backed up today by physicochemical analyses in place, and also by representation of sedimentary sequences or dips.

Geological monitoring of boreholes has led to the development of new stratigraphic analysis methods, by foraminifera and other microfossils to begin with, and also by palynology and nanofauna and nanoflora, sparking a considerable development of these disciplines. The establishment of new stratigraphic scales has permitted the dating or identification of a number of series, previously considered as azoic.

Thanks to petroleum techniques, stratigraphy has also been enriched by a dense body of geophysical and logging correlations, which often served to detect unconformities, transgressions, variations in facies and thickness, considerably augmenting our knowledge of sedimentary volumes. Petroleum exploration activities also helped to discover a number of sedimentary mechanisms unknown, or poorly understood, by surface observations alone, especially with respect to the forms of deposition.

The study of materials extracted from boreholes has often revealed the scale and diversity of sedimentary transformations occurring through the different stages of diagenesis and catagenesis, and the interactions between rocks and fluids in particular. The knowledge of diagenesis mechanisms, especially in the particularly rich and sensitive area of organic matter, has thus received a tremendous impetus. Through their diversity and instability, organic substances, and hydrocarbons in particular, constitute very sensitive recorders of sedimentary, thermal and climatic processes. Through their inherent characteristics, certain chemical substances, designated by the name of biochemical fossils, demonstrate the link between the mineral world and the living world.

Petroleum exploration data have also clarified the role of aquifers in the sedimentary basins, of their movements, and of their more or less continuous chemical action on the rocks, especially when the fossilizing role of the hydrocarbons causes the formation, in some pools, of preserved islets which act as relics and serve as markers.

From the structural standpoint, these investigations enrich our knowledge about certain patterns of deformation, and especially about the geometry and kinetics of salt and undercompacted shale domes, growth faults and roll overs, generally highlighting the scope and variety of synsedimentary deformations, and particularly the extent of tensions and stresses that affect the sedimentary volumes long before they break or fold them. The features of these deformations also provide an excellent model for mass displacements of materials in the deep crustal zones. Faced with the need to model its hidden objectives, and by the confirmation of the well, petroleum exploration has developed a certain form of geological, analytical and concrete reflection, which, without hampering the imagination, focuses it on precise hypotheses and enhances its effectiveness.

On the more general level, petroleum exploration has added a new dimension to the knowledge of sedimentary basins, these privileged libraries of the archives of the Earth's history, and, at the same time, the cradles of tectonic structures which are among the most magnificent embellishments of our planet. Thus the study of the basins emerges as the synthesis of many stratigraphic, sedimentary, geochemical, hydrogeological, geothermal, structural and geodynamic activities, which constitute the sciences of the Earth. The understanding of these large sedimentary volumes leads to a dynamic, and hence living representation of our planet.

Exploration conducted to discover raw materials and energy sources, necessary for the improvement of the living conditions of mankind, enriches our most fundamental knowledge, without which these operations could not culminate successfully. More than any other scientist, the petroleum geologist, ceaselessly faced with the uncertainties of exploration, is fully aware of the risks involved. All his knowledge and his know-how are primarily intended to reduce this risk. But risk itself is a creative factor and the source of progress. As recalled recently by P.A. Bailly, President of the Society of Economic Geologists, a society without risks, and hence without responsibilities, would be a society without hope and without a future, "and we as Economic Geologists, who take the risks and enjoy the benefits of these activities, will agree with Helen Keller that 'life is either a daring adventure or nothing'."

REFERENCES

ADAMS, T. D. & KIRBY M. A. (1975). – Estimate of world gas reserves. – Proc. 9th World Petroleum Congr. (Tokyo), **3**, 3-9, 6 tables.

ADLER, F. J. (1971). – Future petroleum provinces of the Midcontinent. – *Mem. amer. Assoc. Petroleum Geol.,* **15**,2, 985-1120.

ALA, M. A., KINGHORN, R. R. F. & RAHMAN, M. (1980). – Organic geochemistry and source rock characteristics of the Zagros petroleum province, Southwest Iran. – *J. Petroleum Geol.,* **3**, 1, 61-89, 19 fig.

Alberta Society of Petroleum Geologists (1973). – A brief on science and technology in the supply and utilisation of energy in Canada. – Future petroleum provinces of Canada.

ALEVEDO, J. S. (1980). – Giant fields of the southern zone, Mexico. – *Mem. amer. Assoc. Petroleum Geol.,* **30**, 339-385, 43 fig.

ALLEN, G. P., LAURIER D. & THOUVENIN, J. (1979). – Etude sédimentologique du delta de la Mahakam. – Notes Mém., *Comp. franç. Pétroles,* **15**, 156 p.

Amer. Gas Assoc., Amer. Petroleum Inst., Can. Petroleum Assoc. (1975). – Reserves of crude oil, natural gas liquids and natural gas in U. S. and Canada and U. S. production capacity, **29**.

ARPS, J. L. & ROBERTS, T. G. (1958). – Economics of drilling for Cretaceous oil on East flank of the Denver - Julesburg Basin. – *Bull. amer. Assoc. Petroleum Geol.,* **42**, 11, 2549-2566.

ARPS, J. J., MORTADA, M. & SMITH, A. E. (1971). – Relationship between proved reserves and exploration effort. – *J. Petroleum Technol.,* **23**, 671-675.

ASQUITH, D. O. (1966). – Geology of late Cretaceous Mesaverde and Paleocene Fort Union oil production, Birch Creek unit. – *Bull. amer. Assoc. Petroleum Geol.,* **50**, 10, 2176-2184, 9 fig.

AVBOVBO, A. A. & OGBE, F. G. A. (1978). – Geology and hydrocarbon productive trends of southern Nigeria basin. – *Oil and Gas Journ., 27 Nov.,* **76**, 48, 90-93.

BILAL, M., JONES, R. W., SLENTZ, L. W., TARTIR, M. & WILSON, A. O., AYRES, M. G. (1982). – Hydrocarbon habitat in main producing areas, Saudi Arabia. – *Bull. amer. Assoc. Petroleum Geol.,* **66**, 1, 1-9, 14 fig.

BACOLLOLI, G., MORALES, R. G. & CAMPOS, O. A. J. (1980). – The Namorado oil field : A major oil discovery in the Campos basin, Brazil. – *Mem. amer. Assoc. Petroleum Geol.,* **30**, 329-338, 13 fig.

BALDUCCHI, A. & POMMIER, G. (1970). – Cambrian oil field Hassi Messaoud, Algeria. – *Mem. amer. Assoc. Petroleum Geol.,* **14**, 477-488, 19 fig.

BALLY, A. W. (1975). – A geodynamic scenario for hydrocarbon occurrences. – Proc. 9th World Petroleum Congr. (Tokyo), **2**, 33-44.

BALLY, A. W. & SNELSON, S. (1980). – Facts and Principes of world petroleum occurrence : realms of subsidence. – *Bull. canad. Soc. Petroleum Geol.,* **6**, 9-94.

BANDY, O. L. & ARNAL, R. E. (1969). – Middle Tertiary basin development San Joaquin Valley, California. – *Bull. geol. Soc. Amer.,* **80**, 783-820, 27 fig., 1 pl.

BANE, S.C. & CHANPONG, R. R. (1980). – Geology and development of the Teak oil field, Trinidad. – *Mem. amer. Assoc. Petroleum Geol.,* **30**, 387-398, 15 fig.

BARK, E. van den & THOMAS, O.D. (1980). – Ekofisk : first of the giant oil fields in Western Europe. – *Mem. amer. Assoc. Petroleum Geol.,* **30**, 195-224, 26 fig.

BARKER, C. (1972). – Aquathermal pressuring : role of temperature in development of abnormal pressure zones. – *Bull. amer. Assoc. Petroleum Geol.,* **56**, 10, 2068-2071, 3 fig.

BARNARD, P.C. & COOPER, B.S. (1981). – Oils and source rocks of the North Sea areas. – *In :* Petroleum Geol. of the Continental Shelf of NW Europe, 169-175, 4 fig. – Inst. Petroleum, London.

BARSS, D.L., COPLAND, A.B. & RITCHIE, W.D. (1970). – Geology of Middle Devonian reefs, Rainbow area. – *Mem. amer. Assoc. Petroleum Geol.,* **14**, 19-49, 25 fig.

BAZANOV, E.A., GURARI, F.G., KOZYREV, V.D., NALIVKIN, V.D., PRITULA, Yu.A., ROVNIN, L.I., RUDKEVICH, M.Ya., TROFIMUK, A.A. & SARKISYAN, S.G. (1967). – Gas and oil bearing provinces of Siberia. – Proc. 7th World Petroleum Congr. (Mexico), **2**, 190-210, 5 fig.

BHAT, H. (1975). – Les champs de Strachan et Ricinus dans l'Ouest canadien. – *Rev. Assoc. franç. Tech. Pétrole,* **233**, 11-21, 14 fig.

BISHOP, W.F. (1980). – Petroleum geology of Northern Central America. – *J. Petroleum Geol.,* **3**, 1, 3-59.

BLEAKLEY, W.B. (1968). – How an old giant is faring. – Oil and Gas J. 1st July, 101-109, 4 fig.

BODENHAUSEN, J.W.A. & OTT, W.F. (1981). – Habitat of the Rijswijk oil province, onshore, the Netherlands. – *In :* Petroleum Geology of the Continental Shelf of NW Europe, 301-309, 9 fig. – Inst. Petroleum, London.

BOSS, R.F., LENNON, R.B. & WILSON, B.W. (1976). – Middle Ground Shoal oil field, Alaska. – *Mem. amer. Assoc. Petroleum. Geol.,* **24**, 1-22, 13 fig.

BOUILLOT, A. (1971). – Les zones nouvelles en dehors de l'URSS. – Proc. 8th World Petroleum Cong. (Moscow), **2**, 301-311, 8 fig.

BOY DE LA TOUR, BRASSEUR, R. & LE DUIGOU, J. (1976). – Réflexions sur la notion de réserves énergétiques. – *Rev. Inst. franç. Pétrole,* **31,** 5, 741-780, 3 fig.

BRADY, T.J., CAMPBELL, N.D.J. & MAHER, C.E. (1980). – Intisar D oil field Libya. – *Mem. amer. Assoc. Petroleum Geol.,* **30,** 543-564, 25 fig.

BRAUNSTEIN, J. (1976). – North American oil and gas fields. – *Mem. amer. Assoc. Petroleum Geol.,* **24.**

BRINK, A.H. (1974). – Petroleum Geology of Gabon Basin. – *Bull. amer. Assoc. Petroleum Geol.,* **58,** 2, 216-235, 10 fig.

British Petroleum Co. (1956). – Oil and gas in Southwest Iran. – *In :* Symposium sobre yacimientos de petróleo ..., **2,** 33-72. Int. geol. Congr., 20th Sess., Mexico.

BROWN, R.N. (1980). – History of exploration and discovery of Morgan, Ramadan and July oil fields, Gulf of Suez, Egypt. – *Mem. Canad. Soc. Petroleum Geol.,* **6,** 733-764.

BRUCE, C.J. & PARKER, E.R. (1975). – Structural features and hydrocarbon deposits in the Mackenzie Delta. – Proc. 9th World Petroleum Congr. (Tokyo), **2,** 251-261, 8 fig.

BURKE, K. (1972). – Longshore drift, submarine canyons, and submarine fans in development of Niger Delta. – *Bull. amer. Assoc. Petroleum Geol.,* **56,** 10, 1975-1983, 10 fig.

CAFLISCH, L. (1975). – Comment to the paper « l'exploration pétrolière au Gabon et au Congo ». – Proc. 9th World Petroleum Congr. (Tokyo), **3,** 199-200, 2 fig.

CALLAWAY, D.L. (1971). – Petroleum potential of San Joaquim Basin. – *Mem. amer. Assoc. Petroleum Geol.,* **15,** 239-253, 11 fig.

CAUGHLIN, W.G. et al. (1976). – The detection and development of Silurian reefs in northern Michigan. – *Geophysics,* **41,** 4, 646-658.

CELSO PONTE, F., DOS REIS FONSECA, J. & CAROZZI, A.V. (1980). – Petroleum habitats in the Mesozoic-Cenozoic of the Continental margin of Brazil. – *Mem. canad. Soc. Petroleum Geol.,* **6,** 857-886.

CERVERA, E. (1967). – Exploration and production in Mexico. – Proc. 7th World Petroleum Congr. (Mexico), **18,** 3-23. 6 fig.

CHANUT, C., HUBAULT, C. & RICHE, P. (1971). – Oil and gas accumulations in Tin Fouye Tabankort area, (Algeria) (abst.). – *Bull. amer. Assoc. Petroleum Geol.,* **55** , 2, 333.

CHAPMAN, R.E. (1972). – Primary migration of petroleum from clay source rocks. – *Bull. amer. Assoc. Petroleum Geol.,* **56,** 11, 2185-2191, 5 fig.

CHAPMAN, R.E. (1972). – Petroleum and geology : a synthesis. – *J. austral. Petroleum Explor. Ass.,* **12,** 36-38.

CHIARELLI, A. (1970). – L'hydrodynamisme, ses applications dans la recherche pétrolière. – *Rev. Inst. franç. Pétrole,* **25,** 9, 1015-1036, 10 fig.

CHIARELLI, A. (1978). – Hydrodynamic framework of eastern Algerian Sahara. Influence on Hydrocarbon Occurrence. – *Bull. amer. Assoc. Petroleum Geol.,* **62,** 4, 667-685, 16 fig.

CHILDS, F.B. & REED, P.E.C. (1975). – Geology of the Dan field and the Danish North Sea. – *In :* Petroleum and the continental shelf of NW europe, 1, 429-438, 8 fig. – Applied Science Publ., Barking, Essex, England.

CHILINGARIAN, G.N. & YEN, T.F. (ed.) (1978). – Bitumens, asphalts and tar sands. – Elsevier, Amsterdam, 331 p.

CLARET, J. & TEMPERE, C. (1967). – Une nouvelle région productrice au Sahara algérien : l'anticlinorium d'Hassi Touareg. – Proc. 7th World Petroleum Congr. (Mexico), **2,** 1-100.

CLARKE, R.H. (1975). – Petroleum formations and accumulation in Abu Dhabi. – 9th arab. Petroleum Congr. (Dubai).

CLIFFORD, H.J., GRUND, R. & MUSRATI, H. (1979). – Geology of a stratigraphic giant – Messla oil field Libya. – *Bull. amer. Assoc. Petroleum Geol.,* **63,** 3, 433 (abstract).

COLMAN SAAD, SP. (1978). – Fold development in Zagros simply folded belt, southwest Iran. – *Bull. amer. Assoc. Petroleum,* **62,** 6, 904-1003, 10 fig.

COMBAZ, A. & MATHAREL, M. DE (1978). – Organic sedimentation and genesis of Petroleum in Mahakam Delta, Borneo. – *Bull. amer. Assoc. Petroleum Geol.,* **62,** 9, 1684-1695, 11 fig.

COOGAN, A.H., BEDOUT, D.G. & MAGGIO, C. (1972). – Depositional environments and geologic history of Golden Lane and Poza Rica trend, Mexico, an alternative view. – *Bull. amer. Assoc. Petroleum Geol.,* **56,** 8, 1419-1447, 21 fig.

CONYBEARE, C.E.B. (1972). – Petroleum accumulations in ancient river sands : a paleogeographic view. – Int. geol. Congr. 24th Sess., Canada (Montreal), **5,** 59-74, 10 fig.

CORDELL, R.J. (1972). – Depths of oil origin and primary migration : a review and critique. – *Bull. amer. Assoc. Petroleum Geol.,* **56,** 10, 2029-2067, 5 fig.

COUSTAU H., RUMEAU, J., SOURISSE, C., CHIARELLI, A. & TISON, J. (1975). – Classification hydrodynamique des bassins sédimentaires. – Proc. 9th World Petroleum Cong. (Tokyo), **2,** 105-119, 8 fig.

COUSTAU, H. (1976). – Stratégie de l'exploration. – Rapport interne Elf-Aquitaine.

COUSTAU, H. (1979). – Logique de distribution des tailles des champs dans les bassins. – *Pétrole et Technique,* **262,** 23-30, 11 fig.

COUSTAU, H. (1980). – Les ressources en hydrocarbures du bassin aquitain. – *Int. geol. Congr.,* 26th Sess. (Paris), Colloque **7,** 331-334.

CRAM, I.H. (1971). – Future petroleum provinces of the U.S. – *Mem. amer. Assoc. Petroleum Geol.,* **15.**

CRICK, R.W. (1971). – Potential petroleum reserves Cook Inlet, Alaska. – *Mem. amer. Assoc. Petroleum Geol.,* **15,** 1, 109-119, 9 fig.

CROSTELLA, A. & CHANEY, M.A. (1978). – The petroleum geology of the outer Dampier sub-basin. – *Austral. Petroleum Explor. Assoc. J.,* **18**, 13-22, 12 fig.

CURRY, W.H. (1971). – Summary of possible future petroleum potential, N Rocky Mountains. – *Mem. amer. Assoc. Petroleum Geol.,* **15**, 538-546.

DAVIES, H.F. (1934). – Structural history and its relation to the accumulation of oil and gas in the Rocky Mountains. – *In :* Problems of Petroleum Geology. – *Amer. Assoc. Petroleum Geol.,* 679-693.

DEMAISON, G.J. (1977). – Tar sands supergiant oil fields. – *Bull. amer. Assoc. Petroleum Geol.,* **61**, 11, 1950-1961, 9 fig.

DESPRAIRIES, P. (1978). – Les limites de l'approvisionnement pétrolier mondial. Ressources énergétiques mondiales 1985-2020. – 10ᵉ Conf. mond. Energie (Istambul), 1977, 1-47. Technip, Paris.

DICKEY, P.A. (1972). – Geology of oil fields of West Siberian Lowlands. – *Bull. amer. Assoc. Petroleum Geol.,* **56**, 3, 454-471, 9 fig., 2 tables.

DICKEY, P.A. & HUNT, J.M. (1972). – Geochemical and hydrogeologic methods of prospecting for stratigraphic traps. – *Mem. amer. Assoc. Petroleum. Geol.,* **16**, 136-167, 24 fig.

DICKEY, P.A. & ROHN, R.E. (1958). – Facies control of oil occurrence. – *In :* WEEKS, L.G. (ed.) : Habitat of Oil, 721-734, 9 fig. Amer Assoc. Petroleum Geol.

DOTT, R.H., Sr., HALBOUTY, M.T., KING, R.E., KLEMME, H.D. & MEYERHOFF, A.A. (1970). – World's giant oil and gas fields, geologic factors affecting their formation and basin classification. – *Mem. amer. Assoc. Petroleum Geol.,* **14**, 502-555, 9 fig., 15 tab.

DOTT, R.H. & REYNOLDS, M.J. (1969). – Sourcebook for petroleum geology. – *Mem. amer. Assoc. Petroleum Geol.,* **5**, 471.

DOW, W.G. (1978). – Petroleum source bed on continental slopes and rises. – *Bull. amer. Assoc. Petroleum Geol.,* **62**, 9, 1584-1602, 22 fig.

DUFFAUD, J. (1970). – Le bassin de Putumayo. – Rapport ELF inédit.

DUMESTRE, A. (1969). – Relations entre hydrocarbures et environnement évaporitique à Rainbow, Alberta. – *Rev. Assoc. franç. Pétrole,* **194**, 29-46, 15 fig.

DUNN, W.W. (1975). – North sea basinal area, Europe, an important oil and gas province. – *Bull. Norges Geol. Unders.,* **29**, 69-97.

DUNNINGTON, H.V. (1958). – Generation, Migration, Accumulation and Dissipation of oil in Northern Iraq. – *In :* WEEKS, L.G. (ed.) : Habitat of Oil, 1194-1251, 22 fig. Amer. Assoc. Petroleum Geol.

DUNNINGTON, H.V. (1967). – Stratigraphic distribution of oil fields in the Iraq-Iran-Arabian basin. – *J. Inst. Petroleum,* **53**, 520, 129-161.

DURAND, B. & OUDIN, J.L. (1979). – Exemple de migration des hydrocarbures dans une série deltaïque : le delta de la Mahakam, Kalimantan, Indonésie. – Proc. 10th World Petroleum Congr. (Bucharest), **2**, 3-11, 8 fig. (publ. 1981).

EL AYOUTI, M.K. (1980). – Oil and gas prospects in Egypt – *26ᵉ Congr. geol. inter.,* Coll. C2, 231-241.

EL ZARKA, M.H. (1975). – Geochemical relations of fluids in oil fields of Gulf of Suez, Egypt. – *Bull. amer. Assoc. Petroleum Geol.,* **59**, 9, 1667-1675, 4 fig., 6 tables.

Energie, Mines et Ressources Canada (1977). – Sables bitumineux et pétroles lourds : perspectives. – Rapport E.P. 77.2F, Approvisionnements et Services, Ottawa, Canada.

ERDMAN, G. (1972). – Geochemical formation of oil. – Int. geol. Congr., 24th Sess., Canada, **5**, 17 (abstract).

EVAMY, B.D., HAREMBOURE, J., KAMERLING, P., KNARP, W.A., MOLLOY, F.A. & ROULANDS, P.H. (1978). – Hydrocarbon habitat of Tertiary Niger Delta. – *Bull. amer. Assoc. Petroleum Geol.,* **62**, 1, 1-39, 28 fig.

EVANS, H. (1972). – Zama, a geophysical case history. – *Mem. amer. Assoc. Petroleum Geol.,* **16**, 440-452, 22 fig.

FALCON, N.L. (1958). – Position of oil fields of southwest Iran with respect to relevant sedimentary basins. – *In :* WEEKS, L.G. (ed.), Habitat of Oil, 1279-1293. Amer. Assoc. Petroleum Geol.

FISHER, W.L. & BROWN, L.F. (1969). – Delta systems and oil and gas occurrence. – *In :* Delta systems in the exploration for oil and gas. A research colloquium. – Univ. Texas, Austin, 54-77.

FLETCHER, G.L. & SOEPARJADI, R.A. (1977). – The land of plenty : Indonesia's 28. – *Oil and Gas J.,* August 22, 150-156.

FOHS, J. (1962). – Petroliferous provinces of USSR. – *Bull. amer. Assoc. Petroleum Geol.,* **46**, 11, 1973-1989.

FOLINSBEE, R.E. (1977). – World's view from Alpha to Zipf. – *Bull. geol. Soc. Amer.,* **88**, 897-907, 19 fig.

FOWLER, W.A. Jr. (1970). – Pressures hydrocarbon accumulation and salinities Chocolate Bayou field, Brazoria County, Texas. – *J. Petroleum Technol.,* April 1970, 411-423, 7 fig.

FOX, A.F. (1959). – Some problems of petroleum geology in Kuwait. – *J. Inst. Petroleum London,* **45**, 424, 95-110.

FOX, F.G. (1959). – Structure and accumulation of hydrocarbons in southern foothills, Alberta, Canada. – *Bull. amer. Assoc. Petroleum Geol.,* **43**, 5, 992-1025, 11 fig.

FRANKEL, E.J. (1967). – The Niger Delta oil province. – Proc. 7th World Petroleum (Mexico), **2**, 195-209, 8 fig.

FRASER, W.W. (1967). – Geology of the Zelten field. – Proc. 7th World Petroleum Congr. (Mexico), **2**, 259-264, 5 fig.

GAGE, M. (1980). – A review of the Viking gas field. – *Mem. amer. Assoc. Petroleum Geol.,* **30**, 39-57, 16 fig.

GAINETTE, G. (1972). – Le développement du champ d'Emeraude au Congo. *Rev. Assoc. franç. Techn. Pétrole,* 23-28.

GALLEY, J.E. (1958). – Oil and geology in the Permian basin. – *In :* WEEKS, L.G. (ed.), *Habitat of Oil,* 395-446. Amer. Assoc. Petroleum Geol.

GARDET, P.H. (1971). – Petroleum potential of Los Angeles basin. – *Mem. amer. Assoc. Petroleum Geol.,* **15**, 298-308, 8 fig.

GATEWOOD, L.E. (1970). – Oklahoma City field. – *Mem. amer. Assoc. Petroleum Geol.,* **14**, 223-254.

GIBSON, G.R. (1965). – Oil and gas in Southwestern region. Geologic framework. – *Mem. amer. Assoc. Petroleum Geol.,* **4**, 66-100, 17 fig.

GHIGNONE, J.J. & DE ANDRADE G. (1970). – General geology and major oil fields of Reconcavo Basin, Brazil. – *Mem. amer. Assoc. Petroleum Geol.,* **14**, 337-358, 12 fig.

GILL, D. (1979). – Differential entrapment of oil and gas in Niagaran pinnacle reef belt of northern Michigan. – *Bull. amer. Assoc. Petroleum Geol.,* **63**, 4, 608-620.

GILLESPIE, J. (1967). – The geology of the Sarir oil field, Sirte Basin, Libya. – Proc. 7th World Petroleum Congr. (Mexico), **2**, 181-193, 7 fig.

GREIG, D.A. (1958). – Oil horizons in the Middle East. – *In* : WEEKS, L.G. (ed.): Habitat of Oil, 1182-1193. Amer. Assoc. Petroleum Geol.

GRIFFITH, B.R. & HODGSON, E.A. (1971). – Offshore Gippsland basin fields. – *Austral. Petroleum Explor. Assoc. J.,* **11**, 1, 85-89.

GRIMES, W.L. (1970). – Bay Marchand - Timbalier Bay - Caillou Island Salt Complex. – *Mem. amer. Assoc. Petroleum Geol.,* **14**, 277-291.

GROSSLING, B. (1976). – The petroleum exploration challenge with respect to the developing nations - UNITAR-IIASA Conference on future supply of nature-made petroleum and gas. Luxembourg, Austria. – Rep. U.S. Geol. Surv.

GUBKIN, I.M. (1934). – Tectonics of Southeastern Caucasus and its relation to the productive oil fields. – *Bull. amer. Assoc. Petroleum Geol.,* **18**, 5, 603-671.

HAJAVH, G.M. (1967). – The Abu Sheikhdom. The on-shore field history of exploration and development. – Proc. 7th World Petroleum Congr. (Mexico), **2**, 129-139, 14 fig.

HALBOUTY, M.T. (1978). – Acceleration in global exploration. Requirement for survival. – *Bull. amer. Assoc. Petroleum Geol.,* **62**, 5, 739-751, 12 fig.

HALBOUTY, M.T. (1969). – Hidden trends and subtle traps in the Gulf Coast. – *Bull. amer. Assoc. Petroleum Geol.,* **53**, 1, 3-29, 20 fig.

HALBOUTY, M.T. (1972). – Rationale for deliberate pursuit of stratigraphic, unconformity and paleogeographic traps. – *Bull. amer. Assoc. Petroleum Geol.,* **56**, 3, 537-541, 2 fig.

HALBOUTY, M.T. & MOODY, J.D. (1981). – Proved and ultimate reserves of crude oil. – Proc. 10th World Petroleum Congr. (Bucharest); **2**, 291-301, 6 fig.

HACQUEBARD, P.E. (1977). – Rank of coal as index of organic metamorphism for oil and gas in Alberta. – *Bull. amer. Assoc. Petroleum Geol.,* **61**, 5, 791 (abstract).

HAMOUDA , A. (1980). – Petroleum potential – Ouargla region Triassic basin, Algeria. – *Mem. amer. Assoc. Petroleum Geol.,* **30**, 539-541.

HARDING, T.P. (1973). – Newport-Inglewood Trend California – an Example of Wrenching Style of Deformation. – *Bull. amer. Assoc. Petroleum Geol.,* **57**, 1, 97-116, 15 fig.

HAUN, J.P. & KENT, H.C. (1965). – Geologic history of Rocky Mountain region. – *Bull. amer. Assoc. Petroleum Geol.,* **49**, 11, 1781-1800.

HAUN, J.D., BARLOW, J.A. & HALLINGER, D.E. (1969). – Gas potential of Rocky Mountain region. – *Bull. amer. Assoc. Petroleum Geol.,* **53**, 1, 211 (abstract).

HEDBERG, H.D. (1964). – Geologic aspects of origin of petroleum. – *Bull. amer. Assoc. Petroleum Geol.,* **48**, 11, 1755-1803.

HEDBERG, H.D. (1968). – Significance of high-wax oils with respect to genesis of petroleum. – *Bull. amer. Assoc. Petroleum Geol.,* **52**, 5, 736-750, 1 tab.

HEMPHILL, C.R., SMITH, R.I. & SZABO, F. (1970). – Geology of Beaverhill Lake reefs, Swan Hills area. – *Mem. amer. Assoc. Petroleum Geol.,* **15**, 50-90, 27 fig.

HERITIER, F. (1975). – Quinze ans d'exploration pétrolière en Mer du Nord. – *Ann. Inst. belge Pétrole,* **9**, 4, 17-41, 38 fig.

HERITIER, F.E., LOSSEL, P. & WATHNE, E. (1980). – Frigg field – large submarine fan trap in Lower Eocene rocks of the Viking graben. – *Mem. amer. Assoc. Petroleum Geol.,* **30**, 59-79, 18 fig.

HILL, P.J. & WOOD, G.V. (1980). – Geology of the Forties field, continental shelf, North sea. – *Mem. amer. Assoc. Petroleum Geol.,* **30**, 81-93, 16 fig.

HOBSON, G.D. & TIRATSOO (1981). – Introduction to petroleum geology. – Scientific Press, 2nd ed., 352 p.

HOCKING, J.B. (1972). – Geologic evolution and hydrocarbon habitat, Gippsland Basin. – *J. austral. Petroleum Explor. Assoc.,* 132-137, 8 fig.

HODGSON, B. (1978). – Natural gas : the search goes on. – *National Geographic,* **154**, 5, 632-651.

HOLLAND, D.S., NUNAN, W.E., LAMMLEIN, D.R., WOODHAMS, R.L. & HALBOUTY, M.T. (1980). – Eugene Island block 330 field, offshore Louisiana. – *Mem. amer. Assoc. Petroleum Geol.,* **30**, 253-280, 38 fig.

HOLMGREN, D.A., MOODY, J.D. & EMMERICH, H.H. (1975). – The structural settings for giant oil and gas fields. – Proc. 9th World Petroleum Congr. (Tokyo), **2**, 45-54, 6 fig.

HRISKEVICH, M.E. (1980). – Exploration in Western Canada. – *Mem. canad. Soc. Petroleum Geol.,* **6**, 397-420.

HRISKEVICH, M.E., FABER, J.M. & LANGTON, J.M. (1980). – Strachan and Ricinus west gas fields. Alberta, Canada. – *Mem. amer. Assoc. Petroleum Geol.,* **30**, 315-327, 14 fig.

HUFFINGTON, R.M. & HELMIG, H.M. (1980). – Discovery and development of the Badak field, East Kalimantan, Indonesia. - *Mem. amer. Assoc. Petroleum Geol.,* **30**, 441-458, 23 fig.

HUFF, K.F. (1980). – Frontiers of world exploration. – *Mem. canad. Soc. Petroleum Geol.,* **6**, 343-362.

HULL, C.E. & WARMAN, H.R. (1970). – Asmari oil fields of Iran. – *Mem. amer. Assoc. Petroleum Geol.,* **14**, 428-437.

ILLING , L.V. & HOBSON, G.D. (ed.) (1981). – Petroleum Geology of the continental shelf of NW Europe. – Inst. Petroleum London, 521 pp.

IRAQ PETROL CY (1956). – Geological occurrence of oil and gas in Iraq. – *In :* Symposium sobre yacimientos de petróleo..., **2**, 73-101. – Int. Geol. Congr., 2th Sess., Mexico.

IVANHOE, L.F. (1976). – Oil gas potential in basins estimated. – *Oil and Gas J.,* Dec. 6, 154-155 ; Dec. 13, 108-110 ; Dec. 20, 82-84.

IVANHOE, L.F. (1980). – World's giant petroleum provinces. – *Oil Gas J.,* June 30, 146-148.

JOHNS, W.D. & SHIMOYAMA, A. (1972). – Clay minerals and petroleum-forming reactions during Burial and Diagenesis. – *Bull. amer. Assoc. Petroleum Geol.,* **56**, 11, 2160-2167, 4 fig.

KAMEN-KAYE, M. (1970). – Geology and productivity of Persian Gulf synclinorium. – *Bull. amer. Assoc. Petroleum Geol.,* **54**, 12, 2371-2394.

KAPOUNEK, J., KOELBE, L. & WEINBERGER, F. (1963). – Results of new exploration in the basement of the Vienna basin. – Proc. 6th World Petroleum Congr. (Frankfurt), **1**, 205-221, 10 fig.

KAYE, P. *et al.* (1972). – The Rankin trend northwest shelf, Western Australia. – *J. austral. Petroleum Explor. Assoc.* **12**, 1, 3-8, 6 fig.

KENT, P.E. & WALMSLEY P.J. (1970). – North Sea progress. – *Bull. amer. Assoc. Petroleum Geol.,* **54**, 1, 168-181, 11 fig.

KENT, P.E. (1980). – The North Sea – evolution of a major oil and gas play. – *Mem. canad. Soc. Petroleum Geol.,* **6**, 633-652.

KING, R.E. (1971). – Prospects for large increases in world hydrocarbon reserves from stratigraphic traps. – Proc. 8th World Petroleum Congr. (Moscow), **2**, 221-226.

KING, R.E. (1972). – Exploration for stratigraphic traps. – *Mem. amer. Assoc. Petroleum Geol.,* **16**, 643-648.

KIRK, R.H. (1980). – Statfjord field – a North Sea giant. – *Mem. amer. Assoc. Petroleum Geol.,* **30**, 95-116, 20 fig.

KLEMME, H.D. (1971). – Giants, super giants and their relation to basin types. – *Oil and Gas J.,* March 1, 8, and 15.

KLEMME, H.D. (1971). – Trends in basin development : possible economic implications. – World Petroleum, Oct. 1971, **42**, 10, 47-56.

KLEMME, H.D. (1972). – Heat influences site of oil giants. – *Oil and Gas J.,* July 17, 136-144 ; July 24, 76-78.

KLEMME, H.D. (1975). – Giant oil fields related to their geologic setting. – *Bull. canad. Petroleum Geol.,* **23**, 1, 30-66.

KLEMME, H.D. (1977 a). – Giant fields contain less than 1 % of world's fields but 75 % of reserves. – *Oil and Gas J.,* March 7, 164-169.

KLEMME, H.D. (1977 b). – One fifth of reserves lie off-shore. – *Oil and Gas J.,* "Petroleum 2 000", August, 108-128.

KLOVAN, J.E. (1974). – Development of western Canadian Devonian reefs and comparison with Holocene analogues. – *Bull. amer. Assoc. Petroleum Geol.,* **58**, 5, 787-799, 17 fig.

KOESOEMADINATA, R.P. (1969). – Outline of geologic occurrence of oil in Tertiary basins of west Indonesia. – *Bull. amer. Assoc. Petroleum Geol.,* **53**, 11, 2 368-1 376, 5 fig.

LAW, J. (1957). – Reasons for Persian Gulf Oil abundance. – *Bull. amer. Assoc. Petroleum Geol.,* **41**, 1, 51-69.

LAYER, D.B. (1958). – Oil and gas accumulation in Alberta. – *In :* WEEKS, L.G. (ed.) : Habitat of Oil, 113-128. *Amer. Assoc. Petroleum Geol.*

LEHNER, P. & DE RUITER, P.A.C. (1977). – Structural history of Atlantic margin of Africa. – *Bull. amer. Assoc. Petroleum Geol.,* **61**, 961-981, 18 fig.

LEVORSEN, A.I. (1934). – Relation of oil and gas pools to unconformities in the Mid-Continent region. – *in :* Problem of Petroleum Geology. – *Mem. amer. Assoc. Petroleum Geol.,* 761-784.

LEVORSEN, A.I. (1956). – The Mid-Continent region. – *In :* Symposium sobre yacimientos de petróleo y gaz. – Int. geol. Congr., Rep. 20th sess, (Mexico), **3**, 221-233.

LING, H.C. (1975). – The petroleum industry of the People's Republic of China. – Hoover Inst. Press, Stanford, 264 pp.

LINK, T.A. (1949). – Interpretation of foothills structures, Alberta, Canada. – *Bull. amer. Assoc. Petroleum Geol.,* **39**, 9, 1 475-1 501, 29 fig.

LOFTING, M.J.W. *et al.* (1975). – Exploration results and future prospects in the Northern Australasian region. – Proc. 9th World Petroleum Congr. (Tokyo), **3**, 65-81, 8 fig.

LOPATIN, N.V. (1971). – La température et le temps géologique comme facteur de carbonification. – *Izv. Akad. Nauk SSSR, Ser. geol.,* **3**, 95-106 (traduction).

LOUIS, M. & HOMAN, P.T. (1963). – Cost analysis in the petroleum industry. – J. Graduate Research Center. South Methodist Univ. Press, Dallas 22, Texas.

LYTH, A.L. (1971). – Summary of possible future petroleum potential, western Rocky Mountains. – *Mem. amer. Assoc. Petroleum Geol.,* **15**, 406-412.

McCAMIS, J.G. & GRIFFITH, L.S. (1968). – Middle Devonian facies relations Zama area, Alberta. – *Bull. amer. Assoc. Petroleum Geol.,* **52**, 10, 1899-1924, 26 fig.

McKAY, I.H. & NORTH, F.K. (1975). – Undiscovered oil reserves. – *Amer. Assoc. Petroleum Geol.,* Studies Geol., **1**, 76-86.

McCASLIN, J. (1977). – The search goes on for giant fields. – *Oil and Gas J.,* "Petroleum 2 000", August, 99-106.

McGINNIS, C.J. (1958). – Habitat of oil in the Denver Basin. – *In :* WEEKS, L.G. (ed.) : *Habitat of Oil,* 328-843. *Amer. Assoc. Petroleum Geol.*

McGREGOR, A.A. & BIGGS, C.A. (1968). – Bell Creek Field, Montana : a rich stratigraphic trap. – *Bull. amer. Assoc. Petroleum Geol.,* **52,** 10, 1 869-1 887, 23 fig.

McKENZIE, D.B. (1972). – Primary stratigrahic traps in sandstones. – *Mem. amer. Assoc. Petrol. Geol.,* **16,** 47-63, 6 fig;

McNEAL, R.P. (1965). – Hydrodynamics of the Permian Basin. – *Mem. amer. Assoc. Petroleum Geol.,* **4,** 308-320, 16 fig.

MAGARA, K. (1978). – Geological model predicting optimum sandstone percent for oil accumulation. – *Bull. canad. Petroleum Geol.,* **26,** 3, 380-388, 6 fig.

MAGARA, K. (1978*a*). – Geological model predicting optimum sandstone percent for oil accumulation. – *Bull. canad. Petroleum Geol.,* **52,** 10, 1869-1887, 23 fig.

MAGARA, K. (1978*b*). – Compaction and fluid migration. – *Elsevier,* Amsterdam.

MAGLOIRE, P. (1970). – Triassic gas field of Hassi el R'mel, Algeria. – *Mem. amer. Assoc. Petroleum Geol.,* **14,** 489-501, 12 fig.

MAGOON, L.B. & CLAYPOOL, G.E. (1981). – Petroleum geology of Cook Inlet basin – an exploration model. – *Bull. amer. Assoc. Petroleum Geol.,* **65,** 6, 1 043-1 061, 20 fig.

MAGNIER, P. (1975). – The Mahakam Delta, Kalimantan, Indonesia. – Proc. 9th World Petroleum Congr. (Tokyo), **2,** 239-250, 7 fig.

MALLORY, W.W. (1972). – Geologic Atlas of the Rocky Mountains region. – Rocky Mountains Assoc., Denver, 331 p.

MATHAREL, M. de, LEHMANN, P. & OKI, T. (1980). – Geology of the Bekapai field. – *Mem. amer. Assoc. Petroleum Geol.,* **30,** 459-469, 11 fig.

MARTINEZ, A. (1970). – Giant fields of Venezuela. – *Mem. amer. Assoc. Petroleum Geol.,* **14,** 326-336.

McDONALD, D.R. & KAYE, P., MARTINSON, N.W. (1973). – Exploration on the continental shelf off Northwest Australia. – *Bull. amer. Assoc. Petroleum Geol.,* **57,** 6, 972-989, 13 fig.

MENESES DE GYVES & MEJIA DAUTT, O. (1980). – Geología de la Sonda de Campeche. – *In :* Fossil fuels, Hydrocarbons, 47-76. *Technip,* Paris.

MESOLELLA, K.J., ROBINSON, J.D., McCORMICK, L.M. & ORMISTON, A.R. (1974). – Cyclic deposition of Silurian carbonates and evaporites in Michigan Basin. – *Bull. amer. Assoc. Petroleum Geol.,* **58,** 1, 34-62, 24 fig.

MEYER, R.F. (1972). – Statistical correlation of petroleum reservoir data. – Int. geol. Congr., 24th sess., Canada (Montreal), **5,** 92-97.

MEYERHOFF, A.A. (ed.) (1968). – Geology of natural gas in South Louisiana. – *Mem. amer. Assoc. Petroleum Geol.,* **9,** 376-381.

MEYERHOFF, A.A. (1974). – Geological principles of world oil occurrence : a summary. – *Bull. canad. Petroleum Geol.,* **22,** 4, 413-419.

MEYERHOFF A.A. & WILLUMS, J.O. (1976). – Petroleum geology and industry of the People's Republic of China. – U.N. ESCAP - CCOP techn. Bull., **10,** 103-212, 68 fig.

MEYERHOFF, A.A. (1977). – Best chances on-shore are in China and Russia. – Oil and Gas J., "Petroleum 2 000", August, 132-138.

MEYERHOFF, A.A. (1981). – Proved and Ultimate Reserves of Natural Gas and Natural Gas Liquids in the World. – 10th World Petroleum Congress (Bucharest), **2,** 303-311, 11 fig.

MEYERHOFF, A.A. (1981). – Geology and petroleum fields in Proterozoic and Lower Cambrian Strata, Lena-Tunguska petroleum province, Eastern Siberia, USSR. – *Mem. amer. Assoc. Petroleum Geol.,* **30,** 225-253, 17 fig.

MICHELSON, J.E. (1976). – Miocene deltaic oil habitat Trinidad. – *Bull. amer. Assoc. Petroleum Geol.,* **60,** 9, 1 502-1 519., 14 fig.

MILLER, J.B. *et al.* (1958). – Habitat of oil in the Maracaïbo Basin, Venezuela. – *In :* WEEKS, L.G. (ed.) : *Habitat of Oil,* 601-640, 18 fig. Amer. Assoc. Petroleum Geol.

MILLER, J.B. *et al.* (1963). – Medio ambiante el petróleo en la cuenca de Maracaïbo, Venezuela. Aspectos de la industria petrolera en Venezuela. – 1er Congr. venez. Petróleo (Caracas), 5-99.

MILLOT, G. (1964). – Géologie des argiles. – Masson, Paris.

MILLS, H.G. (1970). – Geology of the Tom O'Connor field, Refugio county, Texas. – *South Texas geol. Soc.,* **9.**

MIRCHINK, M.F., TROFIMUK, A.A. & CHEPIKOV, K.R. (1959). – Specific features of the geological structure of platform regions in the Soviet Union. – Proc. 5th World Petroleum Congr. (New York). **1,** 525-538.

MIRCHINK, M.F. *et al.* (1963). – Problems of oil and gas content in Mesozoïc deposits in the South of the USSR. – Proc. 6th World Petroleum Congr. (Frankfurt), **1,** 97-113.

MIRCHINK, M.F., KRYLOV, N.A., LETAVIN, A.I., ROVNIN, L.I., TOKAREV, V.P., FEIGIN, M.V. & ERVIE, Y.G. (1967). – New major oil and gas areas of young platform of the USSR. – Proc. 7th World Petroleum Congr. (Mexico), **2,** 101-108.

MOODY, J.D. (1975). – Distribution and geological characteristics of giant oil fields. – *In :* FISCHER, A.G. & DISDSON, S. (ed.); Princeton Univ. Press, 322 p.

MOODY, J.D. & ESSER, R.W. (1975). – An estimate of the world's recoverable crude oil resource. – Proc. 9th World Petroleum Congr. (Tokyo), **3,** 11-20, 5 tabl.

MOODY, J.D. (1979). – Outlook for world hydrocarbon industry (not published).

MORGRIDGE, D.L. & SMITH Jr., W.B. (1972). – Geology and discovery of Prudhoe Bay. – *Mem. amer. Assoc. Petroleum Geol.,* **16,** 489-501, 18 fig.

MORRISON, R.R. *et al.* (1971). – Potential of Sacramento Valley gas province. – *Mem. amer. Assoc. Petroleum Geol.,* **15,** 329-338, 8 fig.

MOSSOP, G.D. (1980). – Facies control on bitumen saturation in the Athabasca oil sands. – *Mem. canad. Soc. Petroleum,* **6,** 609-632.

MURRAY, G.E. (1956). – Geological occurrence of oil and gas in Gulf Coastal province. – *In :* Symposium sobre yacimentos... Int. geol. Congr., 20th Sess. (Mexico), **3,** 235-290.

MURRAY, G.E., KACZOR, M.J. & McARTHUR, R.E. (1980). – Indigenous Precambrian Petroleum Revisited. – *Bull. amer. Assoc. Petroleum Geol.,* **64,** 10, 1681-1700, 15 fig.

MURRIS, R.J. (1980). – Hydrocarbon habitat of the Middle East. – *Mem. canad. Soc. Petroleum Geol.,* **6,** 765-800.

MYERS, J.D. (1968). – Differential pressure trapping mechanism in Gulf Coast oil and gas fields; – *Gulf Coast Assoc. geol. Soc.,* **18,** 56, 80-81.

NEHRING, R. (1978). – Giant oil fields and world oil resources. – Rand R 2284 Report for CIA.

NEHRING, R. (1980). – The outlook for world oil resources. – *Oil Gas J.,* oct. 27, 170-175.

NELSON, P.H.H. (1980). – Role of reflection seismics in development of Nembe Creek field, Nigeria. – *Mem. amer. Assoc. Petroleum Geol.,* **30,** 565-576, 8 fig.

NORTH, F.K. (1971). – Characteristics of oil provinces : a study for students. – *Bull. canad. Petroleum Geol.,* **19,** 3, 601-658, 2 fig., 7 tabl.

OSMATION, M.F. (1978). – Some fundamental aspects of plate tectonics bearing on hydrocarbon location. – *In :* HOBSON, G.D. (ed.) : Development in petroleum geology, **1,** 53-82, 14 fig. Science Publishers, London.

OSMENT, F.C., MORROW, R.M. & CRAIG, R.W. (1967). – Petroleum geology and development of the Cook Island Inlet basin of Alaska. – Proc. 7th World Petroleum Congr. (Mexico), **2,** 141-150, 4 fig.

OVANESOV, G.P. *et al.* (1973). – Prospects of discovering oil and gasfields in the Baku Archipelago. – *Petroleum Geol.,* **2,** 4, 1-4.

OWEN, E.W. (1975). – Trek of the oil finders : a history of exploration for petroleum. – *Mem. amer. Assoc. Petroleum Geol.,* **6,** 1, 647 p.

PARASCHIV, D. (1975). – Geologia Zacamintelor de hidrocarburi din România. – *Inst. Geol. si Geofi.,* Bucuresti, 363 pp.

PARASCHIV, O. & OLTEANU, G.H. (1972). – Oil fields in Mio-Pliocene zone of Eastern Carpathians (Ploesti). – *Mem. amer. Assoc. Petroleum Geol.,* **14,** 399-427, 13 fig.

PEAKE, E. *et al.* (1972). – The potential of Arctic sediments : hydrocarbons and possible precursors in Beaufort sediments. – *Int. geol. Congr., 24th Sess., Canada (Montreal),* **5,** 28-37.

PEGRUM, R.M., REES, R. & NAYLOR, D. (1975). – Geology of the Northwest European continental shelf. **2** : The North Sea. – Graham Trotman Dudley Ltd., London, 225 pp.

PERRODON, A. (1961). – Sur la notion de province pétrolifère. – *Rev. Inst. franç. Pétrole,* **16,** 6, 659-677, 10 fig.

PERRODON, A. (1965). La persévérance récompensée ou exemples de quelques découvertes pétrolières difficiles. – *Rev. franç. Energie,* **170,** 345-349.

PERRODON, A. (1966). – Géologie du Pétrole. – Presses Univ. France, Paris, 440 p.

PERRODON, A. (1971). – Les provinces pétrolières de type paralique et récifal. – *Rev. Inst. franç. Pétrole,* **26,** 5, 423-438, 8 fig. ; 6, 521-537, 20 fig.

PERRODON, A. (1972). – Essai de classification des bassins sédimentaires. – *Sci. de la Terre,* **16** (1971), 2, 197-227, 28 fig., 2 tabl., 1 pl.

PERRODON, A. (1972). – Provinces pétrolières : approches statistiques et géologiques de quelques types d'habitat de l'huile et du gaz. – Int. geol. Congr., 24th Sess., Canada (Montreal), **5,** 176-186, 6 fig.

PERRODON, A. (1978*a*). – Coup d'œil sur les provinces géantes d'hydrocarbures. – *Rev. Inst. franç. Pétrole,* **33,** 4, 493-513, 9 fig.

PERRODON, A. (1978*b*). – Le pétrole demain ? – Ing. Géol., **28,** 3-8, 3 fig.

PERRY, E.A. Jr. & HOWER, J. (1972). – Late-stage dehydration in deeply buried pelitic sediments. – *Bull. amer. Assoc. Petroleum Geol.,* **56,** 10, 2013-2021, 8 fig.

PHILIPPI, G.T. (1965). – On the depth time and mechanism of petroleum generation. – *Geochim. cosmochim. Acta,* **29,** 9, 1021-1049.

PIPPIN, L. (1970). – Panhandle Hugoton field. – *Mem. amer. Assoc. Petroleum Geol.,* **14,** 204-222.

PORTER, J.W. & McCROSSAN (1975). – Basin consanguinity in Petroleum Resource estimation. – *In* : HAUN, J.D. (ed.) : *Studies in geology.* 1. – Methods of estimating the volume of undiscovered oil and gas resources, 50-75. Amer. Assoc. Petroleum. Geol.

POWERS, M.C. (1967). – Fluid-release mechanisms in compacting marine mudrocks and their importance in oil exploration – *Bull. amer. Assoc. Petroleum Geol.,* **51,** 7, 1240-1254.

PULUNGONNO, A. (1976). – Recent Knowledge of hydrocarbon potentials in sedimentary basins of Indonesia. – *Mem. amer. Assoc. Petroleum Geol.,* **25,** 239-249, 7 fig.

PYE, W.D. (1957). – Habitat of oil in Northern Great Plains and Rocky Mountains. – *In* : WEEKS, L.G. : Habitat of oil, 178-224. Amer. Assoc. Petroleum. Geol.

RAINWATER, E.H. (1972). – Petroleum in deltaic sediments. – Int. geol. Congr., 24th Sess., Canada (Montreal), **5,** 105-113, 6 fig.

RAYNAUD, J.F. & ROBERT, P. (1976). – Les méthodes d'étude optique de la matière organique. – *Bull. Centre Rech. Pau-SNPA,* **10,** 1, 109-127, 1 fig., 2 tabl., 4 pl.

REEVES, F. (1928). – The carbon ratio theory in the light of Hilt's law. – *Bull. amer. Assoc. Petroleum Geol.,* **12,** 8, 795-823, 5 fig.

RENZ, H.H. *et al.* (1958). – The eastern Venezuelan basin. – *In* : WEEKS, L.G. (ed.) : Habitat of Oil, 551-600, 21 fig.

RENZ, H.H. *et al.* (1963). – La cuenca oriental de Venezuela. Aspectos de la industria petrolera en Venezuela. – 1er Congr. venez. Petróleo (Caracas), 100-189.

REYRE, D. (1960). – Particularités géologiques des bassins côtiers de l'Ouest Africain. – *In* : Bassins sédimentaires du littoral africain. Symposium Ass. Serv. Géol. afric., Paris, **1**, 253-304.

REYRE, D. *et al.* (1966). – Evolution géologique du bassin gabonais. – *In* : Bassins sédimentaires du littoral africain. Symposium. Ass. Serv. Géol. afric. Paris, **1**, 171-191, 12 fig.

RICE, D.D. & CLAYPOOL, G.E. (1981). – Generation, accumulation and resource potential of biogenic gas. – *Bull. amer. Assoc. Petroleum Geol.*, **65**, 1, 5-25, 11 fig.

RICHE, Ph. & PRESTAT, B. (1981). – Paléogéographie du Crétacé moyen du Proche et Moyen-Orient et sa signification pétrolière. – Proc. 10th World Petroleum Congr. (Bucharest), **2**, 57-75, 11 fig.

ROBERT, P. (1971). – Etude pétrographique des matières organiques insolubles par la mesure de leur pouvoir réflecteur. – *Rev. Inst. franç. Pétrole*, **24**, 2, 105-135, 15 fig.

ROBERTS, J.M. (1970). – Amal field, Libya. – *Mem. amer. Assoc. Petroleum Geol.*, **14**, 438-448.

ROGERS, M.A. *et al.* (1972). – An explanatory and predictive model for the alteration of crude oils in reservoirs in the Western Canada Basin. – Int. geol. Congr., 24th Sess., Canada (Montreal), **5**, 48-55, 5 fig.

RONA, P.A. (1973). – Relations between rates of sediment accumulation on continental shelves, sea-floor spreading and eustacy. – *Bull. geol. Soc. Amer.*, **84**, 2852-2872, 16 fig.

RONOV, A.B. (1958). – Organic carbon in sedimentary rocks. – *Geochemistry*, **5**, 497-509.

ROUCHET, J. du (1980). – Une explication paléoclimatique des grands gisements superficiels de gaz sec du Nord de la Sibérie Occidentale. – *Bull. Centres Rech. Explor.-Prod. Elf-Aquitaine*, **4**, 1, 119-142, 11 fig.

ROVNIN L.I. *et al.* (1975). – Laws governing oil and gas accumulations in West Siberia. – Proc. 9th World Petroleum Congr. (Tokyo), **3**, 141-147, 4 fig.

RUSSEL, W.L. (1925). – Relation between isocarbs and oil and gas production in Kentucky. – *Econ. Geol.*, **20**, 3, 249-260, 2 fig.

SABINS, F.F. (1972). – Comparison of Bisti and Horseshoe Canyon stratigraphic traps, San Juan Basin. – *Mem. amer. Assoc. Petroleum Geol.*, **16**, 610-622, 10 fig.

SAID, R. (1962). – The Geology of Egypt. – Elsevier, Amsterdam.

SALEHI, I. & PUJOL, P. (1979). – Recent huge gas discoveries in Kangan area, Iran. – *Bull. amer. Assoc. Petroleum Geol.*, **63**, 3, 520 (abstract).

St-JOHN, B. (1980). – Sedimentary basins of the world and giant hydrocarbon accumulations. – *Amer. Assoc. Petroleum Geol.* 1 map.

SANFORD, R.M. (1970). – Sarir field, Libya – *Mem. amer. Assoc. Petroleum Geol.*, **14**, 449-476.

SAPPENFIELD, L.W. (1974). – Indonesian geology and petroleum potential. – Exploration and Economics of Petroleum Industry, **12**, 169-195, Matthew Bender, New-York.

SCHERER, F.C. (1980). – Exploration in East Malaysia over the past decade. – *Mem. amer. Assoc. Petroleum Geol.*, **30**, 423-440, 13 fig.

SELLEY, R.C. (1978). – Deltaic facies and petroleum geology. – *In* : HOBSON, G.D. (ed.) : Developments in petroleum geology, **1**, 197-224. Applied Science Publ., London.

SETUDEHNIA, A. (1978). – The Mesozoic sequence in southwest Iran and adjacent areas. – *J. Petroleum Geol.*, **1**, 1, 3-42, 26 fig.

SHIBAOKA, M., SAXBY, J.D. & TAYLOR, G.H. (1978). – Hydrocarbon generation in Gippsland Basin, Australia. Comparison with Cooper Basin, Australia. – *Bull. amer. Assoc. Petroleum Geol.*, **62**, 7, 1151-1158, 9 fig.

SIMONSON, R.R. (1958). – Oil in the San Joaquim valley. – *In* : WEEKS, L.G. (ed.) : *Habitat of Oil*, 99-112. Amer. Assoc. Petroleum Geol.

SINERIZ, B.G., QUEROL, R., CASTILLO, F. & ARRIBAS, J.R.F. (1979). – A new hydrocarbon province in western Mediterranean. – Proc. 10th World Petroleum Congr., (Bucharest), **2**, 191-197, 8 fig.

SMITH, J.E. (1971). – Relationship between petroleum reservoir rocks and source rocks. – *J. Inst. Ass. Math. Geol.*, **3**, 2, 183-192.

SOEPARJADI *et al.* (1975). – Exploration play concepts in Indonesia. – Proc. 9th World Petroleum Congr. (Tokyo), **3**, 51-64, 8 fig.

SMIRNOV, L.P. (1958). – Oil-bearing basins on eastern of the Russian platform. – *In* : WEEKS, L.G. (ed.) : *Habitat of Oil*, 1168-1181. Amer. Assoc. Petroleum Geol.

SOKOLOV, V.Y. *et al.* (1973). – The Shatlysk giant gas deposit in East Turkmenia. – *Soviet Geol.*, **2**, 66-75.

SPILLERS, J.P. (1965). – Distribution of hydrocarbons in South Louisiana by types of traps. – *Bull. amer. Assoc. Petroleum Geol.*, **49**, 3, 358 (abstract).

STANLEY, T.B. (1970). – Vicksburg Fault zone, Texas. – *Mem. amer. Assoc. Petroleum Geol.*, **14**, 301-308.

STAPLIN, F.L. (1969). – Sedimentary organic matter, organic metamorphism oil and gas occurrence. – *Bull. canad. Petroleum Geol.*, **17**, 1, 47-66, 2 fig.

STEWART, G. (1981). – Mexico's giants enter their second decade. *World Oil*, Feb. 1, 59-66.

TARLING, D.H. (1973). – Continental drift and reserves of oil and natural gas. – *Nature*, **243**, June 1, 277-279.

TATSCH, J.H. (1974). – Petroleum deposits. – Tatsch, Sudbury; 378 p.

TENAILLE, M., BURGER, J.J. & PERRODON, A. (1975). – Algeria - Trek of the oil finders. – *Mem. amer. Assoc. Petroleum Geol.,* **6,** 1450-1471.

TERRY, C.E. & WILLIAMS, J.J. (1969). – The Idriss « A » bioherm and oil field. – *In* : The Exploration for Petroleum in Europe and N Africa, 31-48. The Institute of Petroleum; Adlard and Son Ltd, Dorking, U.K.

THOMAS, T.M. (1975). – Search for hydrocarbons in shelf seas of NW Europe : Progress and Prospects. – *Bull. amer. Assoc. Petroleum Geol.,* **59,** 4, 573-617, 20 fig.

THOMPSON, T.L. (1976). – Plate tectonics in oil and gas exploration of continental margins. – *Bull. amer. Assoc. Petroleum Geol.,* **60,** 9, 1463-1501, 41 fig.

THRALLS, W.H. & HASSON, R.C. (1956). – Geology and oil resources of eastern Saudi Arabia. – *In* : Symposium sobre yacimientos de petróleo..., **2,** 9-32. – Int. geol. Congr., 20th Sess., Mexico.

TIMKO, D.J. & FERTL, W.H. (1971). – Relationship between hydrocarbon accumulation and geopressure and its economic significance. – *J. Petroleum Technol.,* August, 923-933, 19 fig.

TIPSWORD, H.L. *et al.* (1971). – Possible future petroleum potential of Lower Miocene-Oligocene. – *Mem. amer. Assoc. Petroleum Geol.,* **15,** 2, 836-854.

TIRATSOO, E.N. (1976). – Oil fields of the World. 2nd ed. – Scientific Press, Beaconsfield.

TISSOT, B., CALIFET-DEBYSER, Y., DEROO, G. & OUDIN, J.L. (1971). – Origin and evolution of hydrocarbons in early Toarcian shales, Paris Basin. – *Bull. amer. Assoc. Petroleum Geol.,* **55,** 12, 2177-2193, 25 fig.

TRAVES, P.M. (1971). – Stratigraphic traps in the Roma area, Queensland, Australia. – Proc. 8th World Petroleum Congr. (Moscow), **2,** 275-284, 7 fig.

TROFIMUK, A.A. (1955). – Conditions that led to formation of oil fields in the Ural-Volga oil-bearing region. – Proc. 4th World Petroleum Congr. (Rome), **1,** 731-736.

TROFIMUK, A.A. (1974). – Some important lessons from the discovery of the western Siberian oil and gas province. – *Geol. Geofiz.,* **15,** 5, 29-36.

VALITOV, N.B. (1972). – Regional variations in properties of oil, Tataria. – *Int. geol. Rev.,* **14,** 1, 54-64.

VASSOYEVICH, N.B. *et al.* (1972). – Sedimentary basins. – Int. geol. Congr., 24th Sess., Canada (Montreal), **5,** 187-194.

VERDIER, A.C., OKI, T. & SUARDY, A. (1980). – Geology of the Handil Field (East Kalimantan – Indonesia). *Mem. amer. Assoc. Petroleum Geol.,* **30,** 399-421, 23 fig.

VEST, E.L. (1970). – Oil fields of Pennsylvanian-Permian Horseshoe Atoll, W Texas. – *Mem. amer. Assoc. Petroleum Geol.,* **14,** 185-203.

VIDAL, J., JOYES, R. & VAN VEEN, J. (1975). – L'exploration pétrolière au Gabon et au Congo. – Proc. 9th World Petroleum Congr. (Tokyo), **3,** 149-165, 7 fig.

VIDAL, J. (1980). – Geology of Grondin field, Gabon – *Mem. amer. Assoc. Petroleum Geol.,* **30,** 570-590, 11 fig.

VIGRASS, L.W. (1968). – Geology of Canadian heavy oil sands. – *Bull. amer. Assoc. Petroleum Geol.,* **52,** 10, 1984-1999.

VINCELETTE, R.R. & SOEPARJADI, R.A. (1976). – Oil-bearing reefs in Salwati Basin of Irian Jaya, Indonesia. – *Bull. amer. Assoc. Petroleum Geol.,* **60,** 9, 1448-1462, 13 fig.

VINIEGRA, F. *et al.* (1970). – Golden Lane fields. – *Mem. amer. Assoc. Petroleum Geol.,* **14,** 309-325.

VINIEGRA, F. (1981). – Great carbonate bank of Yucatan, Southern Mexico. – *J. Petroleum Geol.,* **3,** 247-278.

VINKOVETSKY, Y. (1979). – Why are larger deposits discovered first ? – (Not published).

WALLIS, W.E. (1967). – Offshore petroleum exploration Gippsland and Bass Basins. – Proc. 7th World Petroleum Congr. (Mexico), **2,** 783-791, 8 fig.

WALMSLEY, P.J. (1975). – Oil and gas developments in the North Sea. – Proc. 9th World Petroleum Congr. (Tokyo), **3,** 131-140, 8 fig.

WALMSLEY, P.J. (1977). – Ten years in the North Sea with B.P. – *Pétrole et Technique,* **241,** 7-20, 13 fig.

WALTERS, L.J. Jr. & GEORGE, D.T. (1979). – Effects of source material and thermal maturation on chemical composition of Gulf Coast crude oils. – *Bull. amer. Assoc. Petroleum Geol.,* **63,** 3, 547 (abstract).

WATSON, J.M. & SWANSON, C.A. (1975). – North Sea - Major petroleum province. – *Bull. amer. Assoc. Petroleum Geol.,* **59,** 7, 1098-1112, 17 fig.

WEBER, K.J. (1971). – Sedimentological aspects of oil fields in the Niger Delta. – *Geol. en Mijnb,* **50,** 3, 559-576.

WEBER, K.J. & DAUKORU, E. (1975). – Petroleum Geology of the Niger Delta. – Proc. 9th World Petroleum Congr. (Tokyo), **2,** 209-221, 7 fig.

WEBER, K.J. *et al.* (1978). – The role of faults in hydrocarbon migration and trapping in Nigerian growth fault structures. – 10th ann. offshore Technol. Conf. (Houston, 8-10 May 1978) (preprint) OTC 3356, **4,** 2643-2653.

WEEKS, L.G. (1952). – Factors of sedimentary basin development that control oil occurrence. – *Bull. amer. Assoc. Petroleum Geol.,* **36,** *11,* 2071-2124.

WEEKS, L.G. (1958). – Habitat of oil and factors that control it. – *In* : WEEKS, L.G. (ed.) : *Habitat of Oil,* 1-61. Amer. Assoc. Petroleum Geol.

WEEKS, L.G. (1961). – Origin, migration and occurrence of petroleum. – *In* : Petroleum Exploration Handbook, 5.1-5.50. McGraw Hill, New York.

WEEKS, L.G. (1969). – Offshore petroleum developments and resources. – *J. Petroleum Technol.,* April, 377-385.

WEEKS, L.G. (1971). – Marine geology and petroleum resources. – Proc. 8th World Petroleum Congr. (Moscow), **2**, 99, 106, 2 tabl.

WEEKS, L.G. (1972). – Critical interrelated geologic, economic and political problems facing the geologist, petroleum industry and nation. – *Bull. amer. Assoc. Petroleum Geol.*, **56**, 10, 1919-1930.

WEEKS, L.G. (1975). – Potential Petroleum Resources. – *In* : HAUN, J.D. (ed.) : Studies in geology. 1. - Methods of estimating the volume of undiscovered oil and gas resources, 31-49. Amer. Assoc. Petroleum Geol.

WENGER, R. (1973). – Le bassin sédimentaire gabonais et la dérive des continents. – 27th Brazilian Geol. Cong. Proc., **3**, 13-52.

WENNEKERS, J.H.L. (1958). – South Sumatra basinal area. – *In* : WEEKS, L.G. (ed.) : *Habitat of Oil*, 1347-1358, 8 fig. Amer. Assoc. Petroleum Geol.

WHITLE, A.P. & SHORT, G.A. (1978). – The petroleum geology of the Tembungo field. East Malaysia. – Offshore S.E. Asia Conf. – SE Asian Petroleum Explor. Soc. Session (preprint).

WIJNE, D.H. van, LUTZ, M. & KASSCHIETER, J.P.H. (1980). – The Rotliegend in the Netherlands and its gas accumulation. – Geol. Mijn, **59**, 1, 3-24, 18 fig.

WILCOX, R.E., HARDING, T.P. & SEELY, D.R. (1973). – Basic wrench tectonics. – *Bull. amer. Assoc. Petroleum Geol.*, **57**, 1, 74-96, 16 fig.

WINNOCK, E. & MERMOD, P. (1975). – Libye : le gisement de Magid A. – *Rev. Assoc. franç. Techn. Pétrole*, **234**, 11-18, 11 fig.

YARBOROUGH, H. (1971). – Sedimentary environments and the occurrence of major hydrocarbon accumulations. – *Trans. Gulf. Coast Assoc. Geol. Soc.*, **5**, 21-82.

YERKES, R.F. *et al.* (1965). – Geology of the Los Angeles Basin, California : an introduction – *U.S. geol. Surv. prof. Pap.*, **420 A**, A 1-A 57.

YOUNG, G.A. (1971). – Exploring for stratigraphic traps in the Oficina area. – Proc. 8th World Petroleum Congr. (Moscow), **2**, 249-257, 8 fig.

ZAMBRANO, E. *et al.* (1970 et 1971). – Paléogéographie et synthèse pétrolière du Vénézuéla occidental. – *Rev. Inst. franç. Pétrole*, **25**, 12, 1449-1492, **26**, 1, 50-72.

ZIEGLAR, D.L. & SPOTTS, J.H. (1978). – Reservoir and source-bed history of Great Valley, California. – *Bull. amer. Assoc. Petroleum Geol.*, **62**, 5, 813-826, 16 fig.

ZIEGLER, P.A. (1977). – Geology and hydrocarbon provinces of the North Sea. *Geol. J.* **1**, 7-31, 15 fig.

ZIEGLER, P.A. (1980*a*). – Hydrocarbon provinces of the North West European basin. – *Mem. canad. Soc. Petroleum Geol.*, **6**, 653-706.

ZIEGLER, P.A. (1980*b*). – Northwestern Europe : subsidence patterns of post variscan basins. – Int. geol. Congr., 26th Sess., Paris, Colloque **C6**, 249-280, 10 fig.

ZIEGLER, P.A. (1981). – Evolution of sedimentary basins in North West Europe. – *In : Petroleum Geology of NW Europe*, 3-39, 17 fig. – Inst. Petroleum, London.

LIST OF FIGURES

LIST OF TABLES

LIST OF PLATES

INDEX

M

N

O

P

IMPRIMERIE LOUIS-JEAN – 05002 GAP

Dépôt légal n° 341 — Juin 1983.